Current Topics in Veterinary Medicine

Volume 2

Patterns of Growth and Development in Cattle

Patterns of Growth and Development in Cattle

A Seminar in the EEC Programme of Coordination of Research on Beef Production held at Ghent, October 11-13, 1977

Sponsored by The Commission of the European Communities, Directorate-General for Agriculture, Coordination of Agricultural Research

Edited by

H. De Boer
Research Institute for Animal Husbandry (Schoonoord)
Zeist

J. Martin
University of Ghent

Martinus Nijhoff - The Hague/Boston/London   1978
for
The Commission of the European Communities

Publication arranged by

Commission of the European Communities,
Directorate-General Scientific and Technical Information
and Information Management, Luxembourg.

EUR 6007 EN

ISBN-13: 978-94-009-9758-5   e-ISBN-13: 978-94-009-9756-1
DOI: 10.1007/978-94-009-9756-1

Table of contents

VIII

List of Publications in the series on Beef Production

EUR 5545 The Early Calving of Heifers and its impact on Beef Production
          A seminar held in Copenhagen, Denmark on June 4-6, 1975. xii+295 pp
          Price: BFR 250   DKR 39.30   DM 16.90   FF 30   LIT 4 550
          HFL 17.30   UKL 3.10   USD 7.20

EUR 5451 Perinatal ill-health in calves
          A seminar held in Compton, England on September 22-24 1975. 200 pp
          Price: BFR 200   DKR 31.50   DM 13.50   FF 24   LIT 3 650
          HFL 13.80   UKL 2.40   USD 5.80

EUR 5491 Egg Transfer in Cattle
          A seminar held in Cambridge, England on December 10-12 1975. 416 pp
          Price: BFR 800   DKR 126   DM 54   FF 96   LIT 14 500
          HFL 55.30   UKL 9.80   USD 23

EUR 5488 Improving the nutritional efficiency of Beef Production
          A seminar held in Theix, France on October 14-17 1975. 402 pp
          Price: BFR 350   DKR 55   DM 23.50   FF 42   LIT 6 350
          HFL 24.24   UKL 4.25   USD 10.10

EUR 5489 Criteria and methods for assessment of carcass and meat
          characteristics in Beef Production experiments
          A seminar held in Zeist, The Netherlands on 9-12 November 1975.
          406 pp
          Price: BFR 600   DKR 94.50   DM 40.50   FF 72   LIT 10 900
          HFL 41.50   UKL 7.30   USD 17.20

EUR 5490 Optimization of cattle breeding schemes
          A seminar held in Dublin, Ireland on November 26-28 1975. 354 pp
          Price: BFR 300   DKR 47.20   DM 20.30   FF 36   LIT 5 450
          HFL 20.70   UKL 3.70   USD 8.60

EUR 5492 Crossbreeding experiments and strategy of breed utilization to
          increase beef production
          A seminar held in Verden, France on February 9-11, 1976. 490 pp
          Price: BFR 900   DKR 140.70   DM 57.70   FF 120   LIT 20 950
          HFL 60.40   UKL 13.90   USD 24

This publication contains the proceedings of a seminar held
in Ghent, Belgium on October 11th-13th 1977, under the auspices
of the Commission of the European Communities, as part of the
EEC programme of co-ordination of research on beef production.

The seminar was initiated by the scientific working group
on 'Carcass and Meat Quality' and attracted the interest of the
working groups on 'Genetics and Selection' and on 'Nutrition and
Management'. Consequently it developed into a multi-disciplinary
programme, organised under the responsibility of the working group
on Carcass and Meat Quality. This group comprised Ir. H. de Boer
(Chairman), Netherlands; Prof. R. Boccard, France; Dr D.E. Hood,
Ireland; Dr R.W. Pomeroy, UK; Dr A. Romita, Italy; Professor
Dr L. Schön, Fed. Rep. of Germany; Mr P. L'Hermite, CEC; and
Dr J.C. Tayler, UK (adviser to the CEC).

In view of the broad scope of the subject, a special planning meeting was organised, involving experts from the different
disciplines involved. The additional participants were:
Dr B. Bech Andersen, Denmark; Dr A.J.H. van Es, Netherlands;
Prof. Dr J. Martin, Belgium; Dr. St.C.S. Taylor, UK.

The multidisciplinary scope of this seminar follows a
series of seminars in 1975-76 on more specific aspects of beef
production research in the individual fields involved. It seems
logical that further seminars should integrate the approaches by
different disciplines in order to achieve a balanced programme
of research on the very complex topic of beef production.

Although a seminar of this type sets high requirements for
the presentation of the specific topics to a multidisciplinary
audience, we feel that very good results have been achieved in
this case. The material brought together in this book could
provide research workers with a valuable view of neighbouring
disciplines, which at the same time retain scientific depth. We
think the purpose of the published contributions may go beyond

the immediate results of the discussions during the seminar itself

The Commission wishes to thank those representatives of the member States who took responsibility in the organisation and conduct of the seminar; notably Ir. H. de Boer (Chairman) and Prof. Dr J. Martin (Local Organiser). We gratefully acknowledge the assistance and the hospitality rendered by the Study Centre for Beef Production of Melle.

Thanks are due as well to the co-ordinators of sessions: Dr R.W. Pomeroy, Prof. R. Boccard and Drs. P.L. Bergström (in specific parts of session 1), Mr R. Jarrige and Mr. C. Béranger (session 2), Dr St.C.S. Taylor (session 3), Ir.H. de Boer (session 4) and Prof. Dr H.J. Oslage (final session).

OBJECTIVES OF THE SEMINAR

To review results of research on the sources of variation
in absolute and differential growth of the main tissue groups in
the bovine body; to quantify the possible effects of genetic and
nutritional factors and their influence on carcass evaluation;
to consider the application of research results in practice in
beef production, and to consider future priorities and directions
for research in order to improve efficiency, quantity and quality
of beef production.

BACKGROUND

The processes of change involved in growth and development
make an impact on the efficiency of beef production and also on
the nature of the final product. Genetic and nutritional factors
are the main sources of variation - influencing protein and fat
deposition - and should be directed towards optimal production
of carcasses of adequate weight, composition and quality, including meat quality. To that end, long term aims in the beef
production industry need to be well defined, taking into account
the impact meat technology is likely to have in the future, and
also possible changes in consumer demand.

Potential possibilities in each part of the chain of beef
production need to be integrated to achieve a high degree of
overall efficiency; this necessitates a multidisciplinary approach.
However, in view of the many uncertainties regarding future
trends in production techniques and requirements for the end
product, the seminar should concern itself not only with the
quantification of relations and trends, but even more with the
reasons why differences in growth rhythm and efficiency occur.
Basic understanding of cell differentiation, neuro-hormonal
associations with growth, metabolic turnover of nutrients etc.
provide the necessary background so that the industry will be in
a position to react appropriately to any unforeseeable changes
which may take place in the future.

SECTION I

PATTERNS OF GROWTH AND DEVELOPMENT
OF BONE, MUSCLE AND FATTY TISSUE

General Co-ordinator:

R.W. Pomeroy

Co-ordinators:

R. Boccard

P.L. Bergström

HISTORICAL AND GENERAL REVIEW OF GROWTH AND DEVELOPMENT.

R.W. Pomeroy.

Agricultural Research Council Meat Research Institute.
Langford, Bristol, BS18 7DY, UK

ABSTRACT

The growth in size and weight of meat animals and the accompanying changes in body form and composition are of great economic significance. It is, therefore, surprising that the systematic study of the changes which occur during growth and development began only about fifty years ago. Early growth studies were largely concentrated on external body measurements and increases in gross liveweight. The quantitative basis of changes in proportions of organs and tissues of the body during growth and the factors like sex and level of nutrition which might influence them were extensively studied in the nineteen thirties by Hammond and his co-workers. An important conclusion from this work was that not only were the proportions of the major tissues of the carcass affected by the level of nutrition but that the distribution of the tissues, particularly muscle, between high and low value parts of the carcass were also affected. This view has been challenged in more recent years particularly by Butterfield and his colleagues who, by using a strictly anatomical dissection technique, showed that the distribution of muscle weight at a given total muscle weight was remarkably constant and little affected by nutrition or breed.

The literature on growth and development of farm animals was thoroughly reviewed by Brody (1945) and by Palsson (1955) and this paper is not intended to be another complete review but rather a re-evaluation of some of the more important earlier work in the light of more recent research.

Growth represented by increase in size and weight with age, and development which consists of the changes in body proportions and compositon as the animal grows from conception to maturity, are of great economic significance. It is therefore, somewhat, surprising that the first major attempt to quantify patterns of growth and development was represented by the work of Hammond and his colleagues at Cambridge as recently as the nineteen-thirties (Hammond 1932; McMeekan 1940; Palsson 1939; Verges 1939; Wallace 1948 and Pomeroy 1941). The qualitative aspects of the anatomy of the animal body had been known for many years previously and quantifying them did not require the development of new and sophisticated techniques - such as were needed for example, to study the role of hormones in controlling growth and development.

Earlier workers like Glättli (1894), Nathusius (1905), Eckles and Swett (1918) had made deductions about differential growth in various species from external measurements taken at successive ages and weights but a technique of carcass dissect-ion was necessary for a quantitative definition of differential growth and of the factors which affect it. The Hammond school developed a dissection technique which was described in detail by Palsson (1939) and by McMeekan (1940) and all the workers in Hammonds department used essentially the same technique. Their work showed that during foetal life the head and lower limbs grow relatively faster than the rest of the body, so that, at birth, the young animal has a relatively large head and long legs while the trunk is short and shallow. After birth the growt rate of the upper limbs and trunk overtakes that of the head and lower limbs and the body lengthens, deepens and widens so that the head and lower limbs become a progressively smaller proportion of the whole. The differential growth pattern was described by Palsson (1955) as :

    1) A primary wave of growth starting in the region of the cranium and moving forwards towards the facial region of the head and backwards towards the lumbar region.

    2) A secondary wave of growth starting from the metatarsal and metacarpal regions and moving down to the digits and upwards along the limbs and trunk to the lumbar region.

Thus, according to Palsson, the lumbar region is the last part of the body to attain its maximum growth rate and hence is the latest developing part of the body.

    A similar pattern was described for the differential growth of the major tissues which is illustrated in Figure 1.

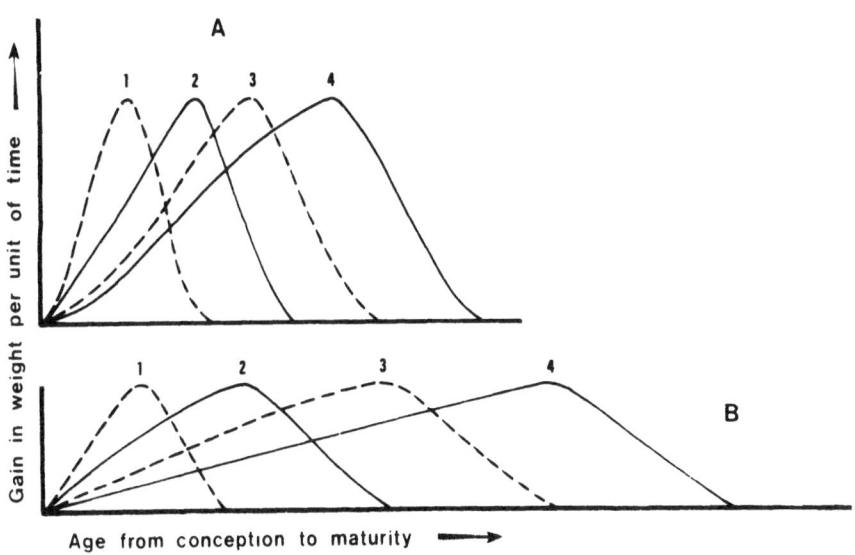

Fig. 1 Order of development of different parts and tissues of the body (from progress in the Physiology of Farm Animals 1955, by courtesy of Butterworths London).

The rate of increase in weight, showing the order of
development of the different parts and tissues of the body and
the way in which the changes in shape and body proportions are
affected by early and late maturity and/or the level of nutrition.

A = early maturity or high plane of nutrition.

B = late maturity or low plane of nutrition.

Curves: 1	2	3	4
Head	Neck	Thorax	Loin
Brain	Bone	Muscle	Fat
Cannon	Tibia-fibula	Femur	Pelvis
Kidney fat	Intermuscular fat	Subcutaneous fat	Marbling fat

The central nervous system is shown as the earliest
developing part of the body followed by bone then by muscle and
lastly by fat. Similarly, within the fat depots the kidney fat
is the earliest developing followed by the intermuscular fat
then the subcutaneous fat and lastly the intramuscular fat.
The succession is the same for all types of animal but selection
for early maturity causes it to proceed more rapidly so that,
as was shown by Hammond (1932), the body proportions of an
improved breed of sheep like the Suffolk are similar at a few
days old to those of an adult of an unimproved breed like the
Mouflon.

Rearing on a high plane of nutrition tends to compress the
developmental changes towards lighter weights and younger ages
while rearing on a low plane of nutrition extends them towards
heavier weights and older ages. In general, a low plane of
nutrition penalises most the later developing parts of the
body and Waters (1908) showed that in cattle kept on a maintenance diet so that there was no increase in live weight, the
early developing bone continued to grow at the expense of the
later developing muscle and fat.

Because the Cambridge workers all used essentially the same dissection technique they all tended to reach the same conclusions from their experimental data. However, the technique contained certain defects which resulted in some misinterpretation of that data.

In retrospect, it must be admitted that the technique was peculiar in that it was neither based on an accepted commercial jointing system nor was it based strictly on the anatomy of the musculature. The carcass was divided into nine pseudo-anatomical regions, head, neck, thorax,'loin', pelvis, two legs and two shoulders. In the process, many muscles were cut and appeared partly in one joint and partly in another. For example, part of the longissimus dorsi muscle was included in the loin and part in the thorax. In addition muscles with very different functions and rates of development were, in some cases, included in the same joint. This was particularly true of the musculature of the loin which included the muscles attached to the lumbar vertebrae and the greater part of the abdominal muscles. This was particularly unfortunate because the muscles attached to the lumbar vertebrae like the longissimus dorsi and psoas muscles are of high commercial value whereas the abdominal muscles are of low commercial value. However, the abdominal muscles are late developing and the inclusion of these with the high value lumbar muscles gave rise to the impression that these were also late developing and could be affected differentially by the plane of nutrition on which the animal was reared.

It is interesting to note that the only member of the Hammond school to doubt the proposition that the growth of muscle could be affected differentially by the plane of nutrition was Wallace (1948) who expressed the view that the weight of muscle in a given region was proportional to the total weight of muscle and that the apparent effect of nutrition on the differential growth of muscle arose from its effect on total muscle weight. In spite of this a further twenty-five years or so were to elapse before Butterfield and his co-workers, Butterfield (1963, 1965), Butterfield and Berg (1966) Butter-

field, et al, (1966), using a strictly anatomical dissection
technique concluded that for a given total weight of muscle the
distribution of individual muscle weights was practically the
same regardless of breed, sex, level of nutrition and so on.
This finding has been confirmed by other workers and it is an
important one in that it implies that there is very limited
scope for changing the distribution of muscle weight and hence
carcass value, by either genetic selection or nutritional
manipulation.

Butterfield's work also implies that the importance of
conformation or shape is much less than had previously been
supposed in that the distribution of muscle weight between the
high and low value parts of the carcass will be the same in
carcasses of good or bad conformation but of the same total
muscle weight. This is not to say that good conformation is
commercially unimportant because thickness of muscle as distinct
from proportionate weight may have considerable commercial
value. Clearly, butchers attach value to good carcass conformation since it is included in most carcass classification schemes
but such value as it has must relate to thickness of muscle or
to muscle to bone ratio and not to distribution of muscle weight.

In interpreting the results of his dissection data Butterfield used the allometry equation devised by Huxley (1932).
This equation is of the form $Y = bX^k$ where Y is the weight of
a part of the body, X is the weight of the whole body or the
whole body less the part and b and k are constants. The
constant b is of no particular biological significance being the
value of Y when X = 1 but the constant k is the ratio of the
specific growth rates of Y and X.

Huxley applied the equation to a range of plants and
animals notably the Fiddler Crab (*Uca pugnax*) in which he concluded that, 'A body of a given weight would have attached to it
a claw whose weight would be the same whether the body had
taken three weeks or three years to reach its present size',
although he qualified this statement by saying that it was

probably only true as an approximation. The Cambridge school
was more explicit and Palsson (1955) stated '.....since the
plane of nutrition affects the various parts and tissues of the
body as well as different parts of the same tissue differentially the theory of a constant specific growth ratio between the
growth of any two parts of the animal, formulated by Huxley
cannot safely be applied to growth in animals of determinate
growth'. These two opinions are in direct conflict and it
is surprising in view of the economic implications, that little
critical experimentation to resolve the conflict was undertaken
for many years.

The answer to the problem appears to lie in the fact that
the Cambridge workers related carcass composition to carcass
weight ie including fat, and the nutritional treatments imposed
by McMeekan, for example, changed the proportion of fat and
hence the proportion of muscle in the carcass. The apparent
effect of nutrition on the distribution of muscle weight observed by the Cambridge workers was thus due to differences in
total muscle in carcasses of the same weight.

Butterfield and his co-workers applied the Huxley
equation $Y = bX^k$ to individual muscles and muscle groups using
the logarithmic form.

$$Log\ Y = log\ b + k\ log\ X$$

which is a straight line with slope k. Muscles with k = 1 grow
at the same rate as the total musculature, those with k less
than 1 at a slower rate and those with k greater than 1 at a
higher rate and he described these muscles respectively as
having average, low or high growth impetus. The Huxley
equation in its logarithmic form is an attractive way of describing growth patterns but there are pitfalls in its use. For
example, in calculating the allometric relationship between a
group of muscles like those of the hind leg and total muscle
it must be remembered that the former comprise about 30% of
the latter. With a part-whole relationship of this magnitude
any factor affecting the growth of the muscles of the hind-leg

will almost certainly produce a high correlated response in the growth of total muscle.

Carcass composition studies in cattle involving complete dissection are expensive in both material and labour. An understanding of the differential growth patterns of the various tissues is, therefore, important in facilitating prediction of carcass composition from less expensive partial dissections. Good quantitative data relating to differential growth in cattle is still somewhat scarce since many workers have had to use rather heterogenous samples of animals in terms of breeds, sexes and nutritional history.

There is, particularly, a dearth of information on the relationship between growth in the late pre-natal and early postnatal periods and subsequent growth to commercial weights although these periods have tended to become a progressively larger proposition of the total lifespan of the animal.

REFERENCES

Brody, S. 1945. Bioenergetics and Growth. Reinhold, New York.

Butterfield, R.M. 1963. Symp. Carcass Composition and Appraisal of Meat Melbourne, C.S.I.R.O.

Butterfield, R.M. 1965. Proc. N.Z. Soc. Anim. Prod. 25. 152.

Butterfield, R.M. and Berg, R.T. 1966. Res. Vet. Sci. 7. 326.

Butterfield, R.M. Pryor, W.J. and Berg, R.T. 1966. Res. Vet. Sci. 7. 417.

Eckles, C.H. and Swett, W.W. 1918. Univ. Mo. Exp. Sta. Bull. 51.

Glättli, G. 1894. Landw. Jahrb. Schweiz. 8. 144.

Hammond, J. 1932. Growth and development of mutton qualities in the sheep. Oliver and Boyd, Edinburgh.

Huxley, J.S. 1932. Problems of relative growth, Methuen, London.

McMeekan, C.P. 1940. J. Agric. Sci. Camb. 30. 276. 387. 511.

Nathusius, S. von, 1905. Arb. d. Deut. Landw. Gesell. H. 112.

Palsson, H., 1939. J. Agric. Sci. Camb. 29. IV. 544.

Palsson, H., 1955. In: Progress in the Physiology of Farm Animals Vol. 2. Butterworths, London.

Pomeroy, R.W., 1941. J. Agric. Sci. Camb. 31. 50.

Verges, J.B. 1939. Suff. Sheep. Soc. Year BK.

Wallace, L.R. 1948. J. Agric. Sci. Camb. 38. 367.

Waters, H.J. 1908. 29th Proc. Soc. Prom. Agr. Sci. N.Y. 71.

BOVINE COMPOSITIONAL INTERRELATIONSHIPS

R.G. Kauffman

University of Wisconsin-Madison, 53706 USA.

ABSTRACT

Composition studies were completed using 49 steers representing variations in muscularity that included double muscled cattle. Results indicated that non-muscular cattle were fatter than anticipated when evaluated subjectively, and the reverse was true for muscular cattle. Muscular cattle contained from 2 to 7% more fat-free muscle and this was attributed to higher muscle-bone ratios; however, fatness was the dominating variable in determining ultimate composition. The proportionate contributions of each muscle to total muscle mass and each bone to total skeletal mass was nearly constant for all cattle, and this was also true for body parts (holding fat content and weight constant). When fat content and live empty body weight were held constant, all cattle converted equal quantities of feed to fat-free muscle. The combination of fat depth, longissimus area and marbling content provided a useful combination of variables to predict quantity of fat-free muscle. Muscular cattle had higher dressing yields which were attributed to their possessing proportionately smaller body cavities. Fatter cattle did not have higher dressing yields when live weight and muscularity were held constant because it appeared that increased quantities of mesentery fat compensated on a proportional basis for increased quantities of carcass fat. One double muscled steer possessed the biological capacity to deposit excess quantities of fat when time on feed was unlimited.

INTRODUCTION

This report provides an overview of compositional inter-relationships that affect the efficient utilisation of cattle for human food. The first portion focuses on a brief review of information already known, but presented from a different prospective. The second part includes a summary and clarification of topological studies conducted at the University of Wisconsin that should stimulate a renewed interest in bovine composition. The third segment projects some speculative thoughts that produce more questions than answers, but ones that could stimulate improvements for the beef industry. Finally, I have listed 17 key references that further elaborate on the concepts presented here.

COMPOSITIONAL CONCEPTS

1. Definition

In a broad sense, composition is defined as the makeup of a substance, or the aggregate of ingredients and manner of their combination or constitution. It is a mixture of various components or a quantitative arrangement of the parts, so as to form a unified, harmonious whole. Chemically, bovine composition is the sum total of the separable components, notably protein, lipid, water, ash and carbohydrate, or in a more basic sense carbon, hydrogen, oxygen, nitrogen, sulphur and the minerals. However, physically separable parts including ingesta, excreta, viscera, skin, fat, bone and muscle may be more meaningful when considering realistic worth of the edible product. Even though muscle, fat and bone will vary greatly in extractable lipid, the edible product is either readily available and acceptable for consumption, or it is not. Figure 1 is included to illustrate one approach in defining bovine composition. It divides the animal into carcass and non-carcass components and then subdivides each into more discrete anatomical units. Each of the units is expressed as a proportion of the other subdivisions of which it is a member. For example, the

Fig. 1 Approximate composition of the bovine animal

scapula represents less than 0.5% of the live animal, 1% of the carcass, 5% of the carcass skeleton, 10% of the appendicular skeleton, 24% of the thoracic limb skeleton and 3% of the total skeleton. The sum of percentages within a subdivision equals 100. This method of presentation provides an expedient approximation of proportionality of one part to another or combination of parts. Since quantities of ingesta, excreta and fat vary, each value serves only as a guide. This same approach could be applied to closely trimmed, boneless cuts of beef or to elemental composition.

2. Expression

When scientists express composition differently, it creates difficulty interpreting and comparing experimental results. In expressing bovine compositional variation there are several methods of choice. Depending on the goal, a componen may be expressed as a portion of the live or carcass weight. Secondly, the more commercially valuable parts (wholesale or retail cuts) of the carcass may be either trimmed of excess fat and deboned or left untouched after separation. Finally, the soft tissues (muscle and fat) may be separated from the skeleton and individually weighed, or homogenised and sampled for chemical analysis.

For the most appropriate, standardised approach, I recommend that live empty body weight be used as the initial weight. If time or finances are limited, then hot carcass weight should be used. The hot carcass is then easily dissected into fat, muscle and skeleton. Each of the soft tissues should be chilled, ground, mixed and carefully sampled for moisture and lipid analysis, depending on the experimental objectives. For most carcass composition studies, expressing composition on a fat-free muscle basis is easily standardised and is a desirable choice. If the fat-free approach is desired, then a correction is necessary. Since lipid extracted from muscle does not represent the connective tissue and water associated with adipose tissue, the lipid percentage should be adjusted as shown in Figure 2 (Fahey et al., 1977). Once the quantity or

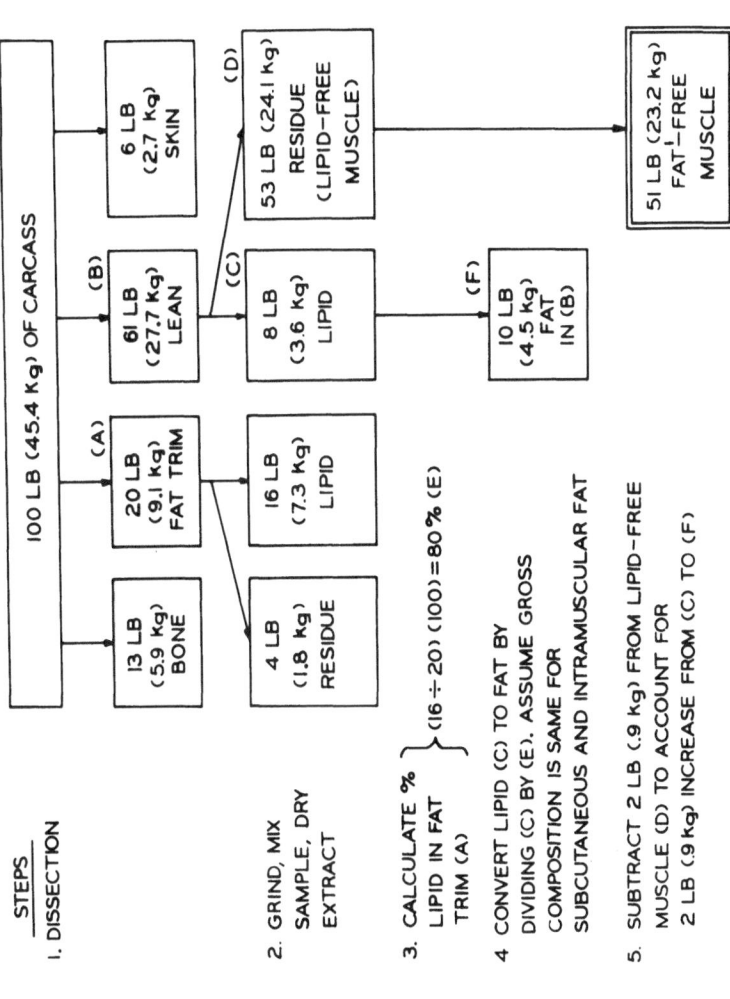

STEPS

1. DISSECTION

100 LB (45.4 Kg) OF CARCASS

13 LB (5.9 Kg) BONE

(A) 20 LB (9.1 Kg) FAT TRIM

(B) 61 LB (27.7 Kg) LEAN

6 LB (2.7 Kg) SKIN

(D) 53 LB (24.1 kg) RESIDUE (LIPID-FREE MUSCLE)

2. GRIND, MIX SAMPLE, DRY EXTRACT

4 LB (1.8 kg) RESIDUE

16 LB (7.3 Kg) LIPID

(C) 8 LB (3.6 Kg) LIPID

(F) 10 LB (4.5 Kg) FAT IN (B)

51 LB (23.2 kg) FAT-FREE MUSCLE

3. CALCULATE % LIPID IN FAT TRIM (A) } (16 ÷ 20) (100) = 80 % (E)

4. CONVERT LIPID (C) TO FAT BY DIVIDING (C) BY (E). ASSUME GROSS COMPOSITION IS SAME FOR SUBCUTANEOUS AND INTRAMUSCULAR FAT

5. SUBTRACT 2 LB (.9 Kg) FROM LIPID-FREE MUSCLE (D) TO ACCOUNT FOR 2 LB (.9 kg) INCREASE FROM (C) TO (F)

[1] FAT IS DEFINED AS A COLLECTION OF ADIPOSE CELLS SUSPENDED IN A LARGER CONNECTIVE TISSUE MATRIX, THAT ARE DISTENDED WITH CYTOPLASMIC LIPIDS, WATER AND OTHER MINOR CONSTITUENTS.

Fig. 2 An example of calculating fat-free muscle

proportion of fat-free muscle is determined, then it may be adjusted to a fat standardised basis by arbitrarily deciding what proportion of fat the muscle should contain, and then dividing fat-free muscle by 1.00 - the fraction of desired fat level (Example: if % fat-free muscle = 54, and the desired standard content of fat is 10%, then fat-standardised muscle = 54% ÷ (1.00-.10) = 60%). This method focuses on the total muscle concept which should be the primary aim of beef production. It expresses the answer in forms, units and language understandable and appropriate for all.

3. Variables

This includes a brief identification of established factors that affect bovine composition (Kauffman, 1971). Detailed discussion is omitted except where there is need for explanation. It is obvious that proportion of fat in the animal is the most important effect that each of the following factors has even though it is recognised that muscle-bone ratio is also involved.

Environmentally, state of nutrition is important in that food deprivation reduces fat content, and low protein diets increase proportions of fat, especially during periods of compensatory growth. However, within practical limits, plane of nutrition will not markedly affect composition if cattle are fed to the same end weight. Extent of exercise slightly increases the proportion of muscle due to the mobilisation of stored lipids and to the stimulation of muscle fibre hypertrophy. The presence of abnormalities due to injury or disease (such as nerve damage) results in muscle degeneration and fatty infiltration. A number of factors affect live weight, especially when empty weight is not used. They include quantities of ingesta and excreta, stage of pregnancy, cleanliness of hide, method of slaughter, and presence of bruises, diseased tissue, reproductive organs and horns.

Genetically, the combination of stage of growth and weight

affects composition. In young, light weight cattle, fat
content and muscle-bone ratios are low and the proportion of
non-carcass is high, and these variables change during growth
(Ellenberger et al., 1950, Hedrick, 1968). Weight is related
to composition; however it is possible to observe two cattle
of similar weights having dissimilar compositions. Preston
(1971) suggested that, instead of believing that live or
carcass weight determines composition, it is more correct to
visualise composition as a function of the proportion of
mature body weight which has been attained at any given live
weight. Secondly, sex influences composition according to
Brannang (1966). The carcass composition between monozygous
twins, in which one twin was castrated, demonstrated that
steers yielded lighter carcasses that possessed 92% as much
muscle and 151% more fat when compared to their intact twins.
It is recognised that these differences are actually reflections
of different stages of growth and development. Finally,
bovine topology as a phenotypical expression of genotype
influences composition as described in the following section.

TOPOLOGICAL STUDIES AT WISCONSIN

1. Materials and Methods

Forty-nine yearling steers weighing an average of 275 kg
were used to represent five muscling groups as determined
subjectively. The most angular shaped group was represented
by Longhorns and the second most angular group included dairy
breeding. The more muscular groups were represented by the
Angus, Hereford and Charolais breeds. The extremely muscular
group was double muscled. Table 1 provides the distribution
of numbers. Seventeen control steers were slaughtered to
determine composition prior to test feeding. Another 31 steers
were individually fullfed a balanced corn, soybean and haylage
ration for 132 days. One double muscled steer was fullfed an
additional 14 months.

TABLE 1

DISTRIBUTION OF 48 CATTLE USED FOR THE TOPOLOGICAL STUDIES

	Muscling Groups					
	Very Angular	Angular	'Average' Musculature	Muscular	Very Muscular	
Controls, N	3	4	4	4	2	17
Fed 132 days, N	7	6	9	7	2	31

After each steer was slaughtered by conventional procedures, the carcass, ingesta, excreta, blood, head, hide, feet, organs, viscera and mesentery fat were weighed separately. During evisceration, the sternum was left intact and the thoracic cavity was filled with water. The water was expressed as a proportion of carcass weight to estimate thoracic capacity.

Each hot carcass was sectioned medially, and one side was dissected in a pre-rigor state to facilitate separation of major muscles as well as the remaining soft tissues from the skeleton. Individual muscles, bones and remaining soft tissues were weighed and chilled and then homogenised for chemical analysis to determine total fat-free muscle as outlined in Figure 2.

The other side was chilled and then measured to confirm muscular dimensions. Other physical and subjective quality observations were made to include as independent variables in stepwise regression equations to predict fat-free muscle. The method used to calculate feed conversion is illustrated in Figure 3.

2. Results

A. __Feed Conversion:__ When live empty body weight and body fat content were held constant by analysis of covariance, muscling did not affect the conversion of feed to fat-free

muscle. As expected, fatter cattle required more feed to
produce fat-free muscle. Muscling *per se* did not affect feed
efficiency, but the advantage of producing muscular cattle was
that they could be fed to heavier weights to produce leaner
carcasses at a given chronological age. For this reason, they
could be more efficient converters of feed to fat-free muscle.
This was especially obvious for double muscled cattle.
(Kauffman et al., 1977).

Example

STEP 1
    Obtain live 24-hr shrunk weight of acclimated
    steer to be fed   280 kg

STEP 2
    Determine dressing percentage of matched control
    steer after slaughter   55%

STEP 3
    Using control dressing percentage, determine
    theoretical carcass weight of steer to be fed   55% x 280 = 154 kg

STEP 4
    Determine percentage fat-free muscle (FFM)
    of control steer carcass   65%

STEP 5
    Calculate theoretical on-test FFM weight of
    steer to be fed   65% x 154 = 100.1 kg

STEP 6
    Determine actual FFM weight of steer at end
    of test   150 kg

STEP 7
    Net gain of FFM/132 days test   150 - 100.1 = 49.9 kg

STEP 8
    Obtain net weight of feed consumed during test   1250 kg

STEP 9
    Calculate kg feed consumed/kg FFM produced   1250 ÷ 49.9 = 25.1

Fig. 3. Method used to calculate feed required to produce one kg
       fat-free muscle.

B. Composition: In the experiment described above and in others (Kauffman et al., 1973, de Boer, 1969), muscular cattle contained approximately 2% more fat-free muscle, even when percentage fat was similar. In more extreme examples as shown in Figures 4 and 5, the difference was 7%. This was true because muscular as compared to angular cattle have more than one unit higher muscle-bone ratios. This suggests that if fat content in the carcass was standardised to 20%, a 550 kg muscular steer would contain about 11 kg more fat-free muscle when compared to a 550 kg angular steer. If the fat content had been standardised to 10%, then the difference would increase to 12 kg, and if fat content was 30%, the difference would decrease to 10 kg. It is because of this that muscle-bone ratios of cattle should be compared only when fat content is known. However these differences are small when compared to the influences that body fatness has on composition.

C. Muscular and Skeletal Proportionalities: On the basis of this experiment as well as others (Berg and Butterfield, 1976 and Charles and Johnson, 1976_2), it is clear that more muscular steers of a given weight yielded shorter and thicker carcasses that had wider, thicker and more bulging thoracic and pelvic limbs. However, when individual muscles were expressed (on a fresh or fat-free basis) as a proportion of the total muscle mass, degree of muscularity did not alter the proportions. When carcass fat was held constant, this phenomenon was also true for wholesale cuts, and for individual bones when expressed as percentages of the skeleton. This suggests that the convex appearance of muscular steers and concave appearance of angular steers is not the result of disproportionate sizes of individual muscles and bones.

D. Fat Deposition: As already demonstrated by Lewis (1967) Johnson et al. (1972) and Charles and Johnson (1976_1), fat deposition was neither proportionally similar at various stages of growth, nor similar at a given stage for various breeds. The more angular cattle (as compared to muscular cattle) had about

LIVE WEIGHT, kg	285	425
CARCASS WEIGHT, kg	165	260
DRESSING YIELD, %	58	61
THORACIC CAPACITY, kg WATER 100 kg CARCASS	13	8
CARCASS FAT CONTENT	TRACES	TRACES
LOIN EYE AREA, cm^2	45	100
CARCASS LENGTH, cm	117	115
BONE WEIGHT, kg	40	40
MUSCLE WEIGHT, kg	128	220
MUSCLE BONE RATIO	3.2	5.5

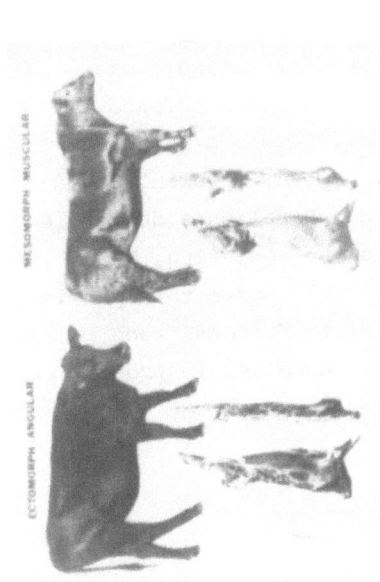

Fig. 4 Extremes in muscling

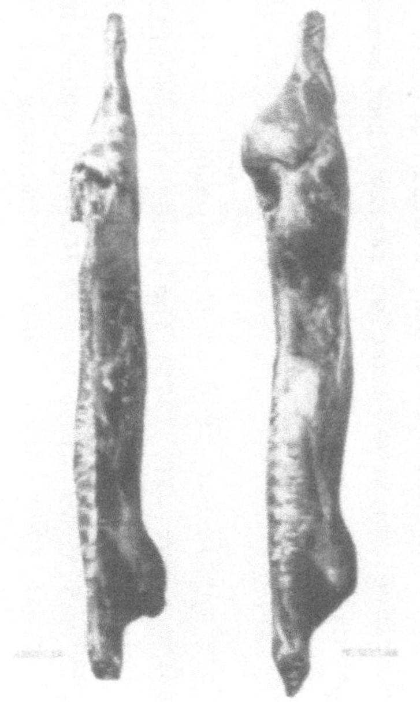

Fig. 5 Two beef sides similar in length, skeleton content and fat deposition, but different in muscling

5% more cavity fat when expressed on a total body fat basis.
Furthermore, from a subjective point of view, live, angular
cattle appeared leaner than they actually were, whereas live,
muscular cattle appeared fatter than they actually were. Since
muscular cattle are wider, deeper and shorter than angular
cattle of a similar weight, and because fatness is generally
associated with these descriptive dimensions, livestock buyers
should recognise that angular cattle may contain obscure fat
deposits. Furthermore, enlarged muscles rather than excess
fat may be responsible for the additional thickness associated
with muscular cattle.

When fat-free muscle was used as the dependent variable,
the results (Kauffman et al., 1975) indicated that marbling
should be used as a predictor. When marbling was included
with fat depth and *longissimus* area in a regression equation,
the standard error was reduced 0.7% and the increase in
percentage of accountable variation exceeded 20% (Table 2).
When marbling was combined singly with other traits, the
relative increase in accountable variation in fat-free muscle
ranged from 7 to 180%. When amount of seam (intermuscular)
fat at the 5th costae was assessed, it too accounted for a
large percentage of variation. For obvious reasons, these two
measures would not be useful if the dependent variable was
trimmed retail cuts. Because carcasses are not sectioned at
an anatomical site that exposes seam fat, it would not be as
appropriate to include in an equation; however marbling could
be easily determined at the 12th costae. Marbling's dual role
as a measure of quality and quantity may be confusing, but in
addition to predicting composition, it could serve to emphasise
that as more of it is preferred for quality reasons, such
improvement would necessarily result in decreases in proportion
of fat-free muscle.

TABLE 2

PREDICTIVE VALUE OF VARIOUS COMBINATIONS OF VARIABLES USED TO ESTIMATE
PERCENT FAT-STANDARDISED MUSCLE[a]

Equation	Variables	Partial r	R^2 x 100 (%)	Standard error of the estimate (%)
1	Adjusted fat thickness, cm	-.57**	32.1	3.50
2	Adjusted fat thickness, cm	-.61**	51.7	3.03
	longissimus area, cm^2	.54*		
3	Adjusted fat thickness, cm	-.62**		
	longissimus area, cm^2	.72**	72.9	2.33
	Marbling score [b]	-.66**		

[a] $N = 22$, * = $P < .05$, ** = $P < .01$.

[b] Devoid = 1, small = 5, abundant = 10.

E. The Fifth Quarter: The results indicated that more
muscular steers yielded proportionately heavier carcasses and
lighter non-carcass components, whereas degree of fatness had
little to do with these yields. Dressing percentage (expressed
on a live, empty body weight basis) was related negatively to
thoracic cavity capacity and positively to muscle-bone ratio
(Kauffman et al., 1976). This information shown in Tables 3
and 4 supports these observations. Preliminary interpretations
suggest that fatness was not directly related to dressing
percentage because the rates of mesentery fat and carcass fat
deposition may be linear (b values were constant). Based on
very limited data, it appeared that for every 1% fat deposited
in the body, about .15% was accounted for by mesentery fat and
the remaining .85% by carcass fat. This assumed that weight
was held constant.

TABLE 3

EFFECT OF MUSCLING ON DRESSING PERCENTAGE AND CAPACITY OF THORACIC CAVITY

Item	Least square means for the five muscling groups[a]				
	I	II	III	IV	V
Dressing percentage[b]	64.8^d	65.2^d	67.2^e	68.1^f	72.5^g
Thoracic capacity[c]	13.7^d	11.2^e	9.8^f	8.4^f	6.2^g

[a] Means in each line having similar superscripts are not different
  ($P > 0.05$).

[b] Expressed on a live, empty body weight basis.

[c] Expressed as kg water held in thoracic cavity/100 kg carcass weight.

TABLE 4

DRESSING PERCENTAGE, THORACIC CAVITY CAPACITY AND CARCASS COMPOSITION
INTERRELATIONSHIPS

N	Thoracic Cavity Capacity[a]	Dressing Percentage[b]	N	Percentage Fat-free muscle[c]	Dressing Percentage[b]
2	16^d	63 low M/B[e]	2	<52	68 high fat
2	12	64	5	53	68
3	12	65	4	55	67
6	11	66	4	57	67
4	9	67	4	59	66
8	9	68	4	61	66
3 .	7	69	6	63	67
1	7	70	2	>64	74 low fat
2	7	>71 high M/B			

[a] kg water/100 kg carcass.

[b] Empty body weight basis.

[c] Carcass basis.

[d] Rounded mean

[e] Muscle/bone ratio

F. Double Muscling: The double muscled cattle used in the project deposited very little fat. As a matter of curiosity we fed an Angus double muscled steer a concentrate ration for 18 months. According to the observations of Novakofski and Kauffman (1977), the carcass possessed excessive fat. Over 2 cm of fat was deposited over the *longissimus* and it contained moderate quantities of marbling. It also contained about 5% cavity fat and excessive quantities of seam fat at the 5th costae. This indicated that double muscled cattle do have the capacity to store fat, but that fattening is delayed because of later maturing characteristics during growth, both sexually and compositionally.

CONCERNS AND OPINIONS

1. Composition vs Quality

The consumable product must be wholesome, nutritious, free of excess shrinkage, attractive and palatable. If beef does not meet these requirements, it is of little value regardless of composition. However, once a minimum of these quality factors are attained, attention must shift toward improving composition. Changes in consumer habits and improved technology of processing will aid in this emphasis.

2. Quantity of Fat

Biologically it is impossible to produce fat-free beef, and if it was, it would not meet the needs of beef consumers. Some fat is essential for processing desirable sausage and ground beef, and there is sufficient evidence that some marbling improves flavour and juiciness. However, it is undesirable for a carcass to contain more than 20% fat. Perhaps 15% is an appropriate goal.

3. Fat Deposition

The growth of adipose tissue is not constant. We need more knowledge about its rate and distribution of postnatal development. We also need to learn more how to control biologically this rate and distribution.

4. Cavity Fat

This specific depot is variable among breeds and we need
to understand why. It should be removed during slaughter since
it would eliminate an unwanted fat depot on the carcass prior
to distribution, and simultaneously it would remain at a point
in processing where it can be best utilised. If it was absent
from carcasses subjected to USDA grading, one subjective
variable would be eliminated.

5. Bulls vs Steers

Male hormones are as effective if not more so than diethyl
stilboestrol in promoting muscular growth. Current experiments
indicate that young bull beef possesses acceptable palatability
characteristics. If management practices can be appropriately
focused on bull production, there is little doubt that faster
growing, more efficient bulls will yield larger, leaner more
muscular carcasses that will help satisfy the demand for both
fresh retail cuts and processed beef, and effectively reduce
energy consumption.

6. Forage vs Grain

Cattle are efficient roughage converters that can utilise
pasture and forages to the best advantage for mankind. They
serve as the last resort in converting waste lands not usable
for crop production, into food. Grain products can best be
processed directly for human consumption, especially in a
starving world and in spite of inefficiencies due to marketing,
distribution, and sociological irregularities. Feeding grain
to cattle should be limited to times of surplus, and only basic
supply-demand relationships should permit this.

7. Weight

Marketing conditions affect final weight more than concerns
for efficiency. Nevertheless, beef carcasses should be as
heavy as possible without jeopardising economic constraints
related to composition, processing limitations, labour
requirements and consumer satisfaction. Heavy cattle (compared

to light cattle) can be slaughtered and processed using similar time, labour, storage and transportation requirements.

8. Double Muscling

For decades, animal scientists have searched for beef cattle that yield proportionately heavy carcasses that are lean and muscular. Double muscled cattle answered this need, but unfortunately the production limitation of low fertility, late sexual maturity, high pre-calving maintenance costs, difficult calving and high death losses at birth have essentially eliminated this genotype from the beef cattle population - except for specialised but rare circumstances. Is it possible that terminal cross breeding programmes might take advantage of the superior carcass traits while avoiding the reproductive hazards?

9. Bovine Excellence

I conclude this paper by describing speculatively the ideal bovine animal that should meet industry goals for it to serve as a source of food for man. The calf will be a twin that was easily calved from a dam weighing less than 550 kg and that provides sufficient milk to grow the calf to weigh at least 350 kg. The polled calf will be an intact male that is late maturing and fast growing to attain a slaughter weight of not less than 500 kg in less than one year of age on a roughage-concentrate diet. He will be docile, healthy and free of any physical abnormalities, and will produce at least 68% of his live empty weight as a muscular carcass possessing less than 0.5 cm subcutaneous fat depth over the loin, less than 2% cavity fat and minimal quantities of intermuscular and mesentery fat. The carcass will be free of abnormalities and the musculature will be firm, free of wateriness, have a cherry red colour and contain small to modest quantities of marbling.

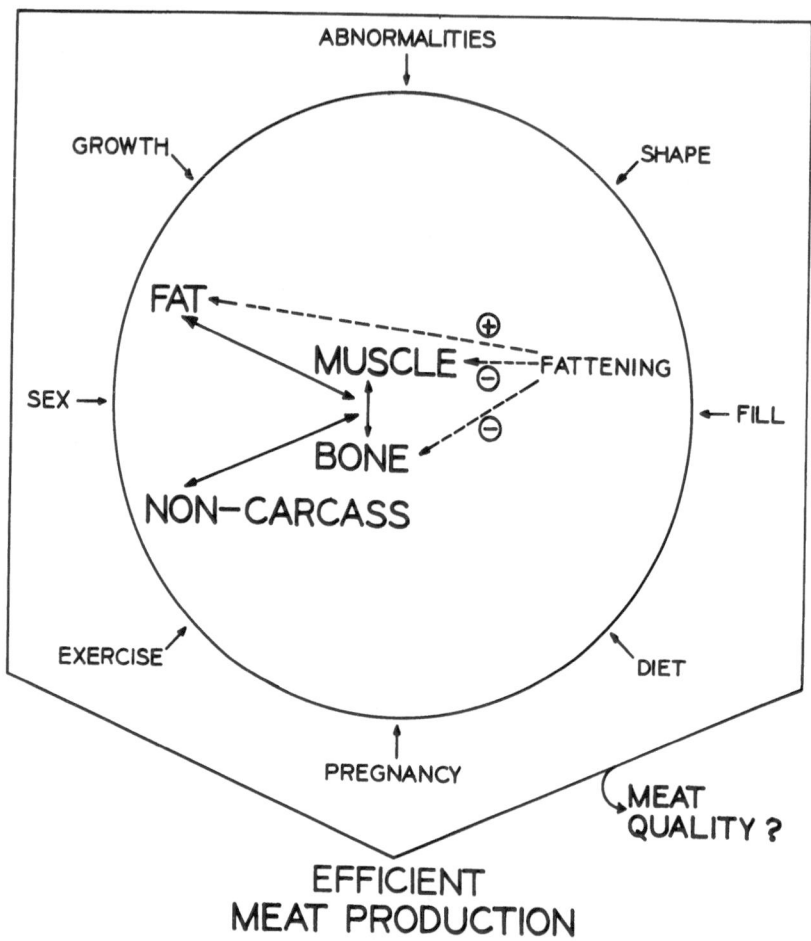

Fig. 6 Interrelationships of bovine composition

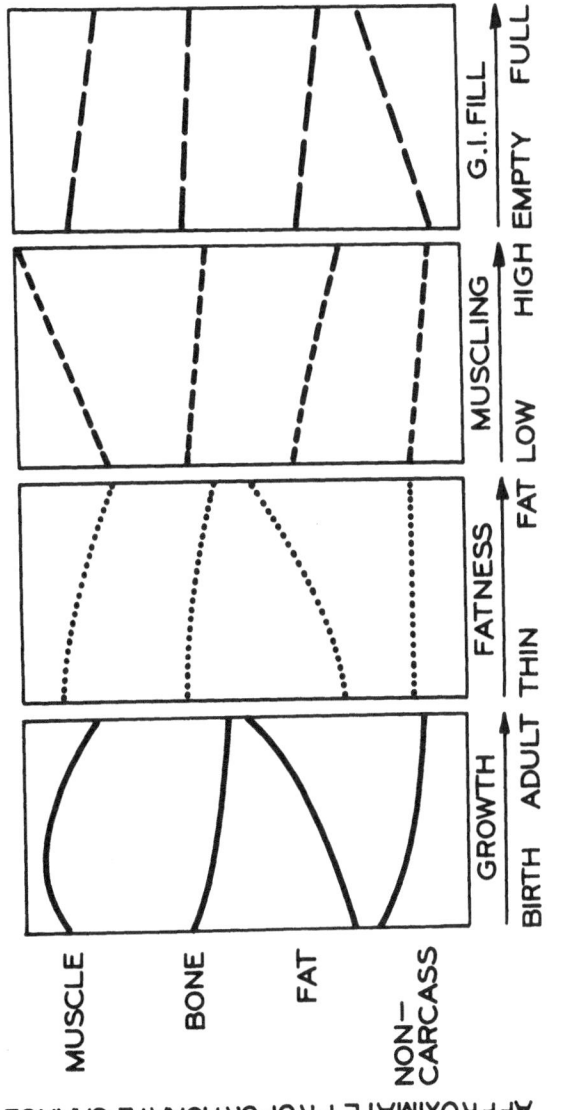

Fig. 7 Factors influencing bovine composition

CONCLUSIONS

Figure 6 emphasises that muscle and bone affect each other's proportions and that this interrelationship is primarily influenced by shape and stage of growth. These two components may increase or decrease together or in opposite directions. Both are affected by fat content and by proportions of the non-carcass portion. However, there is evidence that degree of fatness *per se* does not directly influence the proportions of the non-carcass component. Carcass compositional interrelationships are constantly under the influence of one or more of the eight factors shown outside the circle, but extent of fattening as a result of diet and growth is perhaps the most direct influential factor in increasing fat and decreasing muscle and bone components. These interrelationships directly affect the efficiency in which quantities of meat are produced, yet there is some question as to how they affect product quality. Figure 7 is included to summarise the effects that will occur, on a relative basis, for muscle, bone, fat and non-carcass components, when growth, fatness, muscling, and contents of the alimentary canal change. Even though the four variables are presented collectively to coordinate the interrelationships that exist, each variable affecting relative amounts of each component is to be considered separately, assuming that the other three variables are held reasonably constant. Nevertheless, it is recognised, for example, that as growth proceeds, muscle : bone increases and fat deposition usually occurs; however, you should consider such changes minimal for each circumstance to fully appreciate the relative impact that growth has on composition. Finally, each curve is to be interpreted as a relative change and that absolute changes were not included.

REFERENCES

Berg, R.T. and Butterfield, R.M. 1976.New concepts of cattle growth.
Sydney University Press, Sydney, Australia, p. 29, 178.

Brännäng, E. 1966. Studies on monozygous twins. XVIII. The effect of
castration and age of castration on the growth rate, feed conversion
and carcass traits of Swedish red and white cattle. Part 1.
Lantbrukshogskolans Annaler 32: 329-415.

Charles, D.D. and Johnson, E.R. 1976$_1$. Breed differences in amount and
distribution of bovine carcass dissectible fat. J. Anim. Sci. 42
(2) : 332-341.

Charles, D.D. and Johnson, E.R. 1976$_2$. Muscle weight distribution in
four breeds of cattle with reference to individual muscles,
anatomical groups and wholesale cuts. J. Agric. Sci. 86 : 435-442.

de Boer, H. 1969. Unpublished data. Zeist, The Netherlands.

Ellenberger, H.B., Newlander, J.A., and Jones, C.H. 1950. Composition of
the bodies of dairy cattle. Vermont Agr. Exp. Sta. Bul. 558.

Fahey, T.J., Schaefer, D.M., Kauffman, R.G., Epley, R.J., Gould, P.F.,
Romans, J.R., Smith, G.C. and Topel, D.G. 1977. A comparison of
practical methods to estimate pork carcass composition. J. Anim. Sci.
44 (1) : 8-17.

Hedrick, H.B., 1968. Bovine growth and composition. Res. Bul. 928, U. of
Missouri, College of Agric., Agric. Exp. Sta., Columbia, MO.

Johnson, E.R., Butterfield, R.M. and Pryor, W.J. 1972. Studies of fat
distribution in the bovine carcass 1. The partition of fatty
tissues between depots. Aust. J. Agric. Res. 23 : 381-388.

Kauffman, R.G. 1971. Variation in gross composition of meat animals. Proc.
24th Recip. Meat Conf., Am. Meat Sci. Assn., Lexington, KY. 292-303.

Kauffman, R.G., Grummer, R.H., Smith, R.E., Long, R.A. and Shook, G. 1973.
Does live-animal and carcass shape influence gross composition? J.
Anim. Sci. 37 (5): 1112-1119.

Kauffman, R.G., Van Es, M.D. and Long, R.A. 1976. Bovine compositional
interrelationships. J. Anim. Sci. 43 (1): 102-107.

Kauffman, R.G., Van Es, M.D. and Long, R.A. 1977. Bovine muscularity its
relationship to feed efficiency. J. Anim. Sci. 44 (3): 368-373.

Kauffman, R.G., Van Es, M.E., Long, R.A. and Schaefer, D.M. 1975. Marbling:
its use in predicting beef carcass composition. J. Anim. Sci. 40 (2):
235-241.

34

Lewis, R.W. 1977. Some objective studies related to beef carcass composition.
Ph. D. Thesis. University of Wisconsin, Madison, WI.

Novakofski, J.E. and Kauffman, R.G. 1977. Unpublished data. Dept. Meat
and Anim. Sci. University of Wisconsin. Madison, WI.

Preston, R.L. 1971. Effects of nutrition on the body composition of
cattle and sheep. Georgia Nutrition Conference, University of GA,
Athens. 26-41.

DRESSING PERCENTAGE IN RELATION TO WEIGHT, SEX AND BREED

Y. Geay

Institut National de la Recherche Agronomique
Laboratoire de la Production de Viande
Centre de Recherches Zootechniques et Vétérinaires
THEIX - 63110 Beaumont - France

ABSTRACT

*Several sources of variation of dressing percentage are analysed:
1) when it is related to the full body weight, it depends on the gut content
of the animals and therefore on type of the diet, food intake and length of
starvation period before slaughter; 2) in any case the dressing percentage
increases with fatness. It is higher in bulls than in steers or heifers.
It also varies with breed.*

INTRODUCTION

The dressing percentage is the first criterion of production to be considered at slaughter by producers, scientists,
butchers, economists involved in meat production. However its
reliability is dependent on the mode of expression. Expressed
in relation to empty body weight independently of the gut
content, it may vary according to weight, sex and genotype.

The different ways of expressing dressing percentage

Different ways of expressing dressing percentages are
frequently used according to the need required.

For economists, it is important to be able to link the
live animal's performances and production costs of live weight
gain with slaughter results. On one hand dressing percentage
is usually the relationship between the carcass weight and the
live weight of the animal; the live weight being determined
before the animal leaves the farm.

On the other hand, most buyers, butchers or scientists consider the ratio: carcass weight/live weight, live weight being
measured just before slaughter. In this case the dressing
percentage varies in larger proportions than the previous one
according to the development of gut content. This latter
depends on:

The nature of the diet and especially on the digestibility
and on the rate of passage of food (Table 1). Thus, the gut
content of animals fed with lucerne hay or with lucerne hay +
concentrate is respectively 17.9 and 12.4 per cent of live
weight. Preston and Willis (1969) also observed a difference
of 3.8 per cent in dressing percentage of bulls fed either with
high forage diets or with concentrate diets.

The feeding level: The reduction in food intake by 17 and
27 per cent from ad libitum feeding led to an increase in gut
content of 9 to 15 months old bulls by 15 and 34 per cent

TABLE 1

VARIATIONS OF THE GUT CONTENT WEIGHT PROPORTION IN THE LIVE WEIGHT FOR DIFFERENT DIETS
(Beranger et al. unpublished data)

Diets	Coefficient of Digestibility			Animals			Gut content weight (P.100 live weight at slaughtering)
	Organic matter	Crude protein	Crude fibre	Type	Number	Live weight	
Grass hay	55.7	46.7	58.1	cows	15	549	21.11
Grass hay + barley	65.3	58.5	56.9	oxen	15	576	16.32
Lucerne hay	58.6	73.5	36.6	cows	28	545	17.88
Lucerne hay + concentrate	–	–	–	oxen	8	505	12.42
Lucerne hay + fodder beet	72.6	65.5	52.8	oxen	12	549	14.32
Fresh Ray-Grass	75.5	63.5	74.5	cows	10	587	13.41
Kales	–	–	–	oxen	15	531	13.52

respectively (Table 3). This phenomenon was probably due to
a decrease in rate of passage of food. Consequently for the
same carcass weight the ratio between carcass weight and live
weight was reduced by 0.4 and 1.3 per cent (Geay et al., unpublished data).

The length of the fasting period: When the duration of
fasting increases, the live weight decreases and the dressing
percentage increases, due to the reduction in the weight of gut
contents. Brannang (1966) observed that a short period of
transport (one hour) and yarding (half an hour) increased
dressing percentage by 1.4 per cent. This increase is more or
less rapid according to the digestibility of the food and the
rate of passage of digesta. This latter was low when hay was
fed, and the dressing percentage varied little after feeding;
on the other hand the rate of passage was high with grass (the
gut contents decreased from 4 to 5 per cent in 10 hours) and
the dressing percentage increased during the same period
(Beranger et al., unpublished data).

The age of animal, which influences the feeding system.
Thus the gut contents represented 3% of live weight at birth,
10% two months after weaning, when the animals received roughages (Mathieu, 1961). Later in consequence the dressing
percentage decreased greatly after weaning (Butterfield et al.,
1966, 1971). Finally during the finishing period even though
the composition of the diet remained the same, the proportion
of gut contents in live weight decreased when this live weight
increased: the allometric coefficient (b) of empty body weight,
related to live weight was in fact, 1.05 (unpublished results
obtained by Robelin with 184 animals). Afterwards, the ratio
between carcass weight and live weight increased with the weight
of animals as noted by Field and Schoonover (1967).

Due to these different factors of variations in gut content
which do not depend on the animal directly, it seems difficult
to estimate the ability of animals to produce 'edible meat'
from the ratio between carcass weight and live weight before

TABLE 2

VARIATION OF RELATIVE GROWTH OF BODY COMPONENTS AND VARIATION WITH SEX AND BREED OF BODY COMPOSITION AT THE SAME EMPTY BODY WEIGHT (X = 358 KG)

Dependent variate (Y)	Carcass	Head	Metarcapus + Metatarsus	Fatty tissues (5th quarter)	Rumen	Omasum	Abomasum
Allometric coefficient (b) – 202 animals –	1.02	0.77	0.43	1.84	1.00	1.13	0.75
Adjusted mean of Y (kg)	238.0	10.9	1.9	9.1	6.9	2.7	1.4
Mean related to empty body weight (per cent)	66.5	3.1	0.5	2.6	1.9	0.8	0.4
Deviation to the mean (per cent)							
Bulls:							
Friesian (n = 83)	-5	+7	-5	+21	+30	+42	+21
Salers (n = 30)	-2	+7	+9	-11	+2	+7	0
Charolais (n = 29)	+1	-1	+4	-24	-5	-7	-1
Limousin (n = 42)	+5	-2	-2	-31	-13	-16	-15
Heifers:							
Limousin (n = 8)	+3	-6	-10	+37	-7	-6	-16
Charolais X (n = 10) Salers	-1	-5	+5	+29	-5	-14	+16

TABLE 2 (Cont.)

Dependent variate (Y)	Intestines	Alimentary Tract	Hide	Liver	Heart	Kidney	Lungs
Allometric coefficient (b) – 202 animals –	0.66	0.84	1.07	0.84	0.81	0.65	0.72
Adjusted mean of Y (kg)	9.1	20.3	34.1	5.0	1.5	0.8	3.2
Mean related to empty body weight (per cent)	2.5	5.7	9.5	1.4	0.4	0.2	0.9
Deviation to the mean (per cent)							
Bulls:							
Friesian (n = 83)	+18	+26	-1	+18	+9	+24	+9
Salers (n = 30)	-4	0	+11	+5	+5	+6	+4
Charolais (n = 29)	-2	-3	-2	-6	+2	-1	-5
Limousir (n = 42)	-15	-14	-1	-13	-11	-8	-8
Heifers:							
Limousin (n = 8)	-4	-6	-8	+1	-5	-12	-4
Charolais x Salers (n = 10)	+9	+2	+1	-5	0	-6	+6

slaughter. It would be better to consider the ratio between carcass weight and empty body weight. Its evolution in relation to live weight, sex and genotype will be studied in the following part.

Variations of dressing percentage according to live weight

Generally dressing percentage increases slightly with empty body weight. We have observed that the allometric coeficient of the carcass, related to empty body weight was, on average, b = 1.02 (Table 2). These results have been obtained with 202 animals of different breeds, fed diets containing 75 to 85 per cent of concentrate and 25 to 15 per cent of roughage and slaughtered at 9 or 14-15 months. This increase in carcass weight with the empty body weight was due essentially to fatty tissues, the 'b' value of which was much greater than one (cf. Robelin et al., 1977). Thus dressing percentage increases mainly with fatness of the animal. On the contrary, growth rate of the fifth quarter was relatively lower than that of empty body weight, because the 'b' value of the major part of its components was lower than one (Table 2) except the rumen, the omasum, the hide and the fatty tissues; but their total weight represented only 14% of the live weight. The rumen and the omasum developed later than the abomasum and the intestine, which are already functioning in the pre-ruminant animal. During the finishing period between 9 and 15 months of age, the fatty tissues increased greatly compared to the empty body weight (b = 1.6 to 1.9; cf. Robelin, 1977), especially the fifth quarter fatty tissues but their amount only represented from 20 to 25 per cent of total fatty tissues.

An increase in the quantities of energy intake involves an increase in the proportion of fat, and consequently in the dressing percentage as related by Andersen (1975). Still, in his experiment, the live weight being used as a reference, an increase in the dressing percentage means an increase in the fat proportion and at the same time, a decrease in the relative development of the gut content. Variations in feeding level can affect differently the fatty tissues of the fifth quarter, and

those of the carcass, thus modifying the dressing percentage.
In Table 3 it can be seen that reducing food intake, in young
bulls (Friesian and Charolais X Salers) respectively by 11.5
per cent and 27 per cent from ad libitum feeding, leads to a
decrease in the proportion of fatty tissues in the whole body.
The decrease was much more significant in the fifth quarter
(55 per cent in Friesian and 34 per cent in Charolais X Salers)
than in the carcass (respectively 18 and 34 per cent). So the
dressing percentage was improved from 0.9 to 1.2 points.

Variations of dressing percentage according to sex and genotype

Brännäng (1966) showed that young bulls had a higher
dressing percentage than their castrated twins, when the kidney
fat was removed from the carcass even though the hide weight in
entire males was higher. At the same empty body weight,
Limousin and Charolais X Salers heifers had a lower dressing
percentage by one unit on average than males (Tables 2 and 4).
The 'b' value of their carcass was lower. This difference in
dressing percentage can be explained by a lower muscle weight,
a higher fat weight, in particular in the fifth quarter (6 per
cent additional fat in the carcass of Limousin heifers and 26
per cent in their fifth quarter) and a higher digestive tract
weight. However the weight of their hide was lower (8 per cent).

Many studies have shown differences in dressing percentage
with genotype (Damon et al., 1960; Cole et al., 1964; Fredeen et
al., 1972; Koch et al., 1976). In most of these studies, however, the carcass weight was related to the live weight
measured either at the end of the experiment or before
slaughter. In order to eliminate the influence of gut content,
we have related the carcass weight of young bulls to their empty
body weight (Fig. 1 and Table 2). It can be seen that Limousin
body bulls, are superior to other breeds, especially to Friesian
young bulls. This superiority was confirmed by Koch et al.
(1976) with crossbred Limousin X Hereford and Limousin X Angus
steers and by Andersen (1977) with crossbred Limousin X Red
Danish bulls. Thus, the allometric coefficient of the carcass
of Limousin, Charolais and Salers bulls were identical, and

TABLE 3

VARIATION OF BODY COMPOSITION OF YOUNG BULLS WITH LEVEL OF INTAKE

BREEDS	LEVEL OF INTAKE	LIVE WEIGHT (kg)	EMPTY BODY WEIGHT (kg) (E.B.W.)	CARCASS WEIGHT / LIVE WEIGHT	CARCASS WEIGHT / E.B.W.	GUT CONTENT / LIVE WEIGHT	FATTY TISSUES (kg) TOTAL	5TH QUARTER	CARCASS
Charolais X Salers	100	576	527	60.5	66.1	8.4	71.7	17.3	54.5
	73	557	494	59.7	67.3	11.3	51.0	10.5	40.6
Friesian	100	501	443	55.1	62.3	11.6	63.2	16.6	46.5
	88	499	432	54.9	63.4	13.3	50.3	10.6	39.7

TABLE 4

VARIATION OF BODY COMPOSITION WITH SEX

BREEDS	SEX	EMPTY BODY WEIGHT (kg) (EBW)	CARCASS WEIGHT / E.B.W	MUSCLE (kg)	FATTY TISSUES (kg) CARCASS	5th QUARTER	TOTAL
Charolais X Salers	Bulls (n = 50)	485.7	66.5	236.0	41.3	11.4	52.7
	Heifers (n = 50)	395.6	64.7	170.6	48.1	15.0	63.1
Limousin	Bulls (n = 8)	392.0	69.5	209.0	27.7	7.2	34.9
	Heifers (n = 11)	302.0	67.0	144.0	29.3	9.1	38.4

slightly higher than that of the Friesian bulls. At the same
empty body weight Limousin bulls had a lower fatty tissues weight
in the fifth quarter than the mean of the studied population
(especially the Friesian-75% less) and a lower digestive tract
weight (47% less). Even though they received the same diet as
the Friesian bulls, the Limousin bulls had a lower energy intake and a lower liver weight (36%).

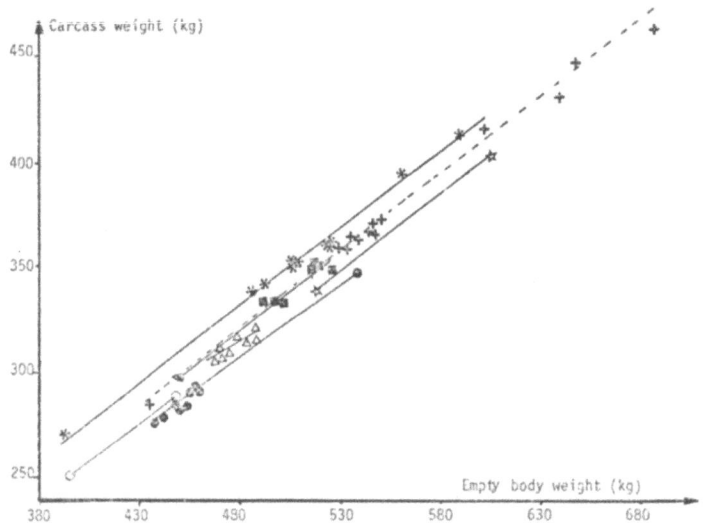

* Limousin ; + Charolais ; ▪ Charolais X Salers ; △ Salers ; ✿ Maine Anjou ○ Hereford ; ● Friesian

Fig. 1. Variation of carcass weight of young bulls with empty body weight
and breed. (Each point represents the mean value of a group of
animals 7 ≤ N ≤ 15).

CONCLUSION

The dressing percentage can be considered as a poor
criterion for evaluating 'edible meat' in animals when carcass
weight is related to live weight (Berg and Butterfield, 1975).
However, when carcass weight is related to empty body weight
the dressing percentage appears to be a useful and cheap
measurement in comparing animals for any given type of production.

REFERENCES

Andersen, B.B., Liboriussen, T., Kousgaard, K., Buchter, L., 1977. Crossbreeding experiment with beef and dual purpose sire breeds on Danish
dairy cows. III. Daily gain feed conversion and carcass quality of
intensively fed young bulls. Livest. Prod. Sci., 4, 19-31.

Andersen, H.R. 1975. The influence of slaughter weight and level of feeding
on growth rate, feed conversion and carcass composition of bulls. Livest.
Prod. Sci., 2, 341-355.

Berg, R.T., Butterfield, R.M., 1976. New concepts of cattle growth. Sydney
University Press.

Butterfield, R.M., Pryor, W.J., Berg, R.T., 1966. A study of carcass growth
in calves. Res. Vet. Sci. 7, 417-423.

Butterfield, R.M. Johnson, E.R., 1971. A study of growth in calves. 1.
Carcass tissues. J. Agric. Sci., Camb. 76, 453-458.

Cole, J.M., Ramsey, C.B., Hobbs, C.S., Temple, R.S., 1964. Effects of type
and breed of British Zebu and dairy cattle on production, palatability
and composition. III. Percent wholesale cuts and yield of edible
portion as determined by physical and chemical analysis. J. Anim. Sci.
23, 71-77.

Damon, R.A. Jr., Crown, R.M., Singletary, C.B., Mc. Craine, S.E., 1960.
Carcass characteristics of purebred and crossbred beef steers in the
gulf coast region. J. Anim, Sci., 19, 820-844.

Field, R.A., Schoonover, C.O., 1976. Equations for comparing longissimus
dorsi areas in bulls of different weights. J. Anim. Sci., 26, 701.

Fredeen, H.T., Martin, A.H., Weiss, G.M., 1971. Characteristics of youthful
beef carcasses in relation to weight, age and sex. II. Carcass measurement and yield of retail product. Can. J. Anim. Sci., 51, 291-304.

Koch, R.M., Dikeman, M.E., Allen, D.M., May M., Crouse, J.D., Campton, D.R.,
1976. Characterisation of biological types of cattle. III. Carcass
composition, quality and palatability. J. Anim. Sci. 43, 48-63.

Mathieu, C.M., 1961. Etude du développement du tractus digestif du veau.
Ann. Nutr. et Alim. 15, 263-266.

Preston, R.R., Willis, M.B., 1970. Intensive beef production. Pergamon,
Oxford.

Robelin, J., Geay Y., Bonatti, B., 1977. Genetic variations in growth and
body composition. Seminar on 'Patterns of growth and development in
cattle'. Gand.

Robelin, J., 1977. Development with age of the anatomical composition of the carcass of bulls. Seminor on 'Patterns of growth and development in cattle'. Gand.

DEVELOPMENT WITH AGE OF THE ANATOMICAL COMPOSITION OF THE CARCASS OF BULLS

J. Robelin

Institut National de la Recherche Agronomique
Laboratoire de la Production de Viande, Centre de Recherches
Zootechniques et Vétérinaires Theix - 63110   Beaumont - France

ABSTRACT

Results of complete dissection of 189 bull carcasses (Friesian, Charolais, Limousin and Salers) from 4 to 19 months of age, weighing from 120 to 650 kg are analysed with allometric relationship.

Relative growth of main carcass tissues, different fatty tissues, and main muscle-groups are successively considered. The main particularities of these continental breed bulls are their low fat and their high muscle content, related to a lower cost of body weight gain.

INTRODUCTION

Development with age and the body weight of main carcass
tissues in cattle (muscle, fat and bone) have mainly been studied
in the past on castrated males (Trowbridge et al, 1919; Callow,
1948; Anon., 1966). A detailed survey of the knowledge has been
given recently by Berg and Butterfield (1976). However, there
have been few results to date, relating to the evolution of the
body composition of entire male animals out of the studies of
Schulz et al. (1974) or Andersen (1975). These works have been
carried out with Friesian type cattle. Similar studies have been
done in our laboratory over the past 10 years on Friesian bulls
as well (Robelin et al.,1974) and also on large beef breed animal
(Charolais, Limousin).

The results presented here represent a synthesis of the
dissection data from 189 male cattle of several breeds (Friesian,
Charolais, Salers and Limousin), slaughtered at different
weights between 120 and 650 kg.

MATERIALS AND METHODS

The bulls were serially slaughtered during experiments
conducted for nutritional purposes, in order to measure the
evolution of body composition. They were generally fed a high
energy diet (75% to 85% concentrate) ad libitum (or slightly
restricted) in order to obtain a high daily gain (1 000 to 1 400
g/day). In each experiment and for each slaughter group, the
animals were randomly chosen on the basis of their weight and
their age.

The distribution of animals per breed and weight is
illustrated in Table 1. The distribution per breed is approximately 40% Friesian (early maturing), 35% Charolais and Limousin
(late maturing) and 25% Salers or Charolais crossbred (intermediate). According to the weight distribution, we can reasonabl
accept the validity of results between 200 and 650 kg body weight

TABLE 1

DISTRIBUTION OF ANIMALS PER BREED AND PER WEIGHT

Breed	Body weight range (kg)							
	<150	150-250	250-350	350-450	450-550	550-650	>650	All
Friesian	3	5	23	21	31	1		84
Charolais	1	1	8	5	4	10		29
Limousin			8	6	11	13	4	42
Salers		1	8		10	11		30
Salers x Charolais			4					4
Total	4	7	51	32	56	35	4	189

The dissections were carried out on the right sides of carcasses two days after slaughter. The procedure was fully described in an earlier paper (Robelin et al, 1974). The muscles were separated on an anatomical basis into 7 parts (Robelin and Geay, 1976): neck muscles, muscles of the thoraxic limb, muscles connecting the thorax to the thoraxic limb, muscles of the abdominal wall, muscles surrounding the spinal column, muscles between the ribs, muscles of the pelvic limb. These groups are not exactly identical to the standard muscle groups used by Butterfield (1963), but they were near enough to make consistent comparisons.

The fatty tissues were classified as subcutaneous, inter-muscular and internal fat. Kidney fat was removed from the carcass at slaughter.

The data were analysed after logarithmic transformations with the allometric relationship:

$$\text{Log } Y = a + b \log X \ (\pm \text{ Syx})$$

The residual coefficient of variation used in the text, refers to RCV = 100 x $(1 - 10^{\text{Syx}})$. The breed effect was not taken into account in this paper (it will be analysed in a later paper in Session 2). According to the distribution of animals

between breeds, the results could be considered as a 'mean' for
a sampling of bulls between early and late maturing breeds.

RESULTS AND DISCUSSION

Evolution of main carcass tissues with carcass weight

As it is shown in Figure 1, the allometric coefficients of
carcass tissues are in the classical hierarchy: 0.66, 1.00 and
1.48 respectively for bone, muscle and fat. They indicate that
the percentage of bone decreases slowly from 17% for 150 kg
carcass weight to 14% for 300 kg carcass weight. The percentage
of muscle remains approximately constant (70%). The percentage
of fat increases from 10% for CW = 150 kg to 14% for CW = 300 kg.
The allometric coefficient of fat is slightly underestimated by
the fact that the proportion in late maturing animals is greater
for heavier weight (Table 1; Figure 1); the common slope of fat
obtained by covariance analysis was 1.65.

The muscle/bone ratio, often used as a criterion of muscle
development, increases with the power 0.34 of carcass weight
from 4.0 for CW = 150 kg to 5.1 for CW = 300 kg. The results
on fat growth differ widely with published data obtained with
Angus or Hereford steers (b = 1.8 - 2.3; Tulloh, 1963; Seebeck
and Tulloh, 1968; Murray et al, 1974), or with Friesian bulls
(1.7 to 2.0: Robelin et al., 1974; Andersen, 1975). They vary
as well, with regard to the amount of fat at a given carcass
weight (say 220 kg): approximately 60 kg in an Angus, Hereford,
or Friesian steer (Anon., 1966; Seebeck and Tulloh, 1968)
against 27 kg in our results (29 kg when kidney fat was not
removed at slaughter).

Evolution of muscle distribution

The variation in relative growth of various parts of
musculature is very narrow, from b = 0.9 to b = 1.2 (Figure 2.
Table 2). The neck and thorax to thoraxic limb muscles grow
more quickly than the whole musculature (b greater than 1.1.),
the muscles of the hind limb grow more slowly (b = 0.91). The
comparison established in Table 2 does not reveal major

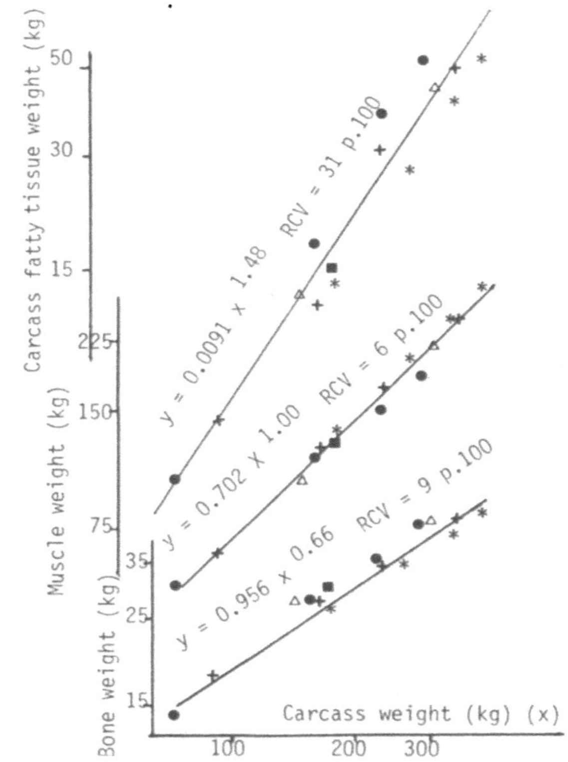

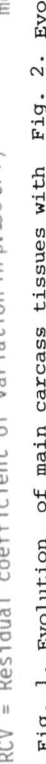

Fig. 1. Evolution of main carcass tissues with carcass weight in bulls.

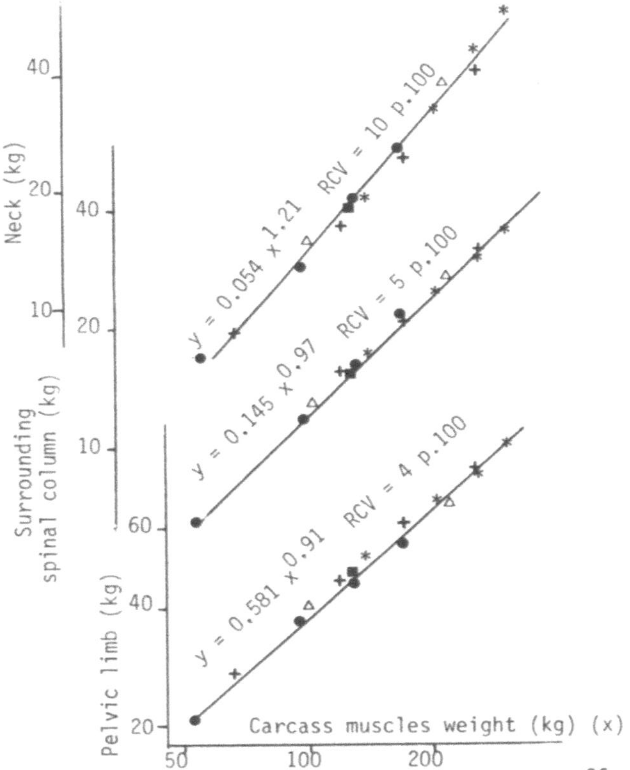

Fig. 2. Evolution of various parts of musculature with the weight of carcass muscles in bulls. (See footnote figure 1).

(N = 189 animals ; ● = FRISIAN ; + = CHAROLAIS ;
* = LIMOUSIN ; △ = SALERS ; ■ = SALERS x CHAROLAIS
(RCV = Residual coefficient of variation in p.100 of Y)

Other parts) Thoraxic limb (kg)=0.167x$^{1.96}$ RCV= 6 %
of the) Abdominal wall (kg)=0.073x$^{0.98}$ RCV= 7 %
musculature) Thorax-Thoracic limb (kg)=0.055x$^{1.13}$ RCV=7 %
) Thorax (kg)=0.055X$^{0.98}$ RCV=10 %

differences between our results and those of Berg and Butterfield
(1976) established with shorthorn crossbred bulls. The relativel
small (lower than 10%) residual coefficients of variation of
muscle groups weight (Figure 2) suggest that there are very
little variations between breeds in muscle weight distribution.
However, it is necessary to compare under similar conditions to
make such conclusions. This subject will be discussed in other
papers (Bergström, 1977; Robelin et al, 1977).

TABLE 2

RELATIVE GROWTH OF VARIOUS PARTS OF MUSCULATURE IN BULLS

Groups of muscles	Allometric coefficient	Allometric coefficient	Groups of muscles
Pelvic limb	0.91	(0.84 (0.92	(Proximal pelvic limb (Distal pelvic limb
Surrounding spinal column	0.97	0.97	Surrounding spinal column
Abdominal wall	0.98	1.12	Abdominal wall
Thoraxic limb	0.96	(1.00 (1.00	(Proximal thoraxic limb (Distal thoraxic limb
Thorax-Thoraxic limb	1.13	1.49	Thorax-Thoraxic limb
Neck	1.21)	(1.28	Neck-Thoraxic limb
M. Between the ribs	0.98)	(1.21	Thorax and neck
Results presented in the text		Berg and Butterfield (1976)	

It is of economical importance to observe that the percentage of expensive muscles (pelvic limb, and surrounding spinal
column) diminishes when the weight of the musculature increases
from 51% of the whole musculature for a muscle weight of 50 kg,
to 48% for 200 kg muscle weight.

Relative growth of fatty tissues

Opposed to muscle groups, the relative growth of fatty
tissues varies widely with their location. The hierarchy in
allometric coefficients, 0.9 for intermuscular fat, 1.0 for
internal fat and 1.4 for subcutaneous fat is comparable to that
observed previously in steers (Anon., 1966; Johnson et al. 1972).
However, the partition of these tissues and particularly the

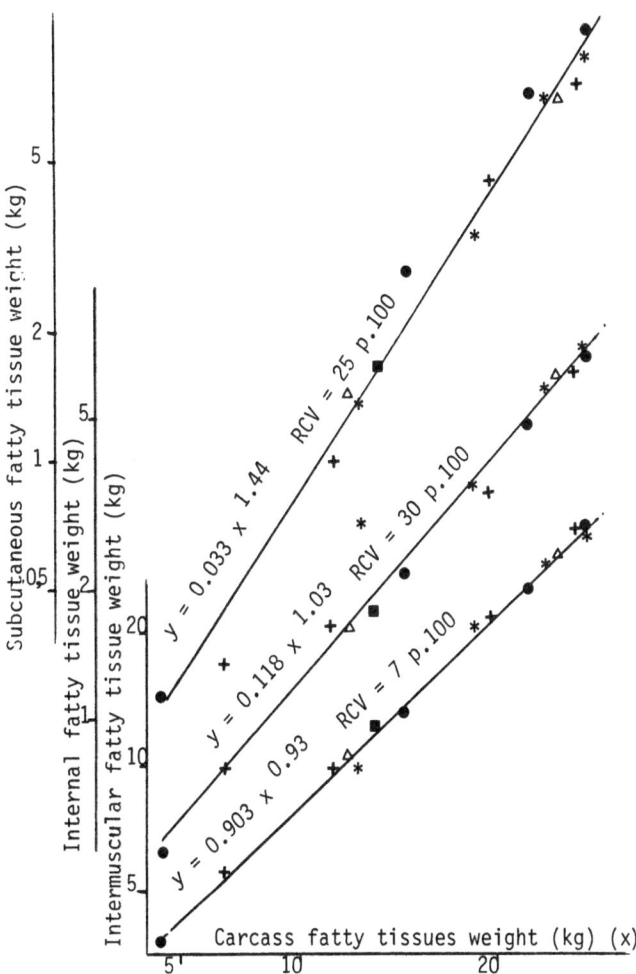

Fig.3. Evolution of various fatty tissues weight with carcass fatty tissues weight in bulls. (See footnote figure 1).

amount of subcutaneous fat seems to be very variable between sexes and breeds. From 17% of carcass fat in our sample of continental breed bulls, it could reach 30% to 37% of carcass fat in Hereford or Angus steers compared with the same weight (40 kg) of carcass fat (Murray et al, 1974; Charles and Johnson, 1976).

CONCLUSIONS

All these results show a certain constancy in relative
growth of carcass tissues of cattle, as well in the relative
increase of their different parts. The number of animals used
in this study and their weight variation (120 to 650 kg) permits
the intrinsic characteristics of continental breeds of bulls
compared to those of Angus, Hereford or Friesian steers to be
known. Their particularities are their high development of
musculature in relation to a low development of fat, particularly
subcutaneous fat. The main practical consequence of this, is
the lower cost of body weight of these animals.

REFERENCES

Andersen, H.R. 1975. The influence of slaughter weight and level of feeding
on growth rate, feed conversion and carcass composition of bulls. Liv.
Prod. Sci., 2, 341-355.

Anonymous. 1966. A comparison of the growth of different types of cattle
for beef production. Report of Major Beef Research Project. The
Royal Smithfield Club, London.

Berg, R.T. and Butterfield, R.M. 1976. Muscle growth patterns in steers.
In 'New concepts of cattle growth' Sydney Press University.

Bergstöm, P.L. 1977. Source of variation of muscle weight distribution.
In 'Patterns of growth and development in cattle'.

Butterfield, R.M. 1963. Relative growth of the musculature of the ox.
In 'Carcass composition and appraisal of meat animals'., 7, 1-14.
(Ed. D.E. Tribe). East Melbourne CSIRO.

Callow, E.H. 1948. Comparative studies of meat. II. The changes in the
carcass during growth and fattening and their relation to the chemical
composition of the fatty and muscular tissues. J. Agric. Sci. 38, 174-199.

Charles, D.D. and Johnson, E.R. 1976. Breed differences in amount and distribution of bovine carcass dissectible fat. J. Anim. Sci., 42, 322-341.

Johnson, E.R., Butterfield, R.M. and Pryor, W.J. 1972. Studies of fat
distribution in bovine carcass. I. The partition of fatty tissues
between depots. Aust. J. Agric. Res., 23, 381-388.

Murray, D.M., Tulloh, N.M. and Winter, W.H. 1974. Effects of three different
growth rates on empty body weight, carcass weight and dissected carcass
composition of cattle. J. Agric. Sci. 82, 535-547.

Robelin, J. and Geay, Y., 1976. Note: Répartition des masses musculaires chez
le jeune bovine mâle entier, et son évolution au cours de la période
d'engraissement entre 8-9 et 16-17 mois. Ann. Zootech., 25, 273-279.

Robelin, J., Geay, Y. and Beranger, C. 1974. Croissance relative des différents tissus, organes et régions corporelles des taurillons Frisons,
durant la phase d'engraissement de 9 a 15 mois. Ann. Zootech., 23 313-323.

Robelin, J., Geay, Y. and Bonaiti, B. 1977. Genetic variations in growth
and body composition of male cattle. CEC Seminar on 'Patterns of
growth and development in cattle', Ghent.

Seebeck, R.M. and Tulloh, N.M. 1968. Developmental growth and body weight
loss of cattle. II. Dissected components of the commercially dressed
and jointed carcass. Aust. J. Agric. Res., 19, 477-95.

Schulz, E., Oslage, H.J. and Daenicke, R. 1974.   Untersuchungen über die Zusammensetzung der Körpersubtanz sowie den Stoff-und Energieansatz bei wachsenden Mastbullen. Fortschritte in der Tierphysiologie und Tierernährung. 4.

Trowbridge, P.F., Moulton, C.R. and Haigh, L.D. 1919. Composition of the beef animal and energy cost of fattening. Mo. Agr. Expt. Station, Research Bulletin number 30.

Tulloh, N.M. 1963. The carcass composition of sheep, cattle and pigs as function of body weight. In 'Carcass composition and appraisal of meat animal'. 5, 1-16. Ed. D.E. Tribe. East Melbourne CSIRO.

BIOCHEMISTRY OF MUSCLE IN RELATION TO GROWTH

R.A. Lawrie

University of Nottingham,
Department of Applied Biochemistry & Nutrition,
Sutton Bonington, Leicestershire, UK

ABSTRACT

This paper deals firstly with the structural and chemical parameters which characterise muscle generally; and with the features which distinguish muscle into two broad types 'red' and 'white'.

An outline is then given of the effect which increasing chronological age in the animal has upon these parameters; and finally consideration is given to the practical significance which certain of these age-related changes have in relation to meat - the commodity which muscle becomes post mortem.

INTRODUCTION

General Nature of Muscle

The light microscope shows muscle to be composed of parallel arrays of fibres which, in turn, are composed of yet smaller parallel entities, myofibrils. The electron microscope reveals that myofibrils are themselves composed of longitudinally linked, repeating structures. These are called sarcomeres and they are the basic units of contractibility. Viewed in longitudinal section, they are bounded, at each end by so-called 'Z-lines'. Their central area is occupied by a parallel array of relatively thick filaments; and relatively thin filaments, originating from the Z-line at each end of the sarcomere, interpenetrate for some distance between the array of thick filaments.

The thick filaments consist mainly of aggregates of tadpole-like molecules of the protein myosin. These are disposed within the thick filament so that the 'heads' of one half of the molecules are directed towards the Z-line at one end of the sarcomere and those of the other half of the molecules towards the 'Z-line' at the opposite end. In the centre of the sarcomere the 'tails' of the two groups of molecules are held by a protein lattice (the 'M' bridge). The 'heads' are also disposed helically around the body of the thick filaments so that each sixth one is opposite one of the surrounding thin filaments. The latter consist of globules of the protein actin disposed along a thread of tropomyosin by which one end of the thin filament attaches to the 'Z-lines'.

In contraction, the myosin 'heads' link with progressively peripheral segments of the actin filaments, so that these interdigitate further inwards between the myosin filaments, bringing the opposing 'Z-lines' to which they are attached closer together and causing the sarcomere to shorten.

This so-called sliding filament hypothesis to explain
muscular contraction was postulated by Huxley in 1960; and has
not been seriously challenged since, although much more detail
is now known. Thus, associated with the 'head' or (H-meromyosin)
of the myosin molecule, are several proteins of relatively low
MW, known as the light-chain components and troponins. The
myosin light chain components are distinguished by how they
are phosphorylated; but their function in contraction has yet
to be elucidated (Perry, Cole, Morgan, Moir and Pires, 1975).
Troponin C binds calcium ions when these are released from the
sarcoplasmic reticulum on receipt of nerve impulses and, as
it does so, it changes shape whereby troponin I no longer
prevents actomyosin ATP-ase from operating. Troponin T binds
to tropomyosin.   There are present also a and B actonins which
control the aggregation of the actin molecules on the tropomyosin thread.

The system of contractile proteins is served by a large
number of water soluble sarcoplasmic proteins which include
the enzymes required for resynthesis, from glycolysis of
carbohydrate, of ATP (whose splitting by actomyosin ATP-ase
yields the energy for contraction), for osmotic regulation and
for many other purposes ancillary to the main contractile
function of the tissue. It is also served by organelle-bound
enzymes which effect ATP resynthesis by respiration. In Table
1 the chemical composition typical of adult mammalian muscle,
after rigor mortis but before degradative changes, is summarised.

POST NATAL CHANGES
In postnatal growth, the observed increase in the diameter of muscle fibres is due to increased numbers of myofibrils
in each; and increase in fibre length is due to an increase in
length of constituent sarcomeres rather than to an increase in
the number of the latter (Goldspink, 1962).

Early postnatal muscle growth is characterised by rapid
radial growth of 'white' myofibrils and slower growth of 'red'
ones. Later postnatal growth is characterised by sustained

TABLE 1

CHEMICAL COMPOSITION OF TYPICAL ADULT MAMMALIAN MUSCLE AFTER RIGOR MORTIS
BUT BEFORE DEGRADATIVE CHANGES POST MORTEM

Components		% Wet Weight
1. WATER		<u>75.0</u>
2. PROTEIN		<u>19.0</u>
(a) Myofibrillar	11.5	
myosin (H and L meromyosins, and several light chain proteins associated with them)	6.5)	
actin	2.5)	
tropomyosins	1.5)	
troponins C, I and T	0.4)	
a and B actinins	0.4)	
M protein etc.	0.2)	
(b) Sarcoplasmic	5.5	
glyceraldehyde phosphate dehydrogenase	1.2)	
aldolase	0.6)	
creatine kinase	0.5)	
other glycolytic enzymes	2.2)	
myoglobin	0.4)	
haemoglobin and other unspecified extracellular proteins	0.6)	
(c) Connective tissue and organelle	2.0	
collagen	1.0)	
elastin	0.05)	
mitochondrial etc. (including cytochrome <u>c</u> and insoluble enzymes)	0.95)	
3. LIPID		<u>2.5</u>
neutral lipid, phospholipids, fatty acids, fat-soluble substances		
4. CARBOHYDRATE		<u>1.2</u>
lactic acid	0.90)	
glucose-6-phosphate	0.15)	
glycogen	0.10)	
glucose, traces of other glycotic intermediates	0.05)	
5. MISCELLANEOUS SOLUBLE NON-PROTEIN SUBSTANCES		<u>2.3</u>
(a) Nitrogeneous	1.65	
creatine	0.55)	
inosine monophosphate	0.30)	
di- and tri-phosphopyridine nucleotides	0.10)	
amino acids	0.35)	
carnosine, anserine	0.35)	
(b) Inorganic	0.65	
total soluble phosphorus	0.20)	
potassium	0.35)	
sodium	0.05)	

continued/

TABLE 1 continued

| magnesium | 0.02) |
| calcium, zinc, trace metals | 0.03) |

6. VITAMINS
Various fat - and water - soluble vitamins, quantitatively minute

[1]Actin and myosin are combined as actomyosin in post-rigor muscle

TABLE 2

COMPARATIVE COMPOSITION OF BOVINE *L.dorsi* MUSCLE AT BIRTH AND 24 MONTHS OF AGE

Parameter	12 day old calf	24 month old steer
Moisture %	78	75
Intermuscular fat : %	0.6	2.0
: iodine No	82.5	58
Nitrogen Total %	3.30	3.52
myofibrillar protein %	1.52	1.61
sarcoplasmic protein %	0.62	0.87
stroma protein %	0.80	0.65
non-protein %	0.36	0.39
Myoglobin %	0.07	0.46

growth of 'red' myofibrils (Swatland, 1976). This would
accord with the observation that myoglobin concentration
appears to increase slowly at first and then more rapidly up
to maturity (cf. Figure 1).

Because of the variability in muscle composition even
within a given species, it is preferable to consider the effect
of animal age/growth within a single muscle; and since by far
the largest muscle in the bovine is *Longissimus dorsi*, it has
been chosen to exemplify the pattern.

A comparison of the composition of the *L. dorsi* of the
newly born calf with that of the mature, 24 month-old steer
(Table 2) shows that there is more moisture and less fat
present at birth; and that the neonatal fat has more unsaturated
since it represents, preponderantly, the lipid of membranes
and other essential muscle structural members. Relative
differences in protein content are much less marked - although
there is rather less myofibrillar and sarcoplasmic protein,
and somewhat more connective tissue (stroma) protein in the
muscle at birth. A particularly striking difference, however,
is shown by the concentration of the muscle pigment myoglobin,
which increases nearly sevenfold between birth and maturity.

Whereas the myofibrillar proteins and those of the
sarcoplasm have attained about 73% of their value at 24 months
by birth, the corresponding value at that time for myoglobin
is only about 13% (Figure 1). Much of the subsequent increase
in myoglobin concentration occurs at a period when no further
change in the major protein composition is taking place.
Concomitantly, however, the gross structure of the muscle is
altering. Thus, Bendall and Voyle (1967) demonstrated that the
number of fibres in bovine *L. dorsi* falls by about 50% from
birth to 24 months of age; whereas their mean fibre diameter
increases. When expressed on the same basis as the other
parameters in Figure 1, the mean diameter closely parallels
the increase in myoglobin.

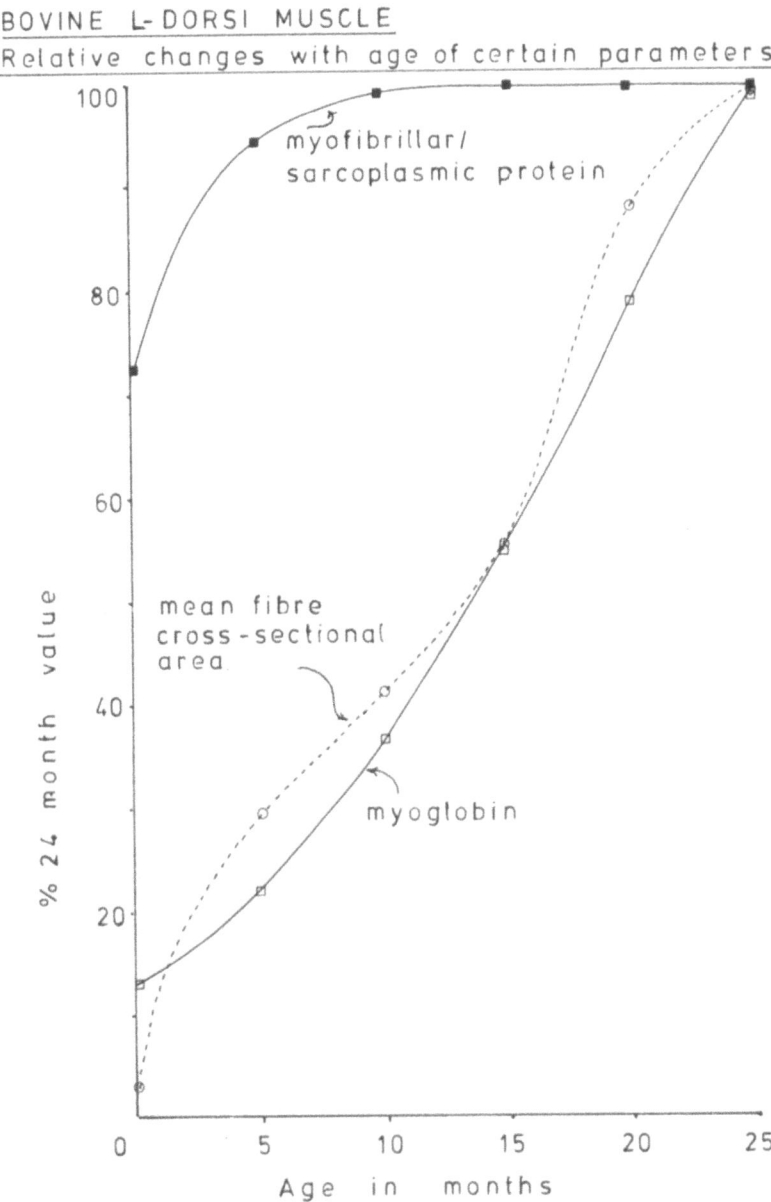

BOVINE L-DORSI MUSCLE
Relative changes with age of certain parameters

myofibrillar/
sarcoplasmic protein

mean fibre
cross-sectional
area

myoglobin

% 24 month value

Age in months

Fig. 1

Myoglobin is a short-term oxygen store elaborated in muscles which require a sustained level of energy provision and which derive it from respiratory mechanisms in mitochondria involving such enzymes as cytochrome oxidase and succinic dehydrogenase. Indeed it can be shown (in equine *L. dorsi*) that the intrinsic activities of these two enzymes increase over the period from birth to 24 months of age in strict parallel with myoglobin concentration (Lawrie, 1952, 1953).

These considerations suggest that a dominant biochemical change associated with growth in bovine musculature is the development of a capacity for aerobic metabolism i.e. a change from a 'white' to a 'red' type of muscle, as the fibre diameter increases to adult dimensions.

Although detailed biochemical information on bovine muscles is less available than that of other species, it would be presumed that general differences which have been demonstrated between 'red' and 'white' muscles would apply. During growth in bovine muscles, therefore, one would anticipate various biochemical changes such as those in Table 3.

SOME PRACTICAL IMPLICATIONS

The biochemical differentiation of the musculature has a direct bearing on meat quality. This has become particularly evident in an era when subdivision of meat before sale, into portions for the individual consumer, is frequent. This practice enhances observation of the effects of muscle differentiation on the quality attributes of meat - tenderness, colour, water-holding capacity and flavour.

When the temperature of prerigor muscles is brought below ca $10^{\circ}C$ whilst the pH is still above ca 6.7 (and they are still physiologically reactive) they 'cold-shorten' (Locker and Hagyard, 1963) due to the release of calcium ions from the sarcotubular system (Cassens and Newbold, 1967) and from the mitochondria (Buege and Marsh, 1975) and to the varying ability of

TABLE 3

SELECTED RELATIVE DIFFERENCES BETWEEN IMMATURE ('WHITE') AND MATURE ('RED')
MUSCLES

Parameter	Immature	Mature	Reference
(a) Enzymic			
Capacity for aerobic synthesis	Low	High	Lawrie (1952)
Respiratory enzymes	Scarce	Plenti-ful	Lawrie (1952, 1953) Bocek et al., (1966)
Glycolytic enzymes	Plenti-ful	Scarce	Lawrie (1953) Scopes (1970)
Control of actomyosin ATP ase (sarcoplasmic reticulum)	High	Low	Gergely et al., (1965)
Calcium activated factor	High	Low	Goll et al.(1971)
Myofibrillar protein synthesising capacity (Polyribosomes)	High	Low	Iyengar & Goldspink (1971)
(b) Structural			
Actin/myosin filament overlap	Little	Consid-erable	Goldspink (1970)
Myosin : 3 MeH content	High	Low	Johnson & Perry (1970)
Troponin I : phosphorylation	Low	High	Cole & Perry (1975)
α/β tropomyosin ratio	High	Low	Cummins & Perry (1973)
Collagen : concentration	High	Low	Lawrie (1961)
: reducible cross links	Plenti-ful	Scarce	Bailey (1974)
Z lines	Narrow	Wide	Gauthier (1969)
Mitochondria	Scarce	Plenti-ful	Lawrie (1952)
Fibre diameter	Narrow	Wide	Bendall & Voyle (1967)

the ATP-energised pump of the sarcoplasmic reticulum to remove
them at lower than in vivo temperatures. Stimulation of the
contractile actomyosin ATP-ase results (Davey and Gilbert, 1974)
leading to shortening and toughening (because in shortened muscles, the actomyosin from adjacent sarcomeres forms a continuum
Locker, 1976). The capacity of the sarcoplasmic reticulum is
less developed in 'red' muscles (Fawcett and Revel, 1961) and
mitochondria are more prevalent in the latter (Lawrie, 1952)
than in 'white' muscles. There is thus a greater tendency for
'red' muscles to toughen when exposed to chilling in the immediat
post mortem period. The increase in myoglobin with increased
age, and the concomitant emphasis on aerobic energy-yielding
mechanisms, to which reference has already been made, is thus
found to be associated with a greater susceptibility to 'cold-
shortening' in the meat of older bovines (Davey and Gilbert,
1975a).

But even for a given degree of shortening post mortem -
whether induced by cold or otherwise - the muscles of young beef
animals are more tender than those of older ones (Davey and
Gilbert, 1975b). During the cooking of veal, its collagen readily dissolves, whereas that of older bovines is insoluble (Carmichael and Lawrie, 1967) and the meat tough. The constituent
polypeptide chains are more highly cross-linked in the collagen
of older bovines (Bailey, 1974) and generate a greater tension
during heat contraction thus binding the denatured myofibrillar
proteins more tightly.

It has long been recognised that when meat is held for
some days at chill temperature it becomes more tender ('conditioning'or 'ageing'); but the increase in tenderness in 'red'
muscles is less marked than that in 'white'. It is now apprec-
iated (Goll et al., 1971) that an important reason for this
difference is the higher concentration in 'white' muscles of a
proteolytic enzyme - calcium - activated sarcoplasmic factor -
which appears to attack the junction between actin filaments
and the 'Z-line' post mortem (Coll, 1970), and the so-called
'gap-filaments' (Locker, 1976; Davey and Graafhis, 1976). It

would thus be expected that the meat from the older animals
would show a less marked improvement in tenderness during conditioning than corresponding muscles from younger ones.

Apart from the increase in myoglobin concentration with
muscle growth, there is another practical implication of the
concomitant development of the cytochrome enzymic system. The
higher residual oxygen uptake by the latter in older animals
(Lawrie, 1952) means that the bright red layer of oxymyoglobin
on the cut surface of meat from older animals is thinner. The
meat colour thus has a darker hue as well as a greater intensity.
On the other hand, because of the lower iodine number of the
intramuscular fat with age in the ruminant (Lawrie, 1961) there
will be less tendency for the synergistic oxidation of fat and
myoglobin; and thus older beef would not be expected to discolour
so readily by the formation of brown metmyoglobin in cold store
(Ledward, 1971).

Veal has a higher water-holding capacity than beef, not-
withstanding the lower concentration of fat and the higher concentration of water, in the musculature. The effect may partly
be due to the somewhat higher ultimate pH found in the musculature of beef animals before maturity (Lawrie, 1961) or to an
intrinsic difference in the protein/water relationship.

The flavour of beef increases up to about 18 months of
age, but remains largely unchanged thereafter (Patterson, 1975).
Its development, therefore, closely parallels the growth of the
animal. Indeed, this is perhaps not surprising since it is
now recognised that meat flavour is mainly due to volatiles
produced during cooking by interactions between fats and proteins
and minor muscle constituents. (Yueh and Strong, 1960; Patterson, 1975) and, as we have seen, these major precursors increase
in quantity, and change in quality, with growth. No doubt differences in metabolism due to the elaboration of sex-linked hormones, and in the absorption and retention of components from
feed, also alter the pattern of flavour volatiles between the
meat from young and mature beef animals. Moreover the develop-

ment of a vigorous rumen microflora as bovines grow, alters the nature of the flavour precursors absorbed from the gastro-intestinal tract; and hence the potential flavour of the meat.

CONCLUSION

In the present brief survey it has not been possible to consider the important field of muscle embryology. Attention has been concentrated on postnatal muscle. This is the form encountered by the consumer of meat.

The biochemical nature of muscular tissue is primarily determined by the tissue's function of contractibility; and growth of the animal affects the biochemical pattern insofar as it determines the nature of the components responsible for contraction.

It is evident that such changes in biochemical pattern must be reflected in various aspects of eating quality; and that this ultimate effect of animal growth must be kept in mind by physiologists and breeders.

REFERENCES

Bailey, A.J. 1974. Tissue and species specificity in the cross-linking of collagen. Path. Biol. 22, 675-680.

Beatty, C.H. and Bocek, R.M. 1970. Biochemistry of red and white muscle. In: Physiology and Biochemistry of Muscle as a Food. Vol. II (Eds. E.J. Briskey, R.G. Cassens and B.B. Marsh). Uni. Wisconsin Press Madison). pp. 155-191.

Bendall, J.R. and Voyle, C.A. 1967. A study of the histological changes in the growing muscles of beef animals. J. Fd. Technol. 2, 259-283.

Bocek, R.M., Basinger, G.M. and Beatty, C.H. 1966. Comparison of glucose uptake and carbohydrate utilisation in red and white muscles. Am. J. Physiol. 210, 1108.

Buege, D.R. and Marsh, B.B. 1975. Mitochondrial calcium and post mortem muscle shortening. Biochem. Biophys. Res. Commun. 65, 478-482.

Carmichael, D.J. and Lawrie, R.A. 1967. Bovine collagen. I. Changes in solubility with animal age. J. Fd. Technol. 2, 299-311.

Cassens, R.G. and Newbold, R.P. 1967. Effect of temperature on the time course of rigor mortis in ox muscle. J. Fd. Sci. 32, 269-272.

Cole, H.A. and Perry, S.V. 1975. The phosphorylation of troponin I from cardiac muscle. Biochem. J. 149, 525-533.

Cummins, P. and Perry, S.V. 1973. The subunits and biological activity of polymorphic forms of tropomyosin. Biochem. J. 133, 765-777.

Davey, C.L. and Gilbert K.V. 1974. The mechanism of cold-induced shortening in beef muscle. J. Fd. Technol. 9, 51-58.

Davey, C.L. and Gilbert, K.V. 1975a. Cold shortening capacity and beef muscle growth. J. Sci. Fd. Agric. 26, 755-760.

Davey, C.L. and Gilbert, K.V. 1975. Cold shortening and toughening of beef J. Fd. Technol. 10, 333-338.

Davey, C.L. and Graafhuis, A.E. 1976. Structural changes in beef muscle during ageing. J. Sci. Fd. Agric. 27, 301-306.

Fawcett, D.N. and Revel J.P. 1961. The sarcoplasmic reticulum of a fast-acting fish muscle. J. biophys. biochem. Cytol. 10, Supple.p.89.

Gauthier, G.F. 1969. On the relationship of ultrastructural and cytochemical features to color in mammalian skeletal muscle. Zellforsch. 95, 462-482.

Gergely, J., Pragey, D., Scholtz, A.F., Seidel, J.C., Srèter, F.A. and Thompson, M.M. 1965. In: Molecular Biology of Muscle Contraction (Igaken Shoin Ltd. : Tokyo). p.145.

Goldspink, G. 1962. Ph.D. Dissertation. Univ. Dublin.

Goldspink, G. 1970. Morphological adaptation to growth and activity In: Physiology & Biochemistry of Muscle as a Food. Vol. II (Eds. E.J. Briskey, R.G. Cassens and B.B. Marsh) (Univ. Wisconsin Press: Madison) pp. 521-536.

Goll, D.E. 1970. Chemistry of muscle proteins as a food In: Physiology & Biochemistry of Muscle as a Food. Vol. II (Eds. E.J. Briskey, R.G. Cassens and B.B. Marsh) (Univ. Wisconsin Press : Madison) pp 755-800.

Goll, D.E., Stomer, M.H., Robson, R.M., Temple, J., Eason, B.A. and Busch, W.H. 1971. Tryptic digestion of muscle components simulates many of the changes caused by post mortem storage. J. Anim. Sci. 33, 963-982.

Huxley, H.E. 1960. In 'The Cell' (Ed. J. Brachet and A.F. Mirskey). Vol. IV (Acad. Press : New York). p.305.

Iyengar, M.R. and Goldspink, G. 1971. Cited by G. Goldspink in: The Structure and Function of Muscle 2nd Ed. Vol. I., p.181. (Ed.G.H. Bourne) (Acad. Press : New York).

Johnson, P. and Perry, S.V. 1970. Biological activity and the 3-methylhistidine content of actin and myosin. Biochem. J. 119, 293-298.

Lawrie, R.A. 1950. Some observations on factors affecting myoglobin concentration in muscle. J. agric. Sci. 40, 356-366.

Lawrie, R.A. 1952. Biochemical differences between red and white muscle. Nature, Lond. 170, 122-123.

Lawrie, R.A. 1953. The activity of the cytochrome system in muscle and its relation to myoglobin. Biochem. J. 55, 298-305.

Lawrie, R.A. 1961. Studies on the muscles of meat animals. I. Differences in composition of beef *longissimus dorsi* muscles determined by age and anatomical location. J. agric. Sci. 56, 249-259.

Lawrie, R.A., Pomeroy, R.W. and Williams. D.R. 1964. Studies on the muscles of meat animals. IV. Comparative composition of muscles from 'doppelender' and normal sibling heifers. J. agric. Sci., 62, 89-92.

Ledward, D.A. 1971. Metmyoglobin formation in beef muscles as influenced by water content and anatomical location. J. Fd. Sci. 36, 138-140.

Ledward, D.A., Chizzolini, R. and Lawrie, R.A. 1975. The effect of extraction, animal age and post-mortem storage on tendon collagen. A differential scanning calormetric study. J. Fd. Technol. 10, 349-357.

Locker, R.H. 1976. Meat tenderness and muscle structure. Proc. 18th Meat Res. Conf. Rotorua pp. 1-4.

Locker, R.H. and Hagyard, C.J. 1963. A cold shortening effect in beef muscles. J. Sci. Fd. Agric. 14, 787-793.

Patterson, R.L.S. 1975. The flavour of meat. In: Meat (Eds. D.J.A. Cole and R.A. Lawrie) (Butterworths : London) pp. 359-379.

Perrie, W.T. and Perry, S.V. 1976. An electrophoretic study of the low-molecular weight components of myosin. Biochem. J. 119, 31-38.

Robbins, S.P., Shimokomaki, M. and Bailey, A.J. 1973. The chemistry of collagen cross-links. Age-related changes in the reducible components of intact bovine collagen fibres. Biochem. J. 131, 771-780.

Scopes, R.K. 1970. Characterisation and studies of sarcoplasmic proteins In: Physiology and Biochemistry of Muscle as a Food. Vol. II (Eds. E.J. Briskey, R.G. Cassens and B.B. Marsh). (Univ. Wisc. Press : Madison) pp.471-492.

Swatland, H.J. 1976. Proc. 29th Ann. Recip. Meat Conf., Provo, Utah. p.86.

Yueh, M.H. and Strong, F.M. 1960. Some volatile constituents in cooked beef. J. Agric. Food Chem. 8, 491-494.

DEVELOPMENT OF CONNECTIVE TISSUE AND ITS CHARACTERISTICS

R. Boccard

Station de Recherches sur la Viande
INRA, THEIX 63110, Beaumont, France

ABSTRACT

Collagens, the main components of muscular connective tissue and tendons, have a special structure and properties. Their characteristics affect meat tenderness and determine its cooking quality.

The distribution of collagens is very heterogeneous, as it is found both within muscles and between muscles.

The quantity of collagen remains almost constant during the ageing of animals and is affected by hormonal and nutritional factors.

The physical properties of collagen (swelling ability, solubility and thermal contraction) exhibit large changes with age. These modifications induce drastic changes in the technological properties of meat.

It is possible by biological means to modify the quantity of collagen in muscle and in the whole body.

The double muscled animals are a typical biological model of this alteration. It is known that certain chemical compounds can modify the properties of collagen but their generalised use for this purpose is not yet practical.

INTRODUCTION

Among body tissues, connective tissues (CT) have a special place. They are omnipresent in the animal kingdom. In the vertebrates they comprise 20 to 30% of the body. Connective tissues contribute to the structure of tissues and organs such as skin, tendons, the cardio-vascular system, lungs and muscles.

From the biological and physiological point of view, CT are especially interesting. Their importance lies in their structural and mechanical properties. They are responsible for the interconnections between tissues, the support of the body, and the transmission of tensile strength.

CT bring an important contribution to homeostasis through the water holding capacity of their mucopolysaccharide compounds and their action on membrane properties. These important tissues are also implicated in many diseases resulting from inborn error, nutrition or ageing.

From the technological point of view, CT have for a long time been shown to be important for meat quality, especially with regard to tenderness.

I - THE CONNECTIVE TISSUES AND THEIR MAIN COMPONENT : COLLAGEN*

In their various forms one may describe connective tissues in a rough way as being composed of three parts. A variable number of cells scattered throughout the whole tissue and generally close to the fibres which comprise a major part of the tissue. The rest of the tissue is a large mass of intercellular and interfibrillar product - the ground substance. The proportion of the above three components varies greatly from one tissue to another, depending on the specific role of each tissue in the body.

* See note at beginning of references

The cells, which consist mainly of fibroblast, are associated with the synthesis of the fibre material and the ground substance. They are very numerous in the loose CT (under the skin, between muscles) but are scarce in the tendon.

Fibres are made up largely of collagen and a small fraction of elastin. The ground substance is generally split into two parts: the glycoproteins (an association of protein with one or several different carbohydrates) and the proteoglycans which comprise mostly polysaccharides called glycosamino glycans, which derive from uronic acids.

The mechanical and biological properties of CT are derived from their two main components glycosaminoglycan and especially collagen.

Collagen

Collagen occurs in numerous conformations such as fibres, laminated sheets, big ropes in tendons or fine filaments spread throughout organs such as muscles. From a structural point of view the different collagens are built up from the same basic unit which is the molecule of tropo/collagen.

The tropo/collagen molecule is a triple coiled strand with a 300A diameter and a molecular weight of 300 000. This molecule contains three polypeptide chains known as α chains which are coiled together in a helix form. The α chains are classified into two family groups: α_1 (4 types) and α_2. Each has a unique amino acid sequence in that glycine takes the third part, and proline (Pro) and hydroxyproline (Hypro) another third. These three AA are mainly in the central part of the polypeptide chain. The presence of Hypro is specific to collagen which allows this protein to be detected by the easy and accurate determination of this constituent AA.

In the vertebrate kingdom the quantity of Hypro per molecule of tropo-collagen is considered to be a biological constant. This enables the quantity of collagen in any constituent or mixture to be determined by the use of a coefficient applied to the quantity of Hypro found in it.

The N and C terminal peptides of the α chains contain a large number of AA (except Try) and among them Lys and Hylys.

The biosynthesis of collagen in fibroblast passes through transcription and translation at the DNA and RNA levels as occurs for other proteins. The production of the long α chain which takes 5 to 10 minutes to be completed needs specific conditions to reach its final step owing to the hydroxylation of some proline and lysins residues to give Hypro and Hylys. This hydroxylation has been shown to take place on the nascent chains prior to their release. Two enzymes, proline and lysine hydroxylase, are involved. They both require molecular oxygen, ferrous ions, ascorbic acid and α ketoglutarate. If only one of these conditions is missing or available at too low a level, the synthesis is disturbed.

At the N terminal of the α chains a small peptide and various carbohydrate groups (glucosyl-galactose and galactose) are covalently bound. These extension peptides and carbohydrates are necessary for the correct alignment to produce the triple helix and to facilitate the extrusion out of the cell.

The extracellular formation of the fibres of collagen by aggregation of tropocollagen becomes possible after the cleavage of the non helical extension peptides by the action of a peptidase. Deficiences of this enzyme produce the dermatosparaxis disease of the calf skin which is characterised by a poor cohesion of the collagen and a low tensile strength (Hanset et al., 1974). Normal aggregation brings about an accretion of the fibre in length and diameter. This polymerisation is followed by a reticulation of the fibre framework into a network pattern where the electrostatic and hydrogen bonds are gradually reinforced by covalent links inside the initial molecule between the α chains, and between the molecules arranged side by side in a quarter stagger array.

A set of possible reactions has been proposed to explain this cross-linkage which changes the mechanical stability of collagen, its resistance to heat and its chemical denaturation as measured by its solubility in different conditions. These cross-linkages occur mainly between lysine, hydroxylysine and histidine residues after alteration by specific enzymes. Copper ions are necessary at this stage. Some chemical compounds such as β amino proprionitrite (β APN) which are found in the seed *Lathyrus Sativum* inhibit the specific enzyme and thus prevent the cross-linking of newly synthesised collagen. The reticulation of the collagen can also be disturbed by penicillamine.

Collagen is subject to turnover, but this phenomenon appears in general to take place very slowly which makes this protein metabolically inert compared to other proteins. Enzymes of the collagenase group have been found in numerous tissues of living animals. Their actions, activated by calcium ions, cleave native collagen. This breakdown of collagen, whether old or newly synthesised, can be traced by the determination of the Hypro content of the blood and urine in which this AA is found free or more often combined in small peptides.

Finally, in live animals, the collagen to be considered in quantity as well as in quality, is the result of the equilibrium of anabolic and catabolic phenomena. It is a mixture of different collagens both newly formed and of different ages with large variations in the degree of polymerisation and reticulation.

Properties of collagen

When a piece of collagen (fibre, sheath, fascia) from anywhere in the body is heated in water, it swells. As the temperature increases, the labile bonds inside the network are broken. When the temperature reaches 60°C the collagen framework begins to contract. The tensile force can be observed either in isotonic or isometric conditions. In the latter case, the collagen develops its tension progressively as the temperature increases above the shrinkage-temperature (Ts). As the heating progresses, the triple helix begins to collapse and portions of collagen are

released into the surrounding media as gelatine. This is the
beginning of its entering into solution which is an important
factor contributing to the technological properties.

It is now well established that the tensile strength is pro
portional to the number of linkages in collagen, while its
solubility decreases with an increase in the number of linkages.
These properties have been largely used as indication of the
ageing of collagen.

Thermal contraction (MacClain et al., 1972) has been extensively studied and it was proposed as a measure of the cross-
linkage and consequently as an indicator of the true physiological age of collagen. Some specific types of apparatus were designed (Kopp et al., 1977) to observe the thermal contraction
aspects, but the relation between the dissolution (in different
chemical and physical conditions) and the degree of cross-linkage
is easier to measure in a non-specialised laboratory. Different
methods (temperature, duration of heating, concentration) have
been used, eg in water only (Goll et al., 1964; Hill, 1965) or
the use of the different pH and ionic strengths to separate collagen into fractions of progressive degrees of polymerisation as
indicated by their solubility in neutral acid or alkaline media
or their insolubility (Valin et al., 1971). They are respectivel
called neutral, acid, alkaline soluble or insoluble collagen.

II - VARIATION WITH AGE

Quality of collagen

The collagen content of the bovine carcass has not yet been
explored in every muscle nor over the animal's whole lifespan,
but considerable data is already available.

The determination of collagen content was obtained some
years ago by the differential extraction of the soluble protein.
The insoluble residue called 'stroma' was considered as the
fibrous part of the connective tissues with small errors resultir
from the soluble part of collagen and insoluble from other

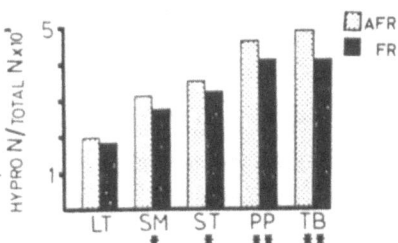

LT: *Longissimus dorsi thoracis* SM: *Semi-membranosus* ST: *Semi-tendinosus*
PP: *Pectoralis profundus* TB CL: *T.B. Caput Laterale*

Boccard et al.,(1977)

Fig. 1. Collagen content in muscles of Friesian and Afpikaner (bulls and steers).

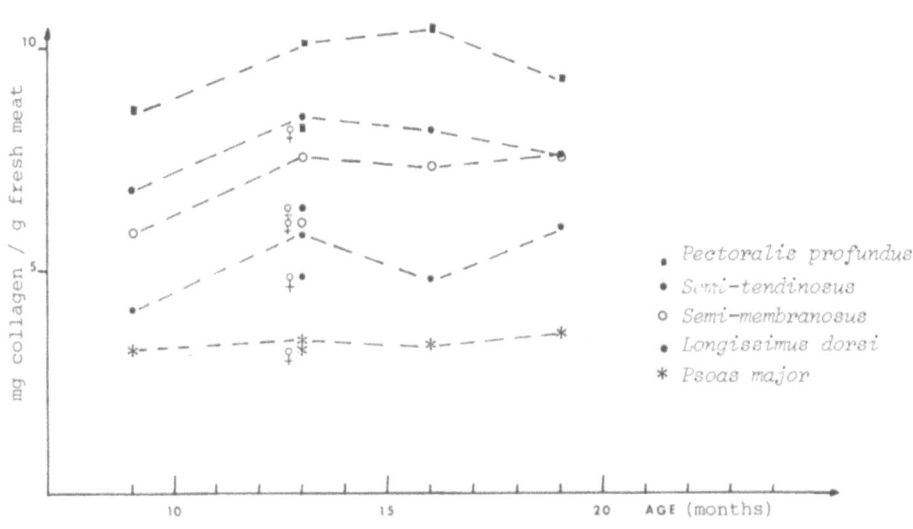

(Kopp, 1977)

Fig. 2. Variation of collagen content with age (Limousin breed).

constituents (Lawrie, 1960). Nowadays the quantity of collagen
is essentially determined by analysing for the Hypro content.
Carrying out the chemical analysis presents no problem but there
remains a lack of uniformity in the presentation of the results
owing to the various coefficients (between 7 and 8) used by
different authors to convert Hypro to collagen, and by the use
of different bases to express it: fresh, dry, fat free, protein
(N x 6.25). Table 1 gives a comparison of some different expressions (Boccard et al., 1976).

In cattle the collagen content varies greatly from one kind
of muscle to another, eg from 1.2 in the *Psoas major* to 5.5 in the
Pectoralis profundus of 24 months old males, Table 2 (Boccard, 1977)
It also varies between individual animals, between breeds -
Figure 1 (Boccard et al., 1977) and with sex. In the same breed,
the Charolais, it has been observed that the genetic type, and
especially the double muscle gene, brings about a large variation
in the collagen content (Table 2). The difference between the
two types seems proportionately larger in the muscles which are
richer in collagen.

During growth from birth to adult the collagen content show
some special change. After birth for two to three months the
collagen content of the muscles decreases, then it is constant
until the onset of puberty. After that it reaches a peak in
entire males (Figure 2). From one year and a half to the adult
stage the proportion of collagen in muscle again remains nearly
constant. A new increase can be detected in advanced age and
especially in milk cows. These variations can be explained by
the relative growth of the connective tissues and the contractile
proteins of muscle. After birth, the diameter of muscle fibre
is far from its maximum owing to the underdevelopment of sarcomere and sarcoplasmic when connective tissue is well organised.

After puberty the increase of collagen in muscle is the
result of a new cycle of growth which can be attributed to the
influence of testosterone at this period (Lund Larsen et al.,1977)
even if it was noted, in rabbits, that this hormone changes the

TABLE 1
DIFFERENT METHODS TO EXPRESS COLLAGEN CONTENT OF MUSCLE (Boccard et al., 1976)

	HYDROXYPROLINE (Hypro)	COLLAGEN
Hypro or collagen content	A_H mg Hypro/g tissue (DM or fresh)	A_C mg coll./g tissue (DM or fresh) = $A_H \times 7.5$
Ratio of Hypro or collagen to N total	B_H mg Hypro/g N total = $\dfrac{A_H}{y\,g\,N\,total}$	B_C mg coll./g N total = $\dfrac{A_C}{x\,g\,N\,total}$
Ratio (‰) of N Hypro or N collagen to N total	C_H mg N Hypro/g N total = $B_H \times 0.1068$	C_C mg N coll/g N total = $B_C \times 1.18$ $\quad B_H \times 1.35$

TABLE 3
COLLAGEN CONTENT VARIATION IN OLD COWS WITH EXTREME FLESHINESS (Boccard et al., 1969)

	Very poor conformation $\dfrac{N\ Hypro}{N\ total}$ ‰	High conformation $\dfrac{N\ Hypro}{N\ total}$ ‰
Internal sample of :		
Triceps brachii longus	5.5	3.0
Teres major	4.5	2.5
Supraspinatus	6.5	4.0
Infraspinatus	6.0	4.5
Triceps brachii laterale	7.0	5.5

	Very poor conformation $\dfrac{N\ Hypro}{N\ total}$ ‰	High conformation $\dfrac{N\ Hypro}{N\ total}$ ‰
Whole muscle :		
Subscapularis	16.5	14.5
Deltoideus	20.0	9.5
Teres major	12.0	8.5
Coraco brachialis	18.0	12.0
Anconaeus	10.0	6.0
Triceps brachii mediale	12.0	6.0
Tensor fasciae antebrachii	14.0	7.0
Brachialis	7.0	5.0
Biceps brachii	15.0	13.0

TABLE 2
COLLAGEN CONTENT IN DIFFERENT MUSCLES FROM NORMAL AND DOUBLE MUSCLED CHAROLAIS ENTIRE ANIMALS AT TWO YEARS OF AGE

N HYPRO / N TOTAL %.

MUSCLES	NORMAL	DOUBLE MUSCLED	MUSCLES	NORMAL	DOUBLE MUSCLED
Psoas major	1.60	1.30	Serratus thoracis	3.34	2.32
Gastrocnemius internus	2.32	2.34	Supraspinatus	3.34	2.65
Sartorius	2.51	1.86	Pectoralis profundus (pars medium)	3.40	2.32
Subscapularis	2.54	1.30	Semimembranosus	3.48	2.42
Vastus medialis	2.55	2.08	Splenius	3.48	2.95
Teres major	2.66	1.50	Infraspinatus	3.53	2.08
Gluteus medius	2.74	1.84	Diaphragma	3.85	4.33
Rectus femoris	2.80	1.76	Biceps femoris (pars inferior)	3.90	2.98
Rectus abdominis	2.92	2.45	Tensor fasciae latae	4.03	2.64
Longissimus dorsi	2.97	1.89	Transversus abdominis	4.30	3.10
Adductor	3.02	2.13	Latissimus dorsi	4.43	2.75
Vastus lateralis	3.06	2.12	Biceps femoris (pars superior)	4.48	2.21
Obliquus abdominis internus	3.06	2.52	Serratus cervicis	4.56	2.58
Triceps brachii caput longum	3.15	2.00	Triceps brachii caput laterale	4.68	2.68
Rectus femoris	3.19	2.34	Biceps brachii	4.90	2.86
Flexor	3.22	2.28	Semitendinosus	5.29	2.72
Gastrocnemius externus	3.24	2.62	Rhomboideus	5.56	3.27
Complexus major	3.31	2.62	Pectoralis (pars sternalis)	5.66	3.83

(Boccard, 1977)

weight of muscle but not its composition (Grigsby, 1976).

In old animals, the synthesis of collagen decreases as the prolyloxydase, responsible for the hydroxylation, falls during ageing, but the proportion of collagen can increase in bad nutrition conditions. In this case, as observed in milk cows at the same age and size (Boccard et al., 1969) the ratio of collagen in their musculature is much higher than in animals which have lost more weight before slaughtering (Table 3).

Recent work suggests that very high as well as very low dietary protein can induce a decrease in the total muscle collagen synthetic activity (McClain et al., 1975) but it was observed with rats, which are generally more sensitive to nutritional deficiencies than cattle, in which the rumen buffers the nitrogen diet variations.

Quality of collagen

As reticulation progresses with the increasing age of the animal, a steady decrease in the solubility of the collagen in the same muscle can be observed (Goll et al., 1964; Hill, 1965) (Figure 3); (Kopp, 1971), but at the same age of the animal the collagen present in skin tendon aponevrosis or inside the peri or endomysium have not got the same structure (Duance et al., 1977) and physiological age and consequently the same solubility. Important differences can also be noted between muscles.

These variations underline the absolute necessity of agreeing upon a standardised representative sample if one wishes to measure the physiological age of meat animals or the collagen solubility. A further complication is that the solubility of collagen is changed by the storage conditions of the samples, even in the deep frozen or the freeze dried state.

In spite of all these difficulties, the measurement of collagen solubility is really important from the biological and technological point of view. With decreasing solubility, the toughness of meat cooked in the same conditions increases -

84

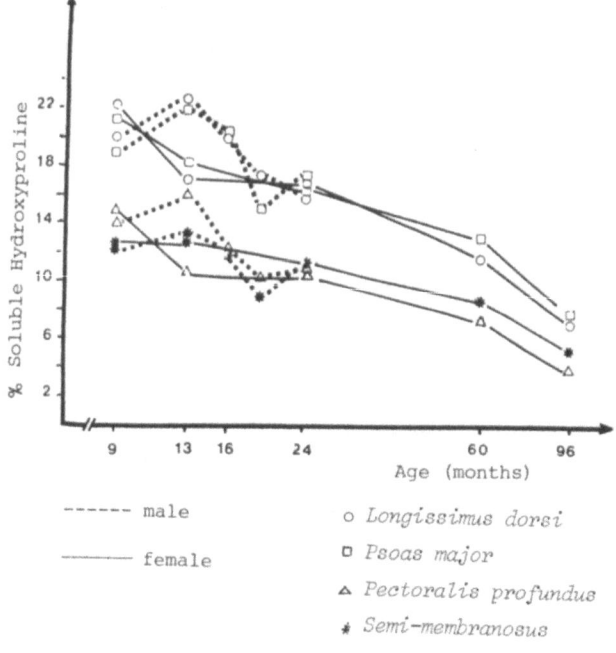

(Kopp, 1976)

Fig. 3. Evolution with age of the solubility of intramuscular collagen
(1h - 70°C).

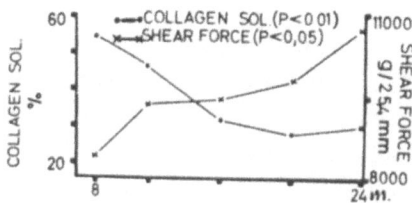

(Boccard et al., 1977)

Fig. 4. Mean age changes in collagen solubility and shear force values of
five muscles of male animals.

(Figure 4)(Boccard et al., 1977). Owing to the variations observed between animals and breeds (Figure 5) (Boccard et al., 1977) it is of major importance to have an indicator of the progress of ageing in the animal bred for meat production.

A more accurate measure of the physiological age can be found in the thermal contraction of the collagen fibre and the value of the maximum tensile strength of the muscular collagen. The isometric tension is of interest for two reasons. Firstly, it varies considerably with the age of the animal (because of the progress of cross-linkage with age) as can be seen in Figure 6 (Kopp, 1976); secondly, the tension is highly correlated with the collagen content of the piece of meat being examined (Figure 7) (Kopp, 1977) and could eliminate the necessity of determining its Hypro content at the same age.

Modification in the quantity and quality of collagen

From the foregoing one may expect that meat quality may be improved by:

1) decreasing its connective tissue and collagen content;
2) increasing the solubility of its collagen whether biologically or by technological processing.

The decrease in tenderness with age has induced breeders to produce fat animals ready for slaughter at much younger ages. But, when considering the economic factors with which they are confronted and the carcass grading system, they are compelled to sell heavy animals. In order to reach a correct weight at a young age, breeders have to change their breeding system, they have to increase the nutritional level of the cattle and frequently have to produce bulls instead of steers. Males have a slightly higher collagen content than steers, but at the same age they have more meat and a better collagen solubility and in the end a meat of about the same degree of toughness.

Since a decrease in the collagen content of muscle cannot be attained by decreasing the age at slaughter, genetic selection may be employed to obtain the desired biological variation of the collagen content. A typical opportunity is presented by

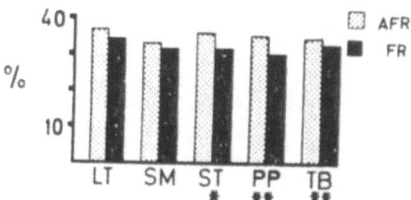

LT: *Longissimus dorsi thoracis* SM: *Semi-membranosus* ST: *Semi-tendinosus*
PP: *Pectoralis profundus* TB: *T.B. Caput Laterale*

(Boccard et al., 1977)

Fig. 5. Variations in muscle collagen solubility in the Afpikaner and
Friesian breeds.

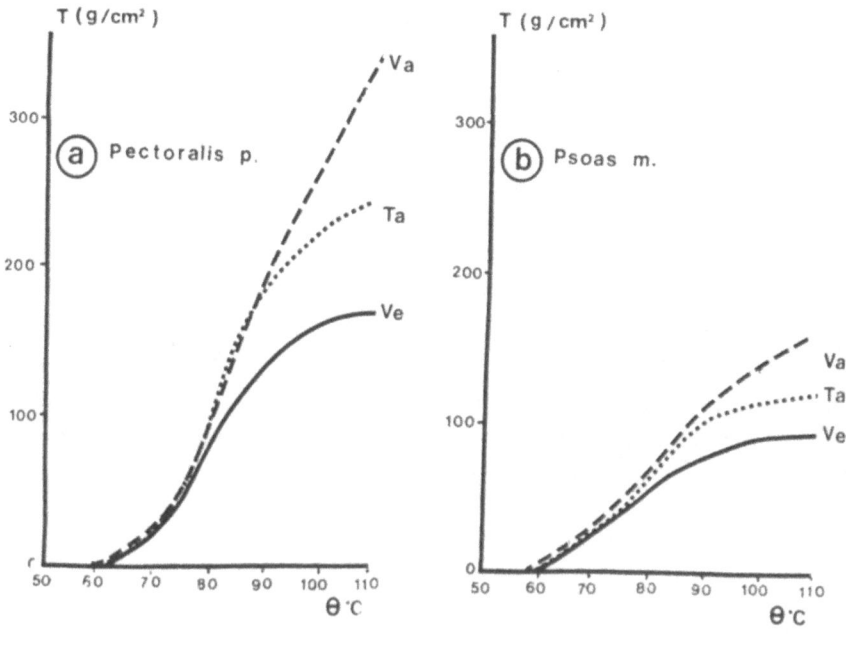

(Kopp, 1976)

Fig. 6. Variations in the thermal tension of intramuscular collagen due to
heating, in cows (Va), young bulls (Ta) and calves (Ve).

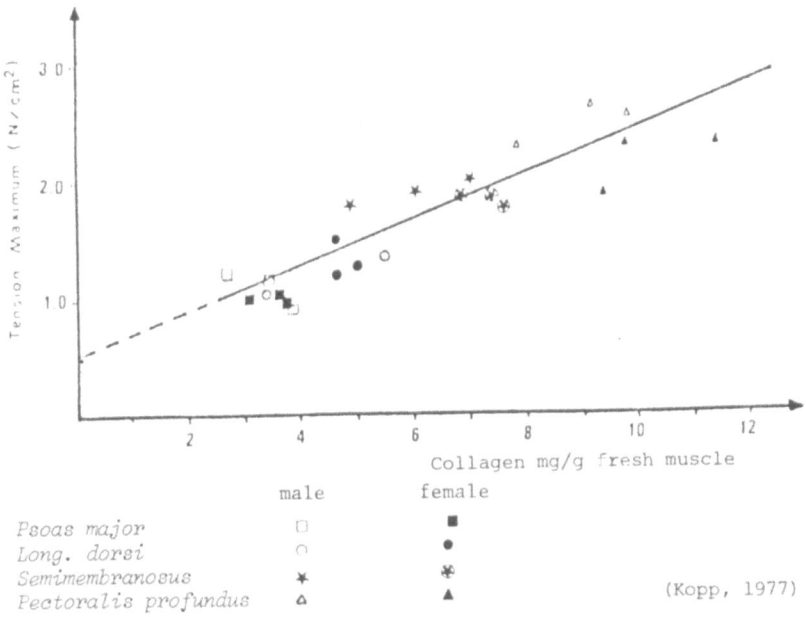

Fig. 7. Relationship between collagen content and maximum isometric strength in Limousin breed at 9 months.

the use of semen of double muscled bulls with normal cows. These animals have a high dressed weight percentage and a high muscle/bone ratio (Dumont et al., 1967). All their muscles are not hypertrophied (Boccard et al., 1974) but all contain less collagen, which offers a very different network in its distribution (Boccard et al., 1967). Both of these characteristics give a better tenderness in the whole musculature.

This genetic approach is certainly the one which is the most promising, because the other methods, which are based on nutrition or on the use of particular compounds, are too expensive and are not yet really suitable for animal husbandry in practice.

REFERENCES*

Boccard R., Dumont B.L. and Schmitt O., 1969 Relation entre la conformation
des carcasses et les caractéristiques de la musculature (sur les vaches
de réforme). Bull. Ac. Vet. <u>52</u> 261-265

Boccard R., Dumont B.L. and Schmitt O., 1967 Relation entre la dureté de
la viande et les principales caractéristiques du tissue conjonctif.
13th Europ. Meet. of Meat Res. Workers, Rotterdam.

Boccard R., and Dumont B.L., 1974 Conséquences de l'hypertrophie musculaire
héréditaire des bovins sur la musculature. Ann. Génét. Sél. Anim.
<u>6</u>, 177-186

Boccard R. and Kopp J., 1976 Tissus conjonctifs et qualité des viandes de
porcs. Bull. Inst. Techn. Porc, 37-47

Boccard R., 1977 In progress

Boccard R., Cronje D.E., Naude R.T. and Smit M.C., 1977 Muscle collagen and
beef tenderness. New Letter S. Af. Soc. An. Prod. <u>16</u> 16

Dumont B.L. and Boccard R., 1967 Le rapport muscle/os critère de sélection
des bovins de boucherie. 2e Symp. Int. Zootech., Milan. 149-155

Goll D.E., Hoekstra W.G. and Bray R.W., 1964 Age associated changes in bovine
muscle connective tissue. II Exposure to increasing temperature.
J. Fd. Sci. <u>29</u> 161

Grigsby J.S., Bergen W.G. and Merkel R.A., 1976 The effect of testosterone
on skeletal muscle development and protein synthesis in rabbits.
Growth <u>40</u>, 303-316

Hanset R. and Lapière Ch., 1974 Inheritance of Dermatosparaxis in the calf:
a genetic defect of connective tissues. J. of Heredity <u>65</u>, 356-358

Hill F., 1965 The solubility of intramuscular collagen in meat animals of
various ages. 11th Europ. Meet. Meat Res. Workers, Belgrade.

Kopp J., 1976 Tendreté de la viande bovine. Principaux facteurs de variation
liés à l'âge des animaux. Bull. Tech. CRZV Theix <u>24</u> 37-46

Kopp J., 1977 Facteurs de variation de la contraction thermique de la viande
liés au collagène. 23rd Europ. Meet. Meat Res. Workers, Moscow.

---

* A very large number of works are published every year on the properties
and biosynthesis of connective tissue. For the general and fundamental
aspects, I would suggest a recent short book which is oriented more
towards the biological and animal husbandry interests of the participants
of this Seminar - Hall, D.A., 1976. The Ageing of Connective Tissue
Acad. Press; London. 204pp.

Kopp J., Sale P. and Bonnet Y., 1977 Contractomètre pour l'étude des propriétés physiques des fibres conjonctives: tensions isométriques, degré de réticulation, relaxation. Can. Inst. Food Sci. Techn. J. 10, 69-72

Kopp J., 1977 In progress

Lawrie R.A., 1960 Analysis of *longissimus dorsi* muscles from cattle implanted with hexoestrol. Brit. J. Nut. 14, 255-258

Lund Larsen T.R., Sunday A., Kruse V. and Velle W., 1977 Relations between growth rate, serum somatamedin and plasma testosterone in young bulls J. Anim. Sci. 44, 189-194

McClain P.E. and Wiley E.R., 1972. Differential scarring calorimeter studies of the thermal transition of collagen. J. Biol. Chem. 247, 692-697

McClain P.E., Wiley E.R. and Beecher G.R., 1975 Influence of high and low levels of dietary protein on the biosynthesis and cross-linking of rat skin and muscle collagen. Nutr. Rep. Int. 12, 317-324

Valin C., Goutefongea R. and Kopp J., 1971 Evolution de quelques caractéristiques physico-chimiques du muscle congelé, conservé à -20°C. Revue Générale du Froid 10, 923-931

SOURCES OF VARIATION IN MUSCLE WEIGHT DISTRIBUTION

P.L. Bergström

Research Institute for Animal Husbandry 'Schoonoord'
Zeist, The Netherlands

ABSTRACT

The weight distribution of the muscles and muscle groups within the total carcass muscle portion is discussed, based on results from literature and own research.

Changes of greater importance are only found in relation to the growth and in particular in the very young growth phases. In general there is a close relationship between the muscle weight distribution and the absolute carcass muscle weight.

The influences of breed, type and quality grade on the muscle weight distribution are very small and here the muscle weight distribution does not reflect the type differences at visual appraisal. The influence of the growth intensity needs further study for some experiments suggest an effect on the muscle weight distribution independent from the absolute muscle weight although not to an important extent.

With regard to the sex the situation in the castrated male is discussed in comparison to entire male and female animals because conclusions concerning the classification of the growth patterns of muscles mainly originate from studies with steers.

Within the various groups of animals of which the group means are not different, an individual variation, which cannot be explained systematically from one of the factors mentioned, can exist.

It is obvious from the preliminary results of an experiment with identical twins that this often rather important individual variation in muscle weight distribution is a genetic variation.

INTRODUCTION

From the literature it can be demonstrated that the weight
ratio between the muscle, fat and bone fractions in the carcass
is the most important factor with regard to differences in carcass composition. Although in the practice of beef production
the apparent differences in muscle weight distribution play a
rather important role with regard to the price differentiation
between beef animals and their carcasses, most of the recent
investigations demonstrate that this factor is not of such importance as is sometimes assumed and expected.

In the present publication we shall restrict ourselves to
the problems of the weight distribution of the components of the
muscle fraction (individual muscles, muscle groups) within the
total carcass muscle fraction. We shall base our discussions
only on data obtained from anatomical dissections with a complete
tissue separation. Only by means of an anatomical dissection
can the problem of muscle weight distribution be studied properly.
Standardised wholesale jointing methods suffer in most cases from
a lack of uniformity concerning the boundaries of the joints or
cuts and in particular the outcomes can be affected by the fat
fraction as a result of incomplete separation between muscles and
fatty tissue. Pioneering work concerning the body composition
was done in the second quarter of the century by research workers
of the Cambridge School of Agriculture. About fifteen years ago
the study of carcass composition was taken up again by several
research workers. Now the changes in carcass composition were
in most cases studied on a 'within tissue' basis and the data
were elaborated according to the principles of allometry as developed by Huxley (1932).

It became quite clear that the muscle weight distribution
is regulated strongly by fixed rules in which functional aspects
can easily be recognised. This results in a far more uniform
pattern of muscle weight distribution as is suggested by differences in conformation and apparent differences in the muscular
development in the various body regions. The only explanation

for this phenomenon can be that, at visual appraisal, factors other than the actual muscle weight distribution, such as thickness of the muscle layer in relation to the dimensions of the skeleton, play an important role.

On basis of data from literature and our own investigations we shall try to discuss the influence of several factors on the muscle weight distribution.

THE MUSCLE WEIGHT DISTRIBUTION IN RELATION TO AGE

As a result of differential growth of the individual muscles, the muscle weight distribution changes in the growing animal. An example for a group of entire male and female Dutch Friesian half-sibs is given in Table 1.

These animals were intensively fed on a weight basis, equal for both sexes, and serially slaughtered partly on an age and partly on a live weight basis. The muscle weight distribution is expressed as percentages relative to the total side muscle weight for the standard muscle groups. These standard muscle groups are basically functional units but their boundaries depend as well on body regions. The standard muscle groups, therefore, reflect at the same time a subdivision of the total side muscle into groups of a different commercial value.

The sequence of the changes in relation to age in Table 1 is in good agreement with the results of Berg (1968), presented in the book by Berg and Butterfield (1976) for shorthorn cross-bred bulls, steers and heifers. This is valid also, with the exception of the muscles of the proximal pelvic limb, for a comparison with the results of Butterfield (1963) for a hetero-geneous material of steers over an age range from birth to over 4 years. The percentages of the individual muscles for three weight groups in the growth experiments with bulls of the Red Danish breed (Refsgaard Andersen, 1975) gives sequences very similar to those in our study. In the studies of Butterfield and Berg, the composition of the standard muscle group varies with our standard muscle group; only the groups of muscles of the limbs

TABLE 1

CHANGES IN MUSCLE WEIGHT DISTRIBUTION IN RELATION TO AGE. DATA OF DUTCH FRIESIAN MALE AND FEMALE HALF-SIBS UNDER INTENSIVE FATTENING CONDITIONS

Standard muscle groups expressed as percentage of the total side muscle weight (means and standard deviation). Data of mature animals of the same breed but not belonging to the progeny group are included for comparison.

| Age or weight group | Standard muscle groups (total side muscle weight = 100%) | | | | | | | | | | Side muscle weight (kg) | n |
	A Prox. pelvic limb	B Dist. pelvic limb	C Prox. thorac. limb	D Dist. thorac. limb	E Neck	F Back and loin	G Sub-lumbar and muscle abdomen	H Thorax and abdomen	I Shoulder girdle	J Skin muscles		
Entire males:												
1 wk 40 kg	31.34 0.20	6.47 0.55	12.20 0.47	4.09 0.20	9.92 0.56	9.95 0.10	2.83 0.15	8.53 0.67	13.02 0.33	1.22 0.12	6.4	10
13 wks 90 kg	34.16 0.78	5.79 0.17	11.28 0.29	3.12 0.21	8.26 0.27	9.95 0.36	3.07 0.13	9.98 0.60	12.59 0.50	1.49 0.13	15.6	10
180 d 200 kg	33.77 0.59	5.37 0.17	11.06 0.27	3.03 0.05	8.02 0.31	10.07 0.11	3.04 0.11	10.50 0.35	12.70 0.36	1.91 0.18	30.6	5
251 d 300 kg	31.95 0.79	4.75 0.25	11.09 0.27	2.71 0.10	9.04 0.40	10.42 0.34	2.93 0.16	11.03 0.51	13.59 0.32	2.23 0.14	49.7	10
337 d 400 kg	30.82 0.92	4.48 0.16	11.46 0.31	2.66 0.07	9.40 0.57	10.78 0.49	2.87 0.14	10.95 0.36	14.25 0.41	2.06 0.10	65.6	5
454 d 500 kg	29.77 0.52	4.15 0.14	11.26 0.27	2.48 0.18	10.67 0.64	10.70 0.25	2.99 0.16	11.46 0.36	14.13 0.31	2.08 0.22	85.3	10
Mature bulls	26.51	3.55	11.63	2.32	13.16	10.94	2.60	11.42	15.93	1.60	170.0	2

For composition of the standard muscle groups compare Appendix

Table 1 continued

TABLE 1 (Cont)

CHANGES IN MUSCLE WEIGHT DISTRIBUTION IN RELATION TO AGE. DATA OF DUTCH FRIESIAN MALE AND FEMALE HALF-SIBS UNDER INTENSIVE FATTENING CONDITIONS

Age or weight group	Standard muscle groups (total side muscle weight = 100%)										Side muscle weight (kg)	n
	A Prox. pelvic limb	B Dist. pelvic limb	C Prox. thorac. limb	D Dist. thorac. limb	E Neck	F Back and loin	G Sub-lumbar muscle	H Thorax and abdomen	I Shoulder girdle	J Skin muscles		
Females												
1 wk 35 kg	32.62 0.62	7.04 0.28	12.51 0.14	4.08 0.13	8.70 0.43	9.91 0.47	2.89 0.15	8.70 0.43	12.18 0.51	0.95 0.27	5.4	5
13 wks 90 kg	34.69 0.75	5.95 0.14	11.30 0.31	2.97 0.31	7.89 0.18	9.47 0.16	3.24 0.13	10.70 0.22	11.93 0.41	1.47 0.18	13.3	5
175 d 165 kg	34.93 0.43	5.70 0.06	10.88 0.29	2.94 0.06	7.45 0.29	10.40 0.31	3.12 0.13	10.47 0.54	11.89 0.19	1.92 0.12	24.9	5
247 d 240 kg	33.92 0.66	5.18 0.10	10.92 0.42	2.87 0.13	7.67 0.35	10.55 0.20	3.15 0.11	10.79 0.28	12.45 0.29	2.07 0.22	36.8	5
335 d 315 kg	32.91 0.25	4.79 0.11	10.88 0.29	2.55 0.05	8.05 0.29	10.63 0.38	3.15 0.10	11.50 0.46	13.28 0.31	2.08 0.11	48.7	5
440 d 390 kg	31.81 0.62	4.47 0.20	11.03 0.35	2.56 0.07	8.42 0.22	11.08 0.38	3.09 0.13	11.90 0.46	13.45 0.37	2.00 0.11	60.1	5
Mature cows	30.31 0.80	4.12 0.20	11.18 0.40	2.30 0.16	9.11 0.45	11.27 0.36	3.07 0.21	13.10 0.21	13.72 0.70	1.77 0.20	91.7	26

For composition of the standard muscle groups compare Appendix

can directly be compared. Nevertheless, the same principles can
be recognised in the several series of data. The rather extreme
position for some of the muscle groups in the 3 months old animals in our material is not indicated in the material of Berg
and Butterfield, apparently because in their study this stage is
not included.

The early maturing muscles of the distal limbs are relativel
well developed at birth. In later life the percentage of these
muscle groups relative to the side muscle weight decreases. In
these muscle groups as well as in the rather high percentages for
the muscles of the neck and shoulder girdle (necessary to support the well-developed head and thoracic limb) in the very young
stage and the low percentage for the muscles of the thorax and
abdomen in the young animal, functional aspects can easily be
recognised. In the later stages the development of the muscles
of the neck and shoulder girdle is relatively strong, in particular in the bulls. It is evident from the data of the mature
animals included in Table 1 that, in comparison to the situation
in the heaviest stages of our study (500 kg for the bulls and
390 kg for the heifers) rather important changes from that stage
to the mature stage in the muscle weight distribution takes
place. The muscles of the proximal pelvic limb increase in percentage relative to the side muscle weight within a short age
period very markedly, whereas, in later stages, the percentage
gradually decreases over a relatively important range. The percentages of the other standard muscle group are, except in the
very young stages for some of these muscle groups, very constant
throughout the growth period of the study, and, as the data of
the mature animal suggests, throughout the total growth period
to maturity. The abdominal muscles in the female animals tend
to increase in percentage in the later stages apparently as a
result of, or as a preparation for, pregnancy. In Table 2 data,
expressed in the same way as in Table 1, are given for a heterogeneous material of the Dutch breeds. The same principles as in
Table 1 can be recognised, although animals which grow relativel
fast to a given muscle weight tend to have in their muscle weight
distribution more 'mature' characteristics compared with those

TABLE 2

THE MUSCLE WEIGHT DISTRIBUTION IN RELATION TO AGE. DATA OF A HETEROGENEOUS MATERIAL OF DIFFERENT CATEGORIES AND FEEDING SYSTEMS OF DUTCH CATTLE BREEDS.

No animals mentioned in Table 1 are included in this Table. The data are presented as in Table 1

Categories and age groups	A	B	C	D	E	F	G	H	I	Total side muscle wt. (kg)	n
Entire males											
Veal calves * 4 mths	32.1 0.8	5.1 0.3	11.3 0.7	2.7 0.2	8.8 0.6	11.1 0.5	2.9 0.1	10.0 0.4	14.2 0.8	32.2	6
Grass fed ** calves 9 mths	32.6 0.8	5.5 0.3	11.2 0.4	2.8 0.2	8.3 0.5	11.1 0.4	3.1 0.1	10.5 0.7	13.4 0.5	40.3	18
Young bulls* 1 yr	30.1 0.9	4.4 0.2	11.2 0.4	2.5 0.1	9.7 0.6	10.8 0.5	3.0 0.1	11.4 0.5	14.3 0.6	79.1	26
Young bulls** 14 mths	30.7 1.2	4.4 0.3	11.7 0.5	2.5 0.2	10.4 0.8	11.0 0.5	2.8 0.1	9.8 0.5	15.2 0.7	77.5	24
Young bulls* 15 mths	29.0 0.4	4.2 0.1	11.6 0.5	2.5 0.1	10.4 0.3	10.8 0.7	3.0 0.2	11.2 0.2	15.2 0.2	97.6	5
Young bulls** 18 mths	30.4 1.2	4.4 0.4	11.5 0.5	2.5 0.2	9.9 1.0	11.0 0.5	2.9 0.1	10.8 0.5	14.9 0.6	83.0	13
Castrated males											
Steers** 2-2½ yrs	32.1 0.6	4.7 0.2	11.6 0.5	2.6 0.2	8.9 0.5	10.6 0.4	3.0 0.2	10.5 0.4	14.7 0.5	87.6	10
Females											
Young heifers* 1 yr	32.2 0.8	4.8 0.2	11.4 0.4	2.5 0.1	8.0 0.4	11.1 0.4	3.1 0.1	11.6 0.7	13.5 0.4	61.4	10
Heifers** 2 yrs	33.0 0.9	4.8 0.3	11.4 0.4	2.4 0.2	8.2 0.4	11.0 0.5	3.3 0.2	11.0 0.5	13.4 0.5	68.2	20

Standard muscle groups (total side muscle weight = 100%)

* Intensively fed: ** Extensively fed

growing more slowly towards that muscle weight.

It is essential for the study of the several factors which
might influence the muscle weight distribution to dispose of a
scheme of the changes which take place in each growing animal.
We have chosen for the present paper to illustrate the influences
on the muscle weight distribution by means of percentages of the
total side muscle weight for the standard muscle groups, because
this procedure demonstrates best the commercial importance. It
is obvious, however, that the results depend largely on the
growth pattern and growth impetus of the muscles included in the
standard muscle groups. For this reason, Butterfield and Berg
(1966$_{a,b}$) and Berg and Butterfield (1976), tried to classify the
individual muscles according to their growth pattern and growth
intensity.

The allometric equation according to Huxley (1932) was used
in its logarithmic form log Y = log a + b log X in which Y is
the weight of the muscle or muscle group and X is the weight of
the total side muscle whereas b is the growth coefficient. They
subdivided a material of 92 calves and steers from birth to over
4 years of age into five growth phases of 10 months each. This
material was extended with data of 4 calves and 12 bulls, 12
heifers and 10 steers serially slaughtered at 3 weights. The
muscles and muscle groups were classified according b-values,
varying significantly or not, among the growth phases (di-phasic
or mono-phasic growth pattern) and according the b-values, varying or not, from 1.0 (high, low or average growth impetus) in
the several growth phases.

This resulted in groups of muscles which could be classified
into low, low/average, average, high, high/average and average/
high, impetus groups whereas a group with about 12% of the total
muscle weight could not be classified satisfactorily for several
reasons. Berg and Butterfield (1976) remark that a grouping of
the muscles into impetus groups is to be preferred in comparison
to functional units or standard muscle groups as inclusion of
muscles with different growth patterns in the same standard muscl

groups can obscure the findings. We can agree with this statement but for the proposed procedure it is a strict necessity
that the classification of the muscles has a very solid basis,
and with the data from literature as well as our own research
taken from somewhat small experiments, including in some studies,
heterogeneous material, we doubt whether this can be realised.

Growth coefficients have also been calculated according to
the allometric equation in its logarithmic form for our Dutch
Friesian half-sibs. It proved necessary to combine at least two
subsequent slaughterings which makes three growth phases for
the bulls for growth periods to 500 kg, and for heifers to 390
kg live weight. In the appendix the b-values with standard
errors are given. The b total values for the total growth
period are given for comparison with other results, although only
if Pa, Pb and Pt are >0.05 can an acceptable total regression
be expected. For comparison with studies which do not include
the very young stages the b_{2+3} - values for the combined growth
phases 2 and 3 are given as well. To give an idea of the muscle
weights the percentages relative to the side muscle weight
are included for bulls in the category of 80 - 89 kg side muscle
and heifers in the category of 60 - 69 kg side muscle. These
were the side muscle weight categories with the highest numbers
of animals in our total material of dissected sides.

We cannot discuss the results in the Appendix in detail and
can only offer some brief comments. We decided not to classify
the muscles according to their growth impetus on basis of a
growth coefficient significantly different or not from 1.0 because, in rather small-scale experiments, the small numbers of
animals in each of the growth phases will influence the results
too much. If a classification according to this procedure were
to be made the results would be the same as those in the study
of Butterfield and Berg (1966[a,b]) for a quantity of the muscles
with a more extreme position; for a relatively large number of
muscles the classification would be different. We must realise,
however, that the material in both studies is different in many
respects (breed, sex, growth intensity and subdivision into

growth phases) so that there can be many reasons for the differences.

The main problems with regard to a classification of the growth patterns of the components of the musculature were that for a relatively important number of muscles, a satisfactory classification in which the slope of the regression as well as the level was taken into account, thus causing great difficulties Moreover, the standard error for the heaviest group in phase 3 was often too high to contribute to the explanation of the growth pattern. The same difficulty was mentioned by Butterfield and Berg (1966a,b) for their heaviest group of growth phase 5. In our study there was less evidence that for an important number of muscles, a clear, di-phasic growth pattern will be found. The growth coefficients tended more to change continuously. It was quite clear that the most important changes in growth intensity take place in the very young animals. A comparison of the b_{2+3} -values for the combined growth phases 2 and 3 and the b-values in the Danish growth experiment (Refsgaard Andersen, 1975) - which cover nearly the same weight range (200 - 500 kg versus 180 - 540 kg) - shows a very close similarity between both series as far as individual muscles are concerned.

We can conclude that the several studies show to a large extent the same changes in muscle weight distribution in relation to age which indicates that these changes are governed in each growing animal by very uniform principles. The results, however, vary too much in details to enable a precise description of the growth patterns of the components of the musculature by means of the data which are now available. Further studies with uniform groups of animals of different genotypes and, above all, under different, but, per experiment, uniform environmental conditions, must be recommended.

THE MUSCLE WEIGHT DISTRIBUTION IN RELATION TO SEX

With regard to the sex influences on muscle weight distribution, Berg and Butterfield (1976) remark that the bull is

TABLE 3

MUSCLE WEIGHT DISTRIBUTION IN RELATION TO AGE.   DATA OF MALE FOETUS (1 day before birth) AND CASTRATED MALES EXTENSIVELY FED.   (After, Butterfield, 1963).

Age groups	Standard muscle groups (total side muscle weight = 100%)									
	1 Prox. pelvic limb	2 Dist. pelvic limb	5 Prox. thorac. limb	6 Dist. thorac. limb	9 Neck and thorax	3 Around spinal column	4 Abdom. wall	7 Thorax to thorac. limb	8 Neck to thorac. limb	n
Male foetus	31.24	6.88	12.53	4.40	9.85	11.98	6.98	8.55	6.98	5
6 - 9 mths	32.62	5.74	11.34	3.06	8.80	12.14	9.06	9.06	6.65	6
10 - 19 mths	32.53	4.98	11.09	2.64	9.43	12.30	10.00	9.13	6.94	7
20 - 29 mths	32.62	5.01	11.38	2.75	9.84	11.90	9.32	9.12	7.05	16
30 - 39 mths	32.34	4.68	11.16	2.55	10.08	12.34	9.45	9.28	7.24	9
> 40 mths	32.34	4.48	11.65	2.63	10.30	12.35	9.03	9.38	7.50	14

The standard muscle groups, 2, 5 and 6 can be directly compared with the standard muscle groups, B, C and D in Table 1. The standard muscle group 1 is only slightly different from the standard muscle group A in Table 1

the only sex which fully utilises its potential of the musculatur
to grow differentially. This means that in entire male and
female animals the changes in the growing animal follow, to a
large extent, the same principles. This is especially evident
in the material of Berg (1968) and our own material as mentioned
in Table 1. In the bull, the development of the muscles in the
neck is more pronounced and this factor is particularly clear in
the mature animals. The development of the abdominal muscles
is more pronounced in the female animals. We shall return to
this subject when we discuss the influence of the absolute muscle
weight on the muscle weight distribution.

In steers the situation seems to be more complicated. In
Table 3 the muscle weight distribution in steers is presented
by Butterfield (1963) and is given for comparison. The composition of the standard muscle groups is different and only the
muscle groups of the proximal and distal thoracic limb and of
the distal pelvic limb can be directly compared. The compositior
of the group of proximal muscles of the pelvic limb is only
slightly different between the experiments. Another comparison
for the muscle weight distribution between the sexes is given
by Berg and Butterfield (1976) based on data of Berg (1968). The
main differences between steers and the other sexes are found
for the sequence of the percentage of the proximal muscles of the
pelvic limb. This percentage remains, in general, very constant
in steers.

A study by Brännäng (1971) with identical twin pairs, was
able to demonstrate a reduction of muscle development mainly of
some neck and shoulder girdle muscles. In our experimental grou
of 28 steers in the side muscle weight range of between 74 and
104 kg, these animals were later castrated at 1 year old, but
when we compared them with 88 bulls in the same side muscle
weight range, a very similar pattern to that found by Brännäng
appeared in respect of differences between both sexes although
less pronounced.

Within this side muscle weight range the changes in relatio

to the muscle weight were different for a number of muscles in
particular in the proximal pelvic limb, thorax, shoulder girdle
and neck. And in some cases the sequence was even reversed very
significantly. In comparison to the bulls important differences
on a side muscle weight basis existed. The effect can be seen
quite clearly in Figure 2 for the muscles of the thorax and
proximal pelvic limb. The situation therefore seems more complicated and cannot be explained solely from a reduction of some
muscles which are responsible for secondary sex characteristics
in the bull and a proportional effect on the other muscles only.
This can have somewhat important consequences for the classification of muscles according to their growth patterns and growth
intensity.

In Table 4 the growth coefficients for the total growth
period in the study of Butterfield and Berg (1966b) for the
standard muscle groups are compared with those of our study of
Dutch Friesian half-sibs. The muscle groups are arranged for
the best comparison between the results of the two studies. The
b-values are very similar and, as far as the sequence of the
changes is much the same, this might be expected; but it is
surprising that for the proximal muscles of the pelvic limb we
see a quite different sequence in the study of Butterfield (1963)
as well as in the study of Berg (1968).

MUSCLE WEIGHT DISTRIBUTION IN RELATION TO PHYSIOLOGICAL AGE AND
ABSOLUTE MUSCLE WEIGHT

It might be expected that where the differential growth of
the musculature proves to be regulated strongly by uniform
principles, the muscle weight distribution depends primarily on
the stage of development on the pathway from birth to maturity
in which the animal is slaughtered. In other words, equal
physiological age would give equal muscle weight distributions.
On the other hand, where functional aspects seem to have a strong
influence on the muscle weight distribution, it seems reasonable
as well that much would depend on the absolute muscle weight.

TABLE 4

GROWTH COEFFICIENTS FOR THE TOTAL GROWTH PERIODS OF THE STUDIES
A comparison of data by Butterfield et al (1966$_b$) and own research (present publication)

DATA OF STEERS PUBLISHED BY BUTTERFIELD AND BERG (1966)

Standard muscle group	b_{total}	SE	Differences between phases	Classification according to growth impetus
Prox. pelvic limb	0.99	0.004	P <0.001	High-average or low
Dist. pelvic limb	0.86	0.006		low
Prox. thorac. limb	0.99	0.004	P <0.01	low-average
Dist. thorac. Limb	0.84	0.006	P <0.01	low-average or low
Surrounding spinal column	1.01	0.004		average
Abdominal wall	1.10	0.001	P <0.01	high-average or high
Neck and thorax	1.01	0.010	P <0.01	low-average
Thorax to thorac. limb	1.03	0.005		high
Neck to thorac. limb	1.07	0.008	P <0.01	average-high

Except for muscles of neck and thorax b_{total} different from 1 (P < 0.01)
Mean total side muscle weight = 47.765 kg

OWN RESEARCH WITH DUTCH FRIESIAN HALF-SIBS

Standard muscle group	b_{total}	Males SE	Differences between phases	b_{total}	Females SE	Differences between phases
Prox. pelvic limb	0.98	0.008	P <0.01	0.99	0.008	P <0.01
Dist. pelvic limb	0.83	0.009		0.82	0.009	
Prox. thorac. limb	0.97	0.005	P <0.05	0.95	0.008	P <0.01
Dist. thorac. limb	0.82	0.010		0.82	0.016	P <0.01
Back and loin	1.03	0.005	P <0.05	1.05	0.009	P <0.05
Sublumbar muscles	1.01	0.008		1.03	0.011	
Thorax and abdomen	1.11	0.008	P <0.05			
Neck	1.03	0.016	P <0.01	0.98	0.014	P <0.01
Shoulder girdle	1.04	0.006	P <0.01	1.04	0.010	P <0.05

Mean total side muscle weight - males = 41.020 kg and females = 31.533 kg

In our material we found only some evidence for an influence on the muscle weight by physiological age for some Jersey bulls slaughtered at 300 kg live weight (Table 9). Although not valid consequently for all muscle groups, those animals were best comparable with Dutch Friesian bulls of over 500 kg live weight. Data presented by Berg and Mukhoty (1970) of Jersey bulls of 294 kg compared with bulls of other breeds or crossings of a mean weight of 434 kg have only significantly higher percentages for the abdominal and neck muscles which may indicate that these Jersey bulls were more mature. The higher percentage of pistol meat relative to the total side muscle weight for the Chianina and Blonde d'Aquitaine crossings and the lower percentage for the Hereford crossings in the Danish crossbreeding experiments (Bech Andersen, 1974) can, in respect of these crossings with bulls of rather extreme types, possibly be explained by differences in physiological age. Apparently only as far as somewhat extreme types are concerned do the results suggest influences due to the physiological age.

In Figure 1 we have compared the muscle weight distribution in the Dutch Friesian half-sibs on a percentage basis relative to the expected mature carcass muscle weight and as a basis of the absolute carcass muscle weight (= 2 x the side muscle weight). From data of the sire, some of his sons and daughters and an average position within the breed with regard to mature size, weight and muscularity, the expected mature side muscle weight may be considered as a reliable estimate for the given type. In the examples on the left hand side of the figure, which is, in fact a comparison on the basis of the physiological age, the differences between the sexes are very important. Compared on the basis of the absolute muscle weight the sex differences are considerably reduced and significant differences are found only for the neck and shoulder girdle muscles and partly for the proximal muscles of the pelvic limb and thorax and abdomen.

A comparison of the Dutch Friesian male half-sibs with Dutch Red-and-White male half-sibs of a sire of a very late maturing and heavy type (for which group of half-sibs a difference in mature weight of 20% above the Dutch Friesian half-sibs

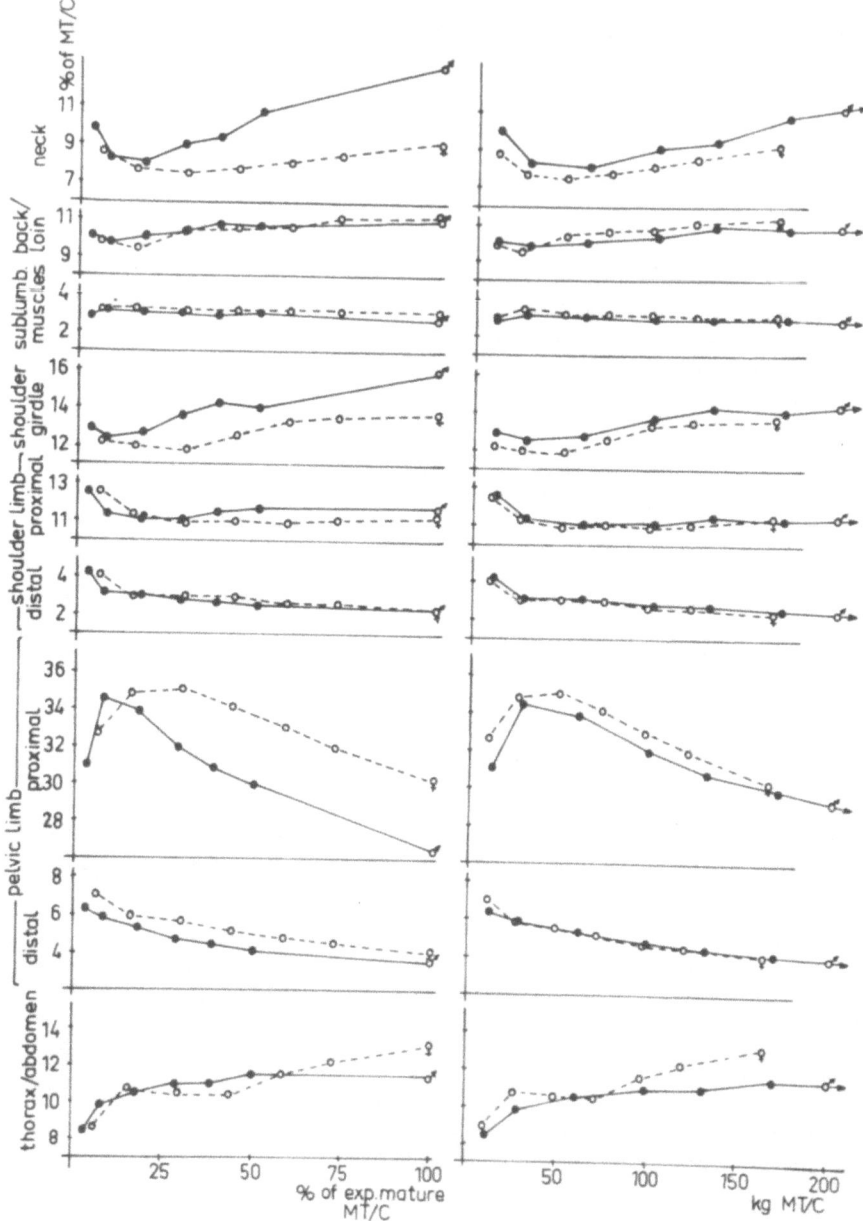

Fig. 1 Muscle weight distribution in Dutch Friesian entire males and female half-sibs expressed as percentage of the total carcass muscle weight (MT/C). Data of mature animals of the same breed are included.

was calculated, based on data of 36 sons of the bull in AI and
a somewhat larger number of daughters) showed very similar results as between the sexes. Here the differences on a muscle
weight basis disappeared nearly completely and significant
differences existed only in the very young stages up to 90 kg
live weight.

In Figure 2, data of a heterogeneous sample of over 300
animals are plotted against the absolute muscle weight subdivided into muscle weight classes of 10 kg. The standard muscle groups are expressed as percentages of the side muscle weight.
The data of the Jersey bulls are not included for these animals
in the side muscle weight class from 40 - 49 kg as the sequence
distortion would be too much. It is obvious that the muscle
weight distribution is influenced to an important extent by the
absolute muscle weight. As far as the age influences are concerned, it is primarily the weight aspect which plays a role and
therefore the muscle weight distribution can never be related
in a very heterogeneous sample to the age alone without regard
to the muscle weight.

THE MUSCLE WEIGHT DISTRIBUTION IN RELATION TO BREED, TYPE AND
QUALITY CLASS

Between breeds and types within breeds, not only differences
in the overall muscularity exist but there seems also to exist
differences in the muscle development in the several body regions.
Studies with animals of different genotypes were seldom able to
confirm differences in weight distribution within the total
muscle fraction in the carcass of any commercial importance.

Data of several breed and crossbred groups presented by
Berg and Mukhoty (1970) demonstrated for only a few standard
muscle groups significant differences. The influences of the sex
were far more important in their material. In a study with four
breed and crossbred groups, Charles and Johnson (1976) found significant differences only for the abdominal muscles and muscles
of the proximal thoracic limb between the genotypes. Robelin

108

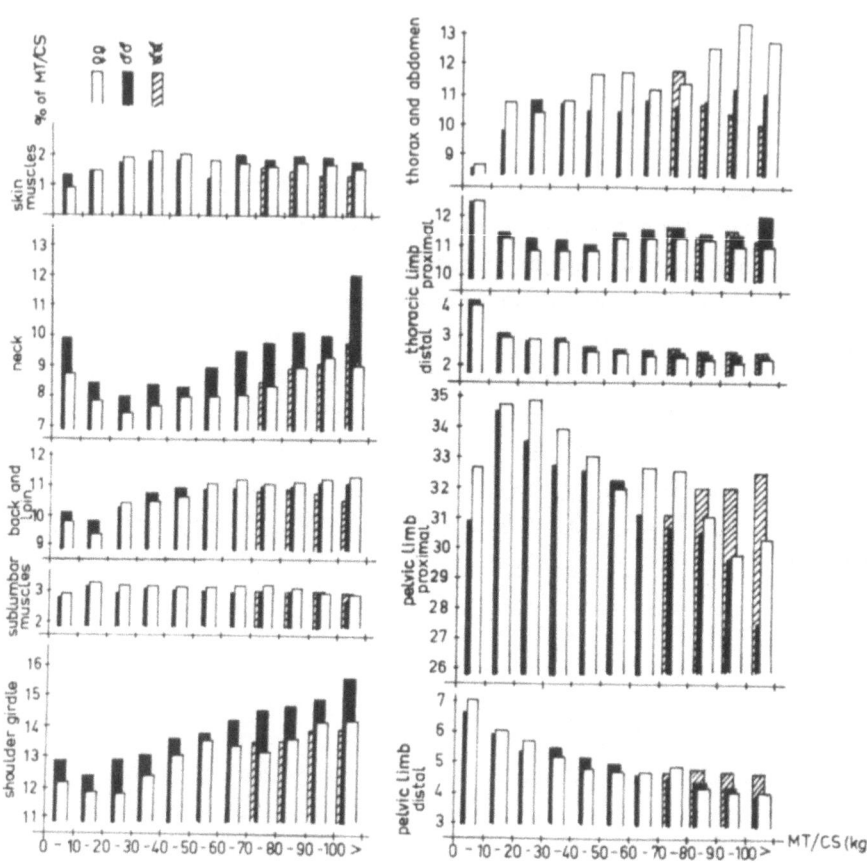

Fig. 2 Muscle weight distribution in a heterogeneous material of 157 bulls,
28 steers and 134 female animals. The percentages relative to the
total side muscle weight (MT/CS) as 100% are plotted against the
absolute side muscle weight, subdivided into classes of 10 kg.

and Geay (1976) studied the muscle weight distribution by means
of the growth coefficients in Charolais and Limousin young bulls.
They conclude that the results were very similar to those of
an earlier study with Friesian young bulls. The results of
anatomically dissected sides by Verbeke (1964, 1965 and 1966)
of animals of the Belgian breeds and by Refsgaard Andersen (1975)
of bulls of the Danish Red breed and our own data of Dutch breeds
and their crossings, do not indicate differences in muscle weight
distribution which are of much commercial importance.

Our own experiments with different genotypes concern animals
of the Dutch Friesian and Dutch Red-and-White breed, crossings
between these breeds and Charolais or Limousin crossings from
Dutch Friesian dams. Results of an experiment with offspring
of different sires but for a large number of half-sibs in the
several sex and treatment categories are given in Table 5.

The data have been adjusted to a common mean side muscle
weight per category. Because of significant differences for the
level and slope of the regression, both groups of heifers cannot
be combined. In each of the categories significant differences
in muscle weight distribution are found between the genotypes,
but differences in the same direction were seldom repeated in
two or more of the categories. Although interactions cannot be
excluded, it might be expected that when breed differences of
any importance might exist, the overall situation would then be
sytematical.

In several studies animals of the Dutch Friesian and Dutch
Red-and-White breeds are compared. Data of young bulls before
and after correction to a common mean side muscle weight are
given in Table 6. The corrected values were significantly different for four standard muscle groups. It may be mentioned
that the influences of the correction on the muscle weight distribution in this heterogeneous sample are well in agreement with
the sequence for the muscle groups as presented in Table 1 for
the Dutch Friesian half-sibs. In an experiment with 36 mature
cows of both Dutch breeds, significant differences were found

TABLE 5

MUSCLE WEIGHT DISTRIBUTION IN DUTCH FRIESIAN (FH) AND DUTCH RED-AND-WHITE (MRIJ) ANIMALS, THEIR CROSSINGS AND CROSSINGS BETWEEN FRENCH BEEF BREEDS AND DUTCH FRIESIAN COWS. DATA OF DIFFERENT SEXES AND TREATMENT GROUPS

The data are per sex/treatment, adjusted to a mean side muscle weight. Significant differences (P < 0.05) are indicated by underlined data

Category and breed or crossing	Standard muscle groups (total side muscle weight = 100%)										n	adjusted to a mean side muscle weight of:
	A Prox. pelvic limb	B Dist. pelvic limb	C Prox. thorac. limb	D Dist. thorac. limb	E Neck	F Back and loin	G Sub-lumbar muscle group	H Thorax and abdomen	I Shoulder girdle	J Skin muscles		
Young bulls intensively fed. 425 kg live weight												
Charolais x FH	31.22	4.44	11.45	2.43	9.28	10.96	3.12	10.86	14.23	1.87	10	
Limousin x FH	31.28	4.38	11.14	2.37	9.77	10.96	3.09	10.64	14.17	2.04	10	80.01 kg
MRIJ x FH	30.09	4.38	11.63	2.47	9.76	10.91	3.05	11.43	14.15	2.01	10	
MRIJ	29.57	4.45	11.92	2.49	10.06	10.64	3.05	11.17	14.57	2.04	5	
FH	30.88	4.51	11.71	2.49	9.50	10.63	3.02	11.04	14.14	1.99	5	
Steers extensively fed, 2½ years												
Charolais x FH	32.44	4.82	11.50	2.60	8.63	10.53	2.99	10.46	14.66	1.23	5	
Limousin x FH	33.10	4.76	11.44	2.50	8.63	10.59	3.04	10.46	14.21	1.14	5	91.60 kg
MRIJ x FH	31.55	4.48	11.30	2.54	9.43	10.95	2.83	10.37	15.16	1.23	5	
MRIJ	31.97	4.85	11.64	2.53	9.17	10.60	3.00	10.11	14.69	1.33	5	
FH	31.99	4.48	11.39	2.52	9.23	10.60	2.84	10.50	14.98	1.36	5	

Many of the animals within breeds or crossings are half-sibs.

Table Continued

TABLE 5 (Cont)

MUSCLE WEIGHT DISTRIBUTION IN DUTCH FRIESIAN (FH) AND DUTCH RED-AND-WHITE (MRIJ) ANIMALS, THEIR CROSSINGS AND CROSSINGS BETWEEN FRENCH BEEF BREEDS AND DUTCH FRIESIAN COWS. DATA OF DIFFERENT SEXES AND TREATMENT GROUPS

The data are per sex/treatment, adjusted to a mean side muscle weight. Significant differences (P < 0.05) are indicated by underlined data.

Category and breed or crossing	Standard muscle groups (total side muscle weight = 100%)										n	adjusted to a mean side muscle weight of:
	A Prox. pelvic limb	B Dist. pelvic limb	C Prox. thorac. limb	D Dist. thorac. limb	E Neck	F Back and loin	G Sub-lumbar muscle group	H Thorax and abdomen	I Shoulder girdle muscles	J Skin muscles		
Young heifers intensively fed. 375 kg live weight												
Charolais x FH	32.58	4.80	11.42	2.51	7.50	11.04	3.33	11.81	13.29	1.63	10	
Limousin x FH	32.82	4.82	10.88	2.32	7.94	11.66	3.05	11.19	13.78	1.70	10	
MRIJ x FH	31.94	4.58	11.23	2.51	8.04	11.20	3.08	11.72	13.49	1.87	10	63.46 kg
MRIJ	32.56	4.84	11.23	2.61	7.97	11.08	3.18	11.10	13.33	1.79	5	
FH	32.97	4.65	11.19	2.42	7.83	11.21	3.06	11.18	13.49	1.91	5	
Heifers extensively fed, 2 years												
Charolais x FH	32.75	4.73	11.17	2.39	8.18	10.84	3.24	11.23	13.84	1.50	5	
Limousin x FH	33.61	4.95	11.46	2.44	8.12	10.66	3.23	10.54	13.27	1.61	5	
MRIJ x FH	32.84	4.94	11.44	2.50	8.06	10.64	3.30	10.92	13.39	1.84	5	70.76 kg
MRIJ	33.04	4.76	11.50	2.29	8.31	10.90	3.21	10.92	13.18	1.68	5	
FH	33.09	4.72	10.87	2.30	7.96	11.32	3.39	11.24	13.31	1.75	5	

Many of the animals within breeds or crossings are half-sibs.

EXPLANATION TO TABLE 5 WITH REGARD TO SIGNIFICANT DIFFERENCES IN MUSCLE WEIGHT DISTRIBUTION IN BREED AND CROSSBRED GROUPS

Category/ breed- crossing	Standard muscles groups									
	A	B	C	D	E	F	G	H	I	J
Young bulls										
Char. x FH	ab		a		a			a		
Lim. x FH	cd		bcd	abc				bc		
MRIJ x FH	ac	N.S.	b	a		N.S.	N.S.	ab	N.S.	N.S.
MRIJ	bde		ac	b	a			c		
FH	e		d	c						
Steers										
Char. x FH	ab	ab			a				ab	
Lim. x FH	bcde				b		a		acde	ab
MRIJ x FH	ac	ac	N.S.	N.S.	ab	N.S.	ab	N.S.	bcf	
MRIJ	d	cd					b		df	a
FH	e	bd							e	b
Young heifers										
Char. x FH		a	a	a	abc	a	abcd	a	a	a
Lim. x FH	a	abc	abc	a	a	abc	ae	b	ab	
MRIJ x FH	ab	b	b		b	b	b	bc		a
MRIJ		c	cd	b	c	c	ce	ac	b	
FH	b		d	b			d			
Heifers										
Char. x FH								a	a	
Lim. x FH			a					a		
MRIJ x FH	N.S.	N.S.	b	N.S.	N.S.	N.S.				N.S.
MRIJ			c				a		a	
FH			abc				a			

Within categories standard muscle groups having the same superscript are significantly different (P <0.05)

TABLE 6

MUSCLE WEIGHT DISTRIBUTION IN DUTCH FRIESIAN AND DUTCH RED-AND-WHITE YOUNG BULLS.
Original data and data after correction to a total mean side muscle weight of a heterogeneous material. (Total side muscle weight = 100%)

Standard muscle groups	FH Mean	FH St. dev.	MRIJ Mean	MRIJ St. dev.	FH Corrected for side muscle weight	MRIJ Corrected for side muscle weight
Prox. pelvic limb	31.59	1.52	30.99	1.43	31.19	31.17
Dist. pelvic limb	4.71	0.45	4.70	0.55	4.63	4.74
Prox. thorac. limb	11.32	0.45	11.45	0.47	11.31	11.45
Dist. thorac. limb	2.69	0.21	2.60	0.22	2.64	2.62
Neck region	9.25	0.99	9.60	1.11	9.44	9.52
Back and loin	10.53	0.46	11.00	0.45	10.53*	11.00
Sublumbar muscles	2.99	0.13	2.92	0.17	2.98	2.92
Thorax and abdomen	10.95	0.59	10.37	0.81	11.09*	10.31
Shoulder girdle	13.77	0.84	14.54	1.04	13.96*	14.46
Skin muscles	1.97	0.21	1.65	0.28	1.98*	1.65
Total side muscles weight (kg)	57.6	21.2	67.9	18.8	Corrected to mean side muscle weight = 64.7	
n	27		65			

* Significant differences (P <0.05) between breeds

only for the proximal muscles of the thoracic limb after correction to a common mean side muscle weight of 91 kg. As far
as groups of Dutch Friesian and Dutch Red-and-White animals are
concerned, significant differences were found for a variety of
standard muscle groups but in practically none of the experiments
for the same muscle groups. In the study of Berg and Mukhoty
(1970) most of the significant differences were found for the
abdominal muscles and muscles of the neck and thorax and in the
study of Charles and Johnson (1976) only for the abdominal muscles and in the proximal muscles of the thoracic limb. In our
study this was valid, among others, for the muscles of the thorax
and abdomen in the young bulls and for the muscles of the proximal thoracic limb in the cows. This can be an indication that
breed differences in muscle weight distribution do not appear
entirely unsystematically.

In order to study weight distribution of the muscles in
animals in which the growth does not play a role and which are,
to an important extent, different in type, we have selected a
group of mature Dutch Friesian and Dutch Red-and-White cows of
different quality classes. The degree of fatness was quite constant except for the 3rd quality and manufacturing quality
animals which were in a poorer condition. To demonstrate that
the differences between quality classes were really important,
the side muscle weight and the muscle to bone ratio is included
in Table 7.

In this material the muscle weight distribution was markedly
constant among the quality classes and only for the muscles of
the thorax/abdomen and shoulder girdle, systematic differences,
partly significant between the quality classes, were found. The
changes in one of these two muscle groups of equal low commercial vale, were almost compensated for by the changes in the
other muscle group.

In Table 7, data of some heifers of different quality are
included. Here the muscle weight distribution was not as constant as in the mature cows.

115

TABLE 7

MUSCLE WEIGHT DISTRIBUTION IN MATURE COWS OF DIFFERENT QUALITY CLASSES. (Means with standard dev.).
Data of some heifers of different quality classes are included in the Table.

| Quality class | Standard muscle groups (total muscle side weight = 100%) | | | | | | | | Side muscle wt.(kg) | Muscle to bone ratio | n |
	A	B	C	D	F	G	H	I			
Extra qual. cows	30.0	4.2	11.4	2.3	11.3	2.8	12.3	14.5	120.7	4.80	3
	0.2	0.1	0.4	0.2	0.2	0.2	1.0	0.5	7.3	0.12	
1st qual. cows	30.4	4.2	11.1	2.3	11.3	3.0	12.7	14.0	95.2	4.12	7
	0.6	0.2	0.3	0.1	0.4	0.3	0.4	0.4	12.3	0.33	
2nd qual. cows	30.4	4.1	10.9	2.3	11.3	3.1	13.2	13.7	87.4	4.05	10
	0.7	0.2	0.4	0.2	0.5	0.2	0.8	0.6	5.9	0.18	
3rd qual. cows	30.2	4.1	11.2	2.3	11.2	3.0	13.2	13.9	92.0	3.56	10
	1.3	0.2	0.5	0.2	0.2	0.1	0.7	0.7	11.7	0.29	
Manufact- uring cows	30.0	4.4	11.8	2.2	11.1	3.0	13.7	13.2	75.1	3.21	3
	1.2	0.1	0.2	0.3	0.6	0.3	0.1	0.9	4.0	0.11	
1st qual. heifers	32.6	4.3	12.1	2.4	11.2	3.1	10.3	13.5	78.7	4.28	3
	0.8	0.3	0.5	0.1	0.5	0.3	0.1	0.3	3.8	0.51	
3rd qual. heifers	31.8	4.8	12.2	2.6	11.2	3.3	10.0	13.3	67.8	3.53	3
	0.3	0.2	0.6	0.2	0.2	0.3	0.3	0.2	1.0	0.05	

standard muscle groups as in Table 1. The standard muscle groups E (neck) and J (skin muscles) are not mentioned here because of the extremely small differences between quality classes. For the neck muscles all means were between 9.0 and 9.3% of the side muscle weight.

TABLE 8

MUSCLE WEIGHT DISTRIBUTION IN SOME DOUBLE MUSCLED AND NORMAL CALVES (INTENSIVELY FED WITH MILK SUBSTITUTES) OF THE BELGIAN WHITE BREED.
Data of individual animals. All animals were entire males. The standard muscle groups not mentioned here differed only very little between the two types.

Standard muscle group	Double muscled type individual animals			Mean	Normal type individual animals			Mean
Prox. pelvic limb	35.64	33.34	31.92	33.63	33.57	32.14	33.52	33.08
Dist. pelvic limb	4.94	4.80	4.47	4.74	5.37	4.72	4.89	4.99
Neck	7.53	8.28	9.27	8.36	8.77	9.46	8.31	8.85
Thorax and abdomen	8.76	9.30	9.38	9.15	10.12	10.53	10.35	10.33
Shoulder girdle	13.73	14.10	15.43	14.42	12.89	13.59	13.68	13.39
Skin muscles	2.04	1.88	2.14	2.02	1.50	1.79	1.57	1.62
Total side muscle weight (kg)	44.7	50.5	68.0	54.4	38.4	38.6	46.9	41.3

TABLE 9

MUSCLE WEIGHT DISTRIBUTION IN A SMALL NUMBER OF JERSEY BULLS, INTENSIVELY FED AND SLAUGHTERED AT 300 kg LIVE WEIGHT.
The animals were rather fat with 22% dissectible fat in the side. (Means with standard deviation)

	Standard muscle groups (total side muscle weight = 100%)									Side muscle	n
	A	B	C	D	E	F	G	H	I		
Jersey bulls 1 yr	28.8	4.1	11.8	2.3	11.5	11.8	3.1	11.3	13.6	47.0 kg	4
	0.5	0.2	0.4	0.2	0.5	0.4	0.2	0.4	0.2	1.7	

Standard muscle groups as in Table 1

Animals of the so-called 'double muscled' type have a very extreme position. We shall not enter into details of the muscle weight distribution in these double muscled animals. From French, Belgian and Italian results of studies on a larger scale it is quite clear that these animals have not the same muscle weight distribution in comparison with normal animals. The influence of a centripetal gradient and a more pronounced hypertrophy of the superficial muscles can be recognised.

An experiment with double muscled veal calves compared with normal calves of the Belgian White breed is worthwhile mentioning (Table 8). For a given age or live weight, double muscled animals can have high muscle weights because of the high dressing percentage and the favourable muscle to bone to fat ratio in the carcass. In our material it was obvious that differences in muscle weight distribution in comparison with normal animals, existed albeit on a quite different level; the sequence of the changes for the standard muscle groups followed very much the same pattern of the changes in relation to the absolute muscle weight as given in Table 1 and Figure 1.

The influence of the physiological age on the muscle weight distribution has already been discussed. In our own material only Jersey young bulls had a rather mature muscle weight distribution at a relatively low live weight. This was not consequently valid for all muscle groups (Table 9) and it therefore can certainly not be considered as a good example of indirect breed influences by the physiological age.

THE MUSCLE WEIGHT DISTRIBUTION IN RELATION TO THE FATNESS

It is essential to know whether the amount of intramuscular fat will influence the muscle weight distribution systematically. In most of the experiments the intramuscular fat is not determined and the eventual effect on the muscle weight distribution is ignored.

Butterfield (1963) showed that the muscle weight distribution

TABLE 1O

MUSCLE WEIGHT DISTRIBUTION IN SOME GROUPS OF ANIMALS WITH DIFFERENT FEEDING INTENSITY. Data of Dutch Friesian heifers and of non-identical twin calves of the Dutch breeds. (Means and standard deviation: side muscle weight = 1OO%)

Standard muscle group	Heifers v. intensively fed	Heifers intensively fed	Heifers extensively fed	Male veal calves	Male grass-fed calves
Prox. pelvic limb	31.8 0.6	32.5 0.7	33.1 1.0	32.1 0.9	33.1 0.4
Dist. pelvic limb	4.5 0.2	4.7 0.3	4.8 0.4	5.1 0.3	5.3 0.4
Neck region	8.4 0.2	7.8 0.3	8.0 0.4	9.0 0.4	8.0 0.1
Shoulder girdle	13.5 0.4	13.5 0.1	13.4 0.5	14.2 0.8	12.8 0.6
Thorax/abdomen	11.9 0.5	11.5 0.6	11.2 0.5	10.0 0.4	12.8 0.6
Total side muscle weight	6O.1 kg	57.4 kg	66.8 kg	32.3 kg	31.5 kg

was influenced by the proportion of dissectible fat in the side. In most of our experiments an analysis of the data indicated significant influences on the muscle weight distribution by the fatness expressed as muscle to dissectible fat ratio, but this does not give an answer to the question whether the changes in muscle weight distribution are caused by a changing level of intramuscular fat. From a study of Johnson et al (1973) it is obvious that among the individual muscles and muscle groups the amount of intramuscular fat varies considerably. In a study with Angus steers in the range from 12.3 to 35.1% dissectible fat in the side, Johnson et al. (1973) found that the partitioning of the intramuscular fat among the standard muscle groups was not influenced by the overall degree of fatness except for the relatively small muscle groups of the distal limbs. No significant differences for the regressions of the weights of the muscle groups on the total muscle weight in the original data and after correction for the intramuscular fat were found.

These findings lead to the conclusion that the muscle weight distribution is not to any great extent influenced systematically by the intramuscular fat content in the muscles. It might be expected that if the fatness did influence the muscle weight distribution, this would be most clearly seen in those muscles in which the fat deposition is relatively important, for example, the muscles of the thorax and abdomen. This was valid for a group of young bulls but, in other experiments, it was mainly the relatively lean groups of muscles of the neck and proximal pelvic limb which seemed to vary in relation to the fatness. These muscle groups are those which show considerably pronounced changes in the growing animal. In the mature cows, significant influences of the fatness were found only for the small group of skin muscles. It seems, therefore, questionable where influences of the fatness on the muscle weight distribution are found; these refer primarily to the amount of intramuscular fat in the muscles.

THE MUSCLE WEIGHT DISTRIBUTION IN RELATION TO THE GROWTH INTENSITY

Butterfield (1963) showed that the most important changes in muscle weight distribution are found in the very young animal in the period in which the birth weight is doubled. This growth phase therefore, was considered the best in which to study the influences of different growth rates.

Butterfield and Berg (1966) showed by means of the group of muscles with a high/average growth impetus, that on an age basis this group of muscles comprised a lower percentage of the total side muscle weight in the calves of the low plane of nutrition. When the high impetus phase ceased in the high plane, calves at 6 and in the low plane at 12 weeks, the impetus group presented a constant proportion of the total muscle weight. If compared on a muscle weight basis, the weight of this impetus group relative to the side muscle weight, was not affected by the plane of nutrition. In an experiment of Butterfield and Johnson (1971) with calves fed at different planes of nutrition significant influeces were found only for the somewhat small group of low/average growth impetus in the calves of the high plane of nutrition. The conclusion is that if compared on an absolute muscle weight basis the relationship between the weights of the muscles is not affected by the growth intensity.

In our opinion this conclusion is too easily extended to other growth phases. In the Danish growth experiment, Refsgaard Andersen, 1975) significant differences were found for 9 muscles or muscle combinations between animals of different feeding level The weight of the expensive muscles in the pistol cut relative to the total side muscle weight was nearly the same in all cases, irrespective of the feeding level.

Our own experiments were not designed to study the effect of the growth intensity on the muscle weight distribution. An experiment with young bulls of the two Dutch breeds started only recently and here, several groups of animals, from intensively fed veal calves, young bulls and young heifers through to extensively fed animals, have been used for comparisons. There was a tendency that animals grown quickly to a given muscle weight would have slightly more 'mature' characteristics in their muscle weight distribution compared with those animals grown more slowly to the same muscle weight.

In Table 10, data of some muscle groups are given for three groups of Dutch Friesian heifers and non-identical male twins of the Dutch breeds fed at different regimes. In some of the muscle groups the values are not in agreement with the sequence in relation to the side muscle weight of Table 1. It must be taken into account that in the calves the abdominal muscles are comparatively strongly influenced by the feeding regime. These aspects of influences of the growth intensity on the muscle weight distribution, independent from the absolute muscle weight, can have consequences for schemes of the changes in muscle weight distribution in relation to age or weight.

Butterfield (1966) studied the effect of weight loss and recovery on muscle weight distribution. The conclusions based on these studies by Berg and Butterfield (1976) are that muscles which are most essential for survival are least affected by nutritional stress. Animals which lose muscle weight tend to revert to the juvenile muscle weight distribution. Recovery after weight loss tended towards a return to the normal muscle weight distribution.

Although the differences indicated in Table 7 between the manufacturing cows (which have certainly lost muscle weight) and the cows of the better qualities, are small, this situation is not in contradiction to the findings of Berg and Butterfield (1976) regarding the effect of muscle weight loss. A study with pairs of identical female twins which were subjected from birth to 9 months to very extreme variations in nutritional regimes, and then, after re-alimentation, slaughtered at 4½ years, has not yet been evaluated completely. The preliminary results, however, show that within pairs, the ultimate muscle weight distribution is somewhat influenced as a consequence of the nutritional regime experienced by the animals between birth and 9 months. The percentage of expensive muscles of the combined muscle groups A, C, F and G relative to the side muscle weight, is highest in the animals of the high plane. This is mainly due to the muscles of the proximal pelvic limb. Between the animals of the very

low and moderate plane, practically no differences were found.
This means that the animals of the very low plane returned completely to a normal muscle weight distribution whereas the
positive influences of the high plane resulted in differences
of 0.5% of the total side muscle weight in the combination of
expensive muscles.

THE INDIVIDUAL VARIATION IN MUSCLE WEIGHT DISTRIBUTION

In all our experiments an individual variation in muscle
weight distribution is found. As far as growing animals are concerned, there is a possibility that this variation can be
attributed to differences in physiological age for even in a
uniform group of half-sibs, quite important differences in ultimate mature weight can be expected. An analysis of data of
mature female animals showed significant influences of the muscle
weight, the fatness, the breed and the age on the muscle weight
distribution. At least a part of the individual variation could
be explained from the factors mentioned.

Our study with female identical twin pairs already mentioned
showed very clearly that, irrespective of the rather small influences by the plane of nutrition in the young stage, the
variation in muscle weight distribution was far greater between
pairs than was valid within pairs. The partners of the same twin
pair had very similar deviations from the normal pattern of
muscle weight distribution in cows. This was valid even for the
small and difficult-to-separate muscles which may indicate that
the technical imperfections at dissection have not influenced
the individual variation to such an extent as we had often feared

Although the twin partners had the same intra-uterine enviro
ment we must accept that the variation in muscle weight
distribution among the twin pairs has a genetic basis. Similar
effects as those found in the identical twin pairs were also
found in non-identical twin pairs.

Functional aspects certainly play an important role with

regard to the muscle weight distribution. If we see, however, the individual variation in the weights of the muscle groups relative to the side muscle weight (eg, in mature cows from 27.3 to 32.4% of the total side muscle weight for the muscles of the proximal pelvic limb or a range of 17% of the mean value), without any evidence that this has influenced the locomotion, we wonder whether 'functional aspects' are not sometimes used too easily to explain some of the findings.

DISCUSSION

It is obvious from several experiments that the differential muscle growth in each growing animal is regulated by very strong and uniform rules. The changes which take place in relation to age and muscle weight are the most important although the most pronounced changes occur in the early growth phases and, very often, before the animal becomes commercially important. For the interpretation of several other factors which may influence the muscle weight distribution, it is a strict necessity to dispose of good schemes in connection with the changes in relation to age and muscle weight. In this respect, further studies, with comparable animals and serial slaughterings, are desirable.

With regard to sex influences in particular, the position of the castrated male animal needs further study because an important part of the conclusions in respect of growth patterns of muscles, is based on data of steers. The results of several studies indicate that the entire male and female animals have more similar muscle growth patterns than is valid for steers versus bulls.

The influences of breed and type on muscle weight distribution are, in most experiments, very small and further studies with animals of breeds which differ in mature weight and muscularity, are needed. The present conclusions are, in a great measure, based on breeds and crossings which vary little in this respect.

The influences of the degree of fatness on the muscle weight distribution needs further attention because the conclusions are based on a very small number of animals. The results thus far do not indicate that the fatness will affect, to a great extent, the muscle weight distribution.

The several studies with regard to the influences of the growth intensity on the muscle weight distribution, are too much limited to the very young growth phases. We fear that the conclusions are too easily extended to other growth phases and that the influence of this factor is often underestimated.

In general, we can say that the influences on the muscle weight distribution as far as influenced significantly by several other factors, are small and in most cases, hardly of commercial importance. It must be clearly understood, however, that we have discussed only the weight ratio between individual muscles and muscle groups. A grouping of muscles according to their commercial value or so-called 'eating quality', as influenced by weight, size, shape of the muscles, sex, breed or type, age and fatness, does not necessarily provide similar results and these factors have been ommitted from this discussion.

It became quite clear that the morphological differences between animals are of a different origin and do not reflect the actual muscle weight distribution. In our material the relationship between the results of the visual appraisal and the weight distribution of the muscles, was invariably extremely poor.

It has often been concluded that the muscle weight distribution cannot be altered by selection because the environmental conditions and the functional requirements remain largely the same. Indeed, many decades of selection in entirely different directions have hardly altered the muscle weight distribution at all and we must realise that a selection, directly on the muscle weight distribution, has never taken place.

REFERENCES

Bech Andersen, B. 1974 Proc. Working Symp. Breed Evaluation and Crossing
    Exp. p 149 Zeist, Netherlands

Berg, R.T. 1968 Proc. 21st Recip. Meat Conf. Chicago

Berg, R.T. and Butterfield, R.M. 1976 New Concepts of Cattle Growth, Univ.
    Press, Sidney

Berg, R.T. and Mukhoty, H.M. 1970 49th Ann. Feeders' Day Report. Univ. of
    Alberta

Brännäng, E. 1971 Swedish J. Agric. Res., 1, 69

Butterfield, R.M. 1963 A Study of the Musculature of the Steer Carcass.
    Thesis, Brisbane

Butterfield, R.M. 1966 Res. Vet. Sci. 7, 168

Butterfield, R.M. and Berg, R.T. 1966 Proc. Austr. Soc. Anim. Prod. 6, 298

Butterfield, R.M. and Berg, R.T. 1966$_a$ Res. Vet. Sci 7, 326

Butterfield, R.M. and Berg, R.T. 1966$_b$ Res. Vet. Sci. 7, 389

Butterfield, R.M. and Johnson, E.R. 1971 J. Agric. Sci. 76, 457

Charles, D.D. and Johnson, E.R. 1976 J. Anim. Sci. 42-2, 332

Huxley, J. 1932 Problems of Relative Growth. Methuen, London

Johnson, E.R., Pryor, W.J. and Butterfield, R.M. 1973 Austr. J. Agric. Res.
    24, 287

Refsgaard Andersen, H. 1975 Beretning fra Statens Husdyrbrugsforsøg 430
    Copenhagen

Robelin, J. and Geay, Y. 1976 Ann. Zootech. 25-2, 273

Verbeke, R. 1964 Mededeling Studiecentrum voor Rundvleesproduktie 2b, Ghent

Verbeke, R. 1965 Mededeling Studiecentrum voor Rundvleesproduktie 2c, Ghent

Verbeke, R. 1966 Mededeling Studiecentrum voor Rundvleesproduktie 7, Ghent

APPENDIX

GROWTH COEFFICIENTS OF STANDARD MUSCLE GROUPS AND INDIVIDUAL MUSCLES OR COMBINATIONS OF MUSCLES FOR DIFFERENT GROWTH PHASES.

Data of Dutch Friesian entire male and female half-sibs (for explanation, see text).

Entire male animals	b_1	SE	b_2	SE	b_3	SE	b_t	SE	P_a	P_b	P_t	b_{2+3}	SE	% of side muscle wt
Prox. pelvic limb	1.12	0.01	0.87	0.02	0.88	0.05	0.98	0.01	2	2	2	1.02	0.01	30.46
m. gluteobiceps	1.17	0.02	0.93	0.03	0.90	0.07	1.03	0.01	2	2	2	0.91	0.01	6.90
glut.medius	1.23	0.03	0.90	0.07	0.69	0.11	1.01	0.02	2	2	2	0.87	0.03	3.58
glut.accessor	1.02	0.14	0.74	0.12	1.18	0.19	0.93	0.03	2			1.00	0.05	0.27
glut.profundus	0.75	0.05	1.08	0.10	1.05	0.10	0.88	0.02		2		0.98	0.03	0.34
semitendinosus	1.21	0.06	1.03	0.04	1.28	0.13	1.07	0.02			1	1.10	0.05	2.71
semimembranosus	1.24	0.02	0.79	0.04	0.66	0.08	0.98	0.02	2	2	2	0.79	0.02	4.97
quadriceps fem.	0.94	0.02	0.75	0.04	0.84	0.09	0.90	0.01	1	2	2	0.80	0.02	5.78
tensor fasciae latae	1.13	0.05	1.04	0.10	0.92	0.10	1.07	0.02	2		1	0.93	0.03	1.34
mm. gemelli+ quadr.	0.83	0.10	0.83	0.12	0.57	0.24	0.88	0.02				0.87	0.05	0.12
m. gracilis	1.03	0.02	0.92	0.06	1.00	0.15	1.00	0.01				0.98	0.03	1.35
sartorius	0.95	0.16	0.87	0.12	1.37	0.22	1.01	0.03				1.00	0.05	0.39
pectineus	1.23	0.07	0.90	0.11	1.06	0.17	0.96	0.02				0.85	0.05	0.57
adductor	1.19	0.06	0.67	0.08	0.92	0.22	0.94	0.02	2		2	0.75	0.04	1.60
obturator ext.	0.85	0.05	0.99	0.07	1.22	0.21	0.92	0.02			1	0.92	0.04	0.54
Dist. pelvic limb	0.82	0.02	0.76	0.02	0.65	0.09	0.83	0.01	1		1	0.75	0.04	4.35
m. soleus	0.93	0.27	0.02	0.90	1.86	0.81	0.69	0.10				1.00	0.25	0.01
gastrocnemius	0.98	0.02	0.75	0.08	0.76	0.11	0.86	0.01	2		2	0.74	0.02	1.86
flexor dig.sup.	0.81	0.05	0.72	0.07	0.92	0.15	0.79	0.01	1			0.82	0.04	0.38
mm. flex.+extens.	0.77	0.03	0.76	0.05	0.62	0.14	0.82	0.01				0.75	0.02	2.10
Prox. thoracic limb	0.89	0.02	1.01	0.03	0.98	0.05	0.97	0.01	2	1	2	1.02	0.01	11.39
m. deltoideus	0.79	0.03	1.15	0.09	1.15	0.18	1.02	0.02	2	2	2	1.18	0.04	0.54
subcapularis	0.75	0.11	1.08	0.09	1.14	0.20	0.93	0.02				1.03	0.04	1.18
infraspinatus	1.03	0.10	1.04	0.09	1.15	0.10	0.99	0.02				1.04	0.03	1.95
supraspinatus	0.91	0.03	1.03	0.05	0.97	0.11	0.99	0.01				1.02	0.03	1.57
teres major	0.99	0.03	1.00	0.07	0.93	0.23	0.99	0.01				1.04	0.04	0.42
teres minor	0.92	0.08	0.89	0.11	0.69	0.14	0.92	0.02				1.01	0.04	0.18

APPENDIX (Cont)

Entire male animals	b_1	SE	b_2	SE	b_3	SE	b_t	SE	P_a	P_b	P_t	b_{2+3}	SE	% of side muscle wt
Prox.														
m. coracobrachial.	0.94	0.11	0.78	0.09	0.55	0.17	0.91	0.02				0.85	0.06	0.14
tensor fasciae														
antebrachii	0.86	0.09	1.26	0.20	1.10	0.16	1.06	0.03				1.12	0.07	0.17
triceps brachii	0.89	0.02	1.01	0.02	0.90	0.09	0.98	0.01	2		2	1.02	0.02	4.18
biceps brachii	0.90	0.04	0.86	0.06	0.85	0.16	0.91	0.01				0.96	0.03	0.62
brachialis	0.76	0.03	0.77	0.09	0.73	0.13	0.85	0.01	2			0.89	0.03	0.43
Dist.thoracic limb	0.67	0.05	0.78	0.03	0.69	0.14	0.82	0.01	2		1	0.81	0.03	2.46
Neck region	0.80	0.02	1.25	0.04	1.50	0.12	1.03	0.02	2	2	2	1.28	0.03	10.09
mm. rect.+obl.cap.	0.97	0.08	1.07	0.20	1.27	0.11	0.91	0.03	2		1	1.16	0.05	0.64
m. splenius	0.58	0.07	1.56	0.16	1.72	0.21	1.13	0.05	2		2	1.69	0.06	1.02
mm. longissimus														
cap. et atl.	0.60	0.08	1.74	0.18	2.06	0.23	1.12	0.05	1	2	2	1.61	0.06	0.42
m. semispinal cap.	0.67	0.03	1.32	0.08	1.68	0.16	0.99	0.03	2	2	2	1.33	0.04	2.01
omohyoideus	0.13	0.50	0.05	0.70	0.62	1.37	0.76	0.12				0.95	0.25	0.03
sternocephal	0.72	0.13	1.14	0.16	1.75	0.72	1.01	0.05				1.23	0.10	0.90
longus colli														
etc.*	1.01	0.24	1.21	0.05	1.40	0.20	1.04	0.04				1.14	0.04	2.21
omotransversar	0.89	0.06	1.48	0.12	1.73	0.13	1.11	0.03		2	2	1.33	0.05	0.62
brachiocephal	0.85	0.07	1.09	0.03	1.18	0.08	1.06	0.02	2		2	1.20	0.02	1.57
subclavius	1.00	0.16	1.20	0.59	1.50	0.58	0.89	0.07	1			1.46	0.15	0.04
mm. scalenus dors+														
ventralis	0.80	0.08	1.29	0.12	1.41	0.37	1.00	0.03		1	1	1.30	0.06	0.65
Back and loin region	0.96	0.01	1.06	0.03	0.97	0.08	1.03	0.01	2	1	2	1.06	0.02	10.93
mm. serratus dors.														
inspir.+expir.	0.87	0.09	1.27	0.16	1.44	0.27	1.12	0.03				1.19	0.06	0.34
iliocostalis	0.71	0.05	1.22	0.13	1.02	0.17	1.00	0.03				1.19	0.04	0.48
longiss.cervic.	0.36	0.09	1.61	0.32	1.27	0.38	0.98	0.06	1	1	2	1.41	0.09	0.35
longiss.thorac.														
+ lumborum	1.15	0.02	1.03	0.02	0.85	0.12	1.07	0.01	1			1.00	0.02	6.50
spinal.et semi-														
spinal.etc.**	0.75	0.02	1.02	0.12	1.10	0.08	0.96	0.02	2		2	1.13	0.03	3.27

APPENDIX (Cont.)

Entire male animals	b_1	SE	b_2	SE	b_3	SE	b_t	SE	P_a	P_b	P_t	b_{2+3}	SE	% of side muscle wt
Sublumbar muscles (= mm. iliopsoas, psoas minor and quadrat.lumborum)	1.13	0.03	0.94	0.06	1.16	0.08	1.01	0.01	2		2	0.99	0.03	2.94
Thorax and abdomen	1.18	0.04	1.09	0.06	1.15	0.08	1.11	0.01				1.08	0.02	10.68
m. transv.thorac.	1.08	0.05	0.94	0.24	1.00	0.30	0.97	0.03	2	1	2	1.01	0.07	0.21
rectus thorac.	0.71	0.07	1.13	0.09	0.98	0.24	0.97	0.02				1.02	0.04	0.11
mm. intercostales etc*)	1.10	0.18	1.15	0.10	1.29	0.10	1.04	0.03				1.18	0.03	2.80
Diaphragm	1.10	0.08	1.17	0.10	1.35	0.24	1.09	0.02				1.10	0.05	0.65
m. obliq.ext.abd.	1.28	0.06	1.15	0.10	1.19	0.16	1.20	0.02	2			1.10	0.03	2.04
obliq.int.abd.	1.30	0.03	0.99	0.07	1.17	0.12	1.17	0.01	1		2	1.05	0.03	1.42
transv.abdomin.	1.29	0.03	0.97	0.11	0.89	0.14	1.13	0.02	2	2	2	0.93	0.04	1.46
rectus abdomin.	1.17	0.04	1.07	0.07	1.02	0.13	1.11	0.01	1	2	2	1.06	0.03	1.99
Shoulder girdle	0.96	0.01	1.14	0.02	0.92	0.06	1.04	0.01				1.10	0.01	14.75
m. trapezius	0.88	0.05	1.47	0.08	0.66	0.59	1.11	0.03	2	2	2	1.27	0.08	1.33
rhomboideus	0.95	0.11	1.45	0.13	1.04	0.12	1.08	0.03	2	2	2	1.26	0.05	1.49
latiss.dorsi.	0.97	0.02	1.12	0.07	1.08	0.08	1.07	0.01	1			1.13	0.03	2.36
serratus ventr.	0.88	0.03	1.09	0.06	1.01	0.09	0.99	0.01	1			1.10	0.02	3.93
pectoralis sup.	0.76	0.04	1.22	0.12	0.95	0.11	1.10	0.02	2	2		1.13	0.03	1.65
pectoral.prof.	1.12	0.03	1.00	0.05	0.72	0.10	1.03	0.01	1		1	0.98	0.03	3.99
Skin muscles	1.08	0.03	1.26	0.11	0.91	0.10	1.21	0.02	2			1.05	0.05	1.78
Female animals														
Prox. pelvic limb	1.06	0.01	0.93	0.02	0.83	0.02	0.99	0.01				0.90	0.01	32.64
m. gluteobiceps	1.07	0.01	0.92	0.04	0.66	0.05	1.02	0.01	2	2	2	0.90	0.02	7.30
glut.medius	1.18	0.02	1.00	0.06	0.77	0.09	1.05	0.01	2	2	2	0.92	0.03	3.92
glut.accessor.	0.87	0.06	1.17	0.19	1.72	0.47	0.97	0.03	2	2	2	1.06	0.10	0.29
glut.profundus	0.80	0.03	0.84	0.12	1.19	0.11	0.87	0.02	1			0.98	0.05	0.34
semitendinosus	1.07	0.05	0.97	0.11	1.11	0.14	1.05	0.01			1	1.04	0.05	2.76
semimembranosus	1.15	0.03	0.96	0.04	0.66	0.16	1.01	0.02	2	2	2	0.86	0.03	5.51
quadriceps fem.	0.93	0.02	0.89	0.07	0.91	0.12	0.91	0.01	2	2	2	0.86	0.03	6.25

APPENDIX (Cont)

Female animals	b_1	SE	b_2	SE	b_3	SE	b_t	SE	P_a	P_b	P_t	b_{2+3}	SE	% of side muscle wt
tensor fasciae														
latae	1.16	0.04	0.88	0.09	0.87	0.26	1.09	0.02	1			0.92	0.05	1.33
mm. gemelli+quadr.	0.83	0.14	0.80	0.09	1.11	0.23	0.82	0.03				0.87	0.05	0.13
m. gracilis	1.03	0.03	0.95	0.04	1.02	0.20	1.01	0.01				0.93	0.04	1.36
sartorius	0.99	0.04	0.75	0.14	1.09	0.42	0.98	0.02	2		2	0.94	0.08	0.40
pectineus	1.27	0.06	0.60	0.19	0.80	0.22	0.96	0.03	2	2	2	0.72	0.07	0.64
adductor	1.14	0.02	0.87	0.08	0.87	0.22	0.95	0.02	2	1	2	0.81	0.04	1.83
obturator ext.	0.88	0.04	1.03	0.10	1.01	0.16	0.91	0.01				1.01	0.04	0.56
Dist.pelvic limb	0.86	0.02	0.75	0.05	0.71	0.09	0.82	0.01	2		2	0.73	0.02	4.68
m. soleus	0.98	0.31	-0.06	1.11	-1.86	1.90	0.53	0.14				0.56	0.47	0.01
gastrocnemius	0.91	0.03	0.72	0.05	0.43	0.14	0.85	0.01	1	2		0.71	0.03	2.03
flexor dig.sup	0.78	0.03	0.66	0.08	0.98	0.18	0.78	0.01				0.80	0.04	0.4C
mm. flex.+extens.	0.75	0.03	0.81	0.05	0.79	0.10	0.81	0.01	2		2	0.73	0.02	2.25
Prox.thoracic limb	0.88	0.01	0.99	0.05	1.10	0.07	0.95	0.01	1	2	2	1.01	0.02	11.28
m. deltoideus	0.87	0.05	0.97	0.11	1.53	0.31	1.00	0.02				1.04	0.07	0.49
subscapularis	0.82	0.09	1.03	0.12	1.01	0.10	0.92	0.02				1.01	0.04	1.13
infraspinatus	0.85	0.03	1.06	0.07	1.28	0.14	0.96	0.02	2	2	2	1.11	0.03	2.06
supraspinatus	1.00	0.08	1.01	0.11	1.27	0.14	1.00	0.02				1.07	0.04	1.68
teres major	0.97	0.05	1.04	0.12	1.02	0.13	0.98	0.02				1.04	0.04	0.41
teres minor	0.77	0.10	0.85	0.15	1.33	0.10	0.94	0.02				0.96	0.05	0.19
coracobrachial	0.78	0.03	1.00	0.09	0.95	0.27	0.89	0.02				0.98	0.05	0.14
tensor fasciae antebrachii	0.93	0.07	0.84	0.17	0.60	0.42	0.99	0.03				0.88	0.09	0.15
triceps brachii	0.90	0.02	0.98	0.04	0.96	0.07	0.95	0.01	2			0.98	0.02	4.00
biceps brachii	0.85	0.04	0.89	0.04	0.90	0.17	0.92	0.01				0.93	0.03	0.61
brachialis	0.85	0.06	0.82	0.10	1.11	0.22	0.81	0.02				0.82	0.05	0.42
Dist.thoracic limb	0.64	0.05	0.91	0.05	1.03	0.06	0.82	0.02	2	2	2	0.82	0.03	2.45
Neck region	0.88	0.02	1.09	0.06	1.24	0.07	0.98	0.01	2	2	2	1.14	0.02	8.07
mm. rect.+obl.cap	0.90	0.07	1.47	0.34	1.38	0.25	0.89	0.04	2	2	2	1.32	0.11	0.63
m. splenius	0.76	0.08	0.95	0.17	0.83	0.19	0.95	0.03	2		2	1.15	0.07	0.53

APPENDIX (Cont)

Female animals	b_1	SE	b_2	SE	b_3	SE	b_t	SE	P_a	P_b	P_t	b_{2+3}	SE	% of side muscle wt
mm. longissimus														
cap. et atl.	0.60	0.11	1.40	0.27	1.36	0.31	0.91	0.04		1		1.19	0.10	0.26
m. semispinal.cap.	0.73	0.02	1.03	0.12	1.38	0.11	0.92	0.02	2	2	2	1.10	0.04	1.53
omohyoideus	1.17	0.20	-0.51	0.70	-0.30	1.04	0.77	0.09		1		0.48	0.29	0.02
sternocephal.	1.45	0.37	0.85	0.29	1.31	0.41	1.32	0.07				1.22	0.11	0.63
longus colli etc.*	0.90	0.04	0.97	0.13	1.18	0.14	0.92	0.02	1			1.03	0.05	1.96
omotransversar	0.98	0.05	0.94	0.10	1.22	0.26	1.05	0.02				1.15	0.06	0.50
brachiocephal.	0.94	0.04	1.33	0.11	1.22	0.23	1.05	0.02		2	2	1.23	0.05	1.39
subclavius	1.13	0.25	1.56	0.55	-0.81	0.61	0.96	0.07				1.04	0.22	0.03
mm. scalenus dors.+ ventralis	0.79	0.06	1.16	0.08	1.37	0.21	0.95	0.02	2	2	2	1.21	0.04	0.58
Back and loin region														
mm. serratus dors. inspir.+expir.	0.94	0.02	1.04	0.03	1.14	0.10	1.05	0.01	2	1	2	1.06	0.02	11.20
m. iliocostalis	1.15	0.09	1.08	0.16	1.48	0.22	1.12	0.02				1.15	0.06	0.27
longiss.cervic.	0.89	0.12	1.65	0.44	1.16	0.12	1.02	0.05				1.39	0.13	0.46
longiss.thorac. + lumborum	0.38	0.14	1.17	0.30	1.26	0.40	0.92	0.05	2		2	1.11	0.11	0.30
spinal.et semi- spinal.etc**	1.06	0.03	1.02	0.07	1.04	0.13	1.11	0.01	1		1	1.04	0.03	7.05
	0.78	0.03	0.99	0.10	1.32	0.14	0.95	0.02	2	2	2	1.08	0.04	3.13
Sublumbar muscles (* mm iliopsoas, psoas monor and quadrat.lumborum)	1.13	0.02	1.02	0.06	0.94	0.11	1.02	0.01	2	2	2	1.00	0.03	3.20
Thorax and abdomen	1.22	0.01	1.06	0.06	1.11	0.11	1.11	0.01	2	1	2	1.14	0.03	11.20
m. transv.abdomen	1.21	0.07	0.70	0.18	1.35	0.40	1.05	0.03		1		0.98	0.09	0.22
rectus thorac.	0.99	0.08	1.13	0.16	0.65	0.22	1.01	0.02				1.09	0.06	0.11
mm. intercostales etc*	0.86	0.03	1.13	0.09	1.44	0.25	0.96	0.03	2	2	2	1.27	0.05	2.89

APPENDIX (Cont.)

Female animals	b_1	SE	b_2	SE	b_3	SE	b_t	SE	P_a	P_b	P_t	b_{2+3}	SE	% of side muscle wt
Diaphragm	1.36	0.08	0.78	0.33	1.42	0.29	1.15	0.04	1		1	1.23	0.12	0.68
obliq.ext.abd.	1.35	0.03	1.08	0.13	1.15	0.15	1.20	0.02	1	1	2	1.18	0.05	2.18
obliq.int.abd	1.44	0.04	1.04	0.13	0.86	0.13	1.22	0.02	1	2	2	1.12	0.05	1.45
transv.abdomin.	1.57	0.03	0.91	0.14	0.75	0.17	1.16	0.03	2	2		0.93	0.05	1.61
rectus abdomin.	1.28	0.02	1.20	0.08	1.01	0.24	1.15	0.02	1		1	1.12	0.05	2.05
Shoulder girdle	0.97	0.02	1.10	0.03	1.08	0.06	1.04	0.01	2	1	2	1.15	0.02	13.46
m. trapezius	0.78	0.05	0.98	0.07	1.38	0.16	1.07	0.03	2	2	2	1.29	0.05	1.12
rhomboideus	0.80	0.16	1.26	0.20	1.10	0.16	1.00	0.04	1			1.29	0.07	1.11
latiss.dorsi	0.95	0.03	1.07	0.09	0.95	0.16	1.06	0.01	2			1.12	0.04	2.20
serratus ventr.	0.96	0.02	1.04	0.06	1.16	0.06	1.01	0.01	2		2	1.15	0.02	3.82
pectoralis sup.	0.89	0.09	1.23	0.06	1.21	0.17	1.11	0.02	1			1.15	0.04	1.42
pectoral prof.	1.12	0.04	1.12	0.05	0.93	0.11	1.05	0.01	1			1.09	0.03	3.81
Skin muscles	1.53	0.13	1.16	0.13	0.87	0.16	1.34	0.03	2		1	1.06	0.05	1.70

* Includes some small muscles along the vertebral column   1 = P $<$0.05

** Includes the mm. multifidi   2 = P $<$0.01

VARIATION AND IMPACT OF MUSCLE THICKNESS

B.L. Dumont

Laboratoire de Recherches sur la Viande INRA
Centre National de Recherches Zootechniques,
78350 Jouy-en-Josas, France

ABSTRACT

Muscles largely differ in thickness due to the variation of their weight during growth, which results mainly from an increase in the diameter of muscle fibres. With increasing thickness, the morphological aspects of the perimysium network is changed. The thickness each muscle may present depends on the dimensions of the skeleton. There is a close relationship between muscle thickness, conformation or fleshiness, meat/bone ratio and meat texture. The thicker the muscles, the better the conformation and the higher the meat/bone ratio and the more tender the meat. All these character-istics are at their maximum development in the hypertrophied type of cattle where muscles show considerable hypertrophy in weight and thickness.

INTRODUCTION

The cattle carcass comprises about two hundred muscles, which form round the skeleton a fleshy 'covering' varying in weight or volume from one part of the body to another. At a given anatomical level this 'covering' will show, between animals, a large variation in thickness. This variation in thickness and its impact on carcass and meat quality will be considered in this paper.

This aspect of musculature growth is still very poorly understood. Anatomical studies up to now have mainly been concerned with aspects of weight evolution rather than with morphological considerations. As we shall see, these represent an interesting subject which is very closely linked with commercial value of carcasses and meat quality.

MUSCLE THICKNESS AND ASSOCIATED MEASURES

Thickness is easy to define when considering one solid with a simple and regular geometrical shape, such as a parallelepiped. Thickness is then one of the three measurements, the other two being length and breadth. The smallest dimension is usually called thickness, whereas the largest is the length, and the third breadth. When the shape of the solid is fairly regular, as with most muscles, it is possible to distinguish, along some arbitrarily defined planes of examination (face, profile etc.) adequate directions corresponding to the dimensions of length, breadth or thickness. It is then necessary to take as a base a reference trihedral (Fig. 1) Maximum thickness can thus be defined and all the intermediate values between this and the zero values at the two ends of the muscle may be recorded.

However, no reference trihedral for any muscle has so far been found which meets with any general degree of acceptance among research workers. This lack of definition can be explained by the difficulty in giving a standardised posture to the various parts of the carcass. It is obvious that the usual position of

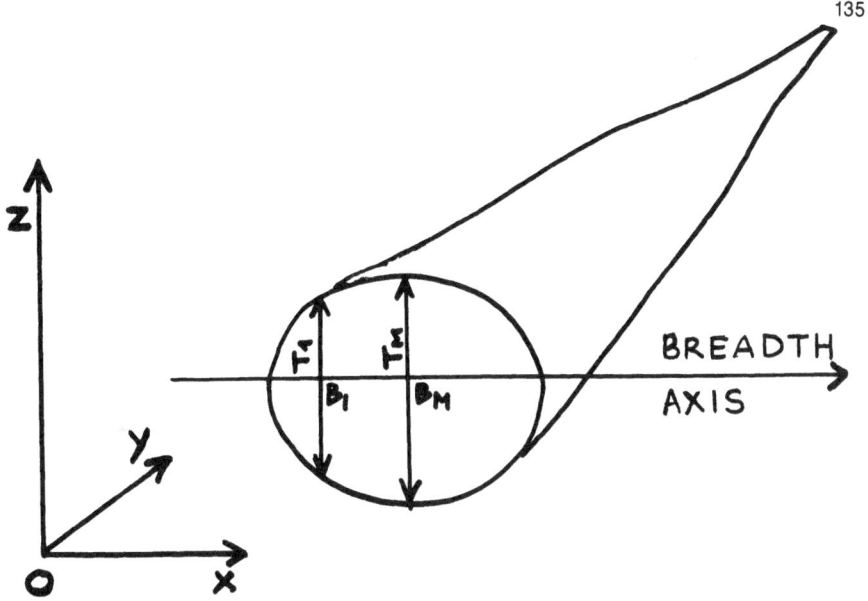

Fig.1 References of muscle thickness measurements

TABLE 1

AVERAGE THICKNESS (T_m) AND FLATTENING RATIO OF BOVINE MUSCLES FROM NINE COWS
OF 252.7 $\pm$ 26.5 kg CARCASS WEIGHT

Muscle	Thickness T_m (cm)	Flattening ratio
Adductor	8.86 $\pm$ 0.76	1.68 $\pm$ 0.24
Longissimus dorsi	4.98 $\pm$ 0.75	3.87 $\pm$ 0.75
Triceps brachii caput longum	8.28 $\pm$ 1.21	1.96 $\pm$ 0.43
Teres major	3.36 $\pm$ 0.37	2.13 $\pm$ 0.29
Semitendinosus	8.27 $\pm$ 0.52	1.30 $\pm$ 0.08
Splenius	2.29 $\pm$ 0.26	4.24 $\pm$ 0.60
Pectoralis profundus	3.03 $\pm$ 0.39	5.23 $\pm$ 0.79
Rhomboideus	2.67 $\pm$ 0.38	2.39 $\pm$ 0.36
Triceps brachii caput laterale	4.95 $\pm$ 0.76	2.01 $\pm$ 0.40

(Thickness T_m is the thickness of muscle at the middle part of the
breadth/length axis (except for *L. Dorsi* where the sample has been
examined in the cross section of the muscle at the level of the third
lumbar vertebra.)

carcasses after slaughter and dressing is very different from
the biological position adopted by live animals. Thickness of
muscles is very largely influenced by carcass position.

It may be useful to consider, in addition to the maximum
thickness, thickness T_1, T_2.....T_n considered at the level
B_1, B_2,...B_n on the breadth scale (eg thickness at ¼ or ⅓ of the
breadth). To help in assessing the important part which thickness plays in shaping a muscle, it is also often necessary to
take into consideration <u>muscle area</u>. This, in fact, is the
integral of thickness. The area/breadth ratio may be used as
an 'average thickness' of the muscle. Finally, one can speak of
a 'flattening ratio' of the muscle, by considering one of the
two ratios -

a) $\dfrac{\text{Maximum Thickness}}{\text{Maximum Breadth}}$ or b) $\dfrac{\text{Thickness } T_m}{\text{Maximum Breadth}}$

where T_m is the thickness of muscle at the middle part of the
breadth axis.

The whole growth of some associated muscles on one anatomical level might be assessed by looking at the thickness of the
various muscles, taking into account the 'puffing up' of the
carcass at this level. In this case, measurement of the thickness involves examination of each of the muscles concerned, from
various angles, taking into account the depth of intermuscular
fat, and possibly also that of the subcutaneous fat - for example
in the case of the thigh depth.

VARIATIONS IN MUSCLE THICKNESS
Nature and amount of variation

Practical experience shows that the different muscles present great variations in thickness. Three types of variation
may be considered:

1) For a given muscle, variation in thickness between
the tendons of origin and insertion;

2) For a specific animal, variation in thickness between
different muscles (Table 1);

3) For a given anatomical level, variation between animals, due to the effects of age, breed, level of nutrition, etc.

Origin of the variations

Muscle thickness is theoretically determined by the cumulative thicknesses of muscle fibres (MFT); the thickness of various sections of the connective tissue network and of the associated fatty tissue (CTT); and the amount of extracellular spaces. The proportion of each of these components is not well established, but it can be assumed that the amount of extracellular spaces is both small and less variable, and that fibres constitute between 75 and 90% of a muscle. In this paper we shall consider only problems relating to muscle fibres and connective tissues, and we shall assume that:-

$$\text{Muscle thickness} \approx \text{MFT} + \text{CTT}$$

MFT depends on number and dimensions of the muscle fibres. CTT is mainly made up of the perimysium.

Muscle fibres

Many factors affect the number and the dimensions of skeletal muscle fibres as shown in Table 2. They have been reviewed, among others, by Hegarty (1971) who has also discussed the importance of physiological conditions of muscle (pre-rigor, rigor) on measurements of fibre dimensions. The general scheme of muscle fibre development is now well understood. Fibre diameter increases with age provided that levels of nutrition are sufficient. The main results are increases in muscle weight and area (Hiner and Bond, 1971). The increase in thickness - sensu stricto - depends on the general morphology of each muscle, but, in general, muscle thickness increases with age. The relationship between thickness and age (or weight) of the different muscles is not known and might well be studied.

On the other hand, it is well established, and agreed, that the number of muscle fibres is determined at birth or thereabouts, so the number of fibres differentiated in the first weeks could directly influence any subsequent increase in thickness during growth. In some cases the fibre number will not be constant

TABLE 2
FACTORS OF VARIATION OF MUSCLE THICKNESS

Component	Character	Factors of Variation
Muscle Fibre	Number	Muscle Breed Sex

	Dimensions	Muscle
		Breed
		Sex
		Age
		Nutrition
		Exercise
		Histological processing
		Physiological state of
		  contraction (methods
		  of slaughter, hanging,
		  storage)

Connective Tissue	Perimysium Amount and distribution	Muscle Breed Age Nutrition
	Intramuscular Fat	Muscle Breed Age Nutrition

during postnatal growth, which could possibly affect muscle
thickness. Thus Bendall and Voyle (1967) found an important
decrease in fibre number (in *L. Dorsi* and *Semitendinosus*) as the
animal grew. Variation in number is also sometimes accompanied
by a reverse variation in the size of fibres.

Little is known of the variation in the number of muscle
fibres between breeds or types. As far as 'hypertrophied'
animals are concerned, there appears to be a close agreement
among researchers that the characteristic muscle enlargement
was mainly due to an increase in the number of myofibres
(Swatland, 1976). Thus the term 'hypertrophied' is questionable.
It is a difficult task (Swatland, 1976) and it is absolutely
ncessary to have precise data on the number of muscular fibres
in the various muscles of different animals. As pointed out by
Swatland, "In the 44 years which have elapsed since (the pioneer
work of Sir John Hammond) no method of selection based on the
histological parameters of muscle mass and growth has been
applied in commercial agriculture..."

Connective tissue

Within each muscle, the perimysium is composed of a very
complex network whose detailed arrangement has not been deeply
investigated up to now, due to the technical difficulties of
such an analysis in the case of large entire muscles. Methods
developed by Schmitt and Dumont (1969) however, enable an approach
to be made to the study of muscle connective tissue network.
At a given anatomical level, the way in which this network is
organised seems to be a characteristic of each muscle (Dumont
et al., 1977).

Evolution of the (approximately 20) characteristics of
the connective network during growth in thickness has been
studied in cattle by considering the *Semitendinosus* (Schmitt and
Dumont, 1972) and *Brachialis* muscles (Dumont and Schmitt, 1975).
In the *Semitendinosus* of a very young animal, connective tissue
seems to be relatively more important than in the older animal.
During growth, muscle enlargement results from a general

distension of the connective tissue network affecting irregularly
each muscular structure unit (major bundles of muscle fibres).
Their sides appeared thin, as if under elastic tension. Their
vertices, corresponding to the nodes of connective tissue framework, were enlarged and look less extensible.

In the *Brachialis* muscle in cattle of various ages and
weights (from birth to 13 years old) it appears that some characteristics of the connective tissue network do not vary. These
are qualitative factors (such as the structure, the shape of the
major bundles of fibres, their localisation within the muscle,
etc.) These characteristics seem to be specific to this type of
muscle and are thus supposed to be fixed at birth and remain
constant during growth. On the opposite side, some characteristics which could be considered quantitative were largely
affected by the increase in thickness of the muscle, for example,
the average size of major bundles of muscle fibres; the number
of nodes at the vertices of the connective network; the relative
thickness of the walls of connective network. It seems that,
during the growth of muscle, the variation of the connective
network was only quantitative, without any remodelling of the
basic plan which fixes at birth the general organisation of the
perimysium framework.

The general impression resulting from histological studies
is of a decrease in the amount per unit of surface of connective
tissue network during growth. At the level of primary bundles
of muscular fibres, Vognarova et al., (1968) found that, though
the size of these bundles was smaller in calves than in older
cattle, the thickness of perimysium is larger in calves. The
decrease in the relative amount of connective tissue in relation
to muscular fibres revealed by the histological analysis can be
compared to the results of chemical analysis which showed
(Vognarova et al., 1968 and other workers) that the proportion
of collagen is higher in veal muscle than in beef.

Variations in muscle thickness resulting from variations
in fleshiness or conformation (see below) are accompanied by

variations in relative amounts of collagen. This has been well established by the works of Boccard, Dumont and Schmitt, who found:-

a) in mature cows with poor fleshiness, the ratio - $\frac{\text{N hypro}}{\text{N total}}$ x 1000 decreased by 25 to 45% between the extreme types considered in the study (Boccard et al., 1968);

b) in double muscled animals (hypertrophied cattle) the amount of collagen is significantly smaller than in the normal type of the same breed (Boccard et al., 1967; Boccard and Dumont, 1974);

c) decreases in collagen content with increasing fleshiness (from the poorest state of fleshiness, ie class 1 of the EAAP system of description of fleshiness (de Boer et al., 1974), to the hypertrophied state, ie class 5 of fleshiness) are very variable between muscles and are the more pronounced as the muscles contain more collagen (Dumont, 1977 unpublished data).

Comparison, by means of histological analysis of the whole muscle, of connective tissue networks in hypertrophied and normal cattle (Dumont and Schmitt, 1973) showed that the perimysium was less abundant in hypertrophied animals, with a very loose network in which it was difficult to distinguish types of muscle fibre bundles other than primary ones.

CONSEQUENCES OF VARIATIONS IN MUSCLE THICKNESS

Variations in muscle thickness greatly affect carcass fleshiness (and conformation), composition and muscle texture. As muscles increase in thickness, fleshiness (and conformation) is improved; anatomical composition is changed (increase in the muscle/bone ratio, and, in the case of double muscled animals, different distribution of muscles); texture is modified (decrease in the connective component of muscle structure) and tenderness is improved; cutting and trimming facilities are higher.

a) Fleshiness and conformation

The concept of muscularity and fleshiness have been defined by de Boer et al., (1974). For a given skeletal support of a particular muscle (or group of muscles) the thicker the muscle, the better the fleshiness. Examination of the profile of muscles in various parts of the body can be used to make a visual assessment of fleshiness (Houdinière, 1957). EAAP standards of reference for the assessment of fleshiness have been established on this basis. An objective judgment can be made by measuring the thickness of a muscle (eg *L. Dorsi m*) or groups of muscles (eg thigh thickness). One can speak of a hypo-myomorphic state when the thickness is small; compared to the average or normal thickness for a given skeleton (eumyomorphic state), whereas the hypermyomorphic condition corresponds to a very important thickness (such as is found in the hypertrophied animals).

The great variation found in muscle thickness between carcasses of the same weight is usually associated with a variation in bone length. The resulting variation in fleshiness gives rise to great differences in carcass type and conformation. Poor fleshiness and long bones are typical of the Ectomorph type, whereas good fleshiness and rather short bones are typical of the Mesomorph type (Kauffman et al., 1970).

Relationship between muscle thickness, muscle development and measurements - in particular length - of the underlying bones should be intensively studied. One may assume as a hypothesis that most of the variation in fleshiness between breeds at the same total muscle weight is due to variations in length of the supporting bones. At similar bone length, the variation in conformation may possibly be due to a considerable variation in muscle thickness and muscle mass, which may explain the large variation encountered, in this case, in carcass weight. Thus, for a given carcass length of 135cm, the average weight of cattle will be about 280 kg in class 2, 350 kg in class 3, 420 kg in class 4 of fleshiness, and over 450 kg in class 5 of fleshiness in the EAAP scale. These figures explain the relevance of 'blockiness' (Verbeke, 1975) or weight to length ratio (Dumont, 1977a) in estimating carcass value.

b) <u>Fleshiness and carcass composition</u>

From the definition of fleshiness a close relationship between muscle thickness and muscle/bone ratio is to be expected, for a given skeletal support. In considering conformation, one may in general expect an increase in the muscle/bone ratio and the 'meat'/bone ratio when the carcass conformation is improved.

Results from comparisons between mesomorph and ectomorph cattle (Kauffman et al., 1970) support this idea. Dumont (1977b) has recently found that there was a high correlation (r = 0.70) between the 'meat'/bone ratio and the fleshiness score in a study of 40 cattle carcasses of various breeds and types. He suggested that the amount of flesh in the carcass, at a given weight and conformation level, would be improved by introducing into the scoring method for carcass fleshiness some additional criteria related to bone development.

It now seems necessary to establish by further studies reciprocal influences of carcass weight, fleshiness, and the different measurements of bones, in order to quantify the relationship of fleshiness to composition. In hypertrophied cattle a different pattern of muscle distribution within the carcass is found, in addition to a higher muscle/bone ratio (Dumont and Boccard, 1967). By comparing cattle carcasses at the same total muscle weight, Boccard and Dumont (1974) found that some muscles were less well developed in weight than others even if their apparent shape was different and they looked hypertrophied. Such muscles may be <u>hyper</u>trophied in thickness, but are <u>hypo</u>trophied in weight. Some muscles are both heavier and thicker than the same type of muscle in normal cattle. These muscles are really hypertrophied, in weight and thickness. Many of the outer muscles and in particular many muscles of the leg, are highly hypertrophied. The so-called 'double-muscled', or hypertrophied, animals might better be given another appellation more appropriate to their morphological or anatomical condition.

However that may be, this type of cattle forms a very interesting
biological model because it shows some of the limits in increase
in thickness that arise during muscle growth, namely: the upper
value of the muscle/bone ratio which can be supported by the skeleton; and the highest degree of thickness each muscle can attain
in a particular direction, owing to restraints imposed on it by
adjacent muscles.

c) Muscle use possibilities

Muscle thickness directly affects the size of meat cuts.
The economic importance of this factor may vary from one country
to another according to the cutting procedures and culinary
customs. In France, where a large part of the meat from cattle
carcasses is consumed as individual steaks, the possibilities
afforded by individual muscles are very important considerations.
The amount of trimming and the time needed to prepare meat cuts
could also be affected by muscle thickness.

d) Meat texture and tenderness

Meat tenderness is largely influenced by the thickness of
the muscle from which the sample is cut. This is a consequence
of the variation of the connective tissue with thickness, as
discussed above. Little work has been done on this question.
It is evident that hypertrophied animals with thicker muscles
give more tender meat than normal types (Boccard et al., 1967).
The opposite is true as well, as is shown by the very poor flesh-
ness and increase in toughness of cow meat (Boccard et al., 1968
The present work, based on a large sample of carcasses from
various types and breeds (Dumont, 1977d) shows that there is a
close inverse relationship between the fleshiness score of
carcasses and the Warner-Bratzler shear index.

CONCLUSION

Muscle thickness is thus a very important characteristic
to be considered in muscle growth studies.

It would seem that an increase in muscle thickness is in favour of most of the desirable attributes of the cattle carcass: fleshiness, conformation, muscle/bone ratio, meat texture etc. - all these characteristics are probably at their maximum development in the hypertrophied type of cattle. At the moment we have only a rough idea about some major aspects of muscle thickness. Further studies need to be done, especially to establish a law regulating growth in thickness of the various muscles, according to breed, sex and level of nutrition. Special attention needs to be paid to the vicinity factors which limit physically the growth in thickness of each muscle which depends mainly on the number of muscle fibres. The influence of the characteristics of adjacent bones on muscle thickness growth should be analysed. Changes in connective tissue characteristics with increasing thickness should be defined for the morphological aspects of the perimysium network, and for the amount, the nature and the structure of its chemical components. It is also necessary to know exactly what influence variations in muscle thickness have on the proportion of the different metabolic types of muscle fibres which affect not only the working performance in live animals but also some other important factors in meat quality such as colour, juiciness and flavour. Finally, variation in muscle thickness must also be analysed in relation to the breeding factors which come first when speaking of muscle growth (meat production): rate of growth, ease of calving, feed intake etc. In this respect it has not yet been proved that the thickest animal is always the best one...

REFERENCES

Bendall J.R. and Voyle C.A., 1967 A study of the histological changes in
    the growing muscles of beef animals. J. Fd. Tech. 2, 259-283.

Boccard R., Dumont B.L. and Schmitt O., 1967 Note sur les relations entre
    la dureté de la viande et les principales caractéristiques du tissu
    conjonctif. 13e réun.eur. Cherch. en Viande, Rotterdam, 20-26 August,
    1967. 22pp

Boccard R., Dumont B.L. and Schmitt O., 1969 Relation entre la conformation
    des carcasses et les caractéristiques de la musculature (Observations
    préliminaires sur les vaches de réforme) Bull. Acad. Vét. 42, 261-265.

Boccard R. and Dumont B.L. 1974 Conséquences de l'hypertrophie musculaire
    héréditaire. Ann. Génét. Sel. anim. 6, (2), 177-186.

de Boer, H., Dumont B.L., Pomeroy R.W. and Weniger J.H., 1974. Manual on EAAP
    reference methods for the assessment of carcass characteristics in
    cattle. Livestock Prod. Sci. 1, 151-164.

Dumont B.L. and Boccard R., 1967   Le rapport muscle/os, critère de sélection
    des bovins de boucherie. 2e Symp. Internl. Zootech. Milan, 14-16th
    April 1967, 149-155

Dumont B.L. and Schmitt O., 1973   Conséquences de l'hypertrophie musculaire
    héréditaire sur la trame conjonctive du muscle de bovin. Ann. Génét.
    Sél. Anim. 5, 499-506

Dumont B.L. and Schmitt O., 1975. Etude de la variation de la trame de tissu
    conjonctif musculaire chez le bovin (Bos taurus). Commun. Réun. Group
    "Développement et croissance" de l'INRA, -INA, Paris-Grignon, 17-18
    April 1975. 4pp.

Dumont B.L., Lefebvre J., Schmitt O. and Barbu S., 1977   Application de
    méthodes d'analyse multidimentionnelle à la différenciation des
    muscles sur la base de l'organisation de leur trame conjonctive.
    10th Eur. Meet. Statist., Leuven, Belgium, 22-26 August 1977

Dumont B.L., 1977a   Variation du rapport poids/longueur des carcasses
    bovines. Ann. Zootech., 26(1), 119-124

Dumont B.L., 1977b   Relations entre la conformation et la composition des
    carcasses de bovins. Ann. Zootech., 26(1), 125-129.

Dumont B.L., 1977c   Relation entre la charnure et la teneur en collagène
    (unpublished data).

Dumont B.L., 1977d   Relation entre la conformation et la dureté de la viande
    (unpublished data)

Houdiniere A., 1957   L'examen des 'profils musculaires' dans l'appréciation
    de la qualité des viandes. Bull. Acad. Vét., <u>30</u>, 51-62

Kauffman R.G., Smith R.E. and Long R.A., 1970   Bovine topography and its
    relationship to composition. Proc. 23rd Ann. Rec. Meat Conf. AMSA.
    100-117

Schmitt O. and Dumont B.L., 1972   Croissance du muscle et tissu conjonctif.
    Ann. Biol. Anim. Biochim. Biophys., <u>12</u>(4), 667-672

Schmitt O. and Dumont B.L., 1969   Méthodes d'analyse de la structure muscu-
    laire. Ann. Biol. Anim. Biochim. Biophys., <u>9</u>(1), 123-134

Swatland H.J., 1976   Recent research on postnatal muscle development in
    swine. Proc. 29th Ann. Recip. Meat Conf. AMSA, 86-104

Verbeke R., 1975   Dissection of the three rib-joint (7th, 8th and 9th) and
    carcass blockiness as estimates of carcass value. EEC Seminar on
    criteria and methods for assessment of carcass and meat character-
    istics in beef production experiments, Zeist, Netherlands, 1975

Vognarova I., Dvorak Z. and Böhm R., 1968   Collagen and elastin in different
    cuts of veal and beef. J. Fd. Sci., <u>33</u>, 339-343

BONE GROWTH AND DEVELOPMENT WITH PARTICULAR REFERENCE TO BREED DIFFERENCES IN CARCASS SHAPE AND LEAN TO BONE RATIO

A.J. Kempster

Meat and Livestock Commission, Bletchley, Milton Keynes, MK2 2EF, UK.

ABSTRACT

Two topics related to bone growth and development are considered:

1) Breed differences in bone weight distribution,

2) The inter-relationships of carcass shape, lean to bone ratio and bone characteristics between breeds.

1) Evidence on bone weight distribution is reviewed and the results of an analysis using MLC dissection files reported. This analysis was based on 753 steer carcasses from 17 breed-type × feeding system groups and is complementary with analyses published recently on fat and lean weight distribution. Small differences of little commercial importance were recorded between groups, a finding which agrees with existing evidence.

2) Breed variation in carcass conformation and its relationship with lean to bone ratio was examined using the data files above and more recent data from the MLC breed comparison trials at Ingliston and Sutton Bonington. Results indicate that, although there is a definite trend for breeds with better conformation to have higher lean to bone ratios at constant carcass subcutaneous fat percentage, certain breeds and crosses do not relate well to this trend. The results are discussed in relation to breed differences in bone characteristics and the genetic history of the various breeds.

BONE WEIGHT DISTRIBUTION

Nearly all bone is removed before retail sale and is of the same low value irrespective of where it comes from in the carcass. Consequently bone weight distribution has less commercial importance than either lean or fat weight distribution except possibly at the wholesale level for trade in carcass quarters or bone-in primal joints. It does have some practical significance in relation to the prediction of carcass bone content from individual bones or the bone content of joints.

There have been few studies of bone weight distribution in cattle and present knowledge of the factors influencing it is limited. Existing evidence is, however, consistent in indicating that bone weight distribution is a relatively inflexible characteristic which varies little over the range of weights at which beef cattle are normally slaughtered.

It has been shown that most bones (or bone weights in different joints) grow at similar rates to total bone over the commercial slaughter weight range (for example, Seebeck and Tulloh, 1968; Seebeck, 1973; Frood, 1976). The leg bones have tended to differ to the greatest degree, growing more slowly than total bone. Bone is an earlier developing tissue than either muscle and fat and these findings are consistent with the growth impetus of bone having nearly run its course leaving bone growth to continue in the proportions established.

Information on the differences between breeds in bone weight distribution is particularly limited. Seebeck (1973) found that Africander cross steers tended to have slightly lighter leg bones and heavier axial skeleton and ribs than Brahman crosses at the same total bone weight. Similar differences were recorded between Friesian and Angus crosses by Truscott et al. (1976), the Friesians having more of their bone weight in the legs and less in the thoracic region. Harte and Conniffe (1967) also reported differences in bone weight distribution between Friesian, Hereford and Angus cross steers.

Extreme differences in plane of feeding have produced only minor changes in bone weight distribution (Seebeck and Tulloh, 1968; Seebeck, 1973; Murray et al, 1974). It is very unlikely, therefore, that differences between more conventional systems of feeding will materially influence bone weight distribution.

Table 1 gives the results of an analysis carried out to compare the bone weight distribution between joints for 17 breed type x feeding system groups. The same sample of carcasses was used by Kempster et al (1976_1) to examine lean weight distribution and details of the groups and jointing and dissection method are given there. There were statistically significant (P < 0.001) but relatively small differences between groups in the proportions of bone in the hind quarter joints and leg joints both at constant total bone weight and constant percentage subcutaneous fat in carcass. Differences were more marked in the case of bone weight distribution in limb joints.

Friesian crosses with the Continental breeds (Charolais, Simmental and Limousin) tended to have more bone in the limb joints at constant total bone weight than Friesian crosses with the traditional British breeds and particularly those of early maturing types. It is likely that these differences are to some extent a reflection of mature size; the Continental breeds are less mature at constant total bone weight and thus have more bone in the early developing leg joints. The results of Callow (1961) and Truscott et al, (1976) can also be interpreted in this way. An interesting question is whether the differences are due entirely to mature size or whether other factors are involved. The results in Table 1 suggest that mature size is not the entire answer. If it were, one would expect the South Devon x Friesian group to have similar results to the Continental breed x Friesian groups since they are of similar mature size. Furthermore, adjustment to constant percentage subcutaneous fat, which would to some extent shift the breed means towards equal degree of maturity (for breeds on

the same feeding system), did not obscure the main differences.
Whatever the conclusion regarding the origin of the variation
in bone weight distribution, it is clear that it is unimportant
commercially. A paper reporting the detailed results of the
analysis and considering them in relation to the problem of
predicting carcass bone content is in press (Kempster et al,
1977).

TABLE 1

GROUP MEANS FOR THE PERCENTAGE OF TOTAL BONE IN THE HINDQUARTER JOINTS AND
THE PERCENTAGE OF TOTAL BONE IN THE LIMB JOINTS* AT CONSTANT TOTAL BONE
WEIGHT (TB)† AND CONSTANT PERCENTAGE SUBCUTANEOUS FAT IN CARCASS (%SF)†

Breed type x feeding system group‡	Number of carcasses	Percentage bone in hindquarter		Percentage bone in limb joints	
		TB	%SF	TB	%SF
Ayrshire (C)	27	44.2	44.4	35.2	35.9
A. Angus x Friesian (G/C)	18	44.9	45.1	35.2	35.8
Hereford x Friesian (C)	31	45.2	45.3	37.0	37.5
Friesian (C)	106	45.2	45.3	37.1	37.2
Limousin x Friesian (C)	25	44.8	44.9	37.4	37.6
Charolais x Friesian (C)	49	46.0	46.1	38.4	38.6
Limousin x Friesian (G/C)	11	45.0	44.9	37.1	37.0
Hereford x Friesian (G/C)	66	44.8	44.9	36.0	36.0
Simmental x Friesian (C)	65	45.5	45.5	38.1	38.0
Miscellaneous commercial cattle	65	44.1	44.1	34.3	34.4
Angus crosses	29	44.6	44.8	34.4	35.5
Friesian (G/C)	72	44.9	44.8	36.4	36.1
Simmental x Friesian (G/C)	51	45.7	45.6	36.9	36.4
Welsh Black and crosses	79	44.0	43.9	33.8	33.2
Friesian x Ayrshire (G/C)	25	43.9	43.8	35.5	35.4
S. Devon x Friesian (G/C)	10	44.8	44.7	35.7	35.3
Simmental x Ayrshire (G/C)	24	44.5	44.4	35.8	35.9
Residual standard deviation		1.16	1.16	1.29	1.32

* Top piece (round), leg (hind shin) and fore shin which include some tarsal
  bones, tibia-fibula, femur, ischium, some coccygeal vertebrae and radius-
  ulna.
† 36.4kg and 7.5% for TB and %SF respectively.
‡ C = cereal feeding and G/C = grass/cereal feeding.

CARCASS SHAPE AND LEAN TO BONE RATIO

Views on the value of carcass conformation as a predictor of beef carcass yields are probably more wide ranging than those on any other topic in animal science. Nevertheless, conformation has been accepted as an essential factor in commercial beef carcass classification schemes in many countries. Its primary role appears to be the identification of differences in lean to bone ratio (and possibly fat distribution) and hence carcass lean content among carcasses of similar levels of subcutaneous fatness. (Adjustment to constant fatness is necessary for interpretation since conformation is defined as the thickness of muscle and fat in relation to skeletal size and ipso facto confounded with fatness). The relationships between tissue percentages and ratios and conformation were considered by Harrington and Kempster (1977) in the context of the classification scheme operated in Great Britain by MLC.

Breed cannot normally be identified at the point of classification and conformation is being used to some extent as a breed indicator. The importance of breed in this relationship is illustrated in Table 2, which is based on 850 carcasses for MLC's breed evaluation trials at Ingliston and Sutton Bonington (the procedure for these evaluations is considered in more detail below). The fat and conformation classes are as in MLC's beef carcass classification scheme (Meat and Livestock Commission, 1975). The results in Table 2 indicate that conformation added usefully to the prediction of percentage saleable meat in the carcass when breed is unknown, but its contribution was much reduced when breed is known. It is also clear that breed explained a substantial proportion of the variation in percentage saleable meat among carcasses of the same weight, fat class and conformation class. This leads us to ask how well particular breeds fit between breed regression of saleable meat to bone ratio (lean to bone ratio) on conformation.

Faced with this question, the scientific literature is unhelpful. Most of the early work on conformation cannot be interpreted in the way required because conformation has been confounded with fatness (for reviews see Oliver, 1967; Barton, 1967). In some of the more recent work specifically designed to look at differences in shape, attempts have been made to control fatness, but experimenters have tended to take cattle of different (usually extreme) shapes with little regard to breed differences (for example, Kauffman et al., 1973; Fredeen et al., 1974). In such work sample sizes have generally been very small. Breed comparisons which are a potentially useful source of information, have rarely been made at constant fatness because of the difficulties encountered in defining slaughter point. Even with serial slaughter experiments, results are seldom reported at constant fatness.

TABLE 2

RESIDUAL STANDARD DEVIATIONS (RSD) FOR THE PREDICTION OF PERCENTAGE SALE-ABLE MEAT IN CARCASS FROM CARCASS WEIGHT, FAT CLASS*, CONFORMATION CLASS †
AND BREED. (BASED ON 850 STEER CARCASSES FROM INGLISTON AND SUTTON
BONINGTON SLAUGHTERED AT SIMILAR LEVELS OF FATNESS)

Standard deviation of percentage saleable meat		2.01
RSD	Weight	2.00 (1)
	Weight + fatness	1.84 (16)
	Weight + fatness + conformation	1.71 (28)
	Breed + weight + fatness	1.56 (40)
	Breed + weight + fatness + conformation	1.50 (44)

Percentage variation explained is given in brackets.

* Subjective assessment of subcutaneous fat percentage in carcass on a 7-point scale.

† Subjective assessment of conformation on a 6-point scale.

Relevant data for a wide range of breed types are available from MLC's breed evaluation trials at Ingliston and Sutton

Bonington and these have been used to examine the between
breed relationship of conformation and saleable meat to bone
ratio (Tables 3 and 4). At the Ingliston Unit, cattle from the
suckler herd are being evaluated on two systems of production
one using autumn born calves (winter finishing) and the other
late winter born calves (summer finishing). The cattle include
crossbred steers by the more important sire breeds out of
Blue-Grey (Shorthorn x Galloway) and Hereford x Friesian dams
(balanced numbers of each), together with purebred steers of
the hill breeds, Galloway, Luing and Welsh Black. The lifetime performance of dairy-bred calves purchased at about 10
days of age is evaluated at the Sutton Bonington unit in two
production systems; one system approximates to the popular
18 month beef system in Britain while the other system involves
slaughter at an average of two years. For each production
system, slaughter takes place between predetermined dates. Within that period cattle are slaughtered when they are predicted, on
the basis of subcutaneous fat area measured by Scanogram ultrasonic machine, to be at a particular point on the MLC beef
carcass classification fatness scale. After slaughter the
left side of each carcass is divided into standardised
commercial joints which are subsequently deboned and subjected
to standardised trimming of excess fat. For the analysis,
results are adjusted to a constant estimated subcutaneous
fat percentage (based on subjective assessment of external fat
cover). Each test group comprises 12 steers and the trials
are being repeated over several years. Tables 3 and 4 include
data for all animals available when the analysis was run.

The results in Tables 3 and 4 indicate a trend for breeds
with better conformation to have higher saleable meat to bone
ratios at the same level of external fatness. The relationship
is, however, particularly dependent on the extreme groups and
several breeds did not relate well to the trend. The major
deviants at Ingliston are the Welsh Blacks, Angus crosses and
Galloways which had higher ratios than would be predicted from
their conformation scores and the Simmental and Charolais crosses

TABLE 3

BREED TYPE MEANS FOR CARCASS WEIGHT, CONFORMATION SCORE AND SALEABLE MEAT/BONE RATIO (M/B) * AT CONSTANT ESTIMATED SUBCUTANEOUS FAT PERCENTAGE** AND DEVIATION OF M/B FROM THE BETWEEN BREED-TYPE REGRESSION ON CONFORMATION SCORE

Breed	Number of carcasses		Carcass weight (kg)		Conformation[‡]		M/B		Deviation	
	S[†]	W[†]	S	W	S	W	S	W	S	W
Luing	19	25	220	198	3.38	3.22	4.34	4.31	-0.05	-0.04
Welsh Black	22	20	241	224	3.29	3.40	4.45	4.52	+0.08	+0.12
S. Devon x	21	24	249	245	3.63	3.37	4.37	4.36	-0.06	-0.03
Lincoln Red x	16	24	246	228	3.81	3.58	4.46	4.36	0	-0.09
Hereford x	20	23	234	223	3.87	3.68	4.38	4.54	-0.09	+0.06
Devon x	20	25	240	218	3.94	3.72	4.52	4.48	+0.04	-0.01
Galloway	24	-	202	-	3.85	-	4.64	-	+0.17	-
Simmental x	24	26	267	271	4.04	3.96	4.37	4.46	-0.13	-0.10
Sussex x	19	26	245	230	4.03	3.98	4.49	4.53	-0.01	-0.03
A. Angus x	17	24	214	201	4.24	3.88	4.64	4.65	+0.11	+0.12
Charolais x	23	25	271	280	4.02	4.41	4.43	4.59	-0.07	-0.09
Limousin x	-	12	-	252	-	4.53	-	4.80	-	+0.09
Residual standard deviation			23.9	22.4	0.76	0.75	0.29	0.29		

** 8.2% and 7.8% for summer and winter finishing respectively

† S = Summer finishing, W = Winter finishing

‡ Defined as in Table 2.

* Saleable meat is deboned primal joints after standardised trimming of excess fat.

TABLE 4

BREED TYPE MEANS FOR CARCASS WEIGHT, CONFORMATION SCORE AND SALEABLE MEAT/BONE RATIO (M/B) * AT CONSTANT ESTIMATED SUBCUTANEOUS FAT PERCENTAGE** AND DEVIATION OF M/B FROM THE BETWEEN BREED-TYPE REGRESSION ON CONFORMATION SCORE

Sire breed	Number of carcasses		Carcass weight (kg)		Conformation†		M/B		Deviation	
	18 mo†	24 mo†	18 mo	24 mo	18 mo	24 mo	18 mo	24 mo	18 mo	24 mo
Friesian	35	29	216	264	2.29	2.34	3.81	3.85	-0.10	-0.10
A. Angus	12	12	176	210	2.73	2.77	4.03	4.25	+0.05	+0.18
Devon	42	37	188	244	2.66	3.02	4.05	4.14	+0.08	0
Hereford	35	29	184	244	2.66	3.26	3.90	4.09	-0.07	-0.12
Sussex	17	16	197	274	2.74	2.98	4.18	4.16	+0.20	+0.03
S. Devon	12	-	240	-	2.77	-	3.89	-	-0.10	-
Simmental	22	12	244	286	3.29	3.02	4.01	4.13	-0.06	-0.01
Charolais	22	13	270	325	3.54	3.54	4.10	4.29	-0.01	+0.01
Residual SD			25.7	21.9	0.68	0.59	0.21	0.24		

* Saleable meat is deboned primal joints after standardised trimming of excess fat.

** 6.9% and 8.8% for the 18 month and 24 month systems of feeding respectively.

† Defined as in Table 2.

‡ 18 month and 24 month systems of production.

which had lower ratios than would be predicted. At Sutton Bonington, the Angus crosses were again shown to have higher ratios and the Simmental crosses lower ratios than predicted although in the case of the latter the differences were less marked. There is some suggestion in these data that differences between breed-type groups in carcass weight are influencing the relationship, positive deviations occurring more frequently for lighter breeds. But, there are some notable exceptions to this for example the Luing, Friesian and Limousin cross results.

The results for some breeds notably the Hereford and Sussex crosses were not very consistent from system to system within unit or between the Ingliston and Sutton Bonington units.

Comparable information for fully dissected carcasses is available from the dissection file referred to in Section A. Results have been computed using data from the recent contemporary breed comparison trials involving Limousin and Simmental breeds (the design of the trials is given by the Limousin and Simmental Steering Committee, 1976). Again, there is evidence from these data that the Simmental crosses in contrast to Limousin crosses in this case have lower lean to bone ratios than would be predicted from their conformation scores (Table 5).

It is relevant to ask why certain breeds differ in characteristic ways from the between breed regression. There are a number of possible explanations among which differences in bone structure appear most promising - higher lean to bone ratios at constant shape and subcutaneous fat percentage could reflect differences in bone density. Breed differences in bone density have been found between Jersey and Hereford cattle at the same age (Fursey, 1975).

An alternative possibility is that breed differences in fat distribution between depots are responsible. In the analyses reported above, it is only subcutaneous fat percentage which has been controlled and it is possible that breeds with a

TABLE 5

BREED TYPE MEANS FOR CARCASS WEIGHT, CONFORMATION SCORE AND LEAN TO BONE RATIO (L/B) AT CONSTANT SUBCUTANEOUS FAT PERCENTAGE*, AND DEVIATION OF L/B FROM THE BETWEEN BREED-TYPE REGRESSION ON CONFORMATION SCORE (LIMOUSIN AND SIMMENTAL TESTS STEERING COMMITTEE DATA)

Breed	Number of carcasses		Carcass weight (kg)		Conformation†		L/B		Deviation	
	C†	G/C†	C	G/C	C	G/C	C	G/C	C	G/C
Friesian x Ayrshire	–	24	–	243	–	2.30	–	4.05	–	-0.04
Simmental x Ayrshire	–	22	–	270	–	2.68	–	4.24	–	+0.12
Friesian	38	10	225	271	2.75	2.91	3.82	4.11	-0.08	-0.03
Hereford x Friesian	30	–	185	–	3.21	–	3.86	–	+0.09	–
Simmental x Friesian	64	47	236	264	3.62	3.41	3.98	4.05	-0.14	-0.13
Limousin x Friesian	25	11	226	239	3.79	3.69	4.35	4.29	+0.15	+0.09
Residual SD			23.1	27.9	0.79	0.60	0.24	0.26		

* Saleable meat is deboned primal joints after standardised trimming of excess fat.

** 7.6% and 6.6% for cereal and grass/cereal systems

† C = cereal feeding, G/C = grass/cereal feeding

‡ Assessed on a 5-point scale.

lower subcutaneous fat/intermuscular fat ratio will appear to have better shape because they have more intermuscular fat. However, examination of the fat distribution results published by Kempster et al. (1976$_2$) and Harrington and Kempster (1977) indicates that there is not a strong case for this proposition.

Unfortunately, we do not have data on bone density for the breeds in the study and the author is interested to know if any other workers at the seminar have relevant data. We do have weights, lengths and circumferences (at the centre of the shaft) of the femur and humerus bones. These have been analysed and preliminary results for the main breeds are shown in Table 6. Weight/length ratios, which are obviously highly weight dependent, are given together with the 'densities' calculated by assuming the bones are cylinders. The mean values for 'density' are obviously much higher than actual values but they could reflect actual differences in bone density between breeds.

It is, in fact, difficult to identify any clear patterns among the means shown in Table 6. The Luing cattle appear to have heavy dense bones and both the Friesians and the Simmental x Friesians above average bone densities. Results for the Limousin crosses were the most inconsistent; they were least dense as judged by the Ingliston winter finishing group but the most dense as judged by the grass/cereal results for the Limousin and Simmental Steering Committee tests. So these results do not provide conclusive evidence that differences in density are involved.

There has been much speculation in this section and some people may disagree with the propositions. However the work does raise some important points and may act as a stimulus for discussion. The points raised are as follows:

1) The results demonstrate that conclusions about the value of conformation will depend on the breed mix in the sample of carcasses examined. Indeed one could postulate

TABLE 6

BREED TYPE MEANS FOR CARCASS WEIGHT AND WEIGHT/LENGTH RATIO (W/L) AND 'DENSITY' (D) OF THE FEMUR AND THE HUMERUS BONES*. MEANS ARE ADJUSTED TO CONSTANT EXTERNAL FATNESS WITHIN FEEDING SYSTEM AND UNIT

	Carcass weight (kg)	W/L (g/mm)	D (g/cm^3)	Carcass weight (kg)	W/L (g/mm)	D (g/cm^3)
Sutton Bonington		18 mo			24 mo	
A.Angus x Friesian	176	4.88	3.63	210	5.22	3.40
Hereford x Friesian	184	5.05	3.64	244	5.69	3.33
Friesian	216	5.75	3.67	264	6.46	3.50
Simmental x Friesian	244	6.02	3.67	286	6.99	3.58
Charolais x Friesian	270	6.35	3.54	325	6.91	3.37
Ingliston		Summer finishing			Winter finishing	
Galloway	202	4.67	3.55	-	-	-
A.Angus x	214	5.06	3.56	201	4.89	3.50
Luing	220	5.21	3.79	198	4.83	3.69
Hereford x	234	5.44	3.43	223	5.27	3.57
Limousin x	-	-	-	252	5.43	3.40
Simmental x	267	6.10	3.56	271	6.02	3.54
Charolais x	271	6.10	3.50	280	6.19	3.53
Steering Committee		Cereal			Grass/cereal	
Hereford x Friesian	185	5.18	3.63	-	-	-
Friesian	225	5.66	3.75	271	5.89	3.50
Limousin x Friesian	226	5.06	3.63	239	5.37	3.68
Simmental x Friesian	236	5.75	3.71	264	5.82	3.54

*An average result for the two bones is given

populations to give almost any conclusion one likes. Care must, therefore, be exercised when drawing conclusions from small samples.

2) When examining the relative precision of different predictors such as conformation, it is useful to look carefully at the sample of animals involved and try to determine in growth terms why particular results are obtained. If this is done it would be easier to develop models which would allow us to generalise results from one set of circumstances to others. This has important implications for the standardisation of carcass evaluation procedures between countries which have widely different cattle populations.

3) Further work is required to determine why certain breeds differ in characteristic ways from the shape - lean/ bone ratio relationship. Are the deviations due to differences in bone density as hypothesised or are there more subtle differences which, for example, influence the subjective assessment of conformation?

4) We need to look critically at the role of conformation in commercial classification and see whether breed itself or a better breed indicator than shape could be substituted in some circumstances.

Finally, as a side issue, it is interesting to look at the shape relationships of the different breeds in terms of their genetic history. For this purpose, the Ingliston results (selected because they include the widest range of breed types) have been combined and are plotted in Figure 1. The breed types show a time series from the more indigenous types, Welsh Black (Celtic in origin) and Luing (Beef Shorthorn x Scottish Highland cattle of ancient origin) with poor conformation through the British red breed crosses, the sires of which are probably descended from cattle introduced during the Anglo-Saxon invasions to the Continental breed crosses selected in more recent times than the British breeds for draught purposes. The Galloway and

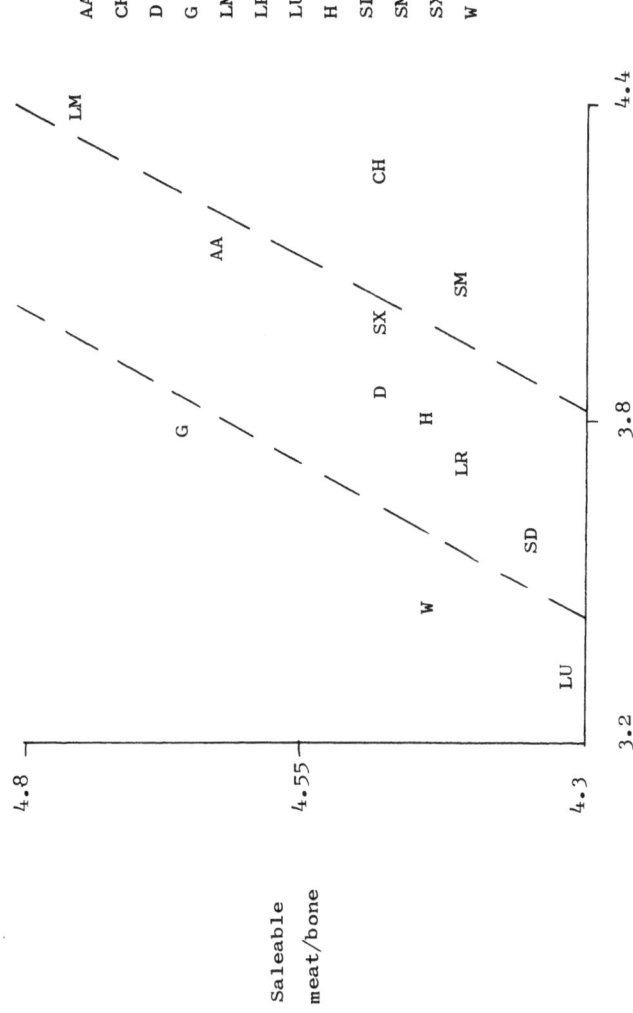

Fig. 1.   Ingliston: breed type means for saleable meat/bone ratio against conformation

Angus crosses stand away from the regression of saleable meat/
bone ratio on conformation and appear different to the Luing
and Welsh Black. The origin of these black polled breeds is
obscure but they are possibly of Nordic Viking ancestry. We
do not, of course, know what the effect of crossing the pure
breeds with Blue-Grey and Hereford x Friesian dams would be and
the Figure is only given as an interesting observation.

ACKNOWLEDGEMENTS

The author would like to thank Mr. G.L. Cook and Mr. D.W.
Jones for their advice on various aspects of this work and
their assistance with the data analysis.

REFERENCES

Barton, R.A. 1967. The relation between live animal conformation and the carcass of cattle. Anim. Breed. Abstr. 35 : 1-22.

Callow, E.H., 1961. Comparative studies of meat. VII. A comparison between Hereford, Dairy Shorthorn and Friesian steers on four levels of nutrition. J. Agric. Sci. Camb 56 : 265-282

Fredeen, H.T., Locking, G.L. and Mc Andrews, J.G. 1974. Carcass conformation as a criterion of yield and value of beef carcasses. Can. J. Anim. Sci. 54 : 551-563.

Frood, I.J.M. 1976. An investigation into the effect of sex and plane of nutrition on the growth performance and carcass quality of British Friesian cattle for beef production. Ph. D. Thesis, University of Reading.

Fursey, G.A.J. 1975. A note on the density of bovine limb bones. Anim. Prod. 21 : 195-198.

Harrington, G. and Kempster, A.J. 1977. Beef carcass yields. Inst. of Meat Bull. No. 95, February 1977 : 2-15. Published by (British) Institute of Meat, London.

Hante, F.J. and Conniffe, D. 1967. Studies on cattle of varying growth potential for beef production. II. Carcass composition and distribution of 'lean meat', fat and bone. Ir. J. agric. Res. 6 : 153-170.

Kauffman, R.G., Grummer, R.H., Smith, R.E. Long, R.A. and Shook, G. 1973. Does live animal and carcass shape influence gross composition? J. Anim. Sci. 37 : 1112-1119.

Kempster, A.J., Cuthbertson, A. and Smith, R.J. 1976$_1$. Variation in lean distribution among steer carcasses of different breeds and crosses. J. Agric. Sci., Camb. 87 : 533-542.

Kempster, A.J., Cuthbertson, A. and Harrington, G. 1976$_2$. Fat distribution in steer carcasses of different breeds and crosses. 1. Distribution between depots. Anim. Prod. 23 : 25-34.

Kempster, A.J., Cuthbertson, A. and Jones, D.W. 1977. Bone weight distribution in steer carcasses of different breeds and crosses, and the prediction of carcass bone content from the bone content of joints . J. Agric. Sci., Camb. 89, 675-685.

Limousin and Simmental Tests Steering Committee, 1976. Report of the evaluation of the first importation into Great Britain in 1970-71 of Limousin bulls from France and Simmental bulls from Germany and Switzerland. 100pp. London : HMSO.

Meat and Livestock Commission, 1975. Progress on Beef Carcass Classification. Marketing and Meat Trade Technical Bulletin, No. 22. Meat and Livestock Commission, Bletchley, Bucks.

Murray, D.M., Tulloh, N.M. and Winter, W.H. 1974. Effects of three different growth rates on empty body weight and dissected carcass composition of cattle. J. Agric. Sci., Camb., 82 : 535-547.

Oliver, W.M., 1967. A review: Shape or weight in cattle for beef. Texas Agricultural Experiment Station, Department of Animal Science, Technical Report No. 6.

Seebeck, R.M. 1973. The effect of body weight loss on the composition of Brahman cross and Africander cross steers. II. Dissected components of the dressed carcass. J. Agric. Sci., Camb. 80 : 411-423.

Seebeck, R.M. and Tulloh, N.M. 1968. Developmental growth and body weight loss of cattle. III. Dissected components of the commercially dressed carcass, following anatomical boundaries. Aust. J. Agric. Res., 19 : 673-688.

Truscott, T.G., Lang, C.P. and Tulloh, N.M. 1976. A comparison of body composition and tissue distribution of Friesian and Angus steers. J. Agric. Sci. Camb. 87 : 1-14.

A NOTE ON CONFORMATION AND MEAT CHARACTERISTICS IN BEEF CARCASSES

R. Verbeke and G. Van de Voorde

Study Centre for Meat Production, Melle, Belgium

ABSTRACT

An objective approach to carcass classification is proposed, parallel to the EAAP standards. An attempt has been made to define conformation subclasses in a mathematical way, by relating changes in carcass weight to carcass length. Results of anatomical dissection and retail cutting are discussed in connection with conformation aspects.

INTRODUCTION

At the agricultural research seminar on criteria and methods
for the assessment of carcass and meat characteristics in beef
production experiments, Zeist, 1975, it was stated that there is
a great diversity in the assessment of beef carcasses because of
the application of different criteria as well as different methods
for similar objectives.

A general conclusion mentioned, as a principal aspect of
agreement on common procedures, the determination of carcass weigh
and the use of EAAP standards (colour slides) for 'fleshiness'
and 'fatness'. (Eur 5489, 1976). Although the value of visual
carcass assessment cannot be ignored, scientists do realise that,
in order to assess biological characteristics subject to continuous variability, the use of objective methods is essential. Only
objective data (dissections, measurement etc.) in connection wit
conformation or other generic characteristics of this kind will
allow correct interpretation by means of statistics, in this case
concerning the relationship of carcass weight to conformation.

MATERIALS AND METHODS

Yeates (1959) observed that at the same weight, the shorter
carcasses had a better conformation than the longer ones. He
defined the 'fleshing index' of a carcass as the weight units
it was heavier or lighter than the average for its length.

Along similar lines the research centre attempted an objective assessment of carcass conformation, resulting in the use
of the nomogram illustrated at Figure 1. The development of the
method may be described as follows:-

1) Data on 'blockiness' (weight to length index) were
collected in experimental and market bulls (Eur. 5489).
The linear regression equation relating carcass length
(y) to carcass weight (X) was given by:

$$y = 95.72 + 0.1040 \, X \quad (r = 0.98; \text{ res.}_{S.D.} = 1.33 \text{ cm})$$

These factors determine conformation also.

2) A practical formula -

$$\frac{carcass\ weight}{10} + 100 - length \gtrless 0$$

similar to Yeates' fleshing index was used to construct a preliminary rectilinear bundle of lines in which five conformation classes at a given carcass length differed in weight by 10%.

3) Superposition of EAAP classification scores for conformation (de Boer et al.) and the indexes obtained by graphical classification, showed clearly that there was no linearity in lower weight ranges. Therefore it seems more correct to express the relation between carcass weight (y) and length (X) by means of an exponential form as given below:

Conformation classes 3-,3,3+ : $y = 20.9127\ e^{0.0207X}$
$$(r^2=0.75,\ n=62)$$

Conformation classes 4-,4,4+ : $y = 21.8779\ e^{0.0211X}$
$$(r^2=0.69,\ n=0.48)$$

The exponents suggest that 15% weight ranges rather than 10% differences between conformation classes should be used in the definite construction of the nomogram.

4) Because the weight of a carcass (of similar density) is proportional to its volume, and because, during the growth process, without any changes in its shape, the volume increases with the cube of linear dimensions (David and Newth, 1970) it may be more accurate to introduce ℓ^3 instead of ℓ.

This mathematical transformation leads to the linear expression employed for the construction of the conformation nomogram:

$$y = 35.29 + (0.1260\ .\ 10^{-3}\ x\ X^3)$$

where y represents carcass weight and X carcass length.

Covering all weights and lengths, a possible mathematical

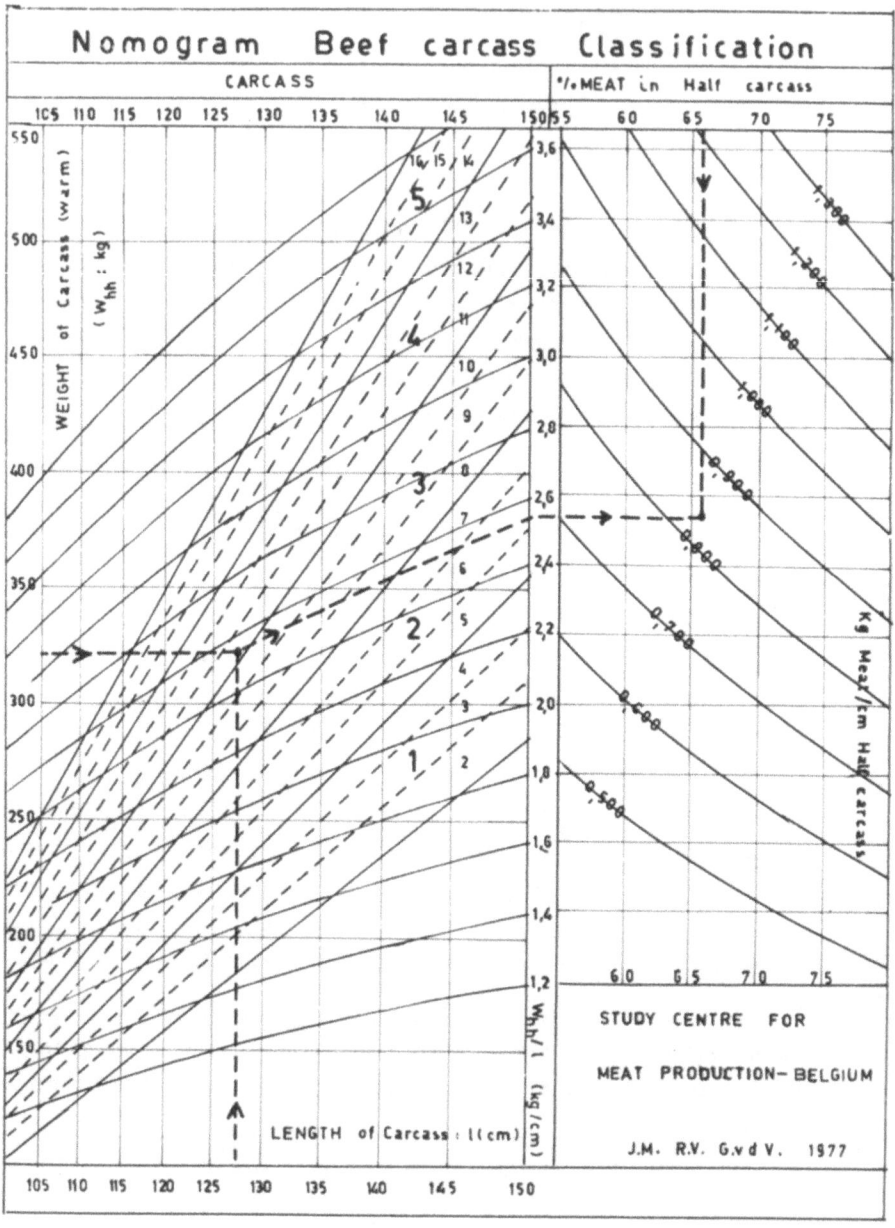

Fig. 1

approximation of the EAAP subclasses (2 up to 16) may be given as:

$$EAAP_{subclass} = 1 + A\ e^{B.X}$$

where: $A = -22.26 + 0.2243\ l\quad (r^2 = 0.9943)$

$B = 0.32 - 0.0017\ l\quad (r^2 = 0.9376)$

$X = $ fleshing index $= \dfrac{carcass\ weight}{10} - (l - 100)$

(in which l represents length)

It should be stressed that this formula should only be applied within certain biological limits - in this case to bull carcasses. In cow carcasses a somewhat different formula may apply.

RESULTS AND DISCUSSION

Since the proposed nomogram.classification is superimposed on the classification with regard to conformation according to EAAP standards, carcass and meat characteristics may be related to both. Weight changes and dissection results of carcasses will be discussed in relation to the conformation classes.

Conformation and weight of carcasses

The formula applied for the construction of the nomogram involves a linear increase in weight (5%) with the sub-class for conformation at a given length. The gain in carcass weight (W_{HH}) in function of the carcass length (l) is written as -

$$\Delta\ W_{HH} = -22.79 + 0.3012\ l\ (r^2 = 0.99)$$

The theoretical increases calculated within some carcass length ranges are given in Table 1.

This means, for example, that a carcass classified as 3.0 (sub-class 9) for conformation at a length of 125 cm and a weight of 280 kg would have been classified as a 5+ (sub-class 16) if the weight of that carcass had been 280 + (14.9 x 7) = 384 kg. Thus the upper limits of the carcass weights may be calculated for all conformation sub-classes at specified carcass lengths (Table 2).

TABLE 1

CHANGES IN CARCASS WEIGHT WITH CONFORMATION CLASSES WITHIN CARCASS LENGTH CLASSES

Carcass Length (cm)	Δ W_{HH} / sub-class (kg)	Σ^{Δ} W_{HH} / sub-class 2 - 16 (kg)
100	7.3	110
105	8.8	132
110	10.3	155
115	11.8	178
120	13.4	200
125	14.9	223
130	16.4	245
135	17.9	268
140	19.4	291
145	20.9	313
150	22.4	336

TABLE 2

UPPER LIMITS OF CARCASS WEIGHT (kg) IN FUNCTION OF CONFORMATION AND LENGTH

Carcass Length	EAAP sub-classes for conformation														
	2 -1	3 1.0	4 +1	5 -2	6 2.0	7 +2	8 -3	9 3.0	10 +3	11 -4	12 4.0	13 +4	14 -5	15 5.0	16 +5
100	107	115	122	129	137	144	151	159	166	173	181	188	195	203	210
105	121	130	139	148	157	166	174	183	192	201	210	219	227	236	245
110	140	150	161	171	181	192	202	212	223	233	243	254	264	274	285
115	155	167	179	191	202	214	226	238	250	262	273	285	297	309	321
120	170	183	197	210	223	237	250	263	277	290	304	317	330	344	357
125	190	205	220	235	249	264	279	294	309	324	339	353	368	383	398
130	210	226	243	259	275	292	308	325	341	357	374	390	406	423	439
135	230	248	266	284	301	319	337	355	373	391	409	427	444	462	480
140	255	274	294	313	333	352	371	391	410	429	449	468	488	507	526
145	280	301	322	343	364	384	405	426	447	468	489	510	531	551	572
150	310	332	355	377	400	422	444	467	489	512	534	556	579	601	623

Dissection data

The Centre for Meat Production studies carcass quality at different weight levels in beef, dairy and crossed types of cattle. For that purpose carcass classification is completed by anatomical dissection and retail cutting, giving information about muscle distribution and cutability. The results so far obtained refer to 30 carcasses and suggest some preliminary but perhaps important conclusions.

Carcass jointing and conformation

The type of carcass does not seem to be related to the differences found in the distribution of wholesale cuts, as shown in Table 3.

TABLE 3

DISTRIBUTION OF CUTS AND INFORMATION

EAAP conformation class	2	3	4	5
% neck + chuck (4 ribs)	12.83	13.77	14.72	14.11
% brisket + rolled rib	17.20	16.42	16.59	17.00
% best rib (5 ribs)	5.27	5.29	5.07	4.66
% sirloin (4 ribs)	3.52	3.59	3.60	3.69
% loin, rumpsteak and rump	13.38	13.28	13.53	13.27
% thick flank	8.74	8.80	8.93	8.23
% shoulder	13.33	12.99	12.97	12.03
% hind leg	25.82	25.87	24.58	27.03

Sizeable differences will occur only in the case of the highest conformation classes. However, only a few really double muscled animals have been dissected up to now, which indicate that there may be an increased percentage of hind leg, and a decreased percentage of thick flank and shoulder.

Carcass lean content and conformation

The variation in the yield of lean meat between the conformation classes is highly significant. In the best classes the percentage of meat is highest, while that of bone is lowest. The relative amount of fat is not associated to such an extent, as can be seen in Table 4.

TABLE 4

CONFORMATION IN RELATION TO ANATOMICAL DISSECTION RESULTS

Conformation class	2	3	4	5
% lean meat ± mean error	65.74 ± 1.97	67.13 ± 1.16	66.55 ± 2.81	70.08 ± 1.05
% fat ± mean error	18.09 ± 1.66	18.86 ± 1.50	20.99 ± 2.55	18.03 ± 1.07
% bones ± mean error	16.17 ± 0.70	14.00 ± 0.65	12.47 ± 0.46	11.90 ± 0.01
Meat/bone ratio	4.07	4.80	5.34	5.89
Meat/fat ratio	3.03	3.55	3.17	3.89
Fat/bone ratio	1.12	1.35	1.68	1.52

These results suggest that only the meat/bone ratio is related to some extent to conformation. However, it should be stressed that for a clear interpretation of the ratio, the absolute percentage of one of the components, meat or bone, should always be considered as well. In this way, the influence of fatness can be taken into account.

Conformation and retail cutting

One side of the dissected carcasses has been subjected to retail cutting. Similar variations in saleable meat cuts occur between the conformation classes, as obtained in anatomical yield Yet 7 to 12% of non lean components other than meat are considere as retail meat, the greatest margin being found in the carcasses with the best conformation.

Conformation class does not significantly affect the

roasting/boiling meat ratio. On the other hand, it was found that a better conformation leads to a slight improvement in the proportion of good quality lean cuts within the roasting joints This, however, may not be realised where there is stereotyped retailing or a lack of professional butchery knowledge.

TABLE 5

RETAIL YIELD AND CONFORMATION

Conformation Class	2	3	4	5
Retail meat (%)	72.66	73.86	74.08	78.33
Retail fat (%)	11.45	12.50	13.68	10.24
Retail bones (%)	15.27	13.01	11.87	11.08
Roasting meat	51.87	50.75	49.74	52.87
Boiling meat	48.13	49.25	50.26	47.13

CONCLUSIONS

The distribution of the weights of carcass joints, which is usually considered to be an important factor in carcass quality, seems not to be affected significantly by conformation. Perhaps one definition of conformation might be, "The relative plus weight of the carcass with regard to the average weight at its slaughter length".

The cutability of a carcass seems to depend on the yield of its component parts and the meat/bone ratio rather than on conformation as such. Consequently, the trading importance of conformation may often be overestimated.

REFERENCES

David W. and Newth R. 1970. Studies in Biology No. 24. Growth and develop-
    ment. Edward Arnold Publishers, London.

de Boer, H. Dumont B.L., Pomeroy R.W. and Weniger J.H., 1974   Manual on
    EAAP reference methods for the assessment of carcass characteristics
    in cattle. Livest. Prod. Sci., 1, 151-164

Verbeke R., 1976.   Eur 5489 - Dissection of the three rib joint (7th, 8th
    9th) and blockiness as estimates of carcass value. 221-226

Yeates N.T.M., 1959   Beef carcass appraisal and grading. J. Aust. Inst.
    Agric. Sci. 25,302. 225

MEAT AMINO ACID COMPOSITION OF CALVES AND STEERS SLAUGHTERED BETWEEN 200 KG AND 500 KG LIVE WEIGHT

G. Piva and D. Guglielmetti

Istituto Scienze della Nutrizione - Facoltà di
Agraria - Università Cattolica S. Cuore - Piacenza - Italy

ABSTRACT

In this paper, the effects of slaughter weight variations on amino acid composition and beef nourishment values are examined in Bavarian Red Pied steers by calculating PDR (Protein Digest Residue) value.

Examination of figures related to ≃ 200 kg lw (live weight), ≃ 300 kg lw, ≃ 400 kg lw, ≃ 500 kg lw slaughtered animals indicates progressive decrease of some amino acids as slaughter weights are increased.

Chemical score is significantly related to slaughter weight which is not however, the case with DPR.

INTRODUCTION

In human nourishment meat provides high value biological proteins. While one cannot ignore the connection between the calorific contribution and the protein content, the connection with the lipidic content (the rate of which varies considerably according to age and fattening condition) is of the highest importance.

The types of protein in muscular meat vary; some are marked by a high and complete essential amino acid composition and therefore have a high biological value; others have low or very low biological value, because they lack some essential amino acids. Collagen, for instance, is mainly made up of glycine, proline and hydroxy proline and it is important to point out the last is plentiful only in collagen (13 - 14%) and is also marked by a high resistance to enzymatic hydrolysis and therefore by a low digestibility.

That means that overall meat feeding value depends on the ratio between the different constituent proteins. This is not a constant ratio, but it is inclined to modify slightly with age, with clear effects on biological value and digestibility.

Bearing these facts in mind, the authors examined meat nourishment peculiarities of steers belonging to the same breed, slaughtered at different weights.

MATERIAL AND METHODS

During the trial, the authors employed steers of the same breed (Bavarian Red Pied steers), of different live weights and fed according to the usual diets.

The authors examined the beef of 16 animals, slaughtered in groups of four at different calf weights as follows:

1) white meat slaughtered at an average 215 kg live weight and fed milk substitutes:

2) steers slaughtered at an average weight of 306.5 kg after having been fed on corn silage, maize meal and protein-extract meal. The protein rate of the ration's dry matter was 14%;

3) steers slaughtered at 420 kg live weight, fed on the same feeds, but with a 13% protein rate on the dry matter;

4) steers slaughtered at 539 kg live weight and fed on a 12% protein rate ration.

From the right half-carcass, at the height of the 10th-11th rib joint, a sample cut of meat was drawn according to Lanari's technique (Lanari, 1973) including portions of the following muscles for each slaughtered animal: supraspinatus, omotransversarius, longissimus costarum, serratus dorsalis anterior, external intercostal, internal intercostal and longissimus dorsi.

These samples were immediately frozen at $-30^{\circ}C$ and then freeze-dried and used as average samples for analytical determinations.

Determinations, besides total nitrogen, included total amino acid content and PDR (Protein Digest Residue), according to Sheffner's method (Sheffner, 1956), modified by Akeson (1964).

Three kind of hydrolysis were carried out to determine total amino acid content:

1) alkaline hydrolysis with Ba $(OH)_2 \cdot 8H_2O$, according to Eggum (1968) for tryptophan;

2) acid hydrolysis with HCl 6N after cystine oxidation to cysteic acid and after methionine oxidation to methionine sulphone with performic acid, according to Moore et al. (1963) and Moore (1958);

3) acid hydrolysis with HCl 6N for all the other amino acids, including hydroxyproline.

An unichrom by Beckman automatic amino acid analyser was employed in the determinations.

For PDR evaluation, the authors carried out a three-hour hydrolysis with pepsin (USP 1/10 000) at $37^{o}C$ in acid ambient with HCl 0.1 N and afterwards a twenty four hour hydrolysis with pancreatin (EUSP) at pH 8 with NaOH 0.2 N.

In the soluble quota, the authors determined released amino acids, after protein precipitation with picric acid 1%, using the above mentioned instrument.

Sheffner's method, modified by Akeson, was followed for PDR calculation. Akeson's variants eliminate tryptophan, which, as already known, is broken down during the protein precipitation stage with picric acid, from the calculation.

RESULTS

1) Essential amino acid summations (isoleucine, leucine, lysine, phenylalanine + tyrosine, methionine + cystine, threonine, tryptophan, valine) expressed as a percentage of total amino acid summation are very close in the different examined samples; only slightly higher for 200 kg lw (live weight) slaughtered animals and a little lower for animals slaughtered at a higher weight with a rather linear decrease:

2) essential amino acid summation expressed as a percentage of total amino acid results closely related to slaughtering weight (r = 0.980, P < 1%);

3) 'chemical score' according to Block and Mitchell (1946), calculated using egg protein as standard protein, shows identical values for group 1 and group 2; slightly decreasing values for groups 3 and 4. Differences between the fourth group and groups 1 and 2 are about 9%. Limiting amino acids are sulphurated amino acids. There is a good correlation with slaughtering weights (r = 0.944, P < 5%);

4) evaluation according to FAO/WHO experts calculation (1965) taking whole egg protein as standard protein (FAO, 1957) renders similar values for the first two groups and decreasing values for groups 3 and 4. Correlation with slaughtering weights is $r = 0.939$ ($P < 5\%$).

5) Hydroxyproline content progressively increases according to slaughtering weights and practically doubles from group 1 to group 4. Correlation with slaughtering weights is highly significant ($r = 0.978$ $P < 1\%$).

6) Examining separately the behaviour of the different amino acids, a significant correlation to slaughtering weights can be pointed out for some of them; for essential amino acids: lysine $r = 0.942$ ($P < 5\%$), leucine $r = 0.978$ ($P < 1\%$), phenylanine $r = 0.964$ ($P < 1\%$), tryptophan $r = 0.882$ ($P < 5\%$); for inessential amino acids: serine $r = 0.995$ ($P < 1\%$), cystine $r = 0.952$ ($P < 5\%$).

7) No significant correlation can be shown between slaughtering weight and PDR, which has a clear meaning of NPU (Net Protein Utilisation) according to Sheffner's and Akeson's researches. The correlation coefficient calculated by Akeson between PDR and BV (Biological Value), even eliminating tryptophan from the calculations, is very high at 0.982 ($P < 1\%$). PDR provides the highest value for the third groups, that is for $\simeq$ 300 kg lw slaughtered steers and the lowest one for steers slaughtered at 500 kg live weight.

CONCLUSIONS

The analytical methods used showed that the nourishment value of the protein share seems negatively influenced by increased slaughtering weight, particularly for animals slaughtered at the heaviest weight.

TABLE 1

AMINO ACID CONTENT, PER CENT OF CRUDE PROTEINS

Samples amino acids	Beef slaughtered at ≃ 200 kg lw	Beef slaughtered at ≃ 300 kg lw	Beef slaughtered at ≃ 400 kg lw	Beef slaughtered at ≃ 500 kg lw
Crude proteins	84.08	86.69	82.77	82.89
Lysine	8.58	8.58	8.26	8.14
Histidine	2.71	2.96	2.58	3.35
Arginine	6.20	6.33	6.16	6.10
Aspartic acid	9.41	8.70	9.17	8.67
Threonine	4.25	4.39	4.08	4.50
Serine	3.95	3.83	3.68	3.51
Glutamic acid	15.51	15.60	15.45	15.53
Proline	4.13	4.50	4.52	4.58
Hydroxy-proline	0.48	0.61	0.93	1.03
Glycine	4.14	4.22	4.06	4.38
Alanine	5.52	5.71	5.40	5.77
Cystine	1.14	1.12	1.01	0.99
Valine	4.59	4.62	4.87	4.46
Methionine	2.62	2.66	2.60	2.46
Isoleucine	4.54	4.65	4.44	4.33
Leucine	7.83	7.72	7.68	7.62
Tyrosine	3.59	3.38	3.38	3.41
Phenylalanine	4.15	3.95	3.87	3.80
Tryptophan	0.92	1.10	1.08	1.16

TABLE 2

ESSENTIAL FREE AMINO ACID % CRUDE PROTEINS, AFTER ENZYMATIC HYDROLYSIS BY AKESON

Samples amino acids	Beef Slaughtered at ≃ 200 kg lw	Beef slaughtered at ≃ 300 kg lw	Beef slaughtered at ≃ 400 kg lw	Beef slaughtered at ≃ 500 kg lw
Lysine	2.46	1.83	0.25	2.13
Histidine	0.13	0.14	0.25	0.43
Threonine	0.33	0.20	0.24	0.15
Cys. ac. +met.	0.19	0.66	0.76	0.82
Valine	0.35	0.39	0.37	0.24
Isoleucine	0.69	0.37	0.41	0.46
Leucine	2.80	1.98	2.29	1.97
Try. +phenyl.	2.91	2.30	2.81	2.10

TABLE 3

EVALUATION INDEX

Samples	Beef slaughtered at ≈ 200 kg lw	Beef slaughtered at ≈ 300 kg lw	Beef slaughtered at ≈ 400 kg lw	Beef slaughtered at ≈ 500 kg lw
$\overline{\Sigma}$ total amino acids	94.25	94.95	92.90	93.79
$\overline{\Sigma}$ essential amino acids	44.02	44.03	42.76	43.06
$\overline{\Sigma}$ not essential amino acids	50.23	50.92	50.14	50.73
$\overline{\Sigma}$ not essential amino acids % $\overline{\Sigma}$ total amino acids	46.70	46.37	46.02	45.91
Chemical score Block & Mitchell	69	69	66	63

PROTEIN SCORES (FAO, 1965)

Reference Hen's egg (whole)	77	78	84	77
PDR	80.72	82.90	80.07	78.08

REFERENCES

Akeson, W.R., and Stahmann, M.A., 1964. A pepsin pancreatin digest index of protein quality evaluation, J. Nutr. 83, 257-261.

Black, R.J., and Mitchell, H.H., 1946. The correlation of the amino acid composition of proteins with their nutritive value, Nutr. Abstr. Rev. 15, 249-257.

Eggum, B.O., 1968. Determination of tryptophan, Acta agric. scand. 18, 127-131.

FAO/WHO, 1957 Protein requirements, FAO Nutritional Studies Series, Nr. 16, Rome.

FAO/WHO, Expert group, 1965. Protein requirements FAO/WHO., Technical Report Series, Nr. 301, Rome.

Lanari, D., 1973. Utilizzazione tagli campione nella stima della composizione della carcassa bovina, Riv. Zoot. Vet. 46, 241-256.

Moore, S., Spackman, D.H., and Stein, W.H., 1958. Chromatography of amino acids on sulphonated solystyrene resin, Anal. Chem. 30, 1185-1190.

Moore, A., 1963. On the determination of cystine as cysteic acid, J. Biol. Chem. 238, 235-237.

MYORHEOLOGICAL, CHEMICAL AND COLOUR CHARACTERISTICS OF MEAT IN WATER BUFFALO AND BOVINE CALVES SLAUGHTERED AT 20, 28 AND 36 WEEKS[*]

D. Matassino[1], A. Romita[2], E. Cosentino[1],
A. Girolami[1] and P. Colatruglio[1]

1 Istituto di Produzione animale, Facoltà di Agraria,
Università degli Studi di Napoli, 80055,Portici, Italy

2 Istituto Sperimentale per la Zootecnia,
via Onofrio Panvinio 11, 00162 Rome, Italy

ABSTRACT

The study was carried out on 13 muscles (Table 1) of water buffalo (Italian) and bovine (Italian Friesian) calves raised under the same micro-environmental conditions and slaughtered at 20, 28 and 36 weeks of age (Table 2). The rheological (hardness, chewiness and water holding capacity, (WHC)), chemical, (dry matter, protein, fat and ash) and colour variables (brightness and spectrophotometric curves) have been evaluated on raw meat. The more significant results were:

1) The muscle confirms its marked individuality by determining independently from age and species all the studied variables.

2) Age is determinant in varying the organoleptic characteristics - at 36 weeks the meat is harder, darker, fatter (positive variation in relation to palatability), requires more chewing energy and has more WHC (positive variation) than at 20 and 28 weeks.

3) The organoleptic characteristics of the meat of the two species vary in a similar way according to the slaughter age.

4) Within the age and considering the qualitative variables all together, the buffalo calves seem superior to the bovine and provide more uniform carcasses.

5) At this stage it seems opportune to integrate these results with

* This work was supported by National Research Council of Italy (Progetto Finalizzato:'Incremento disponibilità alimentari di origine animale').

breeding, slaughtering and cutting data, and to complete the slaughtering ages at 44 and 52 weeks.

6) The results confirm the usefulness of techniques employed for evaluation of organoleptic characteristics of meat in view of its technology, marketing and genetic improvement.

INTRODUCTION

The objective evaluation of meat organoleptic characteristics is a topical problem and improvements have already been
reported on its importance for genetic, technological and commercial aspects of the product. (Bordi et al., 1974; Matassino
et al., 1974, 1975$_a$, 1975$_b$, 1976$_{(a,b,c,d)}$; Matassino and Pilla,
1976; Cosentino et al., 1976$_a$, 1976$_b$, 1977).

The purpose of this investigation is to point out the extent
of meat differences between water buffalo and bovine calves
slaughtered at 20, 28 and 36 weeks of age and to suggest the
best times in order to obtain, qualitatively, the best meat.

MATERIAL AND METHODS

The study was carried out on 13 muscles (Table 1) from 58
sides of water buffalo and bovine calves slaughtered at three
different ages (20, 28 and 36 weeks).

TABLE 1

TESTED MUSCLES

Muscle	Symbol	Function
Caput laterale tricipitis brachii	CLaTB	Extensor
Caput longum tricipitis brachii	CLoTB	Extensor
Supraspinatus	Ss	Extensor
Vastus lateralis	VL	Extensor
Rectus femoris	RF	Extensor
Vastus medialis	VM	Extensor
Gluteus profundus	GP	Extensor
Gluteus medius	GM	Extensor
Pars parameralis gluteobicipitis	PPGb	Flexor
Semitendinosus	St	Flexor
Semimembranosus	Sm	Flexor
Longissimus dorsi	LD	Protractor
Psoas major	PM	Adducent

The fasted and net live weight, as well as that of cold sides, are reported in Table 2. The drawing of muscle samples was carried out on the right side aged 9 days at 3 - 4°C. The techniques of raising, slaughtering and cutting are reported by Romita et al. (1976 and in print). Rheological (hardness, chewiness and water holding capacity (WHC)) chemical (dry matter, protein, fat and ash) and colour (brightness and spectrophotometric curves) variables have been evaluated on raw meat samples as suggested respectively by Matassino et al. (1974; 1975$_c$) Matassino et al. (1976$_{b,d}$) and Matassino et al. (1975$_b$)

TABLE 2

VALUES OF SOME VARIABLES AT SLAUGHTER

Age at Slaughter (weeks)	Buffalo calves						Bovine calves					
	Weight, kg											
	Fasted live		Net live		Cold side		Fasted live		Net live		Cold side	
	Mean	cv %	Mean	cv %	Mean	cv %	Mean	cv %	Mean	cv %	Mean	cv %
20	151.4	7	141.9	6	39.6	9	162.3	6	146.4	6	43.8	6
28	203.4	6	185.3	7	50.3	8	227.8	5	201.1	6	59.2	8
36	229.9	7	210.1	8	57.3	8	291.4	7	260.0	6	74.6	6

RESULTS

We are reporting only the most important results, referring for further details to the original papers (in print).

Water buffalo calves

The muscle is one of the most important factors, affecting the value of all meat characteristics listed in Table 3, with the exception of ash content as already observed by the authors mentioned under 'Introduction'.

The meat hardness tends to be invariant in relation to the age, whether evaluated on the whole side (weighted mean of 13 muscles which represents 87.4, 89.6 and 89.8 percent of the meat

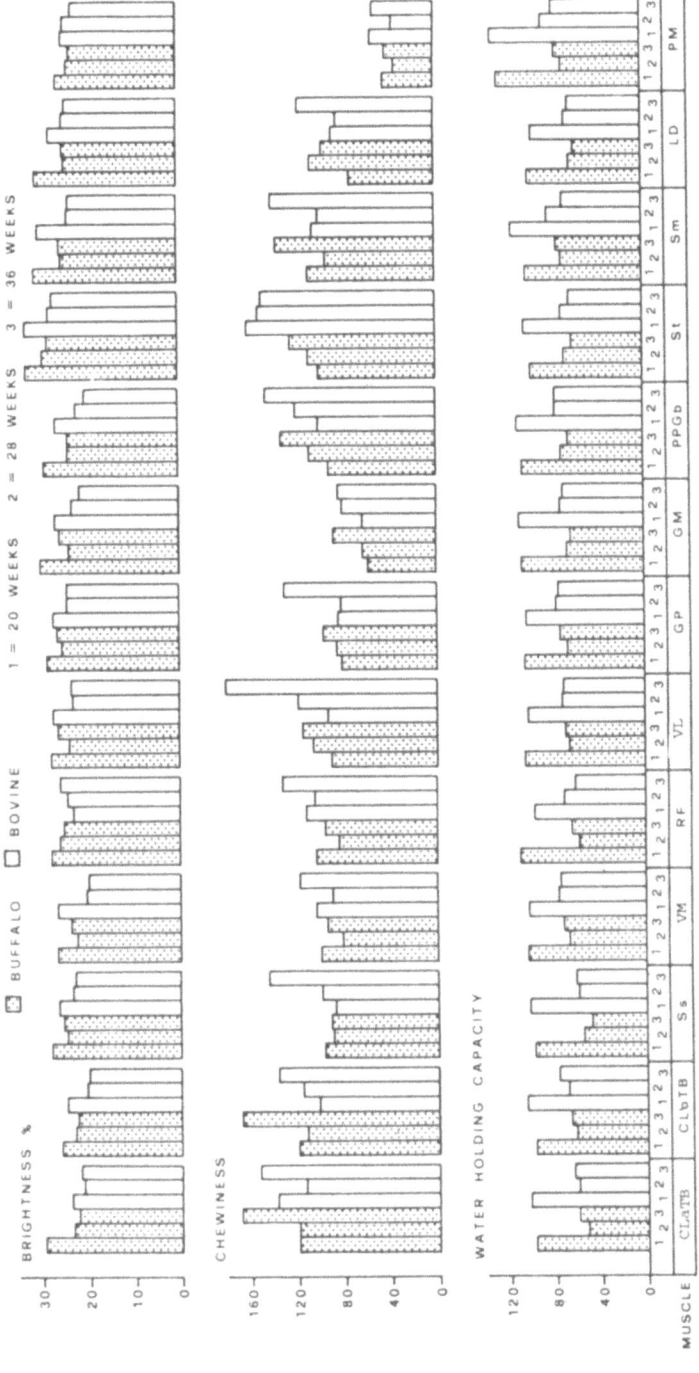

Fig. 1 Mean values of brightness, chewiness and WHC. Chewiness and WHC values were obtained making = 100 the value of water buffalo St muscle at 20 weeks of age

obtained respectively at 20, 28 and 36 weeks, or on the single muscle (Table 3). On the other hand, age (Table 3) is the determining factor of chewiness (rheological variable that synthesises the values of hardness, cohesiveness and springiness) thus suggesting that chewing energy does not increase significantly from 20 to 28 weeks while it increases (about 20%) from 20 - 28 to 36 weeks (P $<$0.01). Within the third stage, the increment of chewing energy, in comparison with the first two stages, reaches about 40% in the following muscles, CLaTB, CLoTB, GM, PPGb and LD (Table 4 and Figure 1).

TABLE 3

RESULTS OF VARIANCE ANALYSIS WITHIN SPECIES

Variable	Buffalo calves F [1]			Bovine calves		
	Age, A	Muscle Mu	Interaction A-Mu	Age, A	Muscle Mu	Interaction A-Mu
Hardness	ns	22^{000}	ns	ns	21^{000}	1^{0}
Chewiness	13^{000}	13^{000}	ns	19^{000}	10^{000}	ns
WHC	378^{000}	6^{000}	ns	247^{000}	6^{000}	ns
Dry matter	104^{000}	24^{000}	4^{000}	129^{000}	13^{000}	2^{000}
Protein	143^{000}	22^{000}	4^{000}	295^{000}	19^{000}	3^{000}
Fat	35^{000}	7^{000}	3^{000}	54^{000}	9^{000}	2^{000}
Ash	6^{00}	ns	ns	15^{000}	ns	ns
Brightness	36^{000}	4^{000}	ns	38^{000}	7^{000}	ns

[1] ns = not significant; ^{0} = P $<$0.05; ^{00} = P $<$0.01; ^{000} = P $<$0.001

The WHC increases markedly (about 40%), P $<$0.01) from 20 to 28 - 36 weeks of age for the single muscle as well as for the whole side (weighted mean of 13 muscles studied: see Tables 3 and 4 and Figure 1).

At this stage, the PM muscle gives meat which is the least hard, requiring the least chewing energy and with the least WHC; the reverse applies to the CLaTB and CLoTB muscles. The absence of significant interaction age/muscle for all the rheological variables means that the rank muscle would tend to be invariant

TABLE 4

RESULTS OF VARIANCE ANALYSIS WITHIN SPECIES WITHIN MUSCLE

Variable	Muscle												
	CLaTB	CLoTB	Ss	VM	RF	VL	GP	GM	PPGb	St	Sm	LD	PM
	F among ages [1]												
Buffalo calves													
Hardness	ns	ns	ns	ns	ns	ns	ns	ns	ns	ns	ns	ns	ns
Chewiness	ns	3^{0}	ns	ns	ns	ns	ns	4^{0}	3^{0}	ns	ns	3^{0}	58^{000}
WHC	33^{000}	24^{000}	45^{000}	25^{000}	36^{000}	29^{000}	27^{000}	23^{000}	25^{000}	20^{000}	21^{000}	28^{000}	ns
Dry matter	ns	18^{000}	5^{0}	ns	12^{000}	4^{0}	8^{0}	12^{000}	11^{000}	44^{000}	18^{000}	23^{000}	ns
Protein	ns	18^{000}	7^{00}	4^{0}	12^{000}	ns	15^{000}	26^{000}	18^{000}	47^{000}	11^{000}	39^{000}	5^{0}
Fat	ns	ns	4^{0}	11^{000}	ns	ns	ns	4^{0}	ns	ns	ns	6^{00}	9^{00}
Ash	ns	ns	ns	ns	11^{000}	ns	ns	ns	ns	ns	ns	ns	ns
Brightness	11^{000}	ns	ns	ns	ns	11^{000}	ns	11^{000}	5^{0}	ns	5^{0}	7^{00}	ns
Bovine calves													
Hardness	ns	ns	ns	5^{0}	ns	9^{00}	ns	ns	ns	ns	ns	4^{0}	ns
Chewiness	ns	ns	ns	3^{0}	ns	8^{00}	4^{0}	ns	ns	ns	ns	ns	ns
WHC	21^{000}	22^{000}	33^{000}	10^{000}	23^{000}	10^{000}	11^{000}	29^{000}	17^{000}	32^{000}	21^{000}	14^{000}	25^{000}
Dry matter	22^{000}	ns	22^{000}	16^{000}	26^{000}	7^{00}	15^{000}	28^{000}	6^{00}	4^{0}	3^{0}	36^{000}	5^{0}
Protein	33^{000}	14^{000}	44^{000}	15^{000}	32^{000}	4^{0}	25^{000}	59^{000}	53^{000}	17^{000}	5^{0}	58^{000}	38^{000}
Fat	7^{00}	5^{0}	12^{000}	4^{0}	ns	ns	ns	7^{00}	4^{0}	6^{00}	ns	11^{000}	13^{000}
Ash	ns	ns	ns	ns	ns	ns	ns	ns	3^{0}	ns	ns	5^{0}	5^{0}
Brightness	ns	5^{0}	ns	8^{00}	ns	4^{0}	ns	4^{0}	5^{0}	6^{00}	12^{000}	ns	ns

[1] ns = not significant; ^{0} = P <0.05; ^{00} = P <0.01; ^{000} = P <0.001.

in relation to the three considered ages.

Variance analysis results (Tables 3 and 4, Figure 2) indicate that age affects significantly the chemical composition of meat: animals slaughtered at 20 weeks of age tend to have a meat with higher dry matter and protein and less fat than at 28 and 36 weeks (P <0.01). At 28 weeks the meat has the lowest dry matter and protein content, while the fat has intermediate values between 20 and 36 weeks.

At a given age, the Ss muscle tends to have the highest content of dry matter, protein and fat, the opposite is true for dry matter and protein content of VM muscle. The significant interaction between age and muscle for the three abovementioned variables would indicate that the muscle does not always reach the same rank with the varying of age.

When all the muscles are considered (Table 3 and Figure 1) meat colour, measured as brightness, ie, the percentage of reflectance compared to white standard or 'visual brightness' (average eye response), tends to become darker with age, showing a brightness decrease of about 23% (P <0.01) between 20 and 28 weeks and of about 13% (P <0.01) between 20 and 36 weeks. The single muscle has the same behaviour with a brightness decrease of 17 - 24%, (P <0.05 - 0.01) between 20 and 28 - 36 weeks: this is especially true for the following muscles: CLaTB, GM, PPGb, Sm and LD. At a given age the PM muscle always gives darker meat whereas the opposite is observed with the St muscle. The absence of significant interaction between age and muscle, suggests that the rank scale of muscles is invariant in relation to age. Within each muscle the reflectance values, measured at 426 - 684 nm are higher at 20 weeks.

Bovine calves

Bovine calves reacted in the same way as water buffalo for numerous characteristics, consequently we report only the results which are specific to this species (Tables 3 and 4; Figures 1 and 2).

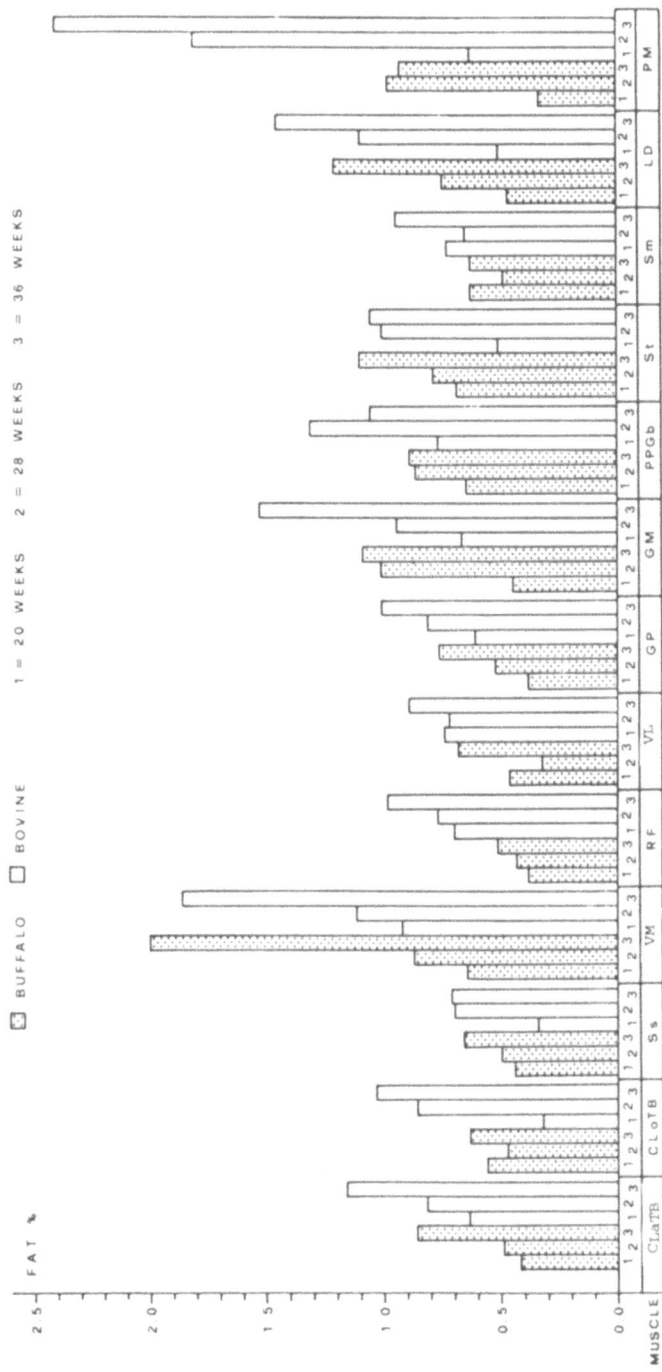

Fig. 2 Mean values of fat content

When the contribution of all muscles to the total meat weight
of the side (which was the same as observed in water buffalo) is
considered, the age increase does not statistically affect the
hardness, while the opposite is observed for the chewing energy
which does not vary between 20 and 28 weeks and increases of
about one-third from 20 - 28 weeks to 36 weeks of age (Tables 3
and 4).

The WHC increases significantly about one-third and two-fifth
from 20 to 28 and from 20 to 36 weeks respectively. The muscles
CLaTB, Ss and LD show the highest WHC.

At a given age, the PM furnishes the most tender meat and
requires the least chewing energy and has the lowest WHC; the
opposite is true for St muscle. The muscle tends to preserve
a marked individuality throughout the growth.

Chemical composition of bovine meat varies significantly in
relation to slaughter age as already observed for the water
buffalo. At a given age, the St muscles have the highest content
of dry matter and protein and the Ss the least of these two
variables as well as of fat (Figure 2). The colour of bovine
meat varies with age as already observed for water buffalo: the
brightness decreases about 22% (P < 0.01) when age increases from
20 to 28 weeks and 14% (P < 0.01) when age increases from 20 to
36; the same results are obtained when the single muscle (and
particularly CLoTB, VM, VL, GM, GP, PPGb, St and Sm) is considere
The rank muscle is invariant in relation to age.

CONCLUSIONS

The results have shown the importance of the age in determining the values of myorheological, chemical and colour characteristics of meat obtained from water buffalo and bovine calves
raised and slaughtered under the same micro-environmental conditions (type of housing, feeding, slaughtering, carcass ageing,
etc). The practical use of the results is without question, of
course, within the three slaughter ages considered.

TABLE 5

RESULTS OF VARIANCE ANALYSIS WITHIN AGE

Variable	Age at slaughter (weeks)								
	20			28			36		
	F [1]								
	Species,S	Muscle,Mu	Interaction S-Mu	Species,S	Muscle,Mu	Interaction S-Mu	Species,S	Muscle,Mu	Interaction S-Mu
Hard-ness	5^{0}	13^{000}	ns	ns	15^{000}	ns	12^{000}	16^{000}	ns
Chewi-ness	ns	10^{000}	ns	ns	10^{000}	ns	7^{00}	6^{000}	ns
WHC	ns	4^{000}	ns	26^{000}	5^{000}	ns	9^{00}	5^{000}	ns
Dry matter	39^{000}	46^{000}	17^{000}	ns	7^{000}	ns	7^{00}	8^{000}	ns
Protein	12^{000}	44^{000}	12^{000}	ns	7^{000}	3^{00}	ns	11^{000}	ns
Fat	46^{000}	11^{000}	8^{000}	29^{000}	6^{000}	ns	21^{000}	7^{00}	2^{0}
Ash	ns	ns	ns	5^{0}	2^{00}	ns	ns	ns	ns
Bright-ness	16^{000}	6^{000}	ns	ns	2^{00}	ns	23^{000}	6^{000}	ns

[1] ns = not significant; ^{0} = P <0.05; ^{00} = P <0.01; ^{000} = P <0.001.

For a better knowledge of meat quality variation with age, and for an objective choice of the best slaughtering age, it is suggested that the study should be completed taking into consideration the ages of 44 and 52 weeks.

All the results show that the two species tend to have the same behaviour for many characteristics. In both species the muscle individuality is the determining factor for all the variables studied, indicating that the muscles tend to have the same rank independently from age and species (Table 5).

By increasing the slaughtering age and considering the weighted mean of 13 muscles, we obtain a meat which: 1) tends to vary very little in its hardness within the two species; 2) in the water buffalo and bovine calves at 36 weeks the meat requires about 20 and 30% more chewing energy respectively than at 20 - 28 weeks; this means that the chewiness is invariant from 20 - 28 weeks; 3) the meat increases its WHC but with differ ent behaviour in the two species: (in water buffalo the variatio (40%) takes place from the first to the second - third slaughter age while in the bovine calves from the first to second (30%) and from the second to the third (14%), with a total (over 40%) between the two extreme ages);4) at 20 weeks, in both species, meat is richer in dry matter and protein and less fat than at 28 and 36 weeks; fat increase is related to the heterogenic growth; 5) the meat tends to become darker, decreasing its brightness over 20% from the first to the second stage and about one-seventh from first to third and this is true within both species. This behaviour is the same for each muscle.

At a given age, in both species, the PM muscle gives meat which is the darkest, the most tender (requiring the least chewing energy) and has the least WHC.

By comparing the two species it is evident that: 1) both the species tend to have the same behaviour in relation to age; 2) at a given age water buffalo, compared to bovine calves, furn ish more tender meat (requiring less chewing energy) with higher WHC, with slightly less dry matter and fat, with more protein content (only at 28 and 36 weeks) brighter (whether measured as brightness or as reflectance at wavelengths from 426 to 684 nm); this is true for each of the 13 muscles as well as for all of them considered together: 3) on the basis of the various variables considered in the present study, the water buffalo meat is of higher quality than the bovine meat, and, moreover, at the muscular level, it is also more uniform. However, to complete these data and for a better evaluation of the matter, further studies should be made in co-operation with those people involved in breeding, slaughtering and cutting.

Finally, the results confirm that the methods used to evaluate the different qualitative characteristics can be considered entirely satisfactory is relation to the genetic improvement, technology and marketing of meat.

REFERENCES

Bordi, A., Colatruglio,P. and Girolami, A. 1974 Composizione chimica di
alcuni muscoli in vitelloni appartenenti a 4 tipi genetici. Prod. Anim.
14, 123-133

Cosentino, E., Colatruglio, P., Girolami, A., Bordi, A., Mammarella, V. and
Matassino, D. 1976a Beef carcass evaluation with 'Texturometer' regard-
ing the meat ageing period and breed. EEC Agric. Res. Seminar, Crit.
Meth. Assess. Carcass Meat Charact. Beef Prod. Exper., Zeist, (1975)
EUR 5489, 339-352.

Cosentino, E., Borghese, A., Rubino, R. and Cavallo, F. 1976b Studio comparativo fra bufali e bovini sulla qualità della carne. III Composizione
chimica all'età di 20 settimane. Atti II Conv. Naz. ASPA, Bari, 401-413

Cosentino, E., Carena, A., Colatruglio, P. and Matassino, D. Valutazione
di alcuni aspetti qualitativi della carne di tacchino. I. Mioreologia
Genet. Agr. (in print)

Matassino, D., Cosentino, E. and Bordi, A. 1974 Quantizzazione di alcuni
aspetti qualitativi della carne bovina con l'impiego della tecnica
tessurometrica. Atti IX Simposio Int. di Zootecnia, Milano

Matassino, D., Bordi, A ., Colatruglio, P., Cosentino, E., Casalini, F. and
Chiericato, G. 1975a Alcune caratteristiche fisiche rilevate con la
technica tessurometrica su muscoli di carcasse di 10 tipi genetici
bovini. Genet. Agr., 29, 11-22

Matassino, D., Cosentino, E., Bordi, A. and Colatruglio, P. 1975b Sulla
composizione chimica di muscoli di carcasse di 10 tipi genetici bovini.
Genet. Agr., 29, 23-32

Matassino, D., Pilla, A.M., Cosentino, E. and Bordi, A. 1976a Utility of
ultrasonic technique 'in vivo' for beef carcasses. EEC Agric. Res.
Seminar, Crit. Meth. Assess. Carcass Meat Charact. Beef Prod. Exper.
Zeist (1975), EUR 5489, 71-80

Matassino, D., Girolami, A., Colatruglio, P., Cosentino, E., Votino, D. and
Bordi, A. 1976b Colour evaluation of muscles in 10 bovine crossbreeds.
EEC Agric. Res. Seminar, Crit. Meth. Assess. Carcass Meat Charact. Beef
Prod. Exper., Zeist (1975). EUR 5489, 285-299

Matassino, D., Romita, A., Colatruglio, P. and Bordi, A. 1976c Studio comparativo fra bufali e bovini sulla qualita della carne. I. Caratteristiche mioreologiche all'eta di 20 settimane. Atti II Conv. Naz. ASPA,
Bari, 373-386.

Matassino, D., Romita, A., Girolami, A. and Cosentino, E. 1976$_d$ Studio comparativo fra bufali e bovini sulla qualità della carne. II. Colore della carne all'età di 20 settimane. Atti II Conv. Naz. ASPA, Bari, 387-400

Matassino, D. and Pilla, A.M. 1976 Genetica e miglioramento degli ovini. Atti II Conv. Naz. ASPA, Bari, 229-262

Romita, A., Borghese, A. and De Maria, C. 1976 Prova comparativa per vitelli bovini e bufalini allevati fino a 20 settimane di età. I. Accrescimento indice di conversione, resa al macello e caratteristiche della carcassa Atti II Conv. Naz. ASPA, Bari, 349-362 Ann. Ist. Sper. Zootec. $\underline{9}$ (1) 79-92

Romita, A., Gigli, S. and Borghese, A. Prova comparativa fra vitelli bovini e bufalini allevati fino a 28 settimane di età. I. Accrescimento, indice di conversione, resa al macello e caratteristiche della carcassa. Ann. Ist. Sper. Zootec. (in print)

Romita, A., Gigli, S. and Borghese, A. Prova comparativa fra vitelli bovini e bufalini allevati fino a 36 settimane di età. I. Accrescimento, indice di conversione, resa al macello e caratteristiche della carcassa. Ann. Ist. Sper. Zootec. (in print)

ASSESSMENT OF CHANGES IN MYOFIBRE SIZE IN MUSCLE

M.J. Clancy and P.D. Herlihy

The Agricultural Institute, Castleknock, Co. Dublin,
and Economics and Rural Welfare Centre, Dublin 4, Ireland

ABSTRACT

This paper reviews and evaluates approaches to the measurement of myofibre cross-sectional area (CSA). We have found that, based on traverse cryosections, accurate estimates of mean myofibre CSA can be obtained using the formulae: $\frac{\pi}{4}D.d$ and $\frac{\pi}{4}D^2_{ea}$, where D is the largest caliper diameter, d the smallest caliper diameter, and D_{ea} the equivalent area diameter. Estimates based solely on the larger or smaller diameter are most inaccurate. The distribution patterns of myofibre CSAs are influenced by the type of measurement employed. The effects due to the angle at which the sections are cut, on estimates of myofibre CSA need not be significant. A large change in myofibre CSA is caused by the shortening or lengthening of the myofibre. Since this effect is considerable and has hitherto been ignored in myofibre sizings, we recommend that, in order to make valid comparisons between the sizes of various myofibres, both the CSA and the sarcomere length be measured and used together to quantify myofibre size in the manner illustrated in this paper.

INTRODUCTION

The essentials of the cellular aspects of muscle growth
and development in meat animals are (a) myofibre numbers,
(b) myofibre cross-sectional area (CSA) and length, and (c) the
differentiation of myofibres into their various morphological
and histochemical types. All these are interrelated and in
particular aspects (b) and (c).

The determination of the absolute number of myofibres in
a muscle is still a major unresolved problem. Swatland (1976)
has recently given an excellent discussion of this area.

There are many approaches to myofibre sizing without any
reliable information on the significance of differences between
the various measurement procedures. This point becomes all the
more important when data obtained by different methods are compared. This paper, therefore, considers firstly, the definition
of myofibre size; secondly, methods for its direct measurement
or estimation; and thirdly, factors which influence the apparent
size of a myofibre.

MATERIAL AND METHODS

Samples of bovine *Longissimus lumbris* and *Psoas major* muscles
were excised 0.5 h and 48 h post mortem, frozen and stored in
liquid nitrogen.

In studying the effects of lengthening and shortening on
myofibre CSA, pairs of bovine *stearnomandibularis* muscles were used.
One of the pair was stretched, either fully or partially, while
its contralateral was shortened by immersion in ice-water for
7 h. Rigor was allowed to develop at 10° for 48 h, when samples
taken at the mid-point of each muscle were frozen and stored as
above.

Transverse and longitudinal cryosections (10 μm thickness)
were cut from all samples, and stained with haematoxylin and
eosin (Dubowitz and Brooke, 1973).

For sarcomere length determinations, the longitudinal cryosections were projected on to white paper at a final magnification of X 1 300 and the number of A-bands per 100 μm were counted along each of 75 myofibres in a muscle sample.

The largest caliper diameter (D, Fig. 1a) and the smallest caliper diameter (d, Fig. 1a) were measured on images of transverse cryosections projected as above. 150 myofibres were measured per sample.

Direct measurements of myofibre CSA were made by the method of cutting out and weighing projected images (Delesse, 1848).

Photomicrographs of transverse sections were taken at either X 50, X 100 or X 200 according to the size of the myofibre CSA, and prints, enlarged to 20.5 x 32 cm, were made. The equivalent area diameter (D_{ea}, Fig. 1b) was measured on these prints using a Model TGZ3 Zeiss Particle Size Analyser (ZPSA).

In measuring the angles of cut, cubic pieces (seven) of frozen muscles were carefully aligned by setting the readily discernible longitudinal fasiculi at right angles to the cutting plane (the x, y plane, Fig. 2). After some transverse sections of the myofibres had been cut, the frozen piece of muscle was re-set at approximately 90° to the first position. Cryosections were next cut along the length of the myofibres, ie in the y, z plane. After another similar resetting, cryosections were cut along the length of the myofibres in the x, z plane. These cryosections were stained as above and projected on to white paper. Using the longitudinal sections, a straight line, representing the x, y plane, was drawn along the edge of the myofibres. Lines were next drawn at three different positions along the length of the myofibres. The angles these lines made with the x, y line are values for θ and ψ in the case of the y, z and x, z cryosections respectively.

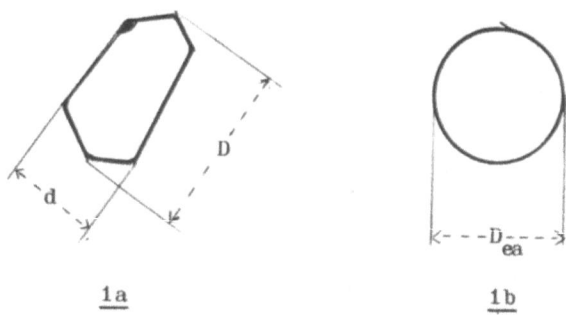

1a   1b

Fig. 1. Dimensions of a myofiber cross-section: 1a, myofiber cross-section:
         D, largest caliper diameter: d, smallest caliper diameter: 1b,
         circle equal in area to 1a: D_{ea}, equivalent area diameter.

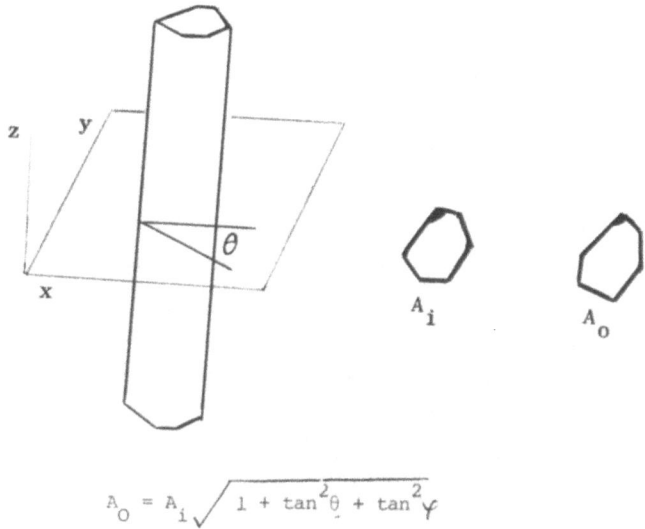

$$A_0 = A_i \sqrt{1 + \tan^2\theta + \tan^2\psi}$$

Fig.2. The effect of obliquity on CSA

RESULTS AND CONCLUSIONS

The measurement of myofibre size

Many workers use transverse sections of muscle as the
basis of their measurement. For this approach cryosections are
to be preferred since the myofibres are not subjected to the
shrinkage effects of dehydration and embedding as is the case
with chemically fixed and embedded muscle. For this reason
cryosections have been used in the present work.

Myofibres have an irregular shape, as is indeed evident
when viewed in transverse section. Therefore, the only satisfactory measurement of radial size is the CSA. Methods for
determining the CSA of a myofibre can be classified into two
main types, ie (a) the area, or some property of it, is directly
measured, and (b) the area is calculated from measured dimensions
of the myofibre. Among the first are the direct area measurement of the projected image of the myofibre using a planimeter
or squared paper (Sissons, 1965), and the calculation of the
area from the weights of the image projected on to uniform paper.
In the second approach, various 'diameters' are measured and
throughout the literature there appears an indiscriminate use
of a variety of 'diameter' types. Furthermore, to many workers
myofibre size and 'diameter' are synonymous. A diameter in the
strict sense of the word can be assigned to a two-dimensional
feature only when it is circular. It should not, therefore, be
used to describe myofibre CSA. Some workers measure what corresponds to the largest caliper diameter (Bell and Conen, 1967)
and others the smallest caliper diameter (Dubowitz and Brook, 1974).
Combinations of these diameters, eg the arithmetic (Edström and
Torlegard, 1969) or geometric mean (this paper) are sometimes
used. The equivalent area diameter, which is the diameter of a
circle having the same area as the figure is another concept
(Fig. 1). In many cases the type of diameter used is not specified. The situation is, therefore, most confusing.

We first studied the problem of measuring myofibre CSA
using a photomicrograph of a typical cross-section of a muscle.

In order to establish a reliable basis with which to make comparisons, the CSAs were determined directly. The largest caliper diameter D, the smallest caliper diameter d, the equivalent area diameter D_{ea} were also measured (Fig. 1). The equivalent area diameter is an extremely useful dimension to measure. It is based on the integrating capacity of the human eye and with a Zeiss Particle Size Analyser (ZPSA), the image of an iris diaphragm can be superimposed on the myofibre cross-sectional area of a suitable enlarged photomicrograph to give accurate areal matching.

TABLE 1

AREAS AND THEIR DISTRIBUTION PATTERNS (N = 76)

	Method	Mean Area (Sq. Units)	CV%	Skewness	Kurtosis
1	$\frac{\pi}{4}D^2$	166.9 a[1]	33.2	1.35***[2]	6.17***
2	$\frac{\pi}{16}(D+d)^2$	108.0 b	24.9	0.80**	4.16***
3	$\frac{\pi}{4}.D_{ea}^2$	102.3 c	21.7	0.97[3]	5.25[3]
4	$\frac{\pi}{4}.Dd$	100.8 c	24.0	0.71**	3.45
5	Direct	100.0 c	22.7	0.72**	3.60
6	$\frac{\pi}{4}.d^2$	62.5 d	33.7	0.97**	3.24

[1] Figures followed by different letters are significantly different at P< 0.05

[2] * P<0.05;   ** P<0.01;   *** P<0.001

[3] Under investigation

Areas based on these dimensions were calculated as set out in Table 1, where the different estimates for the mean value for the myofibre CSA are compared with the value obtained directly. Of these six methods, only two are as accurate as the direct method. The area obtained using the equivalent area diameter is as accurate and as equally precise as the direct method. While the ZPSA has been used in measuring myofibre diameter (Millar, Garwood and Judge, 1975) the present study is, as far as we are aware, the first time such an instrument has

been used and evaluated for determing myofibre cross-sectional
areas. The areas based on the formula $\frac{\pi}{4}D.d$ is also an accurate
estimate of cross-sectional area. Estimates of cross-sectional
areas based solely on the larger diameter or on the smaller diameter are both inaccurate and not as precise as the direct method.

Of particular interest is the effect of mode of measurement
on the distribution characteristics, ie the moments coefficient
of skewness and Kurtosis. It appears that the method used for
measuring or estimating the cross-sectional area influences significantly the distribution pattern. Methods 3, 4 and 5 (Table 1)
which are considered best because of their accuracy and precision,
also lead to an equally significant positive skewness. While
the Kurtosis for methods 4 and 5 are not significantly different
from that of a normal distribution, the peakedness of the result
based on the equivalent area diameter is still under investigation. In view of these observations, caution needs to be exercised in attributing biological signifance to particular distribution patterns. We further suggest that, since histograms
are subjective, distribution data should be presented in terms
of skewness and Kurtosis, which are objective. The coefficient
of variation may be the most useful term of all.

The effect of obliquity

While transverse cryosections appear to be an ideal basis
for estimating mean myofibre CSA, it has been pointed out that
if a myofibre is cut obliquely, the exposed cross-section is
not a true cross-section of the myofibre but a distortion of it.
This distortion could lead to a large error in myofibre sizing.
To correct for this, it has been variously suggested that (a)
the smaller diameter (Dubowitz and Brooke, 1973), (b) the larger
diameter (Bell and Conen, 1967), and (c) the diameter of round
myofibres only (Fenichel, 1963) be measured. None of these
viewpoints has ever been supported by experimental evidence.
Thus, two problems, ie the measurement of CSA and the effect of
obliquity, have been confounded.

Apropos the measurement of CSA, Table 1 shows the extent to which the greater diameter overestimates and the lesser diameter underestimates the CSA, while the measurement of only round myofibres is obviously biased. It remains, therefore, to quantify the effects of obliquity.

In our study of the effect of obliquity on myofibre CSA we used two approaches. In the first, a mathematical model was used in which A_i, the CSA of a myofibre cut at right angles to its length, ie the ideal section, was related to A_O, the CSA of a section cut at an angle θ in the y,z plane (Fig. 2) and at the angle ψ in the x,z plane, ie the oblique section, by the equation:-

$$A_o = A_i \sqrt{1 + \tan^2 \theta + \tan^2 \psi} \qquad (1)$$

Using this equation the effect of angle of cut on the measured CSA of the exposed cross-section can be calculated for different assumed cutting efficiencies - cutting efficiency is defined as the equal likelihood of a section being cut at any angle up to x°. Various cutting efficiencies were investigated. The results are summarised in Table 2.

TABLE 2

EFFECTS OF OBLIQUITY

Cutting Efficiency	Mean Area (Sq. units)	CV%	Skewness	Kurtosis	N
0°	100.00	-	-	-	-
5°	100.13	0.12	0.59	5.69**	21
10°	100.52	0.47	0.64	2.09	41
15°	101.17	1.06	0.65*	2.11	61
20°	102.11	1.89	0.67*	2.15	81
25°	103.35	2.99	0.69*	2.20	101

* $P < 0.05$ ** $P < 0.01$

For a cutting efficiency with a maximum angle of $25°$, a value which is unlikely in practice, the mean CSA had increased by only 3% while the CV was 3%. The mean CSA decreases as the angle of cutting efficiency decreases. Skewness does not vary significantly, and when θ is greater than $7.5°$, Kurtosis stabilises. The distortions are thus very small indeed. Variation in ψ does not change these general conclusions.

In our second approach to the effect of obliquity, we measured the oblique angles θ and ψ. The results of an analysis of variance of these data are given in Table 3.

TABLE 3

ANALYSIS OF VARIANCE FOR OBLIQUE ANGLES θ AND ψ

Source of Variation	DF	Mean Square	F-ratio
Muscle Piece (P)	6	34.39	4.24**
Angle (A)	1	125.15	15.43***
Replication	2	1.01	
Interaction : PXA	6	52.21	6.44***
Error	26	8.11	

** $P < 0.01$ *** $P < 0.001$

θ differs significantly from ψ. In addition there is a significant variation from piece to piece as well as piece X angle interaction. More importantly, the overall mean values for θ and ψ were respectively $2.9°$ and $6.4°$. The angles of cut can thus be small, and, as a consequence, the effects of obliquity are very small indeed. In relation to the large intrinsic variance of myofibre CSAs, the effects of obliquity on the accuracy and precision of estimates of mean myofibre CSA are of no consequence whatsoever when the cryosections are cut properly. There is, however, an effect of skewness which is significant.

Myofibre CSA and sarcomere length

A highly significant negative correlation between myofibre 'diameter' and sarcomere length has been established (Herring et al., 1965). Nevertheless, the implication of this fact on the validity of myofibre CSA measurements, which are made without reference to their sarcomere length, has hitherto been ignored. Indeed, the exact relationship between myofibre CSA and sarcomere length for a given myofibre does not appear to have been formulated.

The volume of a myofibre is related to its length (l) by the formula:-

$$V = CSA \times l$$

which on rearrangement becomes:-

$$CSA = V \times l^{-1} \qquad (2)$$

Thus when a myofibre lengthens or shortens isovolumically the CSA changes hyperbolically with the sarcomere length.

When, for example, the sarcomere length shortens from 2.5 μm to 1.0 μm the cross-sectional area increases 2.5 times.

Equation (2) leads to a family of hyperbolic curves depending on myofibre volume. To illustrate the effect of shortening or lengthening on myofibre CSA, a pair of bovine *stearnomandibularis* muscles were used. One of these muscles was shortened, while the contralateral was lengthened. Fig. 3a and b are photomicrographs, taken at equal magnification, of the transverse cryosections of these muscles. These illustrate very clearly the effect of shortening and lengthening on myofibre cross-sectional area. In addition the shortened myofibres are more dense. This difference in density has important consequence in the interpretation of the results of myofibre histochemistry. Note also the intermyofibre spaces in the shortened but not in the lengthened muscle. Data obtained from the cryosections of these muscles are shown in Table 4, from which it can be seen that the mean myofibre CSA of the shortened muscle is about twice that of the lengthened muscle. This apparent difference in myofibre size in muscles, which are in fact equivalent, can

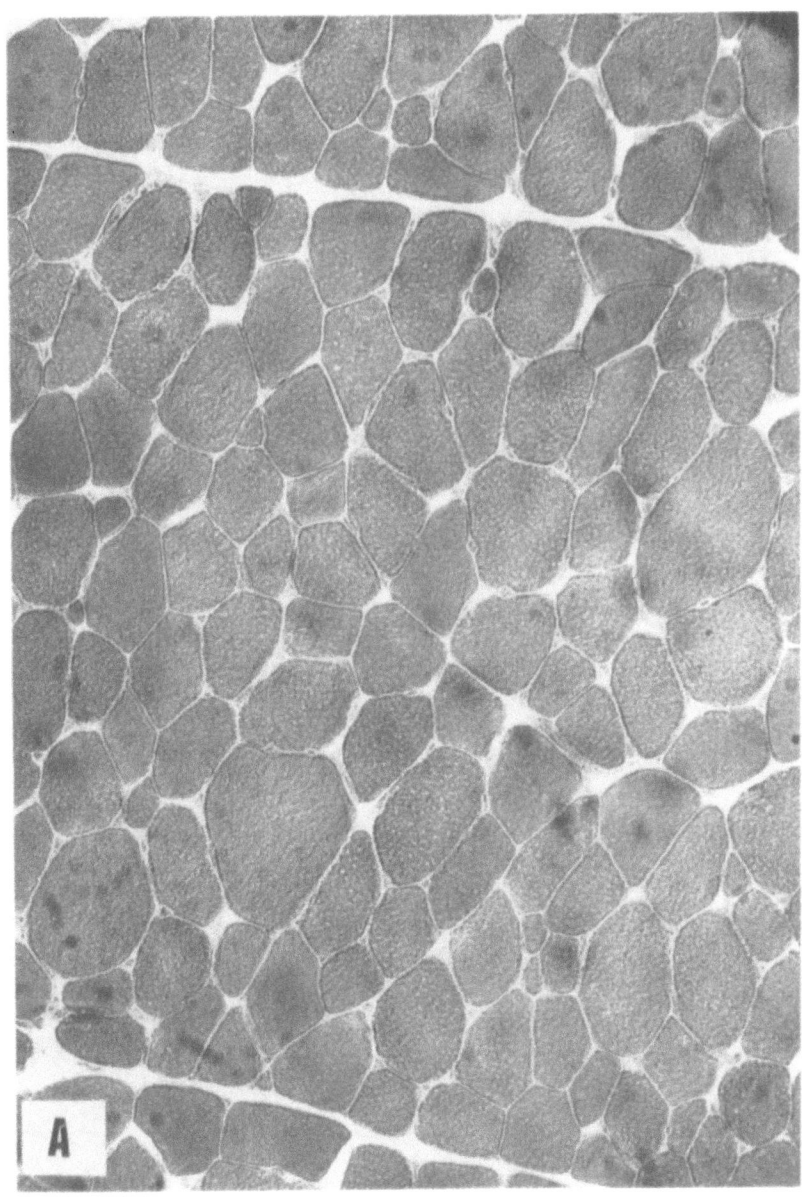

Fig. 3. Transverse 10 μm cryosections of shortened (A) bovine stearnomandibularis muscles from the same animal. Haematoxylin and eosin,
x 75

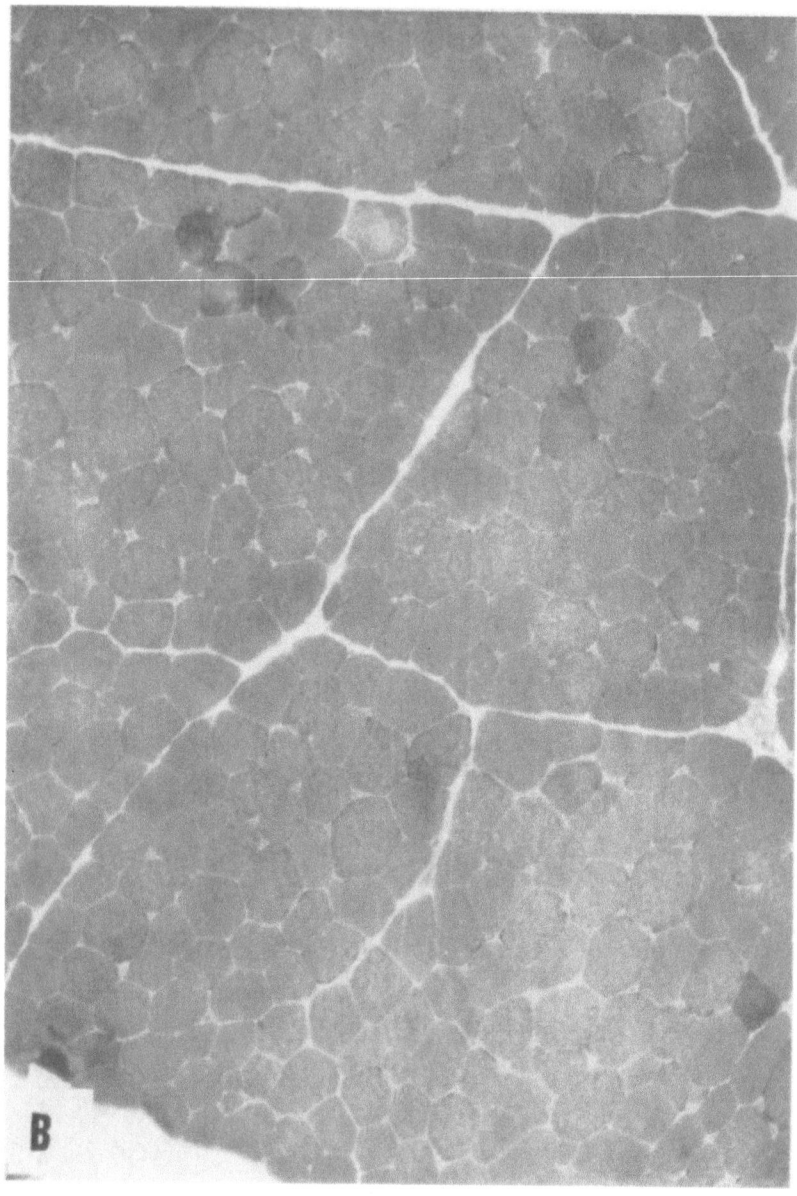

Fig. 3. Transverse 10 μm cryosections of lengthened (B) bovine stearnomandibularis muscles from the same animal. Haematoxylin and eosin,
x 75

be resolved by multiplying the CSA by the sarcomere length (l_s) to give a standard micro-disc volume. The observed mean CSA may also be adjusted to correspond to a standard sarcomere length of say 2.5 μm.

Thus, sarcomere length has enormous but calculable, and therefore manageable, effect on the cross-sectional area of a myofibre, when shortening or lengthening takes place iso-volumically. In cases where shortening or lengthening takes place without isovolumic behaviour, as has been suggested when a myofibre enters rigor mortis, this volume loss, if it occurs, has to be considered. Nonetheless, the fact remains that it is impossible to make valid comparisons of different myofibre CSAs without reference to their sarcomere length.

TABLE 4

EFFECT OF LENGTHENING AND SHORTENING ON MYOFIBRE SIZE

	Means ± SE		
	Shortened muscle	Lengthened muscle	Test of significance
Cross-sectional area (μm^2, N = 150)	2 348.9±106.8	1 021.9±37.2	P <0.001
Sarcomere length (μm, N = 75)	1.51 ± 0.02	3.52 ± 0.03	P <0.001
Standard micro-disc vol. (μm^3)	3 546.8	3 597.1	
CSA adjusted to l_s = 2.5 μm	1 418.7 μm^2	1 438.8 μm^2	

It appears, therefore, that what is regarded as myofibre hyper- and hypo-trophy may not be valid expressions of myofibre size; rather they may be associated with a long or short sarcomere length. The relationship observed between myofibre 'diameter' and meat tenderness (Tuma et al., 1962) may, likewise, be reflections of shortened or lengthened myofibres. Thus, to des- cribe myofibre size fully, both CSA and l_s should be measured.

This concept of micro-disc volume has been used in studying the behaviour of bovine *stearnomandibularis* muscles in rigor mortis. Comparisons were made with each pair of muscles, one of which was lengthened while its contralateral was shortened (Table 5). The micro-disc volume was not affected by the degree of contraction or relaxation in the sarcomere length range 1.22 μm to 3.75 μm. The data of Table 5 give general support to the concept of micro-disc volume as a valid index of myofibre size.

A thaw shortening of longitudinal cryosections, cut from pre-rigor muscle, takes place on the pick-up slide. While this effect can be inhibited by the use of EDTA coated slides (Pearson and Sabarra, 1974), it is not completely eliminated (Beerman, 1976). A similar situation can be expected with transverse cryosections from pre-rigor muscle and the CSA would, therefore, be enlarged. It would thus appear to be extremely difficult, if not pointless, to standardise the degree of stretch in a piece of biopsy muscle before freezing, as is the practice of some workers. It becomes, therefore, all the more important to measure both the CSAs and the sarcomere lengths of biopsy myofibres. Indeed, unless these two measurements are made, there is no means of knowing whether the estimates of myofibre sizes, which are obtained using biopsy material, are valid or not.

TABLE 5

STANDARD MICRO-DISC VOLUMES OF BOVINE *Stearnomandibularis* MUSCLES

| Animal No. | Standard micro-disc volume (μm^3) | | Ratio of left to right |
	Left muscle	Right muscle	
1	1 621.8 (3.22)	1 506.2 (3.75)	1.077
2	2 036.8 (1.98)	2 145.8 (2.71)	0.949
3	3 178.2 (2.73)	3 289.4 (3.38)	1.004
4	3 379.1 (1.22)	3 325.2 (1.94)	1.013
5	3 306.3 (2.31)	3 471.1 (3.64)	0.952
6	3 499.9 (1.49)	3 484.7 (3.41)	1.004
7	3 543.9 (3.11)	3 393.7 (2.37)	1.044
8	3 623.5 (3.30)	3.349.7 (1.23)	1.082
9	4 144.0 (2.70)	4 345.8 (3.47)	0.954
Mean	3 146.9	3 134.6	1.003

Nos. in brackets are the mean sarcomere lengths in μm

REFERENCES

Bell, C.D. and Conen, P.E., 1967. Change in fibre size in Duchenne muscular dystrophy. Nemology 17: 902-913.

Beerman, D.H., 1976. The neural control of skeletal muscle fiber type. Ph.D thesis. University of Wisconsin, USA. p.68.

Delesse, A., 1848. Procédé mécanique pour déterminer la composition des roches. Ann. Mines 13: 379-388.

Dubowitz, V. and Brooke, M.H., 1973. Muscle Biopsy: A Modern Approach. W.B. Saunders Co. Ltd., London, Philadelphia and Toronto.

Edström, L. and Torlegard, K., 1969. Area estimation of transversely sectioned muscle fibres. Z. wiss. Mcks. 69: 166-178.

Fenichel, G.M., 1963. The B fiber of human fetal skeletal muscle. Neurology (Minneap.) 13: 219-226.

Herring, H.K., Cassens, R.G. and Briskey, E.J., 1965. Further studies on bovine muscle tenderness as influenced by carcass position, sarcomere length and fiber diameter. J. Fd Sci. 30: 1049-1054.

Millar, L.B., Garwood, V.A. and Judge, M.D., 1975. Factors affecting porcine muscle fiber type, diameter and number. J. Anim. Sci. 41: 66-77.

Pearson, J., and Sabarra, A., 1974. A method for obtaining longitudinal cryostat sections of living muscle without contraction artifacts. Stain Tech. 49: 143-146.

Sissons, H.A., 1965. Further investigations of muscle fibre size. In: Research in Muscular Dystrophy. Eds. Research Committee of the Muscular Dystrophy Group. Pitman Medical Publishing Co. Ltd., London. 107-118.

Swatland, H.J., 1976. Recent research on postnatal muscle development in swine. Proc. 29th Annual Recip. Meat Conf. 86-104.

Tuma, H.J., Venable, J.H., Wutheir, P.R. and Hendrickson, R.L., 1962. Relationship of fiber diameter to tenderness and meatiness as influenced by bovine age. J. Anim. Sci. 21: 33-36.

PARTITION AND DISTRIBUTION OF FATTY TISSUES

D.R. Williams,

Meat Research Institute, Bristol BS18 7DY. UK.

ABSTRACT

Differences between dairy and beef types in the way in which fat is partitioned into subcutaneous (SCF), intermuscular (IMF) and kidney knob and channel fat (KKCF) in relation to total carcass fat are illustrated. Data from the dissection of 56 Friesian and 63 Hereford beef sides provide the main source of material and the emphasis throughout the paper is upon the assessment of fatness. It is demonstrated that both SCF and KKCF should be taken into account in the assessment of fatness, but that differences in the distribution of SCF and IMF within these two depots are not an important source of breed differences and are relatively unimportant in the assessment of fatness.

INTRODUCTION

It is known that as cattle grow, the order of deposition of fat is kidney knob and channel fat (KKCF) and intermuscular fat (IMF), followed by subcutaneous fat (SCF). Most recently, Johnson et al. (1972) utilised dissection data from four groups of animals which illustrate this principle. The partition of dissectable fat between SCF, IMF and KKCF in these groups of animals is shown in Table 1 and even in this heterogenous material, the general principle is demonstrated. Johnson et al. (1972) conclude their paper with the sentence, 'There is a need for fat distribution studies to be conducted within breeds or strains of cattle'.

TABLE 1

PARTITION OF DISSECTABLE FAT BETWEEN SCF, IMF AND KKCF 1N 4 GROUPS OF ANIMALS (FROM JOHNSON, BUTTERFIELD AND PRYOR, 1972).

Groups	1	11	111	1V
Breeds,	Foetuses	Calves	Steers	Steers
Sexes	3 Friesian females	5 Friesian 3 males, 2 females	7 Angus, 3 Herefords	1 Friesian, 4 Angus
Average age	210 days gestation	28 days	355 days	1056 days
Side weights	3.1 to 3.3kg	8.6 to 22.3kg	43.7 to 80.2kg	119.0 to 179.2k
Dissectable fat as % side weight	3.4	2.5	18.3	27.7
Partition of dissectable fat between 3 depots				
SCF %	7.4	9.5	32.9	32.0
IMF %	48.1	57.5	54.2	56.8
KKCF %	44.4	33.0	12.9	11.2

It is also known that breed differences exist and over 100 years ago Lawes and Gilbert wrote that 'those breeds (of cattle) which have the greater tendency to fatten on the outer frame or carcass, have the less aptitude to do so around the internal organ

(Lawes and Gilbert, 1859). In general, cattle which have been selected for beef production (eg Aberdeen Angus and Hereford) tend to deposit fat subcutaneously (SCF) rather than in the perinephric and retroperitoneal regions (KKCF), while the opposite holds true for cattle selected for milk production (eg Ayrshire, Jersey and Friesian).

In addition to variation in the partition of fat between depots, differences in distribution (in different anatomical regions of the carcass) within the two main depots, SCF and IMF, also arise during growth as animals fatten.

THE ASSESSMENT OF FATNESS

These facts have important implications in the assessment of fatness in classification schemes, where the assessor will make his estimate of fatness on the basis of the visible fat depots. In Britain, classification is based upon the visual assessment of conformation, (which, within a fat class, serves as an indicator of breed type) and SCF alone. This is because it is common practice to remove the KKCF as part of the dressing process immediately after slaughter.

The object of classification for fatness in Britain, is to predict SCF % + IMF % in a mixed population of beef and dairy cattle and crosses between them. However, an estimate of SCF % as an adequate index of SCF % + IMF % relies to a great extent on the accuracy with which SCF % reflects the amount of IMF.

Kempster et al. (1976_1) and Kempster et al. (1976_2) have fulfilled the final appeal by Johnson et al. (1972) in making an important contribution to the study of the partition and distribution of fat in a variety of breeds and crosses under different rearing regimes. In the first of these papers, which deals with the partition of SCF, IMF, KKCF and cod fat as proportions of their total weight, some clear differences between breed/feeding systems are indicated. One example is quoted in Table 2, which serves to illustrate the differences in fat deposition which are characteristic of EXTREME beef or dairy types. The authors point out that the groups of animals came from a number of different investigations and few were compared contemporaneously. However, it is interesting to note that Hereford x Friesian, Simmental x Friesian and Friesian steers from grass/cereal systems had less (mean values) SCF and more KKCF as a proportion of their total fat, than did steers of the same breeds from more intensive cereal systems of rearing.

TABLE 2

EXAMPLE FROM KEMPSTER, CUTHBERTSON AND HARRINGTON (1976)

Breed type	Friesian x Ayshire		Angus Crosses	
Number	25		29	
Feeding regime	Grass/cereal		Suckled/Intensive	
	Mean	SD	Mean	SD
Side weight (ex KKCF) Kg	132.5	11.44	122.1	9.42
Total fat (Kg)	41.5	7.29	44.3	6.26
Total as % side weight	31.2	3.72	36.2	3.27
Depot fats as % total fat				
SCF %	26.4	3.15	36.0	2.42
Cod %	4.0	0.48	4.5	0.64
SCF + cod %	30.4	-	40.5	-
IMF %	48.3	3.21	46.6	2.34
KKCF	21.3	3.71	12.9	1.71

Differences of this kind, especially in the ratio of SCF to IMF, are important in the British classification system, because, as indicated by Harrington and Kempster (1977) of their influence upon saleable meat yield.

DEFINITIONS

It should be noted that in these studies of fat partition and distribution by Kempster et al. (1976), the dissection method adopted was that described by Cuthbertson et al. (1972). By this method, beef sides are separated into fat, lean and bone and also a fourth component, trimmings, which comprises glands, major blood vessels, ligamentum nuchae, tendons in the distal limbs and the thick connective tissue sheaths associated with the diaphragm and the abdominal muscles.

The trimmings correspond to the item labelled Other Tissues (OT) by Pomeroy et al. (1974). These trimmings, or OT, are not included with IMF in the papers by Kempster et al. (1976). However, in addition to the evidence presented by Pomeroy et al. (1974), that the most appropriate grouping of dissected tissues is lean, bone and fat + OT, where OT is included with IMF, dissection data on a further 172 beef sides (unpublished) confirm that OT should be included with IMF. In consequence, this has been done in the brief study of Friesians and Herefords which follows.

FRIESIANS AND HEREFORDS

Dissection data from 56 Friesian and 63 Hereford steers, which came from several different experiments, are used to illustrate differences in fat deposition which might be expected in a typical dairy and beef breed. The minimum side weight (including KKCF) was 85 kg for both breeds and the maximum was 170 kg for the Friesians and 164 kg for the Herefords. The basic compositional data are given in Table 3, and the inter-relationships between the various fats depots are illustrated in Figures 1 to 4.

Figures 1 and 2 show that at given levels of IMF, especially in fatter animals, there is more SCF but less KKCF in the Hereford than the Friesian. This distinction between Herefords and Friesians is illustrated most obviously in Figure 3. In Figure 4, it can be seen that the differences between the two breeds, which are so unmistakable in the fatter animals in Figures 1, 2 and 3 are much less obvious.

224

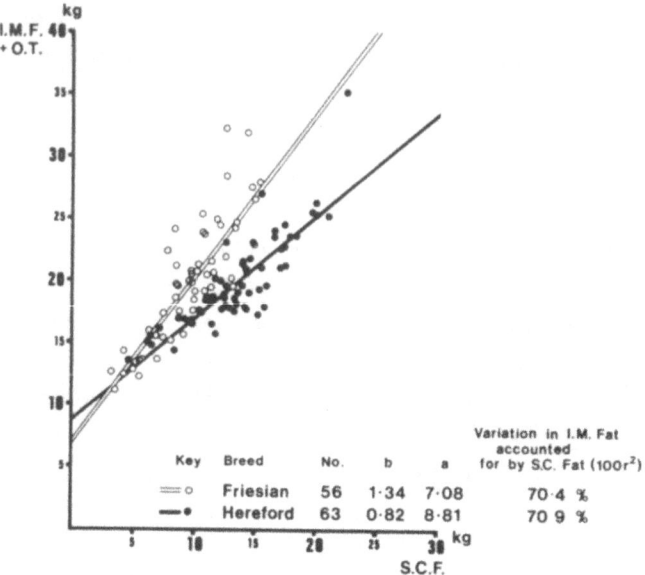

Fig 1.   SCF and IMF

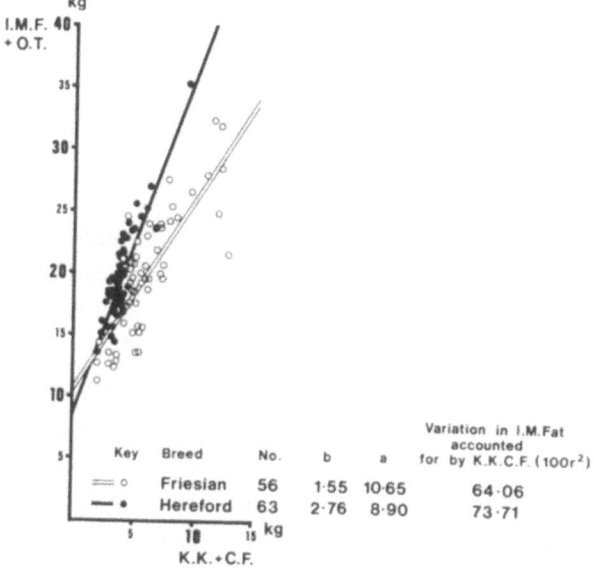

Fig 2.   KKCF and IMF

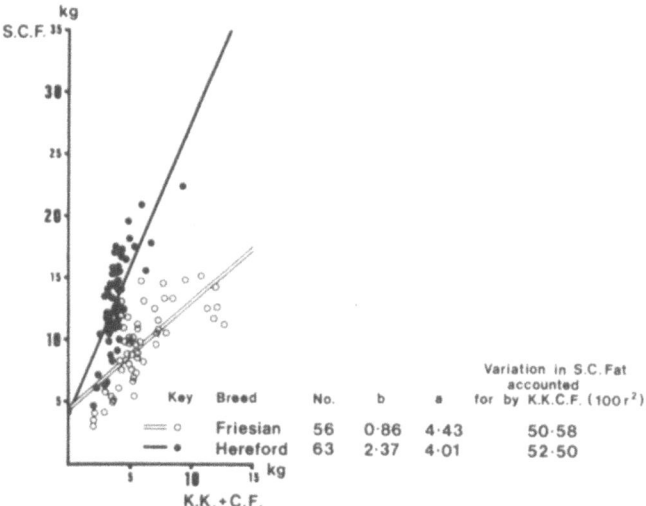

Fig 3.   KKCF and SCF

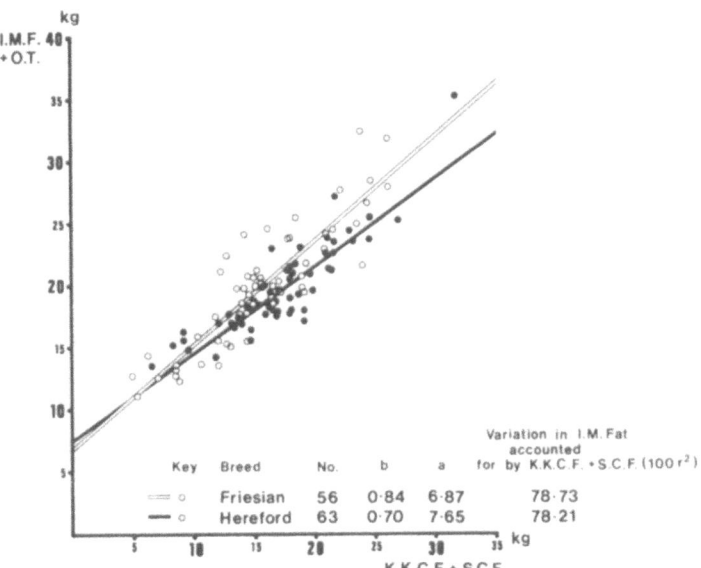

Fig 4.   KKCF + SMF and IMF

TABLE 3

COMPOSITION OF SAMPLE (all weights in kg)

Breed Number	Friesian 56			Hereford 63		
	Mean	SD	CV	Mean	SD	CV
Side weight (incl KKCF)	122.6	22.6	18.4	118.9	14.2	11.9
Total fat as % side weight	28.5	4.3	-	30.5	3.9	-
Depot fats (kg)						
SCF	9.6	3.1	32.4	13.1	3.6	27.5
IMF	19.9	4.9	24.8	19.6	3.5	17.9
KKCF	6.0	2.6	42.8	3.9	1.1	28.3
SCF + KKCF	15.6	5.2	33.7	17.0	4.4	26.0

As has been shown by Pomeroy and Williams (1974) and confirmed by Kempster, Cuthbertson and Harrington (1976) differences in fat partition may produce substantial breed bias when SCF weight is used to predict IMF (or IMF + OT) weight and that the bias is reduced when both KKCF and SCF weights are used as predictors.

EEC STANDARDS

The method of descriptive judgment of the state of fatness of the carcasses of adult cattle developed by Roy and Dumont (1975) takes account of internal (medial surface) fat, especially KKCF development. Their method which, in principle, has been proposed as the EEC standard method for research purposes is illustrated in Figures 5 and 6. The method allows for differences between carcasses in partition of fat, by including the item E. It also allows for possible differences between animals in the distribution of SCF, by including A, B, C and D as separate items. The rates of deposition of SCF (and IMF) are lowest for the distal limbs and increase 'inwards' towards the rib cage and abdominal regions and are highest in the ventral region of the abdomen (the 'flank'). However, breed differences

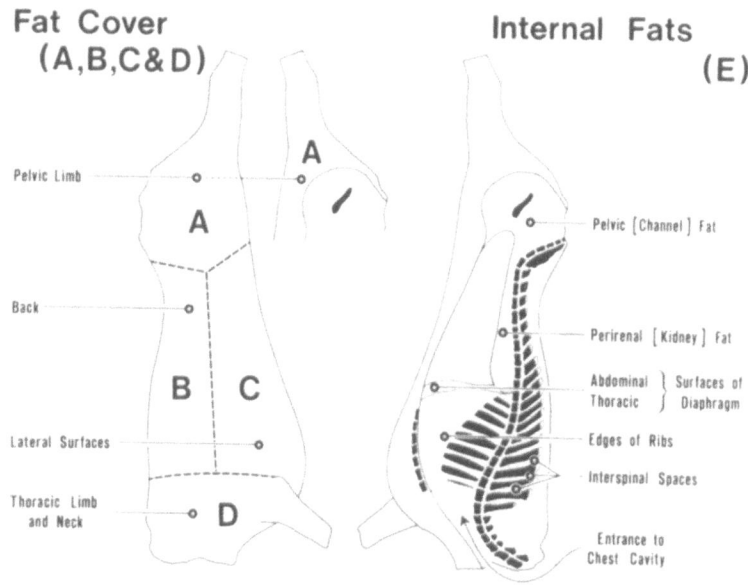

Fig 5. (Adapted from Roy and Dumont, 1975)

ASSESSMENT OF FATNESS

Example Scores

	Region	Fat Class
Fat Cover	A. Pelvic Limb	2+
	B. Back	2+
	C. Lateral Surfaces	3
	D. Thoracic Limb & Neck	2+
	E. Internal Fats	3+

Note that all four regions for Fat Cover, and the Internal Fats, need not be at the same stage of fattening.

Fig 6. (Adapted from Roy and Dumont, 1975)

in the distribution patterns within the two main depots SCF and IMF (or IMF + OT) were found to be small in the 15 breed-type x feeding system groups studied by Kempster et al. (1976). In the assessment of fatness, the differences are certainly of trivial importance compared with breed type differences which exist in the way the depots (SCF, IMF and KKCF) are partitioned relative to total fat.

CONCLUSION

For the assessment of fatness in research programmes, both SCF and KKCF should be taken into account, but differences betweeen breeds in the distribution of SCF are relatively unimportant in this context.

REFERENCES

Cuthbertson, A. Harrington, G. and Smith, R.J. 1972. Tissue Separation -
   to assess beef and lamb variation Proc. Br. Soc. Anim. Prod. (New Series)
   $\underline{1}$, 113-122.

Harrington, G. and Kempster, A.J. 1977. Beef Carcass Yields. Inst. of Meat
   Bulletin, No. 99, February.

Johnson, E.R., Butterfield, R.M. and Pryor, W.J. 1972. Studies of fat
   distribution in the bovine carcass I The partition of fatty tissues
   between depots. Anst. J. agric. Res. $\underline{23}$, 281-288.

Kempster, A.J., Cuthbertson, A. and Harrington, G. 1976$_1$. Fat Distribution
   In Steer Carcasses of Different Breeds and Crosses. I. Distribution
   Between Depots. Anim. Prod. $\underline{23}$, 25-34.

Kempster, A.J., Avis, P.R.D. and Smith, R.J. 1976$_2$. Fat Distribution In
   Steer Carcasses of Different Breeds and Crosses 2. Distribution
   Between Joints, Anim. Prod. $\underline{23}$, 223-232.

Lawes, J.B. and Gilbert, J.H. 1859. Phil Trans R. Soc. 149, 11.

Pomeroy, R.W., Williams, D.R., Harries, J.M. and Ryan, P.O. 1974.
   Composition of beef Carcasses. I Material, measurements, jointing and
   tissue separation. J. Agric. Sci. Camb. $\underline{83}$, 67-77

Pomeroy, R.W. and Williams, D.R. 1974. The partition of fat in the Bovine
   Carcass. Proc. Br. Soc. Anim. Prod. (New Series) $\underline{3}$, 85 (Abstr).

Roy, G. and Dumont, B.L. 1975. Methode du jugement descriptif de l'état
   d'engraissement des carcasses de bovins adultes. Revue Med. vet.,
   $\underline{126}$, 387-400.

FACTORS AFFECTING THE FATTY ACID COMPOSITION OF DEPOT FATS OF CATTLE AND OTHER RUMINANTS

W.M.F. Leat

Biochemistry Department, A R C Institute of Animal Physiology,
Babraham, Cambridge CB2 4AT, UK

ABSTRACT

There are a number of major factors which control the fatty acid composition of bovine depot fat (1) Age: the saturation of internal depot fat increases from birth to about 1 year of age mainly as a result of the deposition of stearic acid formed by hydrogenation in the rumen. (2) Stage of fattening: the depot fats become more unsaturated as the animal enters its fattening phase of development, which is usually at 12 - 18 months of age. (3) Anatomical position: subcutaneous fats are more unsaturated than intermuscular fats which in turn are more unsaturated than internal depots. (4) Diet: high concentrate diets result in the deposition of a more unsaturated depot fat compared to high hay diets which favour the deposition of more palmitic acid.

INTRODUCTION

Beef is a very popular meat accounting for nearly 50% of total meat consumption in the U S A (Hoover, 1973), and contributing about 15% of the total fat intake. Advice and warnings against the consumption of ruminant fat stretch back almost three millenia to Biblical times!: "Ye shall eat no manner of fat, of ox, or of sheep, or of goat" (Leviticus 7.23). However, it is only in the last 10 years since a correlation has been shown between the incidence of coronary heart disease (CHD) and raised plasma cholesterol, and between the intake of saturated fats and plasma cholesterol, that a mounting campaign has been waged against the consumption of saturated animal fats, particularly those of ruminant origin. The pros and cons of the argument on the possible involvement of dietary saturated fat in CHD will not be discussed here, but a consideration of the factors controlling the deposition of fatty acids in the ruminant carcass could benefit producers and consumers in two ways:

1) It may be possible to control the deposition of fat which is energetically and economically expensive;
2) By considering the factors affecting the fatty acid composition of depot fat it may be possible to effect some modification, or at least to define optimum ages for slaughter.

FACTORS AFFECTING THE FATTY ACID COMPOSITION OF RUMINANT DEPOT FAT

(a) General considerations

Animal depot fats are composed almost exclusively of triglycerides which are composed mainly of fatty acids with 16 and 18 carbon atoms. These fatty acids are derived from two sources (a) endogenous and (b) exogenous. Endogenous synthesis of long chain fatty acids occurs in the cytosol of the cell mainly from acetate derived predominantly from rumen fermentation Palmitic acid (C16:0) is the major end product of this *de novo* synthesis, although elongation to stearic acid (C18:0) and desaturation to monoenoic acids can occur subsequently. Exogenous fatty acids are those derived directly or indirectly from the diet. Dietary unsaturated fatty acids are deposited unchanged in the depot fat of non-ruminants such as the pig (Leat et al., 1964), or herbivores having a caecal fermentation, e.g. horse, rabbit (Table 1; Hilditch & Williams, 1964). In ruminant animals, however, dietary unsaturated fatty acids are hydrogenated by rumen microorganisms, mainly to stearic acid (Garton, 1967; Dawson & Kemp, 1970). Ruminant fats are also characterised by the presence of positional and geometric isomers of unsaturated fatty acids, mainly Δ^{11} trans octadecenoic acid, together with variable amounts of branched and odd chain fatty acids formed as a result of microbial activity in the rumen. Although the high content of stearic acid found in many ruminant depots is mainly the result of rumen fermentation, endogenous synthesis of stearic acid can account for at least 16% of the depot fatty acids of sheep (Leat et al., 1977; Duncan et al., 1971).

Palmitic acid and octadecenoic (C18:1) acid are the other major fatty acids of depot fat (Table 1). Hexadecenoic (C16:1) acid is usually a minor component (<5%), but in cattle, particularly in older animals, values up to 20% can be found (Dahl, 1957; Leat, 1975; Table 1). A particularly high value of 33% C16:1 acid has been found in the subcutaneous fat of a red deer stag (W.M.F. Leat, unpublished observations).

(b) Age and development

At birth the calf contains very little adipose tissue (<4%) (Johnson, et al., 1972) derived mainly from endogenous synthesis. The depot fat of the newborn calf is characterised by a high percentage (35%) of palmitic acid (Dahl, 1958; Garton and Duncan, 1969$_1$) the other major acids being C18:1 (40 - 50%) and C18:0 acid (6 - 10%). For comparison the depot fat of the new-born lamb contains less palmitic acid (20 - 23%) and more stearic acid (13 - 17%) (Garton and Duncan, 1969$_2$ Leat, et al., 1977). Virtually no linoleic acid is present in depot fat or

TABLE 1

FATTY ACID COMPOSITION (%) OF SUBCUTANEOUS BRISKET (S) AND PERINEPHRIC (P) DEPOT FATS OF SOME RUMINANT AND NON-RUMINANT HERBIVORES

Fatty Acid	Jersey Cattle ♂0.7y		Jersey Cattle ♀10y		Sheep ♀2y		Goat ♀1.5y		Horse ♀7y		Rabbit ♀-	
	S	P	S	P	S	P	S	P	S	P	S	P
14:0	3.5	1.2	2.7	4.6	2.8	2.8	3.4	2.5	3.4	2.9	2.9	2.6
14:1	1.6	0.3	4.1	0.7	0.9	0.6	2.3	0.6	0.1	0.1	0.2	0.2
16:0	26.8	18.5	19.4	27.5	16.5	21.5	19.6	19.7	20.6	21.8	30.2	28.8
16:1	4.3	1.9	18.1	3.3	6.1	1.3	9.9	1.5	5.6	4.0	2.1	2.0
18:0	14.7	34.3	2.9	17.0	10.1	40.5	4.1	36.3	5.3	5.4	8.1	7.9
18:1	46.1	36.0	47.8	41.5	55.6	24.8	50.2	29.3	34.6	31.4	14.9	14.0
18:2	0.8	1.8	0.8	0.6	1.9	1.2	1.0	2.2	11.0	14.0	13.1	14.4
18:3	<0.5	<0.5	<0.5	<0.5	<0.5	<0.5	<0.5	<0.5	16.9	18.0	26.7	27.9
Other acids	1.7	5.5	3.7	4.3	5.6	6.8	9.0	7.4	2.5	2.4	1.8	2.2

other tissues of newborn ruminants since negligible amounts of this fatty acid cross the placenta (Leat, 1966).

After birth the percentage stearic acid in the perinephric fat of Jersey cattle increases as the rumen begins to function, gradually reaching maximum values at about 1 year of age (Leat, 1975). This is almost certainly due to the deposition of stearic acid formed by hydrogenation in the rumen. However, after 1 year of age the percentage stearic acid gradually decreases being replaced mainly by C18:1 acid. In sheep and goats, however, high levels of stearic acid persist in the depot fat of mature animals (Leat, 1976). This change in fatty acid composition of bovine depot fat coincides with the fattening phase of development suggesting that at this stage the saturated depot fats mainly of exogenous origin are being supplemented with endogenous fatty acids of higher unsaturation.

There seems to be a relationship between the fatness of cattle and the degree of unsaturation of depot fat, the fatter the animal the more unsaturated are the depots (Leat, 1977). There are also breed differences with depot fats of Friesians tending to be more unsaturated than those of Aberdeen Angus at the same degree of fatness.

(c) Diet

Because dietary fatty acids are hydrogenated in the rumen the fatty acid composition of ruminant depot fats is relatively unaffected by alterations in dietary fat. Thus, the feeding of unsaturated oils to cattle has little or no effect on the fatty acid composition of adipose tissue (Thomas, et al., 1934; Tove and Mochrie, 1962). However, if polyunsaturated acids reach the site of absorption they are absorbed and deposited as in non-ruminant animals. This occurs, for example, in suckling animals when reflex closure of the oesophageal groove on sucking allows dietary lipid to bypass the developing rumen and enter the abomasum directly. In adult ruminants under normal husbandry conditions only small amounts of dietary polyunsaturated fatty acids reach the small intestine. If these dietary fatty acids

are protected from hydrogenation by complexing with formaldehyde-
treated protein considerable amounts of unsaturated fatty acids
can be incorporated into ruminant depot fat, particularly in
young animals (see McDonald and Scott, 1977). Whether protected
lipid supplements will gain acceptance as a practical means of
increasing the unsaturation of ruminant depot fat is, however,
debatable.

The proportion of concentrates and hay in the diet can have
an effect on the fatty acid composition of depot fat. Feeding
cereal diets to sheep and goats can result in a softening of
subcutaneous depot fats caused by deposition of unusually high
amounts of branched chain acids, which have a lower melting
point than the usual straight chain acids (see Garton, 1975).
These branched chain acids are derived from methylmalonic acid
formed from the large amounts of propionic acid produced in the
rumen when animals are fed concentrate diets. Cattle, however,
appear to differ from sheep and goats in that feeding predominantly cereal diets results in an increased deposition of
monoenoic acids rather than branched chain acids. Conversely,
the feeding of forage diets tends to increase the saturation of
depot fat, mainly through an increase in the percentage of
palmitic acid (Leat, 1977).

(d) Anatomical position
It is well known that in ruminants, as in pigs, the
unsaturation of depot fats varies with anatomical location,
subcutaneous fats being more unsaturated than intermuscular
and internal depots (see Hilditch and Williams, 1964). In
ruminants the fatty acid composition of subcutaneous brisket
fat varies with the distance beneath the skin (Figure 1) and
usually, but not invariably, there is a gradient in unsaturation
with the deeper fat layer being more saturated than that located
immediately subcutaneously. In cattle, sheep and goats % C18:0
increases and % C16:1 decreases with increasing depth. In
sheep % C18:1 decreases with increasing depth whereas in cattle,
the percentage tends to increase in the deepest layers. The
inner layers of subcutaneous fat are less saturated than

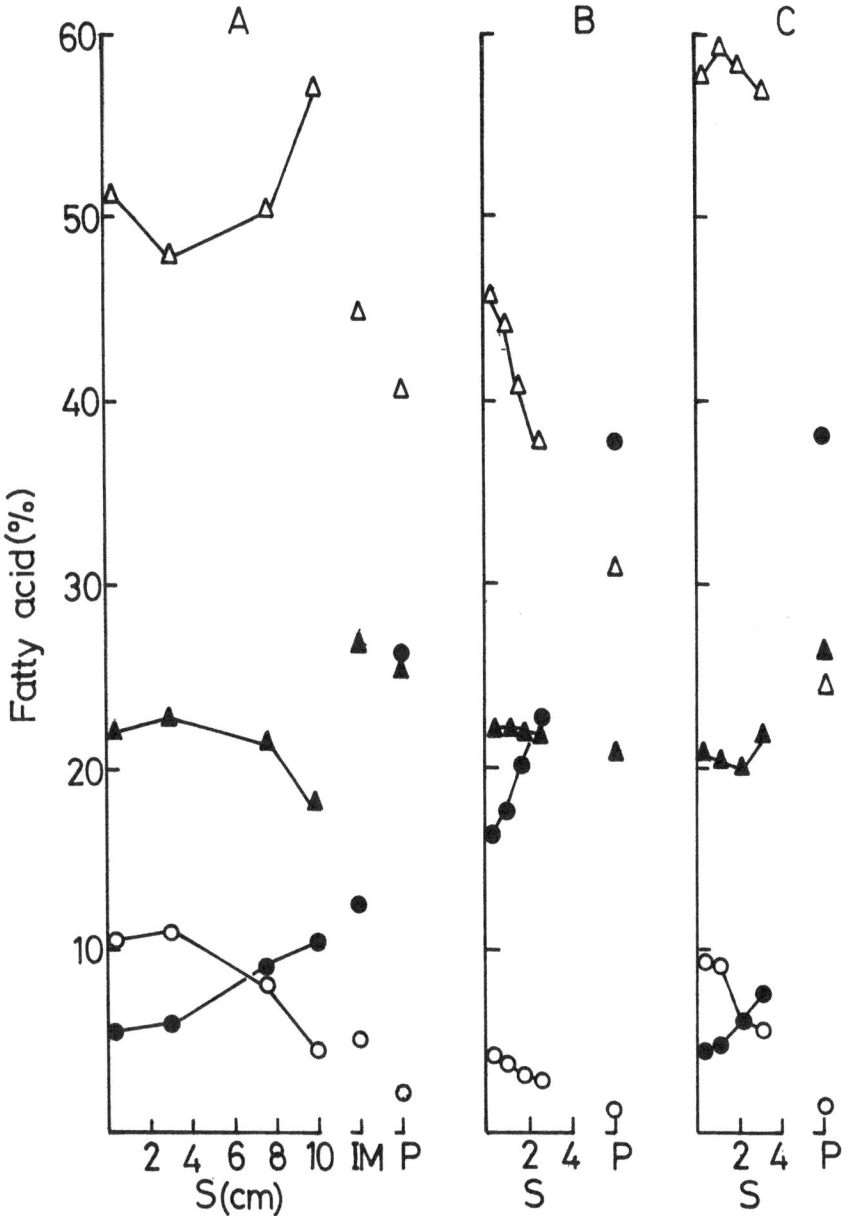

Fig. 1 Fatty acid composition of subcutaneous brisket fat (S) at various
depths (cm) beneath the skin; perinephric fat (P) and intermuscular
fat (IM).

A. Heifer 1.5y   B. Sheep 1.5y   C. Goat 6.3y

△ C18:1   ▲ C16:0   ○ C16:1   ● C18:0

intermuscular fat which is in turn less saturated than perinephric fat. Cuts of meat containing subcutaneous fat will
therefore tend to be less saturated than those containing more
intermuscular fat. This gradient in unsaturation is not universal to all species and some animals, e g the horse, show
virtually no gradient (Dahl, 1958; Table 1). Callow (1958) suggested that the local temperature of the depot, the local rate
of fat deposition and the local level of fatness may be some of
the factors involved in determining fatty acid composition.

(e) Miscellaneous factors
    (i) Season. The unsaturation of bovine depot fat is
subject to seasonal changes with a maximum being found in the
summer and autumn when the animals were on pasture, and a
minimum in the winter (Dahl, 1958). These changes are probably
due to the increased deposition of C18:2 and C18:3 acids of
dietary origin which have escaped hydrogenation in the rumen
possibly consequent upon a change in rumen microflora.

    (ii) Sex. In cattle castration results in an increased
unsaturation of depot fat (Dahl, 1962); and heifers have more
unsaturated fat than steers (Leat, 1977). However, these
differences may not be related to a direct effect of sex hormones
but rather indirectly through the amount of fat in the animal.
In general heifers are fatter than steers and steers fatter than
bulls, and it is known that the unsaturation of depot fat increases with increased fat deposition.

CONCLUSIONS

    Because dietary fats are hydrogenated by micro-organisms
in the rumen it is often assumed that in consequence all ruminant
fats are saturated in character and totally resistant to change.
This assumption is only partly correct. Diet can have an effect
on the fatty acid composition of ruminant depot fat. In cattle,
feeding concentrated diets results in the deposition of a more
unsaturated depot fat compared to forage diets which favour the
deposition of more palmitic acid. The saturation of bovine
depot fat is also controlled by the age and fatness of the

animal. The saturation of internal depots increases to one
year of age and then declines as the animal enters its fattening phase. The fatty acid composition of depot fat is dependent on anatomical position, subcutaneous fats being more
unsaturated than intermuscular, which in turn are more unsaturated than depot fat within the body cavity.

REFERENCES

Callow, E.H. 1958. Comparative studies of meat. VI Factors affecting the
iodine number of fat from the fatty and muscular tissues of lambs.
J. agric. Sci., Camb. 51, 361-369.

Dahl, O. 1957. Hexadecenoic acid as a feature of beef and horse subcutaneous
fat. Acta. chem. scand. 11, 1073-1074.

Dahl, O. 1958. Chemistry of animal depot fats in view of recent investigations. Svensk kem. Tidskr. 70, 43-54.

Dahl, O. 1962. Effect of castration on composition of the depot fat of
monozygous twin cattle. J. Sci. Fd. Agric. 13, 520-524.

Dawson, R.M.C. and Kemp, P. 1970. Biohydrogenation of dietary fats in
ruminants. In: Physiology of Digestion and Metabolism in the
Ruminant. Editor: A.T. Phillipson. Oriel Press, Newcastle-upon-
Tyne. England. pp 504-518.

Duncan, W.R.H., Garton, G.A. and Matrone, G. 1971. Triglyceride fatty
acids of lambs reared on a lipid-free diet. Proc. Nutr. Soc. 30,
48A.

Garton, G.A. 1967. The digestion and absorption of lipids in ruminant
animals. Wld Rev. Nutr. Diet. 7, 225-250

Garton, G.A. 1975. The occurrence and origin of branched-chain fatty acids
in bacterial, avian and mammalian lipids. Rep. Rowett Inst. 31,
124-135.

Garton, G.A. and Duncan, W.R.H. 1969$_1$. Effect of diet and rumen development on the composition of adipose tissue triglycerides of the calf.
Br. J. Nutr. 23, 421-427.

Garton, G.A. and Duncan, W.R.H. 1969$_2$. Composition of adipose tissue
triglycerides of neonatal and year-old lambs. J. Sci. Fd. Agric. 20,
39-42.

Hilditch, T.P. and Williams, P.N. 1964. The Chemical Constitution of
Natural Fats. Fourth Edition. Chapman and Hall, London.

Hoover, S.R. 1973. Research into foods from animal sources. 1. Controlling
level and type of fat. Prev. Med. 2, 346-360.

Johnson, E.R., Butterfield, R.M. and Pryor, W.J. 1972. Studies of fat
distribution in the bovine carcass 1. The partition of fatty tissues
between depots. Aust. J. agric. Res. 23, 381-388.

Leat, W.M.F. 1966. Fatty acid composition of the plasma lipids of newborn
and maternal ruminants. Biochem. J. 98, 598-603.

Leat, W.M.F. 1975. Fatty acid composition of adipose tissue of Jersey cattle during growth and development. J. agric. Sci., Camb. 85, 551-558.

Leat, W.M.F. 1976. The control of fat absorption, deposition and mobilization in farm animals. In Meat Animals: Growth and Productivity. Editors: D. Lister, D.N. Rhodes, V.R. Fowler and M.F. Fuller. Plenum Press, New York and London. pp 177-193

Leat, W.M.F. 1977 (in press) Depot fatty acids of Aberdeen Angus and Friesian cattle reared on hay and barley diets. J. Agric. Sci. Camb.

Leat, W.M.F., Cuthbertson, A., Howard, A.N. and Gresham, G.A. 1964 Studies on pigs reared on semisynthetic diets containing no fat, beef tallow and maize oil: composition of carcass and fatty composition of various depot fats. J. Agric. Sci. Camb. 63, 311-317

Leat, W.M.F., Kemp, P., Lysons, R.J. and Alexander, T.J.L. 1977. Fatty acid composition of depot fats from gnotobiotic lambs. J. Agric. Sci. Camb. 88, 178-179

McDonald, I.W. and Scott, T.W. 1977. Foods of ruminant origin with elevated content of polyunsaturated fatty acids. Wld. Rev. Nutr. Diet 26, 144-207

Thomas, B.H., Culbertson, G.C. and Beard, F. 1934. The effect of ingesting soybeans and oils differing widely in their iodine numbers upon the firmness of beef fat. Rec. Proc. Am. Soc. Anim. Prod., 27th Meeting 193-199

Tove, S.B. and Mochrie, R.D. 1963. Effect of dietary and injected fat on the fatty acid composition of bovine depot fat and milk fat. J. Dairy Sci. 46, 686-689

THE DEVELOPMENT OF ADIPOSE TISSUE IN CATTLE

M. Enser and J.D. Wood

Meat Research Institute, Langford, Bristol, UK

ABSTRACT

A review of the literature on the development of fat cells is presented with particular reference to cattle where possible. A study on fat cell development in Friesian bulls designed to investigate the relationship between age and nutrition is also reported. The overall conclusion is that fat growth up to usual slaughter weight in cattle is accompanied by increases in the number of cells containing fat and in their size. It is clear that only by studying animals much older than those used for beef will it be possible to determine the potential fat cell number and relate this to the distribution of fat within the animal.

INTRODUCTION

Reduction of the quantity of carcass fat in meat animals
is desirable both for economic and health reasons, whilst changes
in the partition of fat between various depots also carries
potential economic advantages. In approaching these problems
it is clear that a detailed knowledge of the growth of adipose
cells within each species is required. This paper attempts to
review our present knowledge of fat cell growth in cattle and
describes work which examines site variations more completely
than is usual in studies of this kind.

STRUCTURE OF ADIPOSE TISSUE

Histological studies of the development of bovine adipose
tissues were reported by Bell (1910) but since then most work
has concentrated on the adipose tissue of humans and laboratory
rodents. Recent interest in meat animals has been confined
mainly to pigs so that, in the absence of specific studies, the
information gained has to be applied by analogy to cattle.

Adipose tissue has a lobular structure (Toldt, 1970) with
the fat cells developing in clusters around a capillary bed,
the whole being supported in a matrix of connective tissue. The
vascular system in adipose tissue is very extensive, similar to
that in skeletal muscle (Gersh and Still, 1945; Hausberger and
Widelitz, 1963). Although fat cells apparently occupy most of
the tissue space they only contain 40 - 70% of the nuclear
material (Rodbell, 1964). The fat, which is mainly triglyceride,
first appears in the cell as a series of droplets which coalesce
to form a single large globule in the centre of the cell, with
the nucleus and cytoplasm pushed to the periphery. Fat cells
have been observed to reach diameters of 250μm (Waters, 1932)
in very fat animals, but are usually smaller in carcasses used
for meat. There is presumably a limiting size for the growth
of fat cells, but restriction of the oxygen-requiring, metaboli-
cally-active elements to the periphery of the cell makes this
more a question of the distance from the blood supply to the

cell rather than a diffusion problem within the cell. The cells at the centre of the lobules, which are nearer the blood supply, tend to be larger than those at the periphery (Bjurulf, 1959; Smith, 1971).

DEVELOPMENT OF FAT CELLS

From his observations on cattle, Bell (1910) suggested that adipose tissue grew through an increase in both the number and size of fat cells. However it is necessary to distinguish between true hyperplasia, the production of new cells by cell division, and the mere appearance of new cells containing fat. Unfortunately the cells which are destined to become fat cells (termed preadipocytes) can, at present, only be recognised after the fat accumulation has begun and fat globules have appeared in the cells. Studies of the incorporation of radioactive thymidine into the fat cells of young adult rats (150 - 175g) showed that there was no significant mitotic production of new fat cells (Hollenberg and Vost, 1968). Thus what is measured as an increase in the number of recognisable fat cells is mainly the result of fat accumulation in existing preadipocytes.

The factors which initiate the expansion of preadipocytes or which cause cells at one site to develop more rapidly or earlier in life than those at another are unknown. Since all tissues are supplied with arterial blood the differences must be the result of inherent responses of cells in each site to the levels of circulating hormones or metabolites. In cattle the perirenal adipose tissue develops early in life compared with the subcutaneous fat whereas in pigs the perirenal depot is relatively late developing. However even within a single defined fat depot the development of cells may not be homogeneous. Clusters of small cells may be present (Kirtland et al., 1975) or there may be size and growth gradients. In the subcutaneous backfat from the shoulder region of the pig we observed a size gradient of cells. The cells under the skin were small but became larger towards the connective tissue which separates the inner and outer fat layers. The largest cells were either side of the connective tissue and the cells then decreased in size toward the underlying muscle (Enser et al., 1976). Studies of the size distribution of cells within a layer suggested that two pupulations of cells were present although they were often poorly separated. Mersmann et al. (1975) observed three populations in their studies. It is probable that irregularities or specific variations in the size distribution of fat cells occur in cattle since in fat cattle the depots are often separated into layers by sheets of connective tissue. Our own studies on 400-day old Friesian bulls have revealed many sites in which the size of the cells was not normally distributed even when 100 contiguous cells were sized by the method of Sjostrom et al. (1971).

The recruitment of preadipocytes to fat cells takes place over a long period of time. In pigs we have observed the hypertrophy of preadipocytes from 3 days of age to 188 days when the pigs weighed 100 kg (Moody et al., 1975; Enser et al., 1976; Wood et al., 1977) and from our results no plateau in cell number could be detected during this period. Anderson and Kauffman (1973) observed that the number of adipocytes in subcutaneous pig adipose tissue plateaued at around 5 - 6 months of age but in intramuscular fat new cells continued to appear even later (Lee and Kauffman, 1974). In cattle Hood and Allen (1973) suggested that cell number had stabilised in the perirenal depot at 14 months of age in steers. In rats, mice and humans, adults have been studied more extensively than in cattle and in that work it appears that the number of adipocytes has become fixed since procedures which caused overeating resulted in an increase in size of existing cells rather than the appearance of new cells (Sims et al., 1968; Johnson and Hirsch, 1972).

Any inbuilt chronology in the development of fat cells will of course be influenced by the supply of nutrients. In order to investigate this relationship we have studied the development of adipose tissue at twelve sites in Friesian bull calves. Nine animals followed a nutritional regime such that at slaughter

at 400 days of age their weights varied from 307 - 467 kg.
Samples were removed from the right side of the carcass and
the left side was dissected by the method of Pomeroy et al.
(1974). The samples for fat cell size determination were taken
from the following sites :

1. Kidney; the anterior limit of the kidney knob.

2. Kidney; halfway along the kidney knob.

3. Channel; from a point situated half way along the line
joining the cranial tip of the pubis, at its symphysis, to
the caudo-ventral corner of the centrum of the last lumbar
vertebra.

4. Thoracic; dorsal to the 5 - 6 sternebrae.

5. Cod; 3 - 4 cm from the ventral border of M. cutaneous
trunci on a horizontal line to the cranial tip of the os
pubis at its symphysis.

6. Brisket; from the point situated on a horizontal line
through the junction of the first and second sternebrae
adjacent to the ventral midline.

7. Crop; on a horizontal line passing through the middle
of the centrum of the 10th thoracic vertebra, 15 cm from the
dorsal midline.

8. Round; on the cranial edge of the tuber ischii.

9. Loin; opposite the midpoint of the centrum of the second
lumbar vertebra, 25 cm from the dorsal median line.

10.Prescapular; from the fat surrounding the prescapular
lymph node.

11. Popliteal; from the fat surrounding the popliteal lymph
node.

12. Omentum.

13. Mesenteric.

The fat depots for the prescapular and popliteal samples
were taken to be either (a) the dissected intermuscular fat
from the neck and thorax or hind limb respectively or (b) the

TABLE 1

CELL SIZE IN 13 BODY SITES AND RELATIONSHIPS BETWEEN CELLULARITY AND DEPOT WEIGHT

Depot+		Cell diameter (μm)	Live weight and depot weight	Cell diameter and depot weight	Cell diameter and depot weight as % of empty body weight	Cell number per depot and depot weight
			Correlation coefficients between:			
Kidney 1	I	108	0.865**	0.693*	0.589	0.595
Kidney 2	I	113	_____	0.704*	0.875**	0.730*
Channel	I	131	0.489	0.184	0.130	0.236
Thoracic	I	118	0.715*	0.411	0.492	0.524
Cod	S	121	0.825**	0.553	0.461	0.011
Brisket	S	100	0.866**	0.317	0.401	0.452
Crop	S	114	0.772*	0.880**	0.884**	0.109
Round	S	101	0.778*	0.819**	0.795*	0.289
Loin	S	121	0.819**	0.821**	0.882**	-0.351
Prescapular (a)	IM	110	0.895**	0.644	0.408	-0.064
Prescapular (b)	IM	-	0.876**	0.521	0.654	0.466
Popliteal (a)	IM	115	0.896**	0.579	-0.008	0.128
Popliteal (b)	IM	-	0.913**	0.459	0.374	0.631
Omentum	I	117	0.622	0.563	0.759*	0.548
Mesenteric	I	116	0.626	0.636	0.419	0.080

+I, internal fat depot; S, subcutaneous fat; IM, intermuscular fat.

* Indicates significance of correlation, $P < .05$, **$P < .01$.

dissected intermuscular fat from the supraspinatus, omo-trans-
versarius and brachiocephalicus for the prescapular or from the
biceps femoris, semitendinosus and gastrocnemius for the popliteal.

Regression analysis revealed significant correlations between live weight and weight of the fat depots except for the
channel, mesenteric and omental depots (Table 1). There were
also some significant positive correlations between live weight
and the weight of the depot expressed as a proportion of empty
body weight, showing that the depots involved were growing more
rapidly than the rest of the body. Three of these sites were
in the subcutaneous fat as would be expected from its relative
late maturity.

Using analysis of variance it was found that the size of
the fat cells did not differ significantly between any of the
depots (Table 1). In particular the classification into internal, subcutaneous or intermuscular did not reveal any consistent
pattern. The mean fat cell diameter ranged from 100 um to 130 µm.
In earlier studies Moody and Cassens (1968) suggested that intermuscular cells were smaller than subcutaneous ones and Hood and
Allen (1973) stated that subcutaneous cells were smaller than
perirenal cells. In the present results there seemed to be
variation in diameter that was unrelated to apparent chronology
e.g. mean kidney fat cell diameter 110 µm, channel 131 µm.
There was also no general relationship between cell diameter and
depot weight expressed either as absolute weight or as a proportion of empty body weight. However in four sites (crop, round,
loin and kidney) there was a significant correlation between
cell size and depot weight. The meaning of such a finding is
difficult to assess since the number of cells per depot was only
significantly correlated with depot weight in the case of the
second kidney fat sample. It appears that in all the sites
except the kidney, crop, round and loin we have a mixed process
of cell recruitment and growth of existing cells, as in the
subcutaneous fat of the pig, with neither process predominant.
However, in the crop, round and loin variations in the size of

the existing cells was responsible for the difference in depot size between animals. Thus the availability of food controls the growth of cells in these depots, and the recruitment of new cells may be limited. In the more developed kidney fat both cell number and cell size are related to the size of the depot, indicating perhaps that cell number is approaching a limiting value. Hood and Allen (1973) reported that the cell number had plateaued in the perirenal fat of steers at 14 months of age. Although the steers were fatter than our bulls, the cell diameters were similar to those reported here and the same as those reported by Moody and Cassens (1968).

REGULATION OF ADIPOCYTE NUMBER

The proliferative stage in the production of preadipocytes ceases early in postnatal life in the species which have been studied. In rats the number of fat cells can be reduced by dietary restriction or exercise but these are only effective if the experimental regime is imposed within two weeks after birth (Knittle and Hirsch, 1968; Hirsch and Han, 1968; Oscai et al., 1972). In pigs, even when food restriction was started 24 hours after birth, only the intramuscular fat which is the latest to develop showed a reduction in cell number (Lee, Kauffman and Grummer, 1973). The calf is born in a more developed state than the pig and it seems likely that the prol-iferative stage will have been completed at birth. There is therefore little hope that postnatal manipulation of food in-take will change the number of potential fat cells in the calf. Manipulation in utero has not been studied but the chances of success seem small. Even the very high fat content of babies of diabetic mothers does not result in a permanent increase in the size and number of fat cells in the adult (Bjorntorp et al., 1974).

Dietary restriction after the first few weeks of life usually only reduces the size of the cells. If the restriction is very severe the apparent cell number is reduced but subsequent feeding at a higher level allows the normal number of cells to be expressed (Haugebak et al., 1974; Lee et al., 1976).

CONCLUSION

Most studies of the growth of fat cells, including this one, have used animals which have not reached their adult state. Only in adult animals fed an adequate diet will it be possible to determine the true relationship of cell number and cell size to the distribution of fat between depots (and this makes the assumption, which still needs confirmation, that fat cell number plateaus in the adult animal). We are currently measuring fat cells in mature bulls and boars in the hope of answering this question.

ACKNOWLEDGEMENTS

We acknowledge the contribution of our collaborators, especially D.J. Restall and H.J.H. MacFie in the pig studies, and Z. Holzer and A.V. Fisher in the cattle studies. For technical assistance we thank P.E. Whittington and Mrs. F.M. Garrard.

REFERENCES

Anderson, D.B. and Kauffman, R.G. 1973. Cellular and enzymatic changes in
    porcine adipose tissue during growth. J. Lipid Res. 14, 160-168.

Bell, E.T. II. 1910. On the histogenesis of the adipose tissue of the ox.
    Am. J. Anat. 9, 412-438.

Bjorntorp, P., Enzi, G., Karlsson, K., Krotkiewski, M., Sjostrom, L and
    Smith, U. 1974. The effect of maternal diabetes on adipose tissue
    cellularity in man and rat. Diabetologia 10, 205-209.

Bjurulf, P. 1959. Human adipose tissue cellularity. Acta. Med. Scand.
    Suppl. 349, 29-54.

Enser, M.B., Wood, J.D., Restall, D.J. and MacFie, H.J.H. 1976. The cell-
    ularity of adipose tissue from pigs of different weights. J. Agric.
    Sci. 86, 633-638.

Gersh, I. and Still, M.A. 1945. Blood vessels in fat tissue. Relation to
    problems of gas exchange. J. Expt. Med. 81, 219-232.

Haugebak, C.D., Hedrick, H.M. and Asplund, J.M. 1974. Adipose tissue
    accumulation and cellularity in growing and fattening lambs. J.
    Anim. Sci. 39, 1016-1025.

Hausberger, F.X. and Widelitz, M.M. 1963. Distribution of labelled eryth-
    rocytes in adipose tissue and muscle in the rat. Am. J. Physiol.
    204, 649-653.

Hirsch, J. and Han, P.W. 1969. Cellularity of rat adipose tissue : Effects
    of growth, starvation and obesity. J. Lipid Res. 10, 77-82.

Hollenberg, C.H. and Vost, A. 1968. Regulation of DNA synthesis in fat
    cells and stromal elements from rat adipose tissue. J. Clin. Invest.
    47, 2485-2498.

Hood, R.L. and Allen, C.E. 1973. Cellularity of bovine adipose tissue.
    J. Lipid Res. 14, 605-610.

Johnson, P.R. and Hirsch, J. 1972. Cellularity of adipose depots in six
    strains of genetically obese mice. J. Lipid Res. 13, 2-11.

Kirtland, J., Gurr, M.I. and Saville, G. 1975. Occurrence of pockets of
    very small cells in adipose tissue of the guinea pig. Nature 256,
    723-724.

Knittle, J.L. and Hirsch, J. 1968. Effect of early nutrition on the
    development of rat epididymal fat pads : cellularity and metabolism.
    J. Clin. Invest. 47, 2091-2098.

Lee, Y.B. Kauffman, R.G. and Grummer, R.H. 1973. Effect of early nutrition
    on the development of adipose tissue in the pig. 1. Age constant
    basis. J. Anim. Sci. 37, 1312-1318.

Lee, Y.B. and Kauffman, R.G. 1974. Cellular and enzymatic changes with animal growth in porcine intramuscular adipose tissue. J. Anim. Sci. 38, 532-537.

Lee, N.S., Dutson, T.R., Lichtenwalner, R.E., Kieffer, N.M., Cross, H.R. and Carpenter, Z.L. 1976. Cellularity of adipose tissue and blood lipid content of double muscled and normal animals on high and low energy rations. Beef Cattle Research in Texas, 1974-75, 295-304.

Mersmann, H.J., Goodman, J.R. and Brown, L.J. 1975. Development of swine adipose tissue : morphology and chemical composition. J. Lipid Res. 16, 269-279.

Moody, W.G. and Cassens, R.G. 1968. A quantitative and morphological study of bovine longissimus fat cells. J. Fd. Sci. 33, 77-80.

Moody, W.G., Enser, M.B., Wood, J.D., Restall, D.J. and Lister, D. 1975. Comparison of fat and muscle development in Pietrain and Large White pigs. J. Anim. Sci. 41, 299.

Oscai, L.B., Spirakis, C.N., Wolff, C.A. and Beck, R.J. 1972. Effects of exercise and of food restriction on adipose tissue cellularity. J. Lipid Res. 13, 588-592.

Pomeroy, R.W., Williams, D.R., Harries, J.M. and Ryan, P.O. 1974. Composition of beef carcasses. 1. Material, measurements, jointing and tissue separation. J. Agric. Sci., Camb. 83, 67-77.

Rodbell, M. 1964. Localisation of lipoprotein lipase in fat cells of rat adipose tissue. J. Biol. Chem. 239, 753-755.

Sims, E.A.H., Goldman, R.F., Gluck, C.M., Horton, E.S., Kellheher, P.C. and Rowe, D.W. 1968. Experimental obesity in man. Trans. Ass. Am. Physicians 8, 153-170.

Sjostrom, L., Bjorntorp, P. and Vrana, J. 1971. Microscopic fat cell size measurements on frozen-cut adipose tissue in comparison with automatic determinations of osmium-fixed fat cells. J. Lipid Res. 12, 521-530.

Smith, U. 1971. Effect of cell size on lipid synthesis by human adipose tissue in vitro. J. Lipid Res. 12, 65-70.

Toldt, C. 1970. Contribution to the histology and physiology of adipose tissue. Sitzber Akad. Wiss. Wien. Math. Naturwiss. KI. 62, 445.

Waters, H.J. 1976. Soc. Promotion Agric. Res. 30, 70, 1909. (Quoted by Hood, R.L. Cellularity of adipose tissue in meat animals. Fed. Proc. 35, 2302-2307)

Wood, J.D., Enser, M.B. and Restall, D.J. The cellularity of backfat in growing pigs. Anim. Prod. submitted, 1977.

THE DEVELOPMENT OF FAT CELLS IN DIFFERENT ANATOMICAL POSITIONS IN CARCASSES OF YOUNG BULLS, HEIFERS AND COW-HEIFERS

Irmgard Schön

Meat Research Institute, Kulmbach, Fed. Rep. of Germany

ABSTRACT

The paper deals with data from literature on the development of fat cells and with own investigations in young bulls, heifers and cow-heifers. Differences in the size of fat cells - μm^2 - between perirenal, subcutaneous, intermuscular and intramuscular fat are significantly influenced by the category, the fattening method and age, and correlate positively with the portion of fatty tissue in the carcasses and the fat content of the specific tissue. As a reference value the mean of 50 single measurements was used each time. A procedure for more exact and quicker measurements of fat cells dimension has been developed.

INTRODUCTION

There is little agreement in literature on the development, numbers and sizes of cells of animal fatty tissue. Wood et al. (1975) suggested that numbers in the perirenal, subcutaneous and intermuscular fat are fixed in early life. Fat cell dimension increases during growth in dependency of the anatomical position; fatty tissues with a higher number of smaller cells develop differently from those with lower numbers of cells and a larger cell area (Wood et al., 1975; Nougues, 1975). Haugebak et al. (1974) and Steele et al. (1974) on the other hand, hold another view on new formation and involution in relation to the feeding level, including transformation of connective tissue cells or muscle fibres to fat cells (Griebel, 1958). Moody et al. (1968) found in bovine fatty tissues located directly to the muscles larger fat cells than those in fatty tissues located further away. In pigs it has been found that in the subcutaneous fat a more rapid enlargement of the fat cells takes place in the inner layer with increasing live weight (Anderson et al., 1972; Moody et al., 1968). While on the one hand differences in cell size have been observed until the final stage of fattening, other authors suggest that cell sizes equalise with increasing live weight, or age (Lee et al., 1974; Nougues, 1975). The increase or decrease of the number and size of the fat cells caused by ad libitum or restricted feeding, is predominantly regarded as reversible (Moody et al., 1968). Martin et al. (1974) restricted this last aspect insofar as by severe feed restriction, existing fat cells disappeared completely after the animals had lost 20% of their body weight. Changes in the structure of cell membranes regenerated again at a normal diet (Desnoyers, 1973).

Results agree on size differences of cells between perirenal subcutaneous, intermuscular and intramuscular fat; in this order the size of the fat cells increases. The development of the intramuscular fat occurs in a later stage of growth; different from other fatty tissues nutritional influences and weight increase do not only affect the size, but also the number of fat cells (Lee et al., 1973, 1974; Martin et al., 1974).

While the amounts of perirenal, subcutaneous and intermuscular fat correlate closely with the sizes of the fat cells
(Evans, 1972; Haugebak et al., 1974; Steele et al., 1974;etc).
an increase of the intermuscular fat has a closer interrelationship with the number of fat cells, according to Lee et al. (1973).
Histologically assessed intramuscular fat is not sufficient as
a measure of the fat content in the musculature, although
significant increases in the fat cell size have been proved
with increasing fat content of the muscles (Moody and Cassens,
1968).

RESULTS OF OWN INVESTIGATIONS

The results refer to size of fat cells from perirenal,
subcutaneous, inter- and intramuscular fat of 160 young bulls,
84 heifers and 69 cow-heifers of different breeds and crossings
of known history. The young bulls and heifers had been
fattened partly indoors and partly on pasture, and included two
age groups. Cow-heifers had been slaughtered 3 and 8 weeks
after calving. The breed spectrum included German Black and
White, German Brown and their crossings.

The sampling of fatty tissue referred to 11 anatomically
defined points. Mentioned cell areas in μm^2 refer to measurements of 50 fat cells of each sample with the ZEISS Research
Microscope Universal.

In the multiple analysis of variance the obtained F-
values and their significance show that the categories,
fattening systems and age of the animals are the most substantial
causes of variance (Table 1). Moreover differences between
various fat locations are existing. The significance of the
F-values is indirectly caused by the importance of the portion
of fatty tissues in the carcasses. Mean values ($\bar{x}$) for the
size of fat cells (Table 2) show clearly these interrelationships.
Besides differences between anatomical sites within groups,

TABLE 1

ANALYSIS OF VARIANCE (MULTIPLE) FOR THE FAT CELL SIZES - μm^2 - OF VARIOUS FATTY TISSUES FROM YOUNG BULLS (YB) HEIFERS (H) AND COW-HEIFERS (C-H) - F-VALUES AND SIGNIFICANCE - $+p = 0.05$; $++p = 0.01$; $+++p = 0.001$

Cause of variance	DF	Peri-renal fat	Subcutaneous fat			Intermuscular fat			Intramuscular fat			
			8th - 11th thor.vert.	round	brisket	8th - 11th thor.vert.	round	brisket	M.long. dorsi	M.adduc-tor	M.psoas major	M.supra-spinam
Total reduction	29	123.8	119.5	125.8	115.0	132.3	88.7	111.3	85.2	84.4	122.0	134.4
Breed	7	++ 3.3	- 1.9	++ 2.9	- 1.0	- 1.8	- 1.2	- 1.1	- 1.2	++ 3.0	- 1.0	- 0.5
Category (YB,H,C-H)	3	+++ 22.0	+++ 30.2	+++ 18.0	+++ 15.1	+++ 15.9	++ 5.3	++ 4.0	+++ 6.8	+++ 6.8	++ 4.6	+ 3.3
System of fattening x age/H	2	+++ 29.2	+++ 24.7	+++ 30.2	+++ 15.3	+++ 28.4	+++ 11.3	+++ 14.9	+++ 16.9	+++ 23.1	+++ 19.0	+++ 31.3
System of fattening x age/C-H [1]	1	+++ 32.8	+++ 18.7	+++ 21.3	++ 6.9	+++ 18.1	++ 9.1	+++ 22.0	+++ 24.3	++ 9.9	- 2.1	- 2.0
System of fattening x age/C-H [2]	1	+++ 13.0	++ 9.0	+++ 14.8	++ 10.3	+++ 13.9	++ 7.8	++ 11.1	+++ 27.5	+++ 22.5	++ 7.3	+++ 30.0
System of fattening x age/YB	6	+++ 13.4	+++ 15.6	+++ 15.0	+++ 11.3	+++ 12.9	+ 2.5	+++ 6.9	+++ 6.8	+++ 6.9	+++ 7.6	+++ 10.7
Breed x category	8	- 1.2	- 1.8	- 1.0	- 0.6	- 0.7	- 0.6	- 1.1	+++ 4.1	+ 2.5	- 0.9	- 1.3

1) 3 weeks after calving   2) 8 weeks after calving

TABLE 2

MEAN VALUES ($\bar{x}$) FOR FAT CELL SIZES – μm^2 IN FATTY TISSUES OF YOUNG BULLS (YB), HEIFERS (H), AND COW HEIFERS (C-H) IN GROUPS OF FATTENING, AGE AND CATEGORY

System of fatten-ing	n – age	Cate-gory	Peri-renal fat $\bar{x}$	Subcutaneous fat 8th-11th thor. vert $\bar{x}$	round $\bar{x}$	brisket $\bar{x}$	Intermuscular fat 8th-11th thor. vert $\bar{x}$	round $\bar{x}$	brisket $\bar{x}$	Intramuscular fat M.long dorsi $\bar{x}$	M.adduc-tor $\bar{x}$	M.psoas major $\bar{x}$	M.supra spinam $\bar{x}$
Total number	321		10193	9194	8996	8618	9560	8684	7250	4782	4846	3988	3544
Indoor Maize silage	18[1]	YB	9097	9873	9581	10067	10052	9076	7957	5197	5148	4413	3929
	24[2]	YB	11321	11284	10629	9774	11447	9588	7956	4906	5802	4694	3817
Indoor Maize silage	29[3]	YB	10795	8985	8986	9422	9767	8524	7437	4744	4818	4236	3760
	30[4]	YB	10205	9181	8000	8207	9048	8443	7286	3900	4431	3863	3496
	31[5]	H	13244	12510	12478	11456	12945	10378	9257	6037	6046	4759	4320
	29[6]	H	15039	13239	12963	11441	13068	11092	8574	6380	6396	5101	4578
Indoor	24	C-H/3[7]	11959	9540	9884	8502	10528	9621	8927	7560	6624	4465	4045
	27	C-H/8[8]	11747	8666	10026	9308	10032	9621	7824	6060	5620	4086	4269
Pasture	13	C-H/3	6027	4503	4252	5101	6021	5588	4216	3352	3655	3504	3178
	13	C-H/8	6379	5400	6427	5672	5636	6126	5237	2797	2871	2903	2461
	24	H	8752	8099	8071	8372	8920	7739	6156	3937	3838	3242	2462
	15	YB	6329	5956	4959	5543	6511	6193	5248	2349	2718	2720	1914
Pasture + supp. food	21	YB	4348	4005	3785	3770	4143	6650	3947	2081	2155	2354	1895
Pasture + indoor finish	83	YB	9056	8426	8541	7925	8459	8091	7145	4086	4397	3690	3544

Age (days) [1]= 510; [2]= 590; [3]= 450; [4]= 520; [5]= 470; [6]= 520; [7]= 3 weeks after calving; [8]= 8 weeks after calving

differences are due to the clear influence of the fattening
system on the portion of fatty tissue in the carcass, eg.
between animals fattened indoors and on pasture. The degree
of fatness in the animals fattened indoors was higher on average,
and therefore also the mean values for the size of fat cells.
This relationship differs to some extent for individual
anatomical positions, eg between intermuscular fat in the dorsal
region (8th - 11th thoracic vertebrae) and the brisket.
Intramuscular fat cells are distinctly smaller and show specific
differences in growth pattern between the muscles. The mean
values in Table 3 illustrate the effect of the fatty tissue
proportion of the carcasses (total dissection) on the cell sizes
at different anatomical locations. At similar level of fatness
the size of the fat cells is smaller in cow-heifers in particular
In subcutaneous and intermuscular fat the cells in the brisket
region are smaller, as is the case in the intramuscular fat of
the M. psoas major and M. supra spinam.

Table 4 gives the correlation coefficients relating the
size of the fat cells (y) to the fatty tissue proportion of the
carcasses (x^1) and the fat content of the specific tissue (x^2).
The x^1 correlations are higher in the perirenal, subcutaneous
and intermuscular fat, but in the intramuscular fat the x^2
correlations are somewhat higher.

The results have been based on the measurement of individual
fat cells. However, within a tissue cell sizes vary widely. Up
to now values on cell size have been calculated as the mean of
50 single measurements. In order to improve random selection
and accuracy of measurement of fat cell areas a modified
technique is used now in which groups of cells are measured. By
automatisation a quicker measurement, involving greater numbers
of cells of a particular fatty tissue could be realised. For
this purpose a line of measurement has been developed, consisting of devices of the Zeiss firm including a Research Microscope
Universal with Compact Camera, Micro-Videomat 2, Macroscope
with marking device (Figure 1), and programmed computing (WANG
720) with automatic output of mean values and deviations of
100 fat cells in µ2 of each fatty tissue sample.

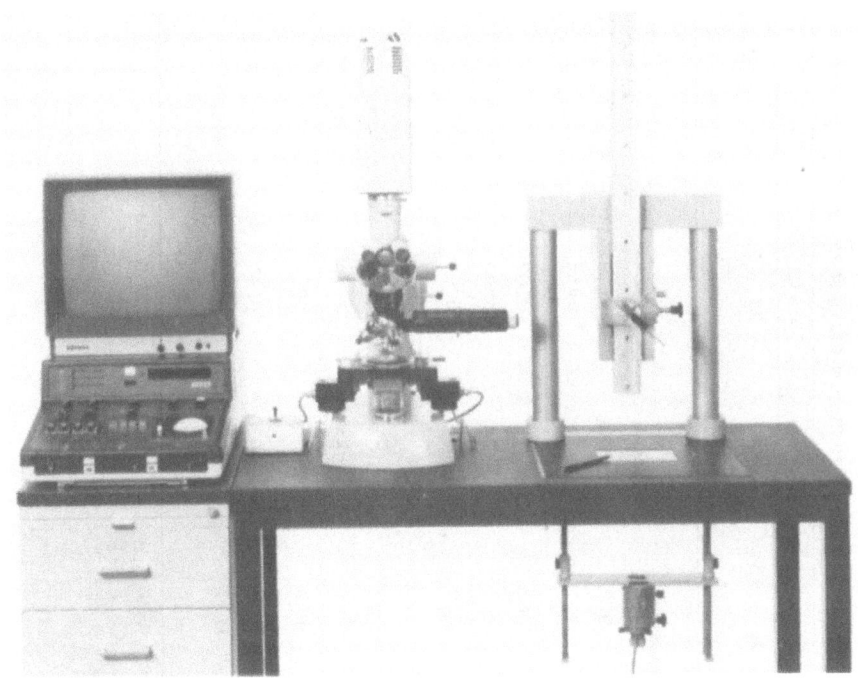

Fig. 1. Part of the line of measurement Zeiss apparatuses from left to
right: Micro-Videomat 2 / Research Microscope Universal with
Compact Camera and Macroscope with marking device.

TABLE 3

CHANGES IN THE FAT CELL SIZES $-\mu m^2$ – WITH INCREASING PERCENTAGE OF FATTY TISSUE IN CARCASSES

Localisation	Category	F-signific.	Fatty tissue proportion as percent of the carcass weight					
			<10 $\bar{x}$ μ^2	10 – 15 $\bar{x}$ μ^2	15 – 20 $\bar{x}$ μ^2	20 – 25 $\bar{x}$ μ^2	25 – 30 $\bar{x}$ μ^2	>30 $\bar{x}$ μ^2
Subcutaneous								
9th – 11th dorsal vert.	YB	+++	4413	8409	10224	11989	–	–
	H	+++	–	6929	9852	13178	15100	15387
	C–H	+++	3501	3711	7512	8835	8742	8396
Brisket	YB	+++	4286	8463	9415	10167	–	–
	H	+++	–	7523	8910	12098	13355	13116
	C–H	+++	3998	3680	7663	8090	8322	14593
Intermuscular								
9th – 11th dorsal vert.	YB	+++	4937	8373	10104	14297	–	–
	H	+++	–	7791	9763	14118	14315	14227
	C–H	+++	3953	3852	7998	9967	10638	9328
Brisket	YB	+++	4217	6824	7963	9241	–	–
	H	+	–	7137	7236	8983	8456	14897
	C–H	+++	3620	3638	7210	7661	8192	11183

Mean values and significance - $^+P = 0.05$; $^{++}P = 0.01$; $^{+++}P = 0.001$

Table continued

TABLE 3 (Cont)

Localisation	Category	F-signific.	Fatty tissue proportion as percent of the carcass weight					
			<10 $\bar{x}$ μ^2	10 - 15 $\bar{x}$ μ^2	15 - 20 $\bar{x}$ μ^2	20 - 25 $\bar{x}$ μ^2	25 - 30 $\bar{x}$ μ^2	>30 $\bar{x}$ μ^2
Intramuscular								
M. long. dorsi	YB	+++	2211	4080	4803	6252	-	-
	H	++	-	3986	4625	6446	6228	8053
	C-H	+++	2182	2669	3859	5615	6938	9673
M. adductor	YB	+++	2342	4487	5238	6029	-	-
	H	-	-	3792	4505	6270	7458	11770
	C-H	+++	3002	2963	4150	4662	6380	8249
M. psoas major	YB	+++	2405	4006	4306	5187	-	-
	H	+++	-	3591	3838	4796	5298	10972
	C-H	+++	2855	2973	3941	4099	4282	3993
M. supra spinam	YB	+++	1958	3261	4010	4789	-	-
	H	+++	-	2976	3250	4365	4956	4893
	C-H	+++	2533	2586	3315	3954	4135	4693

Mean values and significance - $^+$P = 0.05; $^{++}$P = 0.01; $^{+++}$P = 0.001

TABLE 4

CORRELATION COEFFICIENTS (r), RELATING FATTY TISSUE PROPORTION OF THE CARCASS WEIGHT (x^1) AND THE FAT CONTENT OF THE TISSUE (x^2) TO THE SIZE OF THE FAT CELLS (y)

Fat localisation	Fatty tissue % of carcass (x^1) Size of the fat cells (y)			Fat content of the tissue (x^2) Size of the fat cells (y)		
	Young bulls r	Heifers r	Cow-heifers r	Young bulls r	Heifers r	Cow-heifers r
Perirenal	+0.662	+0.658	+0.709	+0.478	+0.455	+0.525
Subcutaneous						
9th-11th dors. vert	+0.654	+0.674	+0.529	+0.278	+0.408	+0.370
Round	+0.708	+0.586	+0.654	+0.348	+0.190	+0.481
Brisket	+0.614	+0.608	+0.618	+0.539	+0.230	+0.507
Intermuscular						
9th-11th dors. vert	+0.676	+0.672	+0.694	+0.556	+0.383	+0.698
Round	+0.373	+0.627	+0.610	+0.365	+0.201	+0.531
Brisket	+0.636	+0.302	+0.604	+0.503	+0.225	+0.448
Intramuscular						
M. long. dorsi	+0.606	+0.459	+0.711	+0.646	+0.559	+0.756
M. adductor	+0.520	+0.631	+0.668	+0.563	+0.731	+0.723
M. psoas major	+0.555	+0.523	+0.364	+0.626	+0.523	+0.466
M. supra spinam	+0.583	+0.501	+0.537	+0.626	+0.601	+0.435

REFERENCES

Anderson, D.B., Kauffman, R.G. and Kastenschmidt, L.L. 1972. Lipogenic enzyme activities and cellularity of porcine adipose tissue from various anatomical locations. Journal of Lipid Research, 13, 593-598.

Desnoyers, F., Vodovar, N. and Durand, G. 1973. Transformation et devenir des cellules adipeuses au cours de l'amaigressement. Etude ultra-structurale. Ann. Biol. Animale Biochim. Biophys. 13, 75-92.

Evans, A.J. 1972. Change in adipocyte size during post-embryonic growth in the female Aylesbury duck. Brit. Poultry Sci. 13, 615-618.

Griebel, C. 1958. Die mikroskop. Untersuchung der fettliefernden tierischen und pflanzlichen Rohstoffe und ihrer Abfallprodukte. Analyse der Fette u. Fettprodukte, Herausgeber: H.P. Kaufmann, Springer-Verlag.

Haugebak, C.D., Hedrick, H.M. and Asplung, J.M. 1974. Adipose tissue accumulation and cellularity in growing and fattening lambs. J. Anim. Sci. 39, 1016-1025.

Lee, Y.B., Kauffman, R.G. and Grummer, R.H. 1973. Effect of early nutrition on the development of adipose tissue in the pig. I. Age constant basis. J. Anim. Sci. 37, 1312-1318.

Lee, Y.B., Kauffman, R.G. and Grummer, R.H. 1973. Effect of early nutrition on the development of adipose tissue in the pig. II. Weight constant basis. J. Anim. Sci. 37, 1319-1325.

Lee, Y.B.and Kauffman, R.G., 1974. Cellular and enzymatic changes with animal growth in porcine intramuscular adipose tissue. J. Anim. Sci. 38, 532-537.

Lee, Y.B. and Kauffman, R.G. 1974. Cellularity and lipogenic enzyme activities of porcine intramuscular adipose tissue. J. Anim. Sci. 38, 538-544.

Martin, R.J., Ezekwe, M., Herbein, J.H., Sherritt, G.W., Gobble, J.L. and Ziegler, J.H. 1974. Effects of neonatal nutritional experiences on growth and development of the pig. J. Anim. Sci. 39, 521-526.

McMeekan, C.P. 1940. Growth and development in the pig, with special reference to carcass quality characters. J. Agr. Sci. Cambridge, 30, 276-343

Moody, W.G. and Cassens, R.G. 1968. A quantitative and morphological study of bovine longissimus fat cells. J. Fd. Sci. 33, 77-80.

Nougues, J. 1975. Adipocyte growth of four adipose deposits in rabbit. Ann. Biol. Animale Biochim. Biophys. 15, 541-546.

Wood, J.D., Enser, M.B. and Restall, D.J. 1975. Fat cell size in Pietrain and Large White pigs. J. Agric. Sci., Cambridge, 84, 221-225.

FATTY ACID COMPOSITION OF FAT IN WATER BUFFALO CALVES AND BOVINE CALVES SLAUGHTERED AT 20 - 28 AND 36 WEFKS OF AGE

A. Borghese, S. Gigli, A. Romita, A. Di Giacomo, M. Mormile
Istituto Sperimentale per la Zootecnia, Rome - Italy.

ABSTRACT

The fatty acid compositions of subcutaneous, intermuscular, intra-muscular, perivisceral and perinephric lipid of 30 bovine and 30 buffalo male calves slaughtered 20 - 28 and 36 weeks are reported.

The 18 : 1 is the fatty acid which is present in a greater degree in both species. Unsaturated fatty acids increase with the age of the animals; 18 : 0 and 18 : 1 occur more in buffaloes rather than in calves. The difference in fatty acid compositions between species increases with the age in all the sites considered.

INTRODUCTION

The composition of animal fat depot changes with the amount and composition of dietary fat, rate of growth of fatty tissues, differential distribution of dietary fatty acids among various fatty tissues, animal species (Shorland, 1953), breed and sex (Gillis et al., 1973), body sites (Dugan, 1952; Cook et al., 1965; Shafrir and Wertheimer, 1965; Marchello et al., 1970; Ingle et al., 1972a, 1972b), seasonal and environmental conditions (Cramer and Marchello. 1964; Marchello et al., 1967; Link et al., 1970b, 1970c). There is also evidence of a relationship between growth and body fat composition (Cramer and Marchello, 1964; Waldman et al., 1968; Terrel et al., 1969; Link et al., 1970a, 1970b, 1970c; Pothoven et al., 1974).

Apart from researches on calves there have been a few researches performed on fatty acid composition. in buffaloes (Intrieri et al., 1967, 1968; Di Lella et al., 1972).

In order to compare chemical characteristics in meat and in lipids of different sites between water buffalo and bovine calves, we have planned a study on fat composition from 20 weeks up to mature age.

The results of the first age are already published (De Maria et al., 1976); in the present report we refer on relationship between some factors as species, age and site and fatty acid compositions up to 36 weeks.

MATERIAL AND METHODS

The trial was carried out on 30 bovine male calves and 30 water buffalo male calves reared under identical breeding conditions and environment and slaughtered at 20 - 28 - 36 weeks of age.

The fatty acid composition of subcutaneous, intermuscular, intramuscular, perivisceral and perinephric fat were analysed.

Perivisceral fat was taken out just after slaughtering,
the other fat samples after 9 days of ageing at $4^{\circ}C$. Subcutaneous fat covering the gluteobiceps, intermuscular fat between
semimembranosus and biceps emoris, intramuscular fat from
longissimus dorsi were sampled. All the fat samples were
stored at $- 25^{\circ}C$ in polyethylene bags.

The procedures for extraction and methylation have been
outlined in a previous paper (De Maria et al., 1976).

The methyl esters were chromatographed on column 2m x ½"
packed with DEGS (diethylene-glycol-succinate on chromosorb
W A W 80 - 100 mesh).

RESULTS AND DISCUSSION

The major fatty acid, both in calves and buffaloes, as is
shown in Tables 1 to 5, for all the sites and the ages
considered, is 18 : 1, in general followed by 16 : 0, 18 : 0,
16 : 1, 18 : 2 and 14 : 0 in accordance with Waldman et al.,
1968; Pothoven et al., 1974. Only in perivisceral buffalo fat
and in perinephric fat of both species the 18 : 0 is the
second and in the intramuscular fat the 18 : 2 is always
higher than 16 : 1.

The percentage of unsaturated fatty acids (UFA) is higher
than that of the saturated fatty acids (SFA); the ratio SFA/UFA
is higher in internal sites (perinephric and perivisceral).

The UFA tend to increase with the age, in fact the values
are highest at 36 weeks, except for perinephric site according
to Pothoven et al., 1974. This trend is generally due to
18 : 1 (as found by Waldman et al., 1968) and to 18 : 2,
while the decrease of SFA is due mostly to 18 : 0, as found by
Link et al., 1970_2 for subcutaneous fat. The 18 : 3 generally
decreases with the age both in calves and buffaloes except for
intramuscular fat where the highest value is found at the
second age. The 14 : 0 content decreases with increasing live

weight in subcutaneous and intermuscular fat as found by
Terrel et al., 1969. The polyunsaturated fatty acids tend to
decline with increasing ages of the animals, except in intramuscular fat; for this site Link et al., 1970$_3$ found a
decreasing trend, perhaps due to the different stage of maturity
of the animals.

TABLE 1

MEANS OF THE FATTY ACID COMPOSITION OF SUBCUTANEOUS LIPID BY AGE OVER
BOTH SPECIES

Age	20 weeks		28 weeks		36 weeks	
Fatty acids	Calv.	Buff.	Calv.	Buff.	Calv.	Buff.
10 : 0	0.24A	0.26a	0.27A	0.16b	0.31A	0.16b
12 : 0	0.41A	0.37a	0.35A	0.26b	0.38A	0.25b
14 : 0	3.49A	2.52a	2.99A	1.67b	3.04A	1.53b
15 : 0	tr.A	tr.a	0.52B	0.21b	0.45B	0.22b
16 : 0	19.93A	17.98a	20.97B	17.17a	20.55B	16.29b
17 : 0	0.22A	0.28a	1.53B	1.18b	1.79C	1.58c
18 : 0	14.90A	17.33a	11.14A	15.08ab	9.99B	11.80b
Satur.	39.19A	38.74a	37.77B	35.73b	36.51B	31.83c
14 : 1	1.59A	0.96a	1.17A	0.40b	1.51A	0.52b
15 : 1	tr.A	tr.a	0.58B	0.34b	0.28C	0.20c
16 : 1	8.50A	6.19a	6.63B	4.48b	7.58A	5.95a
17 : 0	1.10A	1.07a	1.48B	1.03a	1.74C	1.25a
18 : 1	42.68A	46.63a	47.11B	53.95b	45.99A	52.80b
18 : 2	5.35A	5.00a	3.04B	2.57b	5.08A	5.38a
18 : 3	1.59A	1.41a	1.04B	0.96b	0.51C	1.08b
Unsat.	60.81A	61.26a	61.05A	63.73b	62.69A	67.18c

NOTE: Means not underlined are significantly different between species in
the same age. Different letters in the same line mean significant
difference among the ages. (Capital letters for Calves and lower
case letters for Buffaloes).

TABLE 2

MEANS OF THE FATTY ACID COMPOSITION OF INTERMUSCULAR LIPID BY AGE OVER BOTH SPECIES

Age	20 weeks		28 weeks		36 weeks	
Fatty acids	Calv.	Buff.	Calv.	Buff.	Calv.	Buff.
10 : 0	0.31A	0.26a	0.16B	0.20b	0.19B	0.22b
12 : 0	0.47A	0.38a	0.20B	0.20b	0.27B	0.30c
14 : 0	3.47A	2.80a	3.05A	2.08b	3.26A	2.06b
15 : 0	tr.A	tr.a	0.47B	0.29b	0.44B	0.22b
16 : 0	20.24A	19.31a	21.36B	19.36a	20.22A	17.43b
17 : 0	0.32A	0.31a	1.51B	1.45b	1.76C	1.66b
18 : 0	16.67A	19.78a	16.42A	20.43a	13.59B	16.77b
Satur.	41.48A	42.84a	43.17B	44.01a	39.73A	38.68b
14 : 1	1.69A	0.93a	0.79B	0.43b	1.10C	0.62c
15 : 1	tr.A	tr.a	0.34B	0.28b	0.29B	0.23b
16 : 1	7.35A	5.35a	5.29B	3.98b	6.70A	5.20a
17 : 1	1.12A	0.88a	1.13A	1.02b	1.17A	1.17c
18 : 1	41.68A	43.60a	43.76B	42.26a	44.36B	47.65b
18 : 2	5.18A	5.03a	3.02B	2.95b	5.48A	5.50a
18 : 3	1.50A	1.29a	1.17B	1.11a	0.53C	0.43b
Unsat.	58.52A	57.16a	55.50B	55.03b	59.63A	60.79c

NOTE: See Table 1.

There is a large increase in the amount of 17 : 0 between the first and the second age both in calves and in buffaloes for all the sites considered. Generally there is not a definite trend for the fatty acid and therefore they seem to be unrelated to the age in accordance with Link et al., 1970[2], 1970[3].

As is shown in Tables 1 to 5 significant differences between species in all the sites considered are more in the oldest age group, therefore as the age increases there is a differentiation of the fatty acid deposition between species.

TABLE 3

MEANS OF THE FATTY ACID COMPOSITION OF INTRAMUSCULAR LIPID BY AGE OVER BOTH SPECIES

Age	20 weeks		28 weeks		36 weeks	
Fatty acids	Calv.	Buff.	Calv.	Buff.	Calv.	Buff.
10 : 0	0.59A	0.59a	0.31B	0.26b	0.33B	0.29b
12 : 0	0.36A	0.29a	0.45A	0.21b	0.43A	0.23b
14 : 0	2.89A	2.60a	1.76A	1.50b	2.21A	1.79b
15 : 0	tr.A	tr.a	0.49B	0.44b	0.44B	0.42b
16 : 0	19.51A	19.28a	17.66A	17.71b	17.33A	18.48a
17 : 0	0.35A	0.21a	1.32AB	1.24b	1.86B	1.52c
18 : 0	16.72A	18.15a	11.18B	14.86b	11.31B	14.46b
20 : 0	tr.A	tr.a	0.82B	1.36b	0.54C	0.67c
Satur.	40.42A	41.12a	33.99B	37.58b	34.45B	37.86b
14 : 1	1.43A	1.13a	0.66B	0.36b	0.81B	0.47b
15 : 1	tr.A	tr.a	0.77B	0.77b	0.62C	0.64c
16 : 1	7.16A	5.54a	4.76B	3.60b	7.64A	5.43a
17 : 1	1.01A	0.99a	1.55B	1.37b	1.71B	1.39b
18 : 1	42.38AB	42.87a	42.88A	39.13b	40.86B	34.02c
18 : 2	6.07A	6.71a	9.61B	12.01b	10.82B	17.00c
18 : 3	1.53A	1.64a	1.81A	1.89a	0.54B	0.49b
20 : 4	tr.A	tr.a	2.65B	2.37b	1.69C	2.00b
Unsat.	59.58A	58.88a	64.69B	61.50b	64.97B	61.46b

Note: See Table 1.

Another discrepancy between species, is due to 18 : 0, 18 : 1 which generally in all the sites and in the ages considered are significantly more represented in buffaloes rather than in calves.

Frazer (1952) suggested that the reason for more long chain than short chain acids in depot fats, is associated with the method of absorption (short chains via the liver, long chains via the lymphatic system).

TABLE 4

MEANS OF THE FATTY ACID COMPOSITION OF PERIVISCERAL LIPID BY AGE OVER BOTH SPECIES

Age	20 weeks		28 weeks		36 weeks	
Fatty acids	Calv.	Buff.	Calv.	Buff.	Calv.	Buff.
10 : 0	0.27A	0.25a	0.16B	0.15b	0.22A	0.18b
12 : 0	0.36A	0.36a	0.14B	0.15b	0.37A	0.18b
14 : 0	3.43A	2.52a	3.00A	1.55b	3.21A	1.84b
15 : 0	tr.A	tr.a	0.41B	0.17b	0.35B	0.16b
16 : 0	20.75A	17.78a	21.99A	15.18b	21.63A	14.18b
17 : 0	0.32A	0.33a	1.47B	1.41b	1.81C	1.83c
18 : 0	20.26A	22.73a	20.95A	25.34b	16.43B	18.77c
Satur.	45.39A	43.97a	48.12B	43.95a	44.02C	37.13b
14 : 1	1.34A	0.88a	0.63B	0.35b	0.92C	0.45b
15 : 1	tr.A	tr.a	0.29B	0.22b	0.26B	0.17c
16 : 1	5.98A	5.12a	4.51B	3.38b	5.44C	4.29c
17 : 1	0.95A	0.91a	0.96A	0.92a	1.35B	1.27b
18 : 1	39.69A	43.00a	40.55A	46.11b	42.23B	50.63c
18 : 2	5.08A	4.75a	2.88B	3.00b	4.77B	4.78a
18 : 3	1.57A	1.37a	1.11B	1.25a	0.44C	0.85b
Unsat.	54.61A	56.03a	50.93B	55.23a	55.42A	62.43b

Note: See Table 1.

Comparing fatty acid content in different sites of deposition, the percentage of SFA, according to Banks and Hilditch, 1936, Hilditch and Zaky 1941, Dugan et al., 1952, Ostrander and Dugan 1962, Waldman et al., 1968, Pothoven et al., 1974 increased in both species from external to internal sample locations. This trend most likely accounts to 18 : 0 whose content generally increases in the following order: subcutaneous, intramuscular, intermuscular, perivisceral and perinephric. This fact is primarily linked to a replacement of 18 : 1 by 18 : 0 in internal fat depots. The change in saturation is also influenced to a lesser extent by a decrease

TABLE 5

MEANS OF THE FATTY ACID COMPOSITION OF PERINEPHRIC LIPID BY AGE OVER BOTH
SPECIES

Age	20 weeks		28 weeks		36 weeks	
Fatty acids	Calv.	Buff.	Calv.	Buff.	Calv.	Buff.
10 : 0	0.27A	0.26a	0.21A	0.17b	0.24A	0.16b
12 : 0	0.39A	0.37a	0.22B	0.19b	0.37A	0.16b
14 : 0	3.20A	2.49a	2.63B	1.47b	3.32A	1.54b
15 : 0	tr.A	tr.a	0.39B	0.18b	0.47B	0.18b
16 : 0	19.02A	17.24a	21.07B	13.57b	20.95B	14.39b
17 : 0	0.36A	0.29a	1.60B	1.47b	1.83C	1.76c
18 : 0	21.39A	23.56a	20.78B	26.11b	20.37B	28.04c
Satur.	44.63A	44.21a	46.90B	43.16a	47.55B	46.23b
14 : 1	1.24A	0.84a	0.57B	0.29b	1.10A	0.51c
15 : 1	tr.A	tr.a	0.44B	0.34b	0.40B	0.17c
16 : 1	6.07A	4.75a	4.31B	3.16b	5.31C	3.57c
17 : 1	0.93A	0.89a	0.99A	0.84a	1.19A	1.00a
18 : 1	40.37A	42.83ac	42.41B	48.55b	39.54A	45.11c
18 : 2	5.24A	5.09a	3.04B	2.66b	3.63B	2.35b
18 : 3	1.52A	1.39a	0.87B	0.84b	0.43C	0.41c
Unsat.	55.37A	55.79a	52.63B	56.68a	51.61B	53.11b

Note: See Table 1.

in 16 : 1. These observations agree with the findings of Dahl
1962 and Ostrander and Dugan, 1962. When the data of the whole
cycle of the researches now in progress are available we
expect to have a clearer picture of the differences between
species.

REFERENCES

Banks, A. and Hilditch, T.P., 1936. The body fats of the pig. II. Some aspects of the formation of animal depot fat suggested by composition of their glycerides and fatty acids. Biochem. J., 26 : 298.

Cook, C.F., Bray, R.W. and Wechel, R.G. 1965. Variations in the chemical and physical properties of three bovine lipid depots. J. Anim. Sci. 24 : 1192-1194.

Cramer, D.A. and Marchello, J.A. 1964. Seasonal and sex patterns in fat composition of growing lambs. J. Anim. Sci., 23 : 1002-1010.

Dahl, O., 1962. Effect of castration on composition of the depot fats of monozygous twin cattle. J. Agr. Food Chem., 13 : 520.

De Maria Carmela, Romita A. and Borghese, A. 1976. Prove comparative fra vitelli bovini e bufalini allevati fino a 20 settimane di età. II. Digeribilità in vitro delle carni, contenuto in idrossiprolina, e composizione acidica del grasso. Ann. Ist. Sper. Zootec., 9 : 147-161.

Di Lella, T., De Franciscis, G., Rendina, N. and Intrieri F., 1972. Indagini chimico fisico bormatologiche sulle carni di vitellone bufalino di oltre 400 kg. Acta Med. Vet., 18 : 61-75.

Dugan, L.R., Maroney, J.E. and Petheram, M., 1952. Study of carcass fats of beef animals. I. The composition of beef brisket fat. J. Am. Oil Chem. Soc., 29 : 298.

Frazer, A.G. 1952. The mechanism of fat absorption. Symp. Biochem. Soc., 9 : 5.

Gillis, A.T., Eskin, N.A.M. and Cliplef, R.L. 1973. Fatty acid composition of bovine intramuscular and subcutaneous fat as related to breed and sex. J. Food. Sci., 38 : 408-411.

Hilditch, T.P. and Zaky, Y.A.H. 1941. Sheep body fats. II. Component glycerides of perinephric and external tissue fats from same animal. Biochem. J. 35 : 940.

Ingle, D.L., Bauman, D.E. and Garrigus, U.S. 1972[1]. Lypogenesis in the ruminant: in vitro study of tissue sites, carbon source and reducing equivalent generatio for fatty acid synthesis. J. Nutr. 102 : 609-616.

Ingle, D.L., Bauman, D.E. and Garrigus, U.S. 1972[2]. Lypogenesis in the ruminant: in vivo sites of fatty acid synthesis in sheep. J. Nutr. 102 : 617-624.

Intrieri F., Minieri L. and Di Lella, T. 1967. Richerche sul contenuto in acidi grassi delle carni di vitelli bufalini sottoposti a differenti diete alimentari. Atti Soc. It. Sci. Vet., 21 : 448-452.

Intrieri, F., Minieri, L. and De Franciscis, G. 1968. Indagini sul contenuto in acidi grassi delle carni di vitelli bufalini allevati con latte ricostituito. Att. Soc. It. Sci. Vet. 22 : 323-326.

Link, B.A., Bray, R.W., Cassens, R.G. and Kauffman, R.G. 1970_1. Lipid deposition in bovine skeletal muscle during growth. J. Anim. Sci., 30 : 6-9.

Link, B.A., Bray, R.W., Cassens, R.G. and Kauffman, R.G. 1970_2. Fatty acid composition of subcutaneous adipose tissue lipids during growth. J. Anim. Sci., 30 : 722-725.

Link, B.A., Bray, R.W., Cassens, R.G. and Kauffman, R.G. 1970_3. Fatty acid composition of bovine skeletal muscle during growth. J. Anim. Sci., 30 : 726-731.

Marchello J.A., Cramer, D.A. and Miller, L.G. 1967. Effects of ambient temperature on certain ovine fat characteristics. J. Anim. Sci. 26 : 294-297.

Marchello, J.A., Vavra, M., Dryden, F.D. and Ray, D.E. 1970. Influence of sex on certain constituents of bovine muscles. J. Anim. Sci. 31 : 707-712.

Ostrander, J. and Dugan, L.R. Jr. 1962. Some difference in composition of covering fat, intermuscular and intramuscular fat of meat animals. J. Am. Oil Chem. Soc., 39 : 178-181.

Pothoven, M.A., Beitz, D.G. and Zimmeri Ann, 1974. Fatty acid composition of bovine adipose tissue and of in vitro lypogenesis. J. Nutr., 104 : 430-433.

Shafrir, E. and Wertheimer, E., 1965. Comparative physiology of adipose tissue in different sites and in different species. In Hand Book of Physiology; ed. A.E. Renold and G.F. Cahill. Section 5: 417-429.

Shorland, F.B. 1953. Animal fats: recent researches in the fats research laboratory. DSIR. New Zealand. J. Sci. Food Agr. 4 : 447.

Terrel, R.N., Suess, G.G. and Bray, R.W. 1969. Influence of sex, live-weight and anatomical location on bovine lipids. I. Fatty acid composition of subcutaneous and intermuscular fat depots. J. Anim. Sci. 28 : 449-453.

Waldman, R.C., Suess, G.G. and Brungardt V.H. 1968. Fatty acids of certain bovine tissues and their association with growth, carcass and palatability traits. J. Anim. Sci. 27 : 632-635.

EFFECT OF FINNISH LANDRACE AND GALWAY BREEDS ON CARCASS COMPOSITION, FAT DISTRIBUTION AND FATTY ACID COMPOSITION OF DIFFERENT FAT DEPOTS IN LAMBS

J.P. Hanrahan*, P. Allen**, and J.L. L'Estrange**

*Agricultural Institute, Tuam, Co. Galway, **University College Dublin, Ireland.

ABSTRACT

The effect of the Finnish Landrace (Finn) breed on carcass composition was examined by comparing male progeny of Galway ewes which had been mated with either Finn, Galway or Fingalway (½ Finn x ½ Galway) rams. Lambs were slaughtered at two weights (16.0 kg and 19.5 kg carcass weight, approximately) and the chemical composition of the meat from the 7 - 12th rib joint was used as an index of carcass composition. Finn ancestry increased the fat content (P< 0.01). Because of evidence from other workers that breeds can differ substantially in fat distribution a second experiment was conducted to examine the effect of breed on prediction equations relating carcass composition with sample joint composition. Breed differences were significant for equations relating half-carcass composition with the fat content of the 7 - 12th rib joint or with the weight of the kidney knob plus channel fat (KKCF). Using the appropriate within breed prediction equations and sample joint composition data from the first experiment the carcass composition was estimated for all lambs. While the effects of Finn ancestry were reduced the progeny of Finn rams were still significantly fatter.

An examination of breed effects on fatty composition of the fat from three locations (rib, tail and kidney regions) showed that Finn ancestry caused a significant reduction in weight % of C18 : 0 accompanied by an increase in the amount of C18 : 1 and a reduction in fat slip (melting) point (P< 0.05).

INTRODUCTION

The Finnish Landrace breed has been used in Ireland as part of a programme to increase the reproductive performance of native lowland breeds, especially the Galway. Concern about the impact of the Finn breed on live animal conformation and an emerging awareness of the problem of soft carcass fat in lambs led to an investigation of differences between the Finn and Galway breeds in carcass composition and the fatty acid distribution in various fat depots. Evidence of substantial breed effects on carcass fat distribution (McClelland and Russel, 1972) raised serious questions about the usefulness of our breed comparisons using sample joint composition and consequently breed effects on fat distribution were studied.

This paper summarises our experiments involving the effects of the Finn breed on the carcass of lambs at market weight. Information is also included on the effects of Texel ancestry on fat "quality".

MATERIAL AND METHODS

Experiment I

In 1972 lambs born in a large flock of Galway ewes were sired by three ram breeds - Finn, Galway and Fingalway (½ Finn x ½ Galway). Ten rams of each breed were used and a random sample of male (castrated at 3 weeks of age) progeny were assigned to be slaughtered as they reached 39 or 45 kg liveweight. Carcasses were measured, after hanging overnight in a chill, after the description of Timon and Bichard (1965). Each carcass was then divided into three portions

a) Leg and loin (including kidney plus associated fat)

b) Middle (6th to 13th rib)

c) Shoulder

The middle joint was subsequently trimmed in the laboratory to a joint consisting of the 7th to 12th rib. This joint was

weighed, boned and the meat minced and sampled for the
determination of protein, fat and moisture.

Experiment II

Entire male lambs, born in 1975 and representing the Suffolk,
Galway, Fingalway and High Fertility breeds were slaughtered
at approximately 38 kg liveweight and the carcasses were split
down the vertebral column. In addition to determining the
chemical composition of different parts of the side the weight
of the kidney knob and channel fat depots (KKCF) were recorded.
This experiment is described in more detail by Allen (1977).

Fatty acid composition

Lambs born in 1974 and 1975 and slaughtered at market
weight provided fat samples for the determination of fatty
acid composition. Fat from three locations was studied :-

P = Perinephric

R = Subcutaneous - region of the 13th rib

T = Subcutaneous - region of the tail

Analytical methods were as described by L'Estrange and
Mulvehill (1975). Finn and Texel ancestry were compared to
Galway ancestry for effects on the weight per cent of the
various fatty acids as well as on fat slip (melting) point.

RESULTS

Experiment I

Table 1 presents a summary of breed means for carcass
measurements and fat percentage in the meat of the 7 - 12th rib
joint. The Table contains least squares means derived from an
analysis using a model which included slaughter class and
birth - type (single or twin) as main effects and carcass weight
as a covariate. Breed differences in the distribution of
carcass weight are significant with the Finn genes causing a
heavier middle and a lighter leg plus loin. This effect is
reflected in the weight of meat in the 7 - 12th rib joint. There

are no differences in carcass length or bone length but bone circumference is considerably reduced in Fingalway lambs. While breed effects on bone weight parallel changes in bone circumference they are not statistically significant. Finnish Landrace rams increased fat percentage by 5 points compared with Galway rams. The differences between Finn and Galway rams were not changed appreciably when the results were adjusted for differences in the weight of protein in the meat of the 7 - 12th rib joint. The results in Table 2 show that the higher meat weight in lambs sired by Finn rams is due to a corresponding increase in the weight of fat.

TABLE 1

ASPECTS OF CARCASS QUALITY ADJUSTED FOR VARIATION IN CARCASS WEIGHT (EXPERIMENT I)

Trait	Sire Breed			Breed Differences
	Galway	Fingalway	Finn	
No. Lambs	44	41	42	
Leg & Loin wt. (kg)	8.7	8.5	8.5	*
Middle wt. (kg)	2.8	2.9	3.0	**
Shoulder wt. (kg)	6.7	6.8	6.6	n.s.
Carcass length (cm)	98.5	99.6	97.7	n.s.
Leg length (cm)	28.0	28.4	28.2	n.s.
Tibia length (cm)	22.2	23.3	22.4	n.s.
7 - 12th rib joint:				
Meat wt. (kg)	1.48	1.52	1.61	**
Bone wt. (kg)	0.32	0.33	0.31	n.s.
L. dorsi area (cm^2)	13.7	13.2	12.3	**
% fat in meat	28.2	30.8	33.2	**
Cannon bone wt. (g)	53.0	52.0	49.0	n.s.
Cannon bone circumference (cm)	4.8	4.7	4.4	**

n.s. = not significant; *P<0.05; **P<0.01

TABLE 2

DIFFERENCES BETWEEN THE PROGENY OF GALWAY (G) AND FINN (F) RAMS IN THE
COMPOSITION OF THE 7 - 12TH RIB JOINT AND IN CARCASS WEIGHT DISTRIBUTION
(EXPERIMENT 1)

Trait	Difference (G - F)	
	Adjusted to same carcass wt.	Adjusted to same protein wt. in joint
Joint wt. (kg)	-0.12**	-0.08*
Meat wt. (kg)	-0.14**	-0.09**
Bone wt. (kg)	0.01	0.02
Fat wt. (kg)	-0.12**	-0.12**
Leg & Loin wt. (kg)	0.18*	0.11
Middle wt. (kg)	-0.17**	-0.18**
Shoulder wt. (kg)	0.06	0.00

*$P < 0.05$; ** $P < 0.01$

There was no evidence for breed by slaughter class interaction for any trait.

Experiment II

Carcass weight and aspects of fat weight are summarised
in Table 3 which shows that the Fingalway lambs have more KKCF
than Galway lambs as well as having more fat in the meat of the
7 - 12th rib joint. At the same carcass weight the Fingalway
lambs have 10% more fat in the whole carcass compared with 18%
more fat in the 7 - 12th rib joint and 43% heavier KKCF depot.

The linear relationship between fat weight in the half-
carcass and fat weight in the meat of the 7 - 12th rib joint
or weight of the KKCF depot was examined to evaluate the
importance of breed differences in the regression coefficients.
The results of this analysis are summarised in Table 4. There
is significant heterogeneity among breeds in the relationship
between fat weight in the side and depot weight. Thus,
considerable bias may result if breed comparisons rely on the
composition of a sample joint to estimate the carcass composition

unless appropriate prediction equations are derived within breeds.

TABLE 3

COMPARISON OF 4 BREEDS FOR WEIGHT OF FAT (kg) IN THE CARCASS (EXPERIMENT II)

Breed	Carcass wt. (kg)	Fat wt. in side* A	Fat wt. in side* B	Fat wt. in 7-12th rib meat	Wt. of KKCF
Galway	18.5	1.95	1.97	0.44	0.33
Fingalway	17.7	1.95	2.16	0.52	0.47
High Fertility	18.6	2.07	2.07	0.46	0.44
Suffolk	19.9	2.07	2.01	0.41	0.44
Breed differences	-	-	n.s.	n.s.	*

*A = actual wt.; B = adjusted for carcass wt.; *$p < 0.05$

TABLE 4

THE REALTIONSHIP BETWEEN FAT WEIGHT IN THE HALF CARCASS (Y) AND FAT WEIGHT IN THE MEAT OF 7 - 12TH RIB JOINT (X_1) OR WEIGHT OF THE KKCF (X_2) (EXPERIMENT II)

Breed	Regression Coefficients	
	Y on X_1	Y on X_2
Galway	3.32	4.18
Fingalway	2.61	2.33
High Fertility	2.33	2.51
Suffolk	3.44	3.46
F - ratio for heterogeneity	3.65*	2.93*

Fatty acid composition

Analysis of the influence of slaughter age and carcass weight showed significant effects on weight per cent of myristic (C14 : O) and stearic (C18 : O) acids in fat from all three locations studied with the exception of C14 : O in tail fat. Effects on fat slip point were also significant. Consequently, in examining breed effects of fatty acid composition and slip point, carcass weight and slaughter age were inlcuded as covariates. Least squares means for breed groups are given in Table 5 and Table 6. The Tables show values for C18 : 0, C18 : 1, and slip point. There was no suggestion of any breed differences for C14 : 0, C16 : 0, C16 : 1, C18 : 2, total branched chain acids or total odd-numbered C acids (13 : 0, 15 : 0 and 17 : 0). Significant (P < 0.05) breed effects were observed for C18 : 0, C18 : 1 and slip point in at least one location. Averaged over all three locations Finn ancestry increased the weight percent of C18 : 1 and reduced C18 : 0 compared with Galway ancestry. Effects on slip point were consistent with the changes in C18 : 0 and C18 : 1 (l'Estrange and Mulvehill, 1975). There was no evidence for any difference between the effects of Galway and Texel ancestry.

TABLE 5

COMPARISON OF THE PROGENY OF GALWAY (G) AND FINNISH LANDRACE (F) RAMS WITH RESPECT TO ASPECTS OF FAT COMPOSITION AND SLIP POINT

Breed of ram	No. of progeny	Fat depot	C18:0 (g/100g)	C18:1 (g/100g)	Slip point ($^\circ$C)
G	25	R	24.5 ± 1.5	39.1 ± 0.8	41.2 ± 0.6
	11	T	17.6 ± 1.2	39.0 ± 1.4	43.6 ± 0.7
	11	P	30.7 ± 1.2	35.8 ± 1.0	38.9 ± 0.6
F	31	R	19.7 ± 1.4	39.6 ± 0.8	38.9 ± 0.6
	10	T	15.3 ± 1.4	45.7 ± 1.2	37.2 ± 0.8
	10	P	28.4 ± 1.4	36.0 ± 1.6	42.9 ± 0.7
G minus F (averaged over all depots)			$3.2\pm1.1*$	$-2.5\pm1.0*$	$1.5\pm0.5*$

+P < 0.05

TABLE 6

COMPARISON OF THE PROGENY OF FINGALWAY (FG) AND FINN x TEXEL (FT) RAMS
WITH RESPECT TO ASPECTS OF FAT COMPOSITION AND SLIP POINT

Breed of ram	No. of progeny	Fat depot	C18:0 (g/100g)	c18:1 (g/100g)	Slip point ($^{\circ}$C)
FG	26	R	23.9 ± 1.4	37.0 ± 0.8	41.1 ± 0.6
	12	T	17.7 ± 1.1	39.3 ± 1.0	38.8 ± 0.6
	12	P	31.3 ± 1.2	37.0 ± 1.2	44.0 ± 0.6
FT	20	R	22.9 ± 1.8	37.9 ± 1.0	41.0 ± 0.7
	17	T	19.1 ± 1.0	40.5 ± 0.9	38.9 ± 0.5
	17	P	31.1 ± 1.0	35.6 ± 1.1	44.3 ± 0.5

No significant differences between the two sire breeds

DISCUSSION

The results on the effects of Finn ancestry on carcass
composition are consistent with previous reports in showing an
increased fat content (Nitter, 1974; Dickerson, 1974 and
Boylan et al., 1976) and a relatively greater development of
internal fat deposition (KKCF). The results on fat distribution
emphasise the risk of serious bias when sample joints or single
depots are used to rank animals or groups with respect to
carcass fat content. Thus, while Fingalway lambs (at the same
carcass weight) have approximately 10% more fat in the carcass
than Galway lambs, the corresponding figures for fat in the
7 - 12th rib joint and KKCF are 18% and 43%. This bias could
be particularly misleading in calculations of the overall
economic consequences of breed substitution.

Another factor which must be taken into consideration is
the difference between breeds in bone weight. In Experiment II
Fingalway lambs had a significantly (P < 0.01) lower weight of
bone in the leg joint (tibia + femur + ½ pelvis) than Galway
lambs (0.07 $\pm$ 0.02 kg). Breed differences in Experiment I,
while not formally significant (0.05<P<0.10), were in the same
direction. Thus, at the same carcass weight, Fingalway lambs

have more meat. In an attempt to correct for this difference the weight of fat in 7 - 12th rib joint (Experiment I) was adjusted for differences in joint protein weight. The adjustment did not, however, eliminate the significant difference between the progeny of Galway and Finn rams (Table 2).

While differences in degree explain many observations on breed differences in composition (McClelland et al., 1976) there is also evidence of breed (or line) differences in chemical composition at mature weights (Hayes and McCarthy, 1976; McClelland et al., 1976). This point was investigated in the present data by comparing Fingalway lambs slaughtered at the low weight class (16.3 kg carcass) with Galway lambs slaughtered at the heavier weight (19.4 kg carcass). These two groups represent the same degree of maturity at slaughter since Galway ewes are 10 kg heavier than Fingalway ewes (Hanrahan, 1976). The Fingalway lambs had 33.8% fat in the 7 - 12th rib meat compared with 30.1% for the Galway lambs and the difference (3.7 ± 1.84) remains significant. This result may reflect either differences in fat distribution or breed differences in chemical composition at maturity or both.

The results of our examination of breed effects on fat "quality" show that the Finn breed reduces fat slip point relative to the Galway and Texel breeds. This change is accompanied by an increase in the amount of the unsaturated fatty acids. The implication of these changes for meat quality/acceptability are not known.

REFERENCES

Allen, P. 1977. Experimental studies of genetical variation in the pattern
of fat deposition in mice and sheep. Ph.D. thesis, University
College, Dublin (1977).

Boylan, W.J., Berger, Y.M. and Allen, C.E. 1976. Carcass merit of
Finnsheep crossbred lambs. J. Anim. Sci. 42 : 1413-1420.

Dickerson, G.E. 1974. Crossbreeding performance of Finn and domestic (US)
breeds of sheep. In "Proc. working symposium breed evaluation and
crossing experiments" (Research Institute for Animal Husbandry,
Zeist, Netherlands) pp. 421-430.

Hayes, J.F. and McCarthy, J.C. 1976. The effects of selection at different
ages for high and low body weight on the pattern of fat deposition
in mice. Genet. Res. Camb. 27 : 389-403.

Hanrahan, J.P. 1976. Repeatability of ovulation rate and its relationship
with litter size in four sheep breeds. Proc. E.A.A.P. (Zurich)
G30 : 1-8.

L'Estrange, J.L. and Mulvehill, T.A. 1975. A survey of fat characteristics
of lamb with particular reference to the soft fat condition in
intensively fed lambs. J. Agric. Sci. Camb. 84 : 281-290.

McClelland, T.H. and Russel, A.J.F. 1972. The distribution of body fat
in Scottish Blackface and Finnish Landrace lambs. Anim. Prod. 15 :
301 - 306.

McClelland, T.H., Bonaiti, B. and Taylor, St. C.S. 1976. Breed
differences in body composition of equally mature sheep. Anim. Prod.
23 : 281-293.

Nitter, G. 1974. Results of a crossbreeding experiment with sheep for
intensive lamb production. In "Proc. Working Symposium Breed
Evaluation and Crossing Experiments with Farm Animals Zeist". 459-474.

DISCUSSION ON SECTION I

B.L. Dumont *(France)*

When Dr. Kauffman speaks of 'extreme types of double muscling', to what extremes does he refer?

R.G. Kauffman *(USA)*

That would be practical extremes including the long-horned type of animal on the one hand and double muscled on the other.

R. Boccard *(France)*

What is your basis for typifying the double muscled characteristic?

R.G. Kauffman

The characteristics that we use are those that were described by Cartwright, of the general phenotypical expression as shown in the young animal at a stage when there was no fat, a sloping rump, a small head, the very bulging and muscular rumps on the animal corresponding to very high dressing percentages.

A.J.H. van Es *(Netherlands)*

I would like to put a question to Dr. Kauffman. He said that the production of lean tissue at a low age and at a high age, had the same efficiency. Did he include maintenance costs of development of the dam in his figures?

R.G. Kauffman

No.

A.J.H. van Es

It was just the deposition of the metabolisable energy necessary for just lean tissue deposition without any maintenance, without any reproduction, etc?

R.G. Kauffman

The maintenance cost would have to be included in the way that we expressed the data because it was kgs of feed consumed to produce a kg of fat-free muscle. That is the way we expressed it but maintenance costs would be automatically built into that, though no maintenance cost for the dam.

A.J.H. van Es

I think this is still difficult to understand. If you have an animal weighing 200 kg producing about the same amount of lean per day as an animal of, say, 500 kg, that animal of 500 kg requires much more maintenance.

R.G. Kauffman

We were not able to measure that component and I am very concerned about maintenance myself.

I would like to address Mr. Robelin concerning the reason why you have a differential proportion of subcutaneous fat and intermuscular fat in bulls as compared to steers?

J. Robelin *(France)*

When the basis of comparison is the weight of subcutaneous fat plus intermuscular fat, the bulls have less subcutaneous fat than the steers for example - but why? Perhaps it is through selection, for example, the Hereford or Angus have both been selected on their external appearance of fat,

R.G. Kauffman

It seems possible that you are studying animals at different stages of maturity and that the b-values for subcutaneous and intermuscular fat are different at different stages of development and because the bull is later maturing, perhaps you are looking at two different b-values.

J. Robelin

The comparison, at the same weight of subcutaneous plus intermuscular fat takes account of the difference in maturity. I have shown that the corresponding carcass weight of all bulls as 320 kg and the corresponding carcass weight of Friesian and Hereford steers, for example, was approximately 200 kg. I am

not sure that the difference in maturity is not taken into account
in this case - but what exactly is maturity?

J. Kempster (UK)

I think you would solve this problem if you expressed the
fat as a percentage of carcass weight rather than as a percentage
of total fat weight because there is a varying proportion distribution in the carcass. If we were talking about physiological
age, then a percentage of carcass weight would probably be better
than a fixed fat weight as your dependant variable.

J. Robelin

Probably at the same fat percentage of carcass weight the
Hereford steers are younger than the bulls. It will not be a
very good measure of maturity.

J. Kempster

It would be better than expressing it with your dependant
as total fat because that makes the range of maturity wider than
if you express fat as a percentage of carcass weight and make
that your dependent variable.

R.W. Pomeroy (UK)

I think Mr. Robelin is comparing his data from bulls with
some of ours, masquerading under the title 'Anon', for steers,
and Dr. Kempster's. If you are making comparisons of this sort,
you have to be quite sure that the dissection techniques are
strictly comparable - in other words, what you call subcutaneous
fat and what we call subcutaneous fat are the same thing - otherwise differences between bulls and steers become confounded with
differences in dissection techniques. I do not say they were in
this instance, but it is a problem which bedevils all beef cattle
research - the definition of terms and adequacies of samples,
balances of breeds and sexes, proportions of mature weight, etc.
A lot of beef cattle research, in my opinion, has added to our
confusion without substantially increasing our knowledge, also,
because of the very heterogeneous collections of animals with
respect to breeds, sexes and levels of nutrition which are used

in the studies. This is in no way a criticism of Mr. Robelin's
paper. I am simply pointing out some of the defects which are
forced upon us in many cases by lack of money.

J.F. Harte *(Ireland)*
    Dr. Pomeroy might have added that one could attribute the
differences to differences in breed as well, because Callow's
(and the other Cambridge work) was on Herefords and Robelin's
cattle were Charolais crosses.

    In relation to the point that Dr. Kauffman made, I think
where bulls and steer carcasses of the same breed were taken to
quite heavy weights, similar differences were reported. Could
growth physiologists give us reasons why these differences occur?

R. Boccard
    It was a very hard problem to choose a basis to compare
animals. For a long time it was age, now it is weight and carcass
weight. I would propose muscle weight only. The weight itself
is the best basis to compare different breeds and different conditions of production. The physiological age will be discussed
later on and certainly we will need better criteria than only
weight or even fat in the carcass.

V.R. Fowler *(UK)*
    I would like to suggest that some part of this problem is
rooted in the history given to us by Dr. Pomeroy. In fact,
Hammond and Palsson, as I recall, were very aware of Huxley's
equation and they thought that their experiments with differential nutrition actually showed that Huxley's opinion on the
Fiddler Crab did not apply to farm animals because, if you could
produce animals of different fatness, then you could not predict
their composition from a knowledge of their weight. It was
Wallace who pointed out that, if you chose the correct independent variable then many of the differences that were apparently
there when the carcass weight was the independent variable, disappeared when you compared on a 'within skeletal' basis or on a
'within muscle' basis. A lot of the momentum of the Hammond
School failed when they did not appreciate that Hammond's equation

could apply to other independent variates apart from carcass weight.

R.W. Pomeroy

Huxley himself also paved the way for Hammond and Palsson to be misled because, if you remember, that constant in the Huxley equation - the ratio of the specific growth coefficients - Huxley said that it was *a priori* unlikely that this would not be affected by environmental agencies like plane of nutrition, so Huxley really expected, I think, plane of nutrition to affect that constant. He was unduly modest: that constant has proved to be more robust than he anticipated. Nevertheless, as we shall hear later on, it does change during the growth of the animal; it is not nearly as constant as had been implied during recent years.

L. Buchter *(Denmark)*

Professor Lawrie, I would not cast doubts on the New Zealand work showing that you are likely to get more cold shortening in older animals but, when you have an older animal you have a bigger muscle mass and therefore you are less inclined to get cold shortening than with a young animal where you have no protection from fat and only little muscle mass.

You say, 'less conditioning' - I don't know about the amount of cross-linkage broken, but in practice it appears that the conditioning is much faster in the younger animals. For instance, in the six month-old it is finished within one week, while, for a one or 1½ year-old, it takes two weeks; and for the older animals it takes three weeks to reach the maximum amount of conditioning.

Concerning the lower pH in the older animals, this is not in agreement with our findings. When the animals have been treated identically through transportation and slaughtering, and if we speak only of the *l.dorsi* , we tend to get a slightly lower pH in the younger animals compared with the older ones, of the order of 5.45 in the younger animals, 5.50, 5.55 in the older animals - but again, transportation is most important.

I should like to mention a very interesting finding in Denmark, concerning the myoglobin content. In an experiment where we had an animal growing at a maximum rate, an animal growing at 55% feeding level, it was not the age of the animal, it was absolutely the body weight of the animal which determined the myoglobin content and which you implied here.

R.A. Lawrie (UK)

Thank you, Dr. Buchter. May I briefly refer to each of your comments in turn.

I realise that, of course, the dimensions of the younger animals are different from those of the older ones, and that this cuts across the argument here. This is a shorthand situation and I am assuming that, given the same degrees of cooling, then you are going to get more cold shortening with the older animals.

As for the question of conditioning - I think we are on the same wavelength here - I am suggesting that there is less conditioning in the older animal and more in the younger one.

Regarding the question of the pH: this, as I said, was a representation of a reverse, in a sense, of the situation. The higher the pH you have, the more tender it is, so, the reverse then, is that the lower the pH the less one tends to get for that reason. As you rightly point out, the stress susceptibility is a factor in this. This, I would suggest, would be more in younge animals and less experienced animals. The glycogen would tend to be higher for this reason. There would tend to be more glycogen in the older animal because of less bodily movement and a lower pH, and for that kind of reason one would tend to get, other thin being equal, a less tender situation than if you had had stress, with a high pH. I realise that there are two aspects of stress - a fast pH fall and a high pH level. I am talking about the latte

Finally, the point about the myoglobin: I have put down, 'older, redder' but I am implying 'larger' and the dominant situation is that I am presuming that as an animal gets older it gets bigger; but it is the size of the muscle fibre, not the age per se that I am really referring to.

R. Boccard

Thank you. We will continue with the discussion later. Now we come to consider some aspects of connective tissue.

J.D. Wood *(UK)*

Monsieur Boccard, in Table 2 of your paper you show the concentrations of hydroxyproline in normal and double muscled cattle. Can you tell us if those differences are due to different weights of total muscle? Are the animals of the same age? Have they got different muscle sizes and is that the reason why the concentration is lower?

R. Boccard

Animals were killed at the same age and the weights are a little different but in any case we can consider the variation of collagen content as a 'dilution' of the collagen by the hypertrophy of the muscle. The collagen content is depressed in the whole musculature, even in the muscles which are not hypertrophied. It is a specific characteristic of this type of animal, and, in my opinion, the best criterion to make sure that they are double muscled animals.

R. Daenicke *(West Germany)*

What was the meaning of the influence of collagen content in respect of solubility when you referred to broiled or cooked meat instead of the untreated meat?

R. Boccard

When you consider raw meat and collagen content you have a good relationship between the hydroxyproline content and the shearforce value. When you cook your meat you can get variations owing to the content and the solubility of collagen. If you consider muscles with a low collagen content, if you heat them, the toughness will increase because there is shrinkage of the whole network, so, for instance, the *l. dorsi* will be more tough at 25° or 70°C, than at 55°C. If you carry on cooking, especially when using muscles with a very high collagen content, you destroy the network of collagen and the tenderness is increased

R.W. Pomeroy

You showed a slide with, I think, increased solubility of collagen in bulls compared with cows and steers, occurring around puberty possibly. The bull is notoriously more stress-susceptible than the steer, for example, and what I am asking is whether there is an inherent difference in solubility, or were the bulls subjected to more pre-slaughter stress?

R. Boccard

I think collagen is not sensitive to stress. We have, in this case, the production of new collagen in bulls. You have an increase in muscle weight, but it is evident that the first part of the muscle to be synthesised is the framework of connective tissue. Therefore, in young bulls, you have an increase in muscle weight, but first there is an increase in the connective tissue which has a low cross-linkage content. In this case I think we see a dilution of the whole collagen already in the framework. It is mainly this aspect of the growth of connective tissue which can explain the variation in solubility, especially at this period.

Afterwards we have an increase in the myofibrillar and contractile elements of the muscles and we again reach a normal content of collagen. This collagen in its turn is changed by reticulation and new cross-linkage.

D.N. Rhodes (UK)

Could I make a comment on some of the interactions between the two papers we have had, and some factors which have not been mentioned.

First of all, the collagen story, as brilliantly exposed by our friends from Theix, is coherent within itself but cannot be considered except in relation to cooking methods and intended

use of the meat. For example, if you are going to can beef to
be used for stewing, it is no good putting young beef into cans
and heating to 125°C for 90 minutes, because, when it comes out,
it is just pulp. Strong collagen from a 9 year old cow is extremely good material for this purpose. Most of the thinking
about toughness really concerns producing meat for steaks. One
would think, then, that we are in a comparatively simple situation because steaks are eaten from Alabama to Moscow, and
everybody knows what a good steak is. However, even at this
situation, it is very different. Steak in England is a piece of
meat about 1" thick which is heated to an internal temperature
around, perhaps 65 - 70°C, whereas steak in France is a piece
of meat of the same thickness which is heated to an internal
temperature of perhaps 37°C - or even 45°C if you want it well
cooked! These final products are very different from the point
of view of the effect of heat, especially on collagen.

The second point is the interaction between the toughness
due to collagen in the meat and to the myofibrillar proteins.
The points that Dr. Lawrie has elaborated are all pertinent.
Most of them, however, can be eliminated by processing control -
for example, you need not worry about cold shortening if you are
careful not to get cold shortening - and similarly with most of
the others.

If you take a large body of data as we have accumulated in
our work during the last ten years at MRI, one is left with a tremendous variability in toughness, for factors which I must admit
one still cannot explain - and I don't mean small variability.
It is a large variability, and one of the main difficulties
facing the meat industry is to find and eliminate this variability.
When a housewife asks for 'another piece of meat like the piece
she had before' what can the butcher do to produce it? Just
nothing. We can't tell him, even now, how to control breed, sex,
age, slaughter, hanging, pH, cold shortening, and so on. We
are still unable to eliminate this very large variability between
animals.

B. Bech Andersen *(Denmark)*

I have a question for Dr. Bergstrom. Could you please pub-
lish these results on individual variations in muscle distributi‹

P.L. Bergstrom *(Netherlands)*

The data refer to animals of different treatment groups, and
they need further elaboration and completion. However, I can
include these results in the Proceedings, but only as raw materi‹
in the form in which they are given here (see Table 1).

H. de Boer *(Netherlands)*

Perhaps I missed the point in reading the paper: the ratio
of weight to length of bone, its explanatory value - you did
not deal with it, I think, Dr. Kempster?

J. Kempster

It is probably easier to turn to the Table in the text (Tabl
6). We cannot conclude very much on bone density aspects from
erroneous statistical approaches. All that column says is that
density is not a linear function of weight and length. It is th
density factor that we should look at. The results are not
very clear, nor are they very consistent and I don't think that
I would to draw any conclusions from them at all.

H. de Boer

This is not the aspect to which my question referred.

R. Daenicke

I am wondering whether you can make these comparisons withou
knowing the energy and nitrogen intake of these animals, because
we have found that with restricted energy intake in the Simment
nitrogen deposition is quite impaired.

J. Kempster

The first two Tables, the one from Sutton Bonington and
Ingleston, were within system comparisons. They were a contemporary comparison of sire breeds on those animals, compared contemporaneously over two year systems with a year taken out. The

TABLE 1

MUSCLE WEIGHT DISTRIBUTION IN FEMALE MONOZYGOUS TWINS

The data of individual animals are not corrected for eventual influences of different environmental conditions in early life. Absolute weights of two very small muscles are given in g and weights of seven randomly chosen muscles are expressed as percentage of the total side muscle weight.

Twin pairs	musc. subclavius	musc. omohyoideus	musc. brachioceph.	musc. scalenus ventr.	musc. iliocostalis	psoas group	musc. serratus ventr.	musc. gluteobiceps	musc. semitendineus	Total side muscle weight in kg
A	57	13	1.46	0.52	0.44	2.94	4.92	5.83	2.77	103
	48	14	1.54	0.58	0.47	2.84	4.99	6.59	2.69	100
B	64	27	1.51	0.47	0.53	2.77	4.73	7.02	2.62	105
	61	27	1.40	0.46	0.48	2.70	4.54	7.01	2.49	102
C	54	27	1.27	0.40	0.50	2.81	4.91	7.08	2.49	109
	63	31	1.25	0.37	0.50	2.75	4.93	6.94	2.65	114
D	57	66	1.55	0.56	0.54	2.85	4.73	6.64	2.59	129
	70	68	1.50	0.58	0.57	2.71	4.77	6.49	2.59	127
E	42	6	1.60	0.47	0.53	2.44	4.67	7.29	2.63	86
	42	11	1.67	0.47	0.52	2.37	4.69	6.78	2.60	88
F	38	31	1.43	0.45	0.59	2.94	5.24	6.88	1.99	98
	39	38	1.39	0.44	0.59	2.90	5.15	6.68	1.94	91
G	67	46	1.35	0.66	0.53	2.61	5.23	6.23	2.38	129
	62	42	1.37	0.69	0.54	2.40	5.17	5.90	2.53	122
H	55	43	1.35	0.51	0.44	2.85	4.40	6.93	2.37	114
	45	39	1.34	0.49	0.42	2.81	4.22	6.98	2.44	104

Weight in grams   Weight in percentages of the total side muscle weight = 100%

The variation within twin pairs is small in comparison to the variation between pairs.

only results that were not contemporaneous were those shown in the slide for the Limousin and the Simmental Steering Committee tests, and there were quite clear differences between the two systems. Because they were not contemporaneous I showed them in two colours. One of these was cereal fed and slaughtered at much lighter weights - Friesians, Herefords, Simmental, Limousin; the others were grass cereal fed giving a difference in nutrition there, otherwise everything else was contemporaneous with every-thing fed the same diets.

J.D. Wood

Would you like to mention the results of Fursey from MRI who found that there were quite large breed effects on bone density?

J. Kempster

I referred to this in my paper. I could not find anything else in the literature on bone density. His results do not re-late well to ours because he made a comparison on an equal age basis.

D.N. Rhodes

This is a question which I have to put to Dr. Kempster but which might be interesting to consider in this discussion. The correlation that he put on the board between breeds was rather unsatisfactory, and the question I asked him was whether he had a more satisfactory relationship within breeds.

J. Kempster

To some extent we anticipated this question and we brought with us a paper which summarises most of MLC work on conformatio There are a dozen numbers which summarise the whole situation, which are the standard deviations for the percentage lean and the correlations that we get within and between breeds.

If we take percentage lean in carcass, the overall standard deviation is about 1.75. Within breeds the standard deviation is about 1.3. This is over all the data that we have. The overall correlation between conformation and percentage lean in carcass, again at constant external fatness, is 0.35; and the within breed correlation 0.15. The correlations with percentage lean distribution in the high price cuts, which we heard a little about earlier, the standard deviation overall was 1.1, within breeds 1.0, the overall correlation 0.3, and the within breed correlation, 0.1.

That did not exactly answer the question, but it is probably more relevant to look at leanness because there are a lot of obscuring factors. The correlations with saleable meat to bone are somewhat higher - about 0.5 - but they are not very large. Looking at the national slaughter population in Britain, to put the three axes in perspective, we think that among carcasses of different fatnesses, there is a range which is worth about £80, at retail across the counter. The range in conformation mediated through lean to bone ratio is about £20 across the counter, at the same level of fatness; and the range in lean distribution is worth about £5. There is a 5 - 20 - 80 relationship and it is only about £20 that we are talking about in the range.

R.A. Lawrie

Earlier, when we were discussing the implication of biochemical factors in muscle, in older animals versus younger ones, I attributed possibly that some of these biochemical factors would have a bearing on toughness, water holding capacity and other features of eating quality. Dr. Rhodes pointed out that, of course, the method of cooking, and whether cold shortening had been allowed etc., would affect the issue. This is, of course, true. He also talked of the undetermined sources of variation and I just wanted to make clear that one appreciates this very well. In fact, we ourselves had two sibling Friesian/Shorthorn cross heifers, which had been fed the same way and were killed at the same approximate conformation. One of these was commented on very severely by the butcher as being incredibly tough and, when we looked at the under fillet, the Psoas muscle, we found that it contained six times the hydroxyproline content of the sibling. I am only saying that to emphasise that there is a great deal we just do not know. We may be able to control

some factors within limits, avoiding cold shortening, perhaps
by putting some papain in to overcome some of the more severe
aspects of connective tissue, but there is still a lot we do
not know, and I concur with Dr. Rhodes' comment on this. The
only redeeming feature about it is that we all ought to be in a
job for some time to come!

A. Cuthbertson *(UK)*

I thought it might be interesting to tell you a little about
what we are starting to do, to look a bit further than just
straight saleable meat at the primal cut level, which is what
Tony Kempster was talking about. We started to look at the effec
of conformation through to the prepared retail cut stage. The
work is at a very early stage but what we have done is to take a
bunch of cattle in a commercial situation of three weight classe
and of similar fatness but with markedly different conformation
in British terms. In our classification scheme we have 5 basic
classes, and we selected conformation classes 1,3 and 5, and the
5 (which is the best conformation) corresponds, perhaps, to 4 on
the EAAP scale - so we are not dealing with the exaggerated
muscular type of animal, but nevertheless a good conformation
animal by British standards.

What I have done is to take the least squares means, taking
out the effect of weight class and any effect of fatness within
a fat class. We find that there was a difference of weight between the different conformation groups, but this could be taken
out by further analysis. Looking at the area of the *l. dorsi* , a
the 10th rib, we appear to get a marked effect in eye muscle are
but, if we look at it at the last lumbar vertebra and at the
6th rib, we do not get a significant effect due to conformation.
With percent saleable meat, there was a significant effect on
that. The percentage which occurs in the higher price cuts show
a significant effect and there was also a significant effect in
the percentage of lean trimmed from the high price cuts. This i
the amount of lean meat which is trimmed from the high price cut
in the course of preparing them to the primal joint stage, so
what this means is that the animal with good conformation, in th

course of preparing it to primal cut stage, is having to lose
rather more lean meat in trimmings, presumably to face up the
cut to make it saleable. You have to cut through a greater
cross-sectional area.

The next stage is to take these primal cuts and break them
down into prepared retail cuts. We did this work in collaboration
with a large wholesaling firm which prepared meat down to the
prepared retail cut stage. They prepared all this meat for us
and the first set of data relates to the percentage of sirloin
steaks in the trimmed primal cut of sirloin. There was no significant difference due to conformation in the percentage of
steaks that one got from the sirloin. The same goes for the
thick flank, that is the quadriceps group of muscles in the hind
limb. There was no effect due to conformation. Indeed, if you
go through the other cuts we have been unable to detect any effect
of conformation. We now need to look at the effect of eye
muscle area as such rather than at conformation to see what effect
different ranges of eye muscle area may have on these results.
This is a point which is worth making in relation to the other
analysis which Tony Kempster mentioned, where we showed an important effect of breed on saleable meat yield. It could be that
adding eye muscle depth or area instead of conformation might
explain more than conformation.

I would be interested to know if other people have evidence
of this sort, particularly going to the prepared retail cut stage.

R. Boccard
Certainly it is beginning to be well known now that the influence of conformation on the distribution of meat inside the
carcass is not so large as formerly thought. We have to consider
more and more the quality aspects, in particular in relation to
tenderness.

A.J.H. van Es
A short question to Mr. Dumont. You mentioned that the number of fibres might change during growth, and might even be lower.

How can he understand that from a point of view of histology?

B.L. Dumont

I said that according to the results from Bendall and Voyle, it could be that the number of myofibres decreased during growth. At birth the number is quite fixed, it should remain constant for up to one year and thereafter it would probably decrease. I have said, "according to Bendall and Voyle". I don't agree entirely with them, because, from our own experiments, we have not found such large variations between animals of various ages, and I was speaking earlier of some experiments made on the *Brachialis* muscles, and we found only a very small decrease in the number of myofibres according to age by comparing the animals of 2 years, 6 years, 7 years, 10 years and upwards. However, it is important to know if the number of myofibres which we find in the young animals remains the same during growth or whether it changes, in particular for muscle growth studies. It is an open question at the present time.

A.J.H. van Es

Could it be that Bendall has measured the thickness of the number of myofibres in an incorrect way?

R. Boccard

I think it was obtained by calculation, the whole area of the muscle, divided by the area of individual fibre.

B.L. Dumont

The reduction was very important, even drastic. If we consider that at birth the number of muscle fibres for one muscle is fixed, the only way in which we can increase muscle thickness is to increase the diameter of each fibre. In addition, we would have to fight against the diminution of fibres, which is a very difficult problem.

M.J. Clancy *(Ireland)*

We could regard muscle as a dynamic situation, where you have wastage and renewal of myofibres and, in this connection,

satellite cells could be very important because it is thought
that these are the basis of new myofibres that take place, and,
where you see reports of a decrease with age in myofibres, it
could be that the dynamic equilibrium is disturbed and we no
longer have a supply of satellite cells to produce new myofibres.
I agree with Dr. Dumont that this is a very important question
in meat production.

B.L. Dumont

It is important in view of selection of animals on the number
of myofibres. It is clear that the number in a given muscle
differs from animal to animal, so, if we want larger muscles, we
have to select for animals with a high number of muscle fibres.
We must hope that they will still have the number of fibres when
they are grown.

R. Boccard

Have you any method to propose for measuring the number of
fibres at different ages?

B.L. Dumont

I have no special method to recommend but I can tell you
what I do. It is very simple: I take a cross section of the
muscle and make a photographic enlargement of this and put the
photographs on the wall; then I use a pencil and I count - un,
deux, trois .....

We have studied the *Brachialis* muscle in different types of
animal. It is a very tedious task; a very important paper on
this question was delivered by Swatland in the last Reciprocal
Meat Conference which can be referred to.

J.D. Wood

It is a fairly important point, whether muscle fibre number
is fixed. In Goldspink's work with pigs, he assumed that muscle
fibre number was fixed in early life and he showed quite good
correlations between the number of fibres estimated in early and
in late life, and total muscle mass. This does suggest that the

number of fibres is constant. That is the assumption in the
Irish work too, is it not?

R.B. Thiessen *(UK)*

Could I make a remark on Dr. Kauffman's paper when he was
talking about the wide variation in muscle to bone ratio? Apparently one of the animals was a double muscled animal and this
greatly enlarged the range in muscle to bone ratio.

On the subject of using a double muscled sire, which Dr. Boccard remarked on this morning, and the crossing with normal
animals to improve the tenderness by lowering the collagen content, I was associated with a programme in California where you
also got an advantage of about 3 or 4% in dressing percent, about
3 or 4% in increased muscle yield, and about 3% increase in growt
rate, so that the overall improvement in retail cuts per day of
age was about 10%. We also tried to fatten a double muscled stee
and had about the same success as Dr. Kauffman, although we had
to keep it on about another year.

D.N. Rhodes

Could I ask Dr. Thiessen if he in fact got any change in
tenderness?

R.B. Thiessen

As I recall it, there was an interaction between bulls and
heifers but I think that the bulls were more tender than the
normals and the heifers were less tender, though whether that
was a true interaction I could not be sure; but when you pooled
them, the overall result was that you got an improvement in
tenderness.

R. Boccard

The relationship between tenderness and collagen content in
double muscled animals is very important. Many more muscles can
be used for quick cooking compared with normal animals.

D. Lister *(UK)*

I want simply to add a cautionary note to the proceedings.
We talked a little about the effects of breed, and I suspect we
shall talk more about this. If you are comparing dairy breeds
with beef breeds, then I think comparisons are reasonable on this
basis, but, as Dr. Kempster brought out this morning, in the
sort of muddy, grey area which exists between these extremes, a
lot of problems start to emerge. This has been fairly clear in
pigs for a long time, where it is now impossible to say on the
basis of the appearance of an animal whether it is physiologically even a Pietrain.

There is the same situation with sheep as well, and I should
perhaps add a note about a recent experience of ours. We wanted
to select some breeds to be characteristic of certain features,
for instance, large size or small size at maturity, or a physiological type which restricted the deposition of fat, or was very
good at depositing fat. In some respects we had no problems and
our preliminary analyses had shown that, for instance, if you
compared Hampshire sheep, which are large and early maturing,
with Southdown sheep, then physiologically they were identical
although they grew to quite different sizes.

Now, it just so happened that our supplier of Southdown
sheep was unable to provide any more, and we assumed that Southdown sheep were fairly rare and, consequently, might be of a
fairly uniform physiological type, so we based our experimental
procedure on this understanding. However, when we got our Southdowns analysed over the following two years, we discovered that,
physiologically, they were more similar to Cheviot sheep which
one expect to be completely different. Nevertheless, they still
looked like Southdowns, though physiologically they behaved like
a fairly distantly related animal.

Now, if you are talking about breeds of cattle, then I suspect that comparisons on the basis of Hereford v Friesians are
probably all right, but, when you have animals in the middle of
the spectrum, such as Aberdeen Angus and even Charolais and

perhaps Shorthorns and some of the red varieties that we have
in the UK, then the situation becomes very difficult to assess,
and you have the whole spectrum of physiological types represente
within this carefully defined range of types.

Perhaps what I am asking for is much greater definition of
what we mean by the phenotype we are actually working with. It
isn't enough to say that we are working with Aberdeen Angus or
with Southdown sheep. They must be characterised in a comprehensive range of ways. I do not quite know how we should go
about this, but it is no longer adequate to say that we are
working with one breed or another because, certainly in the case
of pigs, there is no relationship to speak of between external
appearance and physiology and I suspect that the same is now becoming true within the range of cattle types.

R. Boccard

This is an important problem. It is true we have a very
large variation in physiological characteristics inside breeds;
this is fortunate, because it is a source for possible improvement in the future, but it is very strange, nonetheless. For
instance, in our work on the Afrikaner and on the Friesian, we
found that for pigment evolution the Afrikaner was the earlier
to mature for this characteristic, whereas, for evolution of
solubility of collagen the contrary was true.

J. Kempster

The important point is to sample the breed properly and to
get a true breed mean. Perhaps a lot of the clouding in the
middle is a reflection of the fact that, even with our data,
where we sample the whole national population in a fairly rigid
way, our sampling errors were large in all our work, particularly
in some of the earlier papers, dealing with very small numbers
of carcasses. Sampling error is much more important in the leve.
that we are working at than people imagine, much more so than
accuracy of the dissection techniques used.

A. Cuthbertson

This is not strictly relevant, but, if there is time, perhaps you could outline how you would assess cross linking of
collagen in a simple way which one could apply for monitoring
meat, for example, under commercial conditions. What would your
recommendation be?

R. Boccard

We have two possible ways. For routine work we can eliminate
the determination of this cross linkage by biochemical determination - according to Bailey - which is too complicated for
that purpose. It is possible to determine the heat soluble collagen content by heating in water and weighing the residue of the
solution in which a hydroxyproline determination is to be made.
Perhaps in the future, some physical apparatus could be developed
to measure the tension, because physical measurements are easier
to make on a large scale.

D.N. Rhodes

I presume, Mr. Chairman, that Dr. Cuthbertson is thinking of
predicting tenderness from whatever measurement and, at that
point, it is easier to measure tenderness directly. There are
simpler methods for measuring tenderness after cooking than
attempting to measure any part of the connective tissue structure.

R. Boccard

It is difficult to include all the factors which could affect
tenderness, and different cooking methods. Perhaps as a first
approach we should consider the biological variation of the factors which induce interesting characteristics from the technological point of view. Variations in collagen content, or number
of fibres, or the ageing of collagen, because there is a large
variation in the scale of the various breeds and animals, must
be considered first from the animal husbandry point of view.
It will be too difficult to fix technological conditions to obtain general agreement and to effect improvements in these
characteristics.

R.G. Kauffman

I would like to ask Dr. Boccard why he thinks that the double muscled animal, architecturally speaking, has a different connective tissue component.

R. Boccard

I think the connective tissue content in the muscles of double muscled animals is different from normal. The content is only one aspect. In the past we tried to determine whether we had different distribution of cross linkage of collagen in normal and double muscled animals but we did not come to any difference. Perhaps in the future we shall carry out experiments to determine the different types of collagen and in this way we may find some differences in the distribution of this type of collagen, but for the moment there is no evidence on this.

The framework of the connective tissue, the network inside the muscle, is different in double muscled and in normal animals. Mr. Dumont showed some pictures where it is very easy to see the difference in structure; even for the same amount and the same nature of collagen there is a change in the distribution inside the muscle. All these characteristics are connected with tenderness and the behaviour of the meat during cooking. From the consumption point of view it is important to have the maximum of meat with low collagen content and not to use strong heating - only to 60 or 65°C.

R.G. Kauffman

But why would you expect the connective tissue structure to be different? The fibre size, and perhaps length also, may be quite similar according to California work. Your work is suggest ing that the thickness of the endomysium is involved. Is this correct? And why should it need to have a thinner structured wall?

M.J. Clancy

Would you care to consider the idea that the number of myo-fibres being greater cause a greater stretch of the connective tissue. The Australians have shown that when connective tissue is stretched it becomes thinner.

R. Boccard

I think that the low content in collagen in muscle has to do with the trouble in synthesis of the protein in the double muscled type. It is not a question of dilution or increase of tension inside the muscle. You get a deficit of collagen formation in double muscled animals.

A.J.H. van Es

Couldn't it be a question of the undernourishment of a double muscled animal because it is difficult for the animal to ingest a sufficient amount of food? It has to produce a lot of protein but maybe it does not get enough energy to produce all that kind of connective tissue needed.

R. Boccard

I do not think it was in connection with the intake of animals. It will be interesting to ask research workers concerned with fat cells to consider these aspects in double muscled animals. For the moment we have no answer. I think the deficit of the production of this connective tissue can be connected with the deficit of connective tissue where fat can be deposited also. But we can say that there is a set of evidence showing that animals finding it difficult to produce enough collagen, are low in myoglobin content; they have respiration difficulties, their lungs are small, their heart is light; in general they have trouble with their respiratory systems, so perhaps if we look inside the cells at the mitochondria level, we may find some difficulty at this level as well.

All these aspects have to do with hydroxylation of the proline inside the collagen chain. It is only a hypothesis, but it may be the way along which we can work to explain the reason why double muscled animals are hypertrophied and have a low collagen content.

St.C.S. Taylor

I wanted to make a comment about the relation of one variable to another, in particular, conformation, breed type and
the like. I find the subject much too difficult and the ground
much too shifting to be able to say anything very sensible and
the best I have been able to do in the last few minutes is to
say that you can get relationships which are 'population dependent'. As a very simple example, you can pick Jerseys and
Friesians and you will get a perfect correlation between colour
and size. Of course, if you pick Charolais and Angus, you also
get a perfect correlation between colour and size. These are
what I call 'population dependent' relationships.

Now, you can never get rid of this complexity and this complication: it is always with us. Sometimes, however, it may
disappear, but only if you have a large enough population where
all these trivial examples no longer appear. Dr. Kempster suggested that, if he got rid of the Limousin at the top and another
breed at the bottom, he would change his correlation. I thought
that this was a slightly negative reply and my own instincts were
to say that, if he had put a few more breeds up at the top, and
a few more down at the bottom, then he would have stabilised the
relationship, if it had existed, or, if there had been no relationship, those extra breeds would have been all over the place
What is required, therefore, is more and more breeds to answer
the kind of question he is asking.

That is only the beginning, however. There is a second kind
of relationship which I might call 'experimentally dependent'
which means that it depends entirely on you what the answer is.
This is a well known example: what is the correlation between
the body weight of a set of animals and their food intake? Normally one says that it is very high, around 0.8 or 0.9. This is
a genetic correlation and, because it is genetic, it is fundamental, it is basic, it is built in. But, in actual fact, if you
decide to feed all those animals exactly the same amount of food,
(which is a very simple thing to do experimentally), that basic
genetic correlation suddenly jumps from 0.9 to become exactly

zero; so here again we are in a very difficult field where we have to realise that we are responsible for our own genetic correlations and that, to try and get rid of that responsibility is irresponsible, scientifically.

There is a third kind, I think, though I am not very sure. I think there is a series of gradations. This third kind are what I have called 'biologically dependent' relationships, physiologically dependent relations, or some such word. Here we are dealing with things like Krebs' cycle which is the same whether it occurs in a Charolais or in a rabbit and if we happen to be examining a biologically basic relationship, then I don't think it matters what breeds we choose. I don't think it matters whether we work between breeds or within breeds. I think we are always going to get the same answer.

That is a very poor attempt to break down different kinds of correlations and enter a very difficult field but I felt I would like to bring it a little further out into the open than it had been.

J. Kempster

I did start off with a basic question this morning, and that was related to commercial population and the point that I was making at the end was that, if you deal with small samples, with a few breeds, you can put a line anywhere you like. If you want to sample a population to answer a practical question such as the one we have on commercial classification, then it is important to sample the population and not get a few double muscled animals at one extreme and a few Holstein at the other and assume that the trend in those relationships will answer the practical question.

V.R. Fowler

May I ask a question to Dr. Clancy. I am rather worried by the idea that one may be able to select by number of muscle fibres and then predict what the breeding value of the animal will ultimately be and yet, listening to the discussion, I get the feeling

that the justification for counting the muscle fibre numbers
very often is that one day we will be able to select animals
almost at birth so that we need not go through this tedious business of performance testing. How it works out I don't know
because we do not really know the correlation between muscle
fibre number in one muscle and the number of fibres in other more
important muscles - I cannot see a digital flexor predicting the
number of fibres in the *biceps femoris* , for example, which might
be very much more relevant to the breeding value of the animal.
I think there is a difficulty that once we see a correlation,
we always assume that it will then be useful predictive material.
Even in the case of quite a high correlation, of 0.6 or 0.7, when
you actually come to predict what the breeding value of an animal
will be where differences of 5% are the difference between an
animal that is good and should be selected and one that is to be
rejected, I can't see muscle fibre number entering into the picture. However, I hope Dr. Clancy is going to respond to this
gauntlet and tell us a little bit more why muscle fibre number
is of such paramount importance.

M.J. Clancy

I do not know, scientifically speaking, why muscle fibre
number is important - it is supposed to be related to meat
quantity, which is extremely difficult to prove. I agree with
a lot of what you are saying, that essentially we come down to
talk about the histology and histochemistry of muscle. We are
talking about ideas and concepts that have been around for a
long time but never genetically or economically proved. To use
a trite phrase, it seems like a good idea to try out. It has
never been put on a firm basis.

R. Boccard

I think the number of fibres is a characteristic of meat
quality also. If for the same area or the same weight you have
more fibres inside the muscle, the meat is more tender.

D. Lister

It seems to me that, by and large, through the mammalian

kingdom, muscle fibre is about the same diameter and certainly
sarcomere length is about the same. If you measure the number
of fibres in a sample of muscle, you are likely to pick out
those animals that are going to grow to a bigger size, not
necessarily those that are going to have a greater proportion
of muscle to bone and I wonder whether those animals that have
a large number of fibres are simply going to grow bigger - not
necessarily more efficiently.

M.J. Clancy

What you say is true but we are talking about beef production and there are enormous problems in counting fibres. People
have measured what I would call apparent numbers, but Mr. Dumont
and other speakers referred to the review by Swatland which
appeared this year, and he has very eruditely discussed this
question of numbers. There is certain circumstantial evidence
about numbers and meat production, but it still needs a good
basis.

H.J. Langholz *(West Germany)*

I would like to ask Dr. Williams, as a breeder, does he think
it is sufficient to work with the kidney knob and channel fat
plus the subcutaneous fat, or should you also add intermuscular
fat (or adjustment for that) from the breeding point of view?

D.R. Williams *(UK)*

As this demonstrates, if you have a very accurate measure of
subcutaneous fat and kidney knob and channel fat, or a very
accurate visual assessment of it, then you have automatically
made a pretty good assessment of the intermuscular fat.

H.J. Langholz

Will this hold for the beef production system on the Continent with a much lower level of fatness than you have?

D.R. Williams

I must admit that all these breeds were British, and we have
not surveyed the Continental breeds, but I think that the same
would be true.

V.R. Fowler

I see a conflict in some of the data presented in the paper by Enser and Wood. First of all, we cannot see the pre-adipocyte cell and yet we know that its number is stable soon after birth, which would seem to me a logical impossibility. I feel that they have perhaps changed their minds over the last years as to what the story is. Five years ago everybody talked about 'cellularity' as if it were a defined concept - that is, that one could speak of the number of nucleii in an adipose depot, relative to its weight. I think a lot of the difficulty is that we do not know where to define the boundary of an adipose depot. All the action is round the outside and in the past we have taken all our samples from the middle. If we had taken our samples from all the outside edges, right across, I wonder whether the story you are now beginning to tell would have been told perhaps twenty years ago. It has really been a sampling problem.

M.B. Enser *(UK)*

I think that what Dr. Fowler said is to some extent quite right. It is a sampling problem and, in fact, we are only now getting down to doing this sort of study. As you know, we have done it with the pig but we haven't done the histology there. We hope in the study which is now going on to do the histology in Hereford and Friesian cattle.

I don't think you are quite right about counting the nucleii because it has been at least eight years since it was demonstrate that only about 30% of the nucleii were actually present in fat cells in adipose tissue. More recent studies have suggested that this may only be 10% and, again, this was never really on, because other cells, such as mast cells, can change quite readily and we certainly cannot differentiate them, just for measurement of nuclear quantity.

With regard to your other point, I think it is quite true that the pre-adipocyte is a hypothesis. The one thing that you can show is that incorporation of radioactive thymidine into nucleii which subsequently appears in recognisable fat cells,

takes place a considerable time before we can recognise those
fat cells through the uptake of fat - in other words because
they have become fat cells. So, the fat cell, which we term the
pre-adipocyte, can be there at least for a fair time before it
takes up fat.

One thing I didn't mention because time was running short,
is that it would be very nice to do some of the studies that
were done by Lister and McCance where you certainly get a very
rapid development of adipose tissue in an animal which has been
so restricted that it has deposited no fat and made very little
bodily growth for a whole year. As soon as it is re-alimented
it deposits fat at a very high rate. We still do not know whether,
if we looked at the tissue in that sort of animal, we would find
potential fat cells or not but we do know from histological
studies that the tissue is recognisable in the foetus as clusters
of cells of some sort around a capillary bed and it is a very
recognisable feature.

St. C.S. Taylor
Is there any question of turnover of fat cells?

M.B. Enser
Fat cells may be lost but there is no great loss and replacement of fat cells. When you starve somebody - and this has been
done in several experiments - the fat cells appear to remain,
albeit as cells very much shrunken with a very crenulated cell
membrane, and they reappear very quickly. This has been done on
rats and the animals were previously labelled with thymidine at
a young age and the starvation is then done before those labelled
nucleii appear as fat cells. The first expansion takes place in
the cells which were there previously, before the starvation and
the labelled nucleii appear very much later. They do not appear
as soon as you re-feed the animal.

D.E. Wood
I would like to address a question to Dr. Hanrahan. Do you
think there is a physiological link between the proportion of

KKCF of total fat and prolificacy, because you certainly find a
high proportion in the Finn, which is a very prolific breed?
The reason why I ask is that in some carcasses which we recentl
dissected similar differences between some prolific breeds compared well at the same level of total fat for instance.

J.P. Hanrahan *(Ireland)*

Although one has to be very careful in looking at the relatio
ships among traits that you measure in different breeds, this
breed here *(slide)* of the high 'prolificacy' type was developed
from very heterogeneous material in Ireland in the early sixties.
Practically no Galway went into the development of that animal,
no Finn either, but the prolificacy here is about the same as in
the Finn cross Galway - they both have prolificacy in the region
of 2 - and they are very similar when it comes to the
position of the fat in terms of kidney knob and channel fat depot
Looking at these two breed types and looking at fat distribution
in more detail, they are so similar that you can hardly distinguish them, but I wouldn't care to comment any more than that.

R.G. Kauffman

I would like to ask Mr. Williams if there was any work done
at all with intestinal fat? I noticed that most of the comments
have been directed towards carcass fat.

D.R. Williams

No, we have not done any detailed work on this, but it is
being done now, in some of the later work.

R.W. Pomeroy

I think the reason for that was that at the time we were doing with the MLC some preliminary studies in carcass classification, and were not really interested in the intestinal fats,
but only in the carcass fats at that time.

P.L. Bergstrom

Now we can move on to questions in regard to fat cell growth
and development.

A.J.H. van Es

There was a remark by Dr. Leat and also by Dr. Enser about the healthiness of fat. I think fat is not very unhealthy, if you are not eating too much energy of fat. In fact, for the liver, carbohydrate is more harmful than fat energy. The danger is to say that fat is unhealthy whereas it is really too much intake of energy which is unhealthy.

W.M.F. Leat *(UK)*

I was only quoting what certain vociferous people have declared, that saturated fats may be dangerous to health. One should take this into account and try to answer their criticisms.

A.J.H. van Es

Human nutritionists at the moment are changing their minds and stressing too much energy intake. That is the important thing just now.

W.M.F. Leat

I agree entirely with that; a varied diet is probably the best thing.

J. Martin *(Belgium)*

I would like to ask whether there is any difference between saturated and unsaturated in relation to energy intake?

A.J.H. van Es

I think you have to discern between those people who are simply eating too much, in which case unsaturated fats will be better for them, and lean people, for whom it perhaps does not matter.

J. Robelin

We have measured the cell size of various adipose tissues of Friesian and Charolais bulls at different weights which are approximately 150, 250, 350 kg. We have represented the increase in cell size on a log scale of cell diameter against the log of the weight of the fat. The higher the slope in this figure, the

higher the impact of cell enlargement on the growth of fatty
tissue.

If we look only at the Friesian bulls we can see that there
are wide differences between depots with regard to their evolution of the size of their cells with the weight of their depots.
For intermuscular and peritoneal fat we have a continuous increase
of cell diameter against the weight of fat depots. On the other
hand, on the late maturing subcutaneous fat depots we have - between 100 and 250 kg - practically the same cell diameter and,
after this 250 kg (seven months approximately) an increase in
cell diameter.

Secondly, between breeds, we find that, for the earlier maturing depots and the later maturing depots, there are practically
no differences in the shape of the curve, but in the intermuscular
fatty tissue the Charolais are earlier in their development than
the Friesian cattle.

D. Lister

I wonder if we could persuade Dr. Leat to say a little more
about his view that reducing the amount of fat in a carcass will
cause an increase in saturation because I think that the evidence
suggests that the saturation of fat depends on the net rate of
formation. In animals in which the amount of fat is reduced,
there is usually an accompanying reduction in the net rate of
formation and an increase in unsaturation, not saturation.

W.M.F. Leat

I take the opposite view. I predicted from my results that,
if the amount of fat decreased, then it would increase in saturation on the basis that fat is made up of what comes from the
diet, which in this case is what comes from the rumen, which is
highly saturated, and that which is synthesised in the tissue,
which is more unsaturated. So, if the animal becomes less fat,
then the dietary fat, which will be more saturated, will make an
increasing contribution to the depot, and this is what I predicted - that the saturation should go up as you decrease the

amount of fat in the carcass. I may be wrong, but it is a hypothesis for testing.

H.J. Langholz

I wonder if we have any more information about the genetic control of the cell numbers.

W.M.F. Leat

I think there may well be genetic control of cell number. The trouble with cattle is that most of the research is done with animals up to meat weights, in other words, animals slaughtered to produce carcasses of 400 - 450 kg. These animals are not mature and I think we will only find out the potential cell number by feeding animals on a high quality diet in ad lib conditions until they have expressed their fat cell number. This has been done with experimental animals. If you take a rat or a mouse and force feed it by various techniques such as damage to the feeding centre of the hypothalmus, this animal will become grossly obese and there is no increase in the number of fat cells. In other words, the animal under normal growth conditions appears to have expressed its fat cell number, but I think there are as yet no studies on cattle which have done this.

We are currently attempting to find out the cell number by looking at animals under these conditions. It is quite possible that there may be a limit in number. Whether this is significant in the quantity of fat deposited of course is still hypothetical since we have this great flexibility in the size of the fat cells. We do not know that the size of the cell is related to many metabolic phenomena. Certainly, in humans and rodents, plasma insulin concentrations are related to size of fat cell, and not fat cell number, but there are situations where one can get changes in number of cells. I think Dr. Kauffman could talk about the miniature pig in this respect, which has a small number of fat cells, and they grow exceedingly large. There has been a change, but I am not quite sure what effect this has had on appetite or any metabolic parameters.

R.G. Kauffman

This is a subject of great interest to me and I thought that
the paper by Enser and Wood was very appropriate though perhaps
it did not answer Dr. Fowler's question entirely.

I am not sure how important mytotic activity has been. We
have studied it in the pig for the last four years and found
that it does occur in postnatal circumstances but it also appears
that it is the size of the cell that is important from the standpoint of fat production, and not the number of cells. I say this
not on the basis of our work but on that of Allen and Hood of
Minnesota, in which they again studied both cattle and pigs. In
the case of pigs they found that the very muscular breeds, including the Pietrain, had many more cells at any given weight
than did the fat Minnesota Number Ones, Twos or even the miniature pigs - and it did turn out that it was the size of the cell
in all cases that was important. I might add that our studies
very clearly indicate that there is mytotic activity in the pig
long after the postnatal circumstances and we don't know when that
stops. It does seem to stop around birth for a moment, but only
for a moment's breath, until it gets started again.

I would ask two other questions: if there is mytotic division
before or after birth, I would have to ask the question, "So
what?" Secondly, I would have to ask the question, "Well, if
we know there is a differential between partitioning of fat in
different parts of the body - again, so what? (in relationship
to total fat production in the animal body).

A.J.H. van Es

My colleague, S. Metz, has measured cell size and cell number in pigs of 20, 50 and 90 kg, and he not only determined the
size of the cells but also the numbers in a given group. The
smallest size he could measure, I think, was about 30 micrometres.
He tested distribution for its normality. At 50 kg he found that
the whole thing seemed to shift to a larger size, and everything
seemed to shift to a larger size. The situation was the same
at 90 kg. This was done within a given layer. From the fact

that no skewness was found I would extrapolate the number of fat cells being constant, only their size being increased. If you test for skewness, especially in the material of Mrs. Schön, it will be possible to see whether in a given layer of fat cells this is the case or not.

L. Schön *(West Germany)*

I would like to ask Dr. Enser about measurement of fat cell size.

M. Enser *(UK)*

The diameter of all the cells was measured in one axis only which corresponded to the direction of the scale on the eyepiece micrometer of the microscope. This procedure was used, rather than a measurement of the long and short axes of the cells, because the preparations which we use show a random distribution of the long axes and, in fact, most cells are almost spherical. We believe that this appearance of the cells is a result of the use of a frozen cut section which is 200 - 250 μm thick. After cutting the section warms up and since it is only two cells thick there is little restraint to maintain the cells in the elongated regular array seen in standard fixed preparations.

P.L. Bergstrom

Well, thank you all very much.

SECTION 2

GROWTH AND NUTRITIONAL EFFICIENCY AND THE EFFECTS
OF GENOTYPE, SEX, HORMONES AND THEIR INTERACTIONS

Co-ordinators:

R. Jarrige

C. Béranger

BIOLOGICAL MODELS OF QUANTIFYING GROWTH AND EFFICIENCY

V.R. Fowler

Rowett Research Institute, Bucksburn, Aberdeen AB2 9SB Scotland

ABSTRACT

Some relationships between input and output are considered in the particular case of the growing pig. Input is restricted to one factor, the metabolisable energy of the feed eaten by the animal (ME). The outputs are assessed both in terms of chemical entities and also as skeletal muscle or lean tissue. Various approaches are discussed which apportion total heat loss from the animal to different physiological functions. Models are presented suggesting the quantitative manner in which ME is partitioned between heat loss, lipid accretion and protein accretion from birth to slaughter, and also the response of pigs weighing 60 kg to changes in the daily supply of ME.

Calculations are made of the relative importance of several factors contributing to the efficiency of utilisation of ME for the production of lean tissue. The most important single component, in terms of the physiology of the animal, is shown to be the growth rate of the lean.

INTRODUCTION

The study of growth and development provides considerable
scope for generating biological models. These may range from
descriptive and hypothetical representations to compact
mathematical equations. The diversity reflects the enormous
volume of biological information associated with growth and the
difficulties which one encounters in attempting to build
observations into a logical framework.

The objective of this paper is to consider a particular
approach to the subject of growth which is mainly associated
with my own field of pig nutrition. It is hoped that the more
general applications of these specific examples will be readily
seen.

GENERAL PRINCIPLES OF INPUT/OUTPUT RELATIONSHIPS

The inputs of animal production can be broadly identified
as capital, manpower and raw materials (feedstuffs). This
bears much similarity to the input side of a conventional
industry. The similarity with industry becomes more obscure
when the productive process itself is considered, that is
growth. The engineer of a factory completely understands the
machinery and each component can usually be tested in terms of
its efficiency and productivity. With animals there are few
direct techniques for establishing the relative importance of
the productive processes and even fewer means of changing them.

There are difficulties too in defining the output of the
process of growth. The product is not a commodity but an animal.
Its structure is complex and the important parameters are
difficult to measure. To some extent the economist can overcome the problems by generating models which relate to the
market prices of inputs and outputs. This is of relatively
little value, however, in providing an understanding of the
biological principles involved and possible ways of modifying
the system.

There have been many attempts to model what can be called 'biological efficiency' As discussed by Spedding (1976) the choice of inputs and outputs depends on the reasons for the calculations. Furthermore, he points out that many models may require exploration merely to gain insights which allow better models to be constructed for specific purposes.

THE UTILISATION OF METABOLISABLE ENERGY FOR GROWTH

One particular concept has proved especially valuable in demonstrating some principles of biological efficiency. It is the utilisation of the energy yielding fraction of the diet for growth. A simple model is to partition metabolisable energy (ME) between the energy retained in the body and the heat lost from the animal by evaporation, convection, conduction and radiation.

$$I_{ME} = R + H$$

Where I_{ME} = Intake of metabolisable energy
R = Energy retained as body tissues
H = Total heat loss from the animal

In the growing pig as much as 60% of ME is lost as heat. A further step is to assign this heat loss to different physiological processes. The so-called ME system for describing the energy requirements of ruminants explained in detail by Blaxter (1967) attributes heat losses to three components, the fasting metabolism, the heat produced when dietary ME is utilised to provide the energy for maintaining the animal, and the heat associated with the growing or fattening process itself.

$$I_{ME} = F/k_m + R/K_f$$

Where I_{ME} = Intake of ME (MJ/day)
F = Fasting metabolism (MJ/day)
k_m = Efficiency of utilisation of ME for F expressed as a decimal fraction
R = Retention of energy in body tissue (MJ/day)

$$k_f = \text{Efficiency of utilisation of ME for R}$$
$$\text{expressed as a decimal fraction.}$$

To expand the model further, attempts have been made to distinguish between the heat production associated with protein accretion and that of lipid accretion. Kotarbinska and Kielanows (1969) and Kielanowski (1972) did this with pigs growing between 20 and 90 kg. Body composition was varied by adjustments to the composition and amount of the diet. Carcass analysis gave the gains of lipid and protein and determination of the energy retained in the carcass allowed the heatloss to be calculated. Using multiple regression equations, the following relationship was determined.

$$I_{ME} = 425M + 67P + 54L$$

Where I_{ME} = Daily intake of ME in kJ

M = Mean metabolic body weight (live weight kg$^{0.75}$)

P = Daily accretion of protein (N x 6.25)

L = Daily accretion of lipid

The efficiency of utilisation of ME for protein accretion k_p from this equation was 0.36 and for lipid accretion k_1 was 0.72. The partial regression coefficient, 425, provided an estimate of the utilisation of ME for maintenance expressed as kJ per kg of live weight to the power 0.75.

This model provided no information concerning the utilisati of ME below maintenance. This is not, however, of particular practical interest in the case of the growing pig. The approach although useful has several widely recognised statistical short-comings. First, it is difficult to ensure that the independent variables are not in fact closely correlated with each other. For example, daily lipid accretion is usually inversely co-related with daily protein accretion. Secondly, the model assumes that the efficiencies of protein and lipid deposition are constant over a wide range of stages of maturity and regard-less of the rate of accretion; in other words, it is essential a linear model.

The basic concept has been followed in many further investigations using either calorimetry or comparative slaughter. Kielanowski (1976) and Pullar and Webster (1976) have reviewed much of the relevant literature. Values for k_p ranged from 0.57 to 0.36 for pigs with estimates for other species also falling within this range. For lipid accretion estimates of k_l usually lay between 0.65 and 0.83 across species but values for the pig remained fairly consistently in the narrower range of 0.77 to 0.70.

Although attempts to differentiate between the costs of lipid and protein accretion are theoretically extremely interesting, the gain in precision is small compared with models using only one term, k_f, for the efficiency of retention of energy above maintenance. In growing pigs the rate of lipid accretion usually exceeds that of protein by about twofold. Even when pigs are growing quite slowly the ratio of the rates of accretion of protein and lipid is rarely below 1 : 1. This means that the most k_f could vary from this cause is from about 0.57 to 0.63, or approximately 10%.

PARTITION OF METABOLISABLE ENERGY FROM BIRTH TO SLAUGHTER

One limitation of the data used by Kotarbinska and Kielanowski is that the values they provide are integrated over one particular period of growth. A model predicting the partition at any stage of growth would be a further major step forward. Unfortunately, comprehensive data covering the whole range are not available. In Figure 1, however, I have attempted a synthesis using data described by Fowler and Livingstone (1972), Oslage and Fliegel (1965), Close and Mount (1975) and Thorbek (1975). The cumulative metabolisable energy is partitioned between the cumulative protein and lipid accretion from birth to 140 kg and the total heat loss from the animal. The inflexion in the curve of protein accretion occurs at about

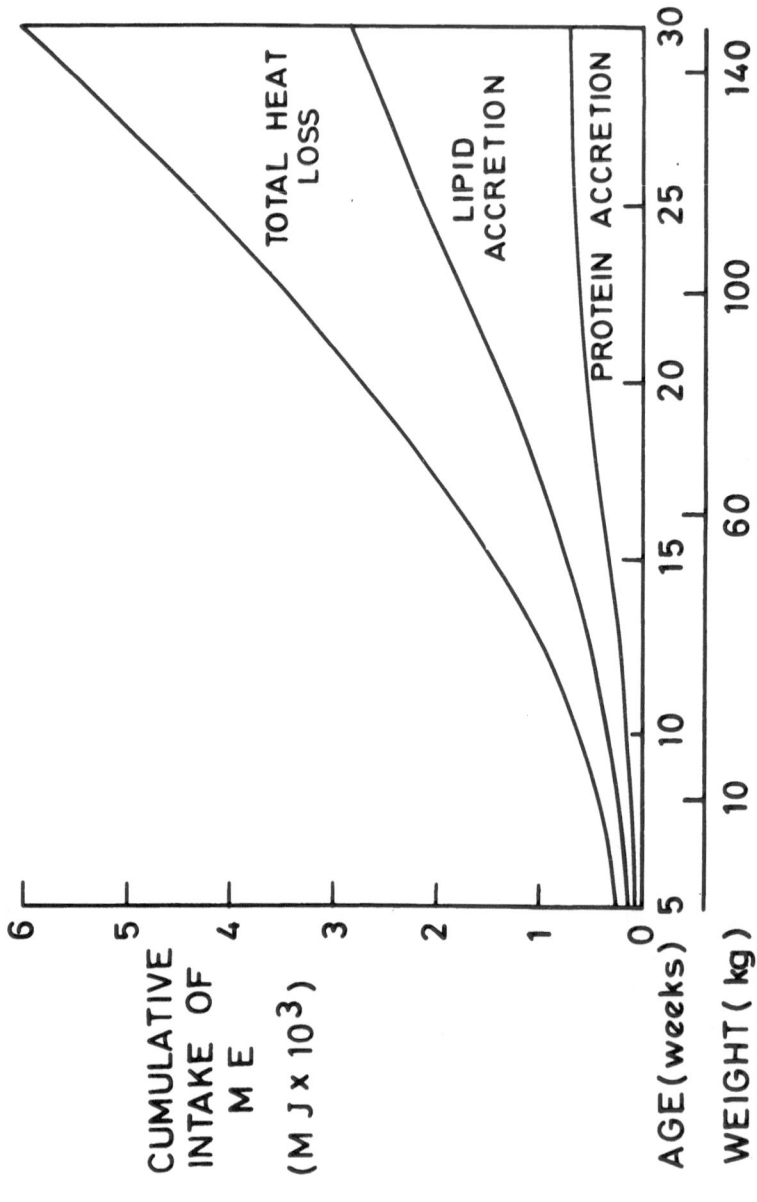

Fig.1 Partition of cumulative ME during growth between heat loss and the accretion of protein and lipid

65 kg live weight and that in the curve of fat accretion at
about 105 kg live weight.

It is immediately obvious from the graph that the energy
retained as protein becomes progressively smaller as the
animal matures. Even though the model gives a description of
energy partition with changing time it is still relatively
static.

A more dynamic model would accommodate changes in the
rate of feed intake. Only limited data exist and these refer
largely to the period of growth from 30 - 90 kg. It does
seem possible, however, to construct a tentative model for
changes in daily intake of ME at 60 kg live weight. To make
the concept easy to grasp, a graphical model is used, based
partly on the excellent and thought-provoking presentation of
Black (1974), who attempted a similar exercise in relation to
the daily ME intake of lambs. The result is shown in Figure
2. To construct the model it has been assumed that a pig at
60 kg given a balanced diet with adequate amino-acid composition
and eating close to its appetite at a rate of 31.5 MJ per day,
could accrete 137.5 g of protein per day. This value corresponds
to 22 g of N retention per day which is at the higher end of
the values in the literature reviewed by Thorbek (1975). Data
from the same source were used to set the daily N retention at
zero energy balance to 2 g and the initial N loss when fasting
to 4 g per day. The maintenance requirement was taken as 0.44
MJ per kg $W^{0.75}$ Using simplified values for k_1 and k_p as 0.75
and 0.40 respectively the partition above maintenance was then
calculated. An arbitrary value of 0.8 for k_m was taken.

To examine whether such a hypothetical model could be
validated by direct experimental data, I have used the data of
Fuller and Boyne (1971 a and b). Data were considered only for
the pigs kept in a thermoneutral environment and the nitrogen
balance data were adjusted to take account of the discrepancies
found by these authors between balances determined in metabolism

332

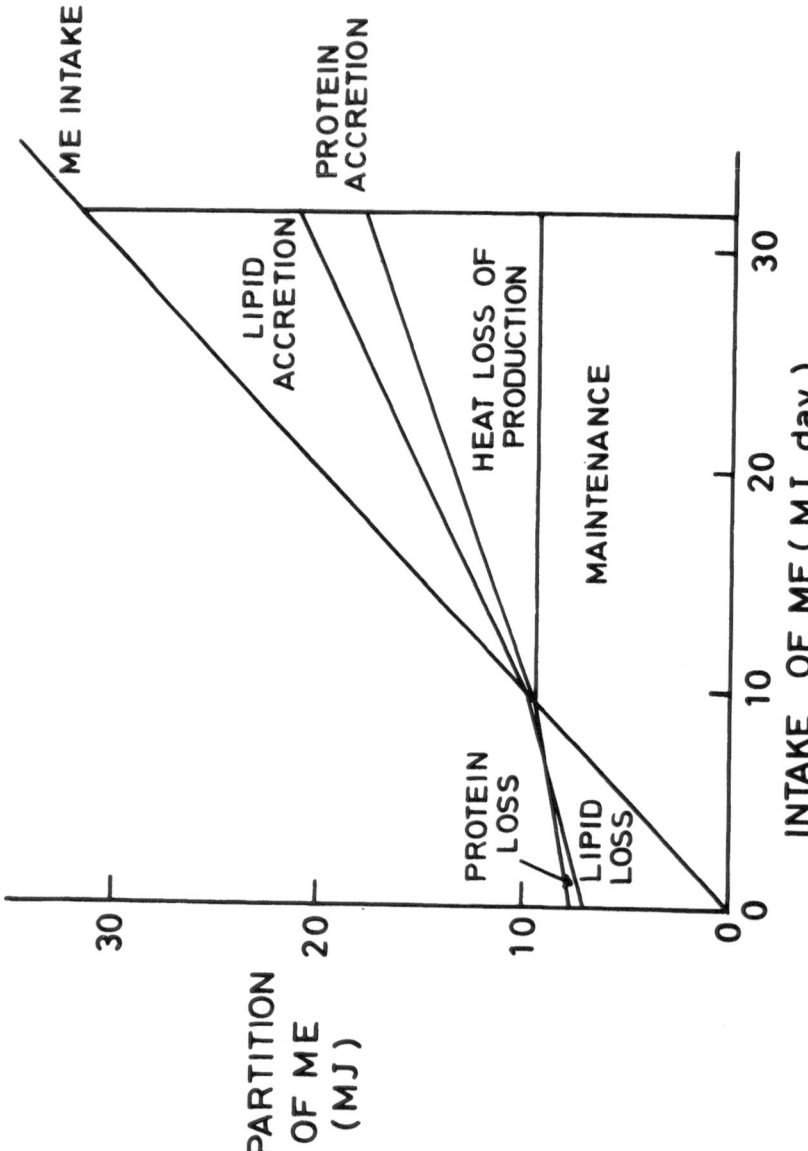

Fig.2 Partition of ME at different rates of daily intake for a pig of 60 kg liveweight

cages and those determined by comparative slaughter. The comparison is shown in Table 1.

TABLE 1

COMPARISON OF RESULTS PREDICTED FROM THE MODEL SHOWN IN FIGURE 2 AND ACTUAL RESULTS CALCULATED FROM THE DATA OF FULLER AND BOYNE (1971 a AND b)

| | | Intake MJ of ME per day | | |
		19.6	23.8	28.8
Heat loss per day (MJ)	(Model	13.4	14.7	16.1
	(Actual	12.2	13.4	14.8
Energy retained as protein per day (MJ)	(Model	1.7	2.3	3.0
	(Actual	2.4	2.9	3.5

It can be seen that although the slopes of the responses to increasing amounts of ME are similar, the intercepts are clearly different. The reasons for the discrepancies may lie to some extent in the circumstances of the experiment of Fuller and Boyne. Heat production was measured with pigs lying in a very confined cage in calorimeters, with little opportunity for activity. For these pigs the temperature was well within the range of thermalneutrality. The maintenance heat production, under these circumstances, may have been well below the value of 0.44 MJ per kg $W^{0.75}$ assumed in the model. In the future, as more data become available it should be possible not only to construct models such as that in Figure 2 with more precision but to have them for a wide range of body weights.

THE RELATIVE CONTRIBUTION OF BIOLOGICAL PARAMETERS TO THE EFFICIENCY OF PRODUCTION OF LEAN TISSUE

Considerable attention has been given to obtaining more precision in the estimates of maintenance and the energy requirements of specific productive processes. The purpose of this section is to attempt to put into perspective the possible consequences of change in some of the biological parameters.

To bridge the gap between chemical entities and meat as it is understood by the customer, it is convenient to assume that the particular component of the animal body which is valued by the consumer is the skeletal muscle.

In Table 2, data from an experiment described by Fowler and Livingstone (1971) are presented in a form assigning the retained energy to different anatomical components of the empty body during the period of gain from 25 - 90 kg live weight.

It can be immediately seen that the recovery of energy as lean tissue is very inefficient and even worse if one considers that the protein energy recovered in lean tissue may be as little as 5% of the original metabolisable energy fed. Considerable nutrition inputs are incurred before the pig reaches 30 kg live weight.

TABLE 2

COMPOSITION OF GAIN FROM 25 - 90 kg LIVE WEIGHT

Component	Gain in weight of tissue (kg)	Energy in tissue (MJ)	Percent of retained energy	% of ME intake
Skeletal muscle	21.5	253	22.8	9.1
Subcutaneous fat + skin, perinephric and intermuscular fat	19.2	640	57.8	23.1
Skeletal tissue including hocks	5.4	48	4.3	1.7
Non carcass head, viscera etc.	12.0	167	15.1	6.0
Total empty body	58.1	1108	100.0	40.0

In Table 3, data from the experiment of Fowler and Livingstone (1971) have been combined with the energy cost of producing a pig weighing 30 kg.

TABLE 3

THE OVERALL COST OF METABOLISABLE ENERGY (ME) OF PRODUCING LEAN TISSUE IN THE CARCASS OF THE GROWING PIG

Live weight (kg)	Cumulative ME intake (MJ)	Weight of lean tissue (kg)	ME cost per kg lean tissue (MJ)
30	1026	11.1	92
40	1289	15.6	83
50	1574	20.5	77
60	1893	25.8	73
70	2258	28.8	78
80	2645	31.2	85
90	3067	34.2	90

To simplify the calculations, it was assumed that the feed cost of rearing the mother could be disregarded since at the end of her reproductive life the sow is also usually sold for meat. Annual production from the sow was set at 20 live pigs weaned and her feed consumption at 1000 kg/year. As can be seen from Table 3, early growth appears inefficient due to the high overhead cost of feeding the dam, whilst later on, growth becomes inefficient because of the increasing increment of fatty tissue and maintenance expenditure associated with each unit of lean-tissue gain. In terms of this calculation the optimum occurred at 60 kg live weight.

It is now possible to examine the relative contribution of components of the biological models discussed so far, to the overall efficiency of utilisation of ME for the production of lean tissue. For comparative purposes, the hypothesis is explored that it might prove possible, by genetic selection or manipulation of the environment, to change a component in a favourable direction, by 10%, without simultaneously altering the contribution of other components.

The results of this exercise are given in Table 4.

TABLE 4

CHANGES IN THE ME REQUIRED TO PRODUCE LEAN TISSUE (MJ/kg) ARISING FROM A 10% CHANGE IN A FAVOURABLE DIRECTION OF THE COMPONENTS OF EFFICIENCY IN A PIG SLAUGHTERED AT 90 kg LIVE WEIGHT

		MJ of ME per kg lean tissue
1	Initial value	90
2	10% reduction in heat lost during lipid accretion	89
3	10% reduction in heat lost during protein accretion	89
4	10% increase in pigs/sow/year	88
5	10% reduction in maintenance	87
6	10% reduction in lipid accretion	87
7	10% reduction in weight at slaughter	85
8	10% increase in growth rate of lean tissue	81
9	Reduction of killing weight to 60 kg	73

These results require some qualification. Items listed 2 to 6 were considered only in the range 30 - 90 kg since satisfactory values at low weights are difficult to obtain.

Despite all the limitations of the approach, the ranking of the relative importance of the factors contributing to efficiency is remarkable. In pig production, the elusive factors of heat production k_p and k_1 taken individually have surprisingly little impact on overall efficiency. It appears relatively of little consequence whether as a correlated response to genetic manipulation they either increase or decrease by a factor of 10%. A 10% reduction in maintenance costs is clearly useful, and the lower the rate of feeding the more critical such a change would become. Similarly a reduction in the lipid associated with each increment of lean deposition would be of value if the energy so saved could be directed proportionately towards the other energy consuming processes. Merely losing it as heat would, of course, achieve nothing.

The most important factor relating to the physiology of the animal is the 10% improvement in the growth rate of the lean tissue. The corollaries of genetic improvement by selection for higher rates of gain of the lean tissue has been discussed in depth by Fowler et al. (1976). It is clear from the table that substantial advantages would accrue if energy used for lipid accretion could be even partially diverted to the deposition of lean tissue.

Equally striking, is the fact that the weight at slaughter has a profound effect on efficiency. Reduction by 10% gives a greater response in terms of efficiency than any single factor apart from the growth rate of the lean tissue. Even more remarkable is that by the simple expedient of slaughtering at 60 kg rather than at 90 kg live weight the improvement is virtually equal to all the other favourable changes of 10% occurring simultaneously. If, as this study suggests, the weight at slaughter is so critical in determining biological efficiency in pigs how much more so must it be in determining that of ruminant species where the maternal overheads are so much greater?

CONCLUSIONS

It is not possible within the framework of one short paper to consider all the possible types of quantitative models of growth. Only one or two aspects of a simple theme have been considered, that is the relationship between a single input metabolisable energy and output considered in crude chemical terms or in rather arbitrary anatomical divisions. The inadequacies are very real, but the frustrations of attempting to use data to construct quantitative models may point the way to designing improved experiments in the future. With all their shortcomings, new insights suggested by quantitative biological models may indicate different priorities for future work, even if that means extrapolating beyond the immediate framework of the model or even from one species to another.

REFERENCES

Black, J.L. 1974. Manipulation of body composition through nutrition. Proc.
    Aust. Soc. Anim. Prod., 10, 211-218.

Blaxter, K.L. 1967. "The Energy Metabolism of Ruminants". 2nd impression.
    (Hutchinson, London).

Close, W.H. and Mount, L.E. 1975. The rate of heat loss during fasting in
    the growing pig. Br. J. Nutr., 34, 279-290.

Fowler, V.R., Bichard, M. and Pease, A. 1976. Objectives in pig breeding.
    Anim. Prod. 23, 365-387.

Fowler, V.R. and Livingstone, R.M. 1972. Modern concepts of growth in pigs.
    In "Pig Production" (Ed. D.J.A. Cole). (Butterworth, London)
    pp. 143-161.

Fuller, M.F. and Boyne, A.W. 1971. a. The effects of environmental
    temperature on the growth and metabolism of pigs given different
    amounts of food. 1. Nitrogen metabolism, growth and body composition.
    Br. J. Nutr., 25, 259-272.

Fuller, M.F. and Boyne, A. 1972 b. The effects of environmental temperature
    on the growth and metabolism of pigs given different amounts of food.
    2. Energy metabolism. Br. J. Nutr., 28, 373-384.

Kielanowski, J. 1972. Energy requirements of the growing pig. In "Pig
    Production") (Ed. D.J.A. Cole). (Butterworth, London) pp. 183-201.

Kielanowski, J. 1972. Energy cost of protein deposition. In "Protein
    Metabolism and Nutrition" (Ed. D.J.A. Cole, K.N. Boorman, P.J.
    Buttery, D. Lewis, R.J. Neale and H. Swan). (Butterworth, London)
    pp. 207-215.

Kotarbinska, M. and Kielanowski, J. 1969. Energy balance studies with
    growing pigs by the comparative slaughter technique. In "Energy
    Metabolism of Farm Animals" (Ed. K.L. Blaxter, J. Kielanowski and
    Grete Thorbek). (Oriel Press, Newcastle-upon-Tyne) pp. 299-310.

Oslage, H.J. and Fliegel, H. 1965. Nitrogen and energy metabolism of
    growing fattening pigs with an approximately maximal feed intake. In
    "Energy Metabolism" (Ed. K.L. Blaxter). (Acad. Press, London : New
    York) pp. 298-306.

Pullar, J.D. and Webster, A.J.F. 1977. The energy cost of fat and protein
    deposition in the rat. Br. J. Nutr., 37, 355-363.

Spedding, C.R.W. 1976. The relevance of various measures of efficiency. In "Meat Animals, Growth and Productivity" A.8 (Ed. D.Lister, D.N. Rhodes, V.R. Fowler and M.F. Fuller). (Plenum Press, New York) pp. 29-41.

Thorbek, G. 1975. Studies on energy metabolism in growing pigs. 424. Beretning fra Statens Husdyrbrugs Forsøg, Copenhagen.

SOME NEUROENDOCRINE ASPECTS OF GROWTH

D. Lister

Agricultural Research Council Meat Research Institute.
Langford, Bristol, BS18 7DY, UK.

The amount and composition of the food an animal eats
determines, for the most part, how it will grow. If an animal
eats too much it gets fat and if it eats too little it gets
thin. But animals, like men, tend not to be grossly overweight
nor excessively thin even when they are allowed to eat their
fill. The long term control of energy input and expenditure
which brings this about is a delicate balance which is rarely
out by more than a few megajoules over a lifetime. The
mechanisms responsible for this control provide a popular topic
for discussion and research. There are enormous problems of
interpretation and extrapolation of results between mice and
men and as yet there seems to be no common denominator by
which the many isolated observations can be assessed. Farm
animals have been extensively investigated from the nutritional
standpoint but only recently have they been examined in terms
of the neuroendocrine phenomena which probably provide the
ultimate control mechanisms. This paper tries to draw some of
these together.

The thyroid was one of the first endocrine glands to
receive the attention of physiologists. Horsley (1886) examined
it in relation to growth and development and Hertoghe (1896)
began to assess its role in milk production which has continued
to be an important topic in dairy research. Hertoghe (1896)
found that when lactating cows were fed preparations of thyroid
gland, their yield of milk increased. His findings were sub-
stantiated much later by Graham (1934) and Folley and White
(1936). When satisfactory techniques became available for the
assay and synthesis of the active hormones, thyroxine and triiodothyronine, it was soon discovered that their glactopoietic
action included the stimulation of lipolysis and led to greater
amounts of fat being secreted into milk (Bartlett et al.,
1954). These and other findings prompted others to investigate

thyroid function in relation to milk production. Pipes et al. (1963) soon established that thyroid activity was higher in dairy than in beef cattle and Sørensen (1962) showed that it was directly associated with the daily yield of milk fat.

Variations in thyroid activity can be seen in animals whose physiology is uncomplicated by lactation or even the requirement to produce milk in the quantities achieved by today's dairy cows. Pigs show quite wide variation in rates of hormone production, the highest being found in the least fat breeds and types. The Pietrain, Poland China and some Landrace pigs belong in this category (Romack et al., 1964; Bodart and Francois, 1972; Palludan, 1972; Lister, 1972, 1976). In these cases, the suggestion is that thyroid hormones are acting to modify energy metabolism and possibly heat production (Whitton, 1970; Chaffee and Roberts, 1971), certainly it seems in the Pietrain which can achieve its leanness by limiting fat deposition rather than by 'the abundant deposition of lean (Lister et al., 1974; Lister, 1976).

Catecholamines, like thyroid hormones are known to be involved in thermogenesis and have well established lipolytic properties (Harrison, 1964; Himms-Hagen, 1967, 1975). There is also evidence for their mutual substitution and synergism with thyroid hormones. Melander et al. (1974), for example, consider that the actions of catecholamines probably depend on the associated levels of thyroid activity. The interaction of catecholamines and thyroid hormones may complicate the assessment of thyroid function in animals.

Measurements of the circulating levels of hormones indicate the balance between their secretion and utilisation and are affected by exercise, emotional state, environmental temperature, age and other hormones (Werner and Nauman, 1968). Some recent experiments in our laboratory have helped to unravel some of the problems of interpretation.

The syndrome known as Malignant Hyperthermia (MH) can develop in pigs like the Pietrain, some strains of Landrace and Poland

China and some people when the relaxant drug suxamethonium or
the anaesthetic halothane are administered to them (Britt and
Kalow, 1970). The condition develops dramatically and usually
fatally and is characterised by intense muscle stimulation,
acidosis and rise in body temperature (Lucke et al., 1976). A
reaction, indistinguishable from MH can also develop spontaneously
in conscious stressed pigs (Lister et al., 1975) and as 'Capture
Myopathy' in some game animals (Harthoorn et al., 1974) during
their chase and capture.

One indicator of the development of this profound metabolic disturbance is the fall in the Free Thyroxine Index (FTI -
Clark and Horn, 1965) in the serum. This is due primarily to
a rise in the triiodothyronine uptake ratio (Herbert et al.,
1965) which denotes an increase in the number of available
binding sites on thyroxine binding proteins and coincides with
a dramatic rise in the circulating levels of catecholamines
predominantly noradrenaline (Lister et al., 1974; Lucke et al.,
1976; Gronert and Theye, 1976). Other signs of the influence
of the sympathetic nervous system and the adrenal medulla on the
reaction are that reserpinisation will prevent a reaction (Lister
et al., 1974; Hall et al., 1975) as will adrenalectomy together
with sympathetic block with bretylium though sensitive pigs
remain so if one or other is left intact (Lucke et al., 1977 -
in preparation; van den Hende, 1977 - personal communication).
α -adrenergic blockade with phentolamine will prevent the development of the usual sequelae after suxamethonium (Lister et al.,1964,
1976) and there is no fall in the FTI. β -blockade with
propranolol modifies some aspects of the reaction but not the
fatal outcome.

A short period of exercise one hour before being given
halothane will increase the proportion of reactors in a population
of sensitive pigs (van den Hende et al., 1976). On the other
hand, neuromuscular blockade of an anaesthetised pig with massive
doses of curare or pancuronium will modify the MH reaction and
in some cases will prevent its fatal development (Hall et al.,
1976). These suggest that muscular activity or 'stress' can
prime a reaction in some pigs and that MH is part of a general

stress syndrome (Wingard, 1974; Lister et al., 1975). Whatever the ultimate mechanism the importance of catecholamines
and especially their α -adrenergic properties is clear.

MH is an extreme and relatively uncommon syndrome, but it
seems to share many of the physiological and biochemical features
of the severe muscle stimulation of exhaustive exercise (Pernow
and Saltin, 1971). Furthermore there are signs that a more
moderate but comparable reaction can be seen among a variety
of pigs and cattle in far less traumatic states.

Pigs are not good candidates for venepuncture for they have
few superficial vessels. The Protein Bound Iodine (PBI) or FTI
of the serum obtained from restrained conscious pigs or taken
at slaughter tend to be lower than they are in samples taken
from pigs fitted with in-dwelling catheters. Low values are
particularly common in pigs such as the Pietrain in which thyroid
activity is known to be raised (Lister, 1972, 1976). The same is
true in cattle where pure dairy breeds tend to have lower values
for FTI than those of beef types (Table 1) which is the contrary
of what might be expected from the known differences in thyroid
activity.

The lowered FTI values suggest a greater rate of utilisation of hormone during stress, especially at slaughter when
the plasma levels of catecholamines are commonly in excess of
40 µg/l (Lister - unpublished observations), comparable to that
occurring during MH. In cattle the circulating levels of thyroid
hormones, or some index of them, fall in response to environmenta
stress which can be associated with a rise in catecholamines,
again, predominantly noradrenaline (Johnson, 1972).

All of these observations support the notion of the
inter-relationships between thyroid hormones, catecholamines
and sympathetic nervous activity. The observed involvement of
α adrenergic responses lends support to the view of Melander
(1971) and Melander et al. (1974) that sympathetic nervous
activity and especially α adrenergic effects are important

regulators of thyroid activity. Our work goes even further to suggest that animals with high thyroid activity show greater sensitivity to the effects of sympathetic stimulation.

TABLE 1*

RANKINGS OF FTI VALUES IN SAMPLES ** OF SERUM FROM VARIOUS BREEDS OF CATTLE

Breed	Serum Free Thyroxine Index (FTI) ± s.d.	
Aberdeen Angus	7.4	1.48
Hereford	6.4	1.96
South Devon	6.3	1.72
Galloway	6.2	1.30
Kerry	6.1	2.17
Guernsey	5.5	0.35
Ayrshire	5.3	1.14
Friesian	5.2	1.00
Luing	4.4	1.66
Jersey	4.3	1.21

* Unpublished observations of D. Lister and D. Lister, St. C.S. Taylor and R. Thiessen.

** Minimum sample 10 animals.

Lean pigs and dairy cattle might in part owe their metabolic and somatic characteristics to the interactions of thyroid hormones and catecholamines and it should not be surprising to discover that they play important roles in animals which need to mobilise fat for energy purposes or for milk production. Similarly, animals which conserve fat might show increased production of, or sensitivity to, lipogenic hormones such as insulin.

A recent paper by Wood et al. (1977) attempted to unravel some of the particular difficulties of the relationships between lipolysis, lipogenesis and the deposition and composition of fat in growing pigs. They examined the effects of feeding, fasting and noradrenaline on fat mobilisation in Pietrain and Large White

pigs. Fatty acids are more readily mobilised in Pietrains
and their faster entry rate into body tissues suggests a
greater reliance on fatty acids for energy purposes. β
-adrenergic blockade is more effective in blocking lipolysis in
Pietrains than it is in Large Whites. Under the same conditions,
Large Whites mobilise more glucose and the effect is blocked more
easily by α -adrenergic blocking agents than it is in the
Pietrains.

Obese people tend to have high levels of plasma insulin in
the fasting state (Paulsen and Lawrence, 1968) and fasting insuli
levels have been used to predict the body composition of human
beings (Bagdade et al., 1967). Using Pietrain and Large White
pigs, Gregory (1976) and Wood et al. (1977) in their paper
demonstrated statistically significant correlations (0.7 - 0.9)
between proportions of muscle and fat in the carcass or some
index of it and the fasting insulin levels. 'Pietrainness' compared to 'Large Whiteness' is typified by reduced fat deposition
which results from enhanced lipolysis under the influence of β
-adrenergic agonists and reduced circulating levels and
sensitivity to the antilipolytic action of insulin.

Hart et al.(1975) and Hart et al. (1975) have studied
Friesian and Hereford x Friesian cows during lactation and the
dry period. During lactation, insulin, which has been shown
to be negatively correlated with milk yield in cows (Koprowski
and Tucker, 1973) was higher in animals containing Hereford
genes. Growth hormone, free fatty acids and β -hydroxybutyrate
were higher and indicative of lipolytic potential in the pure
dairy animals. In the dry period both types of animals showed
similar circulating levels of plasma constituents which resulted
from striking falls in growth hormone, FFA and β -hydroxy-
butyrate and rises in insulin and glucose in the dairy cows.
Total plasma thyroxine was lower in Friesians than Hereford
crosses and lower in both during lactation. Neither of the
differences, however, were statistically significant. The
lower values in Friesians and during lactation could result
from greater rates of hormone utilisation as they do in Pietrain

pigs.

It might be the same phenomenon which provides the physiological basis for the use of thyroxine degradation rates in bulls as a predictor of the milk producing potential of their daughters (Joakimsen et al., 1971).

Changes in metabolic type induced by selection and breeding of pigs are also associated with changes in hormonal balance. Norwegian Landrace pigs selected to grow faster and leaner take on some of the endocrinological features shown in extremes by dairy cattle or Pietrain pigs. The activity of their thyroids increases, fasting levels of FFA and growth hormone in the plasma are raised and glucose is lowered (Bakke, 1975; Bakke and Tveit, 1977).

We clearly need to know a great deal more about neuro-endocrine relationships in regard to growth and development. But even the limited observations recorded here suggest the beginnings of a pattern for the control of energy balance in farm animals. A growing predisposition towards fatness seems to be accompanied by and perhaps attributable to a change in the balance of hormones away from a lipolytic to a lipogenic character ie from that predominant in thyroid, sympathetic and growth hormones to others allowing the lipogenic and, perhaps, the protein anabolic (Cahill and Aoki, 1976) roles of insulin to emerge. The means whereby this progession is achieved is not clear and it is unlikely that a single mechanism is responsible. The sympathetic nervous system comes closest of any for not only is it involved in the regulation of thyroid activity, but also in insulin secretion (Mayhew et al., 1969; Porte and Robertson, 1973) and the interaction of one with the other.

REFERENCES

Bagdade, J.D., Bierman, E.L. and Porte, D. 1967. The significance of basal insulin levels in the evaluation of the insulin response to glucose in diabetic and non-diabetic subjects. J. Clin. Invest. 46, 1549.

Bakke, H. 1975. Serum levels of non-esterified fatty acids and glucose in lines of pigs selected for rate of gain and thickness of backfat. Acta. Agric. Scand. 25, 113.

Bakke, H. and Tveit, B. 1977. Serum levels of thyroid hormones in lines of pigs selected for rate of gain and thickness of backfat. Acta. Agric. Scand. 27, 41.

Bartlett. S., Burt, A.W.A., Foley, S.J. and Rowland, S.J. 1954. Relative galactopoietic effects of 3:5:3 -triiodo-L-thyronine and L-thyroxine in lactating cows. J. Endocrin. 10, 193-201.

Bodart, C. and Francois, E. 1972. Etude comparée du metabolisme thyroidien des porc belge et de Pietrain. In: Isotope studies on domestic animals. International Atomic Energy Agency. Vienna p. 219.

Britt, B.A. and Kalow, W. 1970. Malignant hyperthermia:aetiology unknown. Canad. Anesth. Soc. J. 17, 316.

Cahill, G.F. and Aoki, T.T. 1976. Protein-fat interactions. In: Meat Animals: Growth & Productivity. Editors Lister, D., Rhodes, D.N., Fowler, V.R. and Fuller, M.F. Plenum Press, New York and London p. 221.

Chaffee, R.R.J. and Roberts, J.C. 1971. Temperature acclimatisation in birds and mammals. Ann. Rev. Physiol. 33, 155.

Clark, F. and Horn, D.B. 1965. Assessment of thyroid function by the combined use of the serum protein-bound iodine and resin uptake of [131] I-triiodothyronine. J. clin. Endocr. 25, 39.

Foley, S.J. and White, P. 1936. Effect of thyroxine on milk secretion and on the phosphatase of the blood and milk of the lactating cow. Proc. Roy. Soc. B. 120, 946-965.

Graham, W.R. jr. 1934. The action of thyroxine on the milk and milk fat production of cows. Biochem. J. 28, 1368.

Gregory, N.G. 1977. A physiological approach to some problems in meat production. PhD Thesis. University of Bristol.

Gronert, G.A. and Theye, R.A. 1976. Halothane induced porcine malignant hyperthermia: Metabolic and hemodynamic changes. Anaesthesiology 44, 36.

Hall, G.M., Lucke, J.N. and Lister, D. 1975. Treatment of porcine malignant hyperthermia. Anaesthesia 30, 308.

Hall, G.M., Lucke, J.N. and Lister, D. 1976. Porcine malignant hyperthermia (4) Neuromuscular Blockade. Brit. J. Anaesth. 48, 1135.

Harrison, T.S. 1964. Adrenal Medullary and thyroid relationships. Physiol. Rev. 44, 161.

Hart, I.C., Bines, J.A., Balch, C.C. and Cowie, A.T. 1975. Hormone and metabolite differences between lactating beef and dairy cows. Life Sci. 16, 1285.

Hart, I.C., Bines, J.A., Cowie, A.T. and Balch, C.C. 1975. A comparison of hormones and metabolites in the circulation of beef and dairy cattle. J. Endocr. 65, 3P.

Harthoorn, A.M., van der Walt, K. and Young, E. 1974. Possible therapy for Capture Myopathy in captive wild animals. Nature 247, 577.

Herbert, V., Gottlieb, C.W., Kam-Seng Lau, Gilbert P. and Silver, S. 1965. Adsorption of ^{131} I-triidothyronine (T_3) from serum by charcoal as an in vitro test of thyroid function. J. Lab. Clin. Med. 66, 814.

Hertoghe, E. 1896. Bull. Acad. Med. Belg. (4^{em} serie) 10, 381.

Himms-Hagen, J. 1967. Sympathetic regulation of metabolism. Pharmocol. Rev. 19, 367.

Himms-Hagen, J. 1975. Role of the adrenal medulla in adaptation to cold. In: Handbook of Physiology. Section 7: Endocrinology. Vol VI. Adrenal Gland. American Physiological Society. Washington p. 637.

Horsley, V. 1886. Further researches into the function of the thyroid gland and into the pathological state produced by removal of the same. Proc. Roy. Soc. 40, 6.

Joakimsen, Ø., Steenberg, K., Lien, H. and Theodorsen, L. 1971. Genetic relationship between thyroxine degradation and fat-corrected milk yield in cattle. Acta. Agric. Scand. 21, 121.

Johnson, H.D. 1972. Environmental temperature effects and hormonal control on heat production of cattle. In: Isotope studies on the physiology of domestic animals. International Atomic Energy Agency, Vienna. p. 153.

Koprowski, J.A. and Tucker, H.A. 1973. Bovine serum growth hormone, corticoids and insulin during lactation. Endocrinology 93, 645.

Lister, D. 1972. Hormonal function in relation to adaptation and stress In: Proc. European Association of Animal Production. 23rd Meeting. Verona.

Lister, D. 1976. Hormonal influences on the growth, metabolism and body composition of pigs. In: Meat Animals: Growth and Productivity. Editors Lister, D., Rhodes, D.N., Fowler, V.R. and Fuller, M.F. Plenum Press. New York and London. p. 355.

Lister, D., Hall, G.M. and Lucke, J.N. 1974. Catecholamines in suxamethonium induced hyperthermia in Pietrain pigs. Brit. J. Anaesth. 46, 803.

Lister, D., Hall, G.M. and Lucke, J.N. 1975. Malignant Hyperthermia: a human and porcine stress syndrome? Lancet 1, 519.

Lister, D., Hall, G.M. and Lucke, J.N. 1976. Porcine malignant hyperthermia (3) Adrenergic Blockade, Brit. J. Anaesth. 48, 831.

Lister, D., Lucke, J.N. and Hall, G.M. 1976. Pale Soft, Exudative (PSE) meat, stress susceptibility and MHS in pigs - endocrinological and general physiological aspects. In: Proceedings of the third international conference on production disease in farm animals. Pudoc. Wageningen p. 144.

Lister, D., Wood, J.D. and Perry, B.N. 1974. Towards a physiological basis for animal testing. In: Proceedings of 25th Meeting of European Association of Animal Production. Pigs Commission, Copenhagen.

Lucke, J.N., Hall, G.M. and Lister, D. 1976. Porcine malignant hyperthermia (1) Metabolic and physiological changes. Brit. J. Anaesth. 48, 297.

Mayhew, D.A., Wright, P.H. and Ashmore, J. 1969. Regulation of insulin secretion. Pharmac. Rev 21, 183.

Melander, A. 1971. Interactions of adrenergic blocking drugs with the in vivo release of thyroid hormone induced by thyrotrophin and the long acting thyroid stimulator. Acta Endocr. 66, 151.

Melander, A., Ericsson, L.E. and Sundler, F. 1974. Sympathetic regulation of thyroid activity. Life Sci. 14, 237.

Palludan, B. 1972. Studies of the thyroid function in pigs. In: Isotope studies on the physiology of domestic animals. International Atomic Energy Agency. Vienna p. 199.

Paulsen, E.P. and Lawrence, A.M. 1968. Glucagon hypersecretion in obese children. Lancet 2, 110.

Pernow, B. and Saltin, B. 1971. Muscle metabolism during exercise. Plenum Press, New York and London.

Porte, D. jr. and Robertson, R.P. 1973. Control of insulin secretion by catecholamines stress and the sympathetic nervous system. Fed. Proc. 32.

Pipes, G.W., Bauman, T.R., Brooks, J.R., Comfort, J.E. and Turner, C.W. 1963. Effect of season, sex and breed on the thyroxine secretion rate of beef cattle and a comparison with dairy cattle. J. Anim. Sci. 22, 476.

Romack, F.E., Turner, C.W., Lasley, J.F. and Day, B.N. 1964. Thyroid secretion rate in swine. J. Anim. Sci. 23, 1143.

Sørensen, P.H. 1962. Studies of thyroid function in cattle and pigs. In: Use of radioisotopes in animal biology and the medical sciences. Academic Press, London and New York. p. 455.

van den Hende, C., Lister, D., Muylle, E., Ooms, L. and Oyaert, W. 1976. Malignant Hyperthermia in Belgian Landrace pigs rested or exercised before exposure to halothane. Brit. J. Anaesth. 48, 821.

Whitton, G.C. 1970. In: Comparative Physiology of Thermoregulation. Academic Press, London. Chap. 6.

Wingard, D.W. 1974. Malignant Hyperthermia: a human stress syndrome? Lancet 4, 1450.

Werner, S.C. and Nauman, J.A. 1968. The thyroid. Ann. Rev. Physiol. 30, 213.

Wood, J.D. Gregory, N.G., Hall, G.M. and Lister, D. 1977. Fat mobilisation in Pietrain and Large White pigs. Br. J. Nutr. 37, 167.

ANABOLIC AGENTS IN BEEF PRODUCTION:
THEIR ACTION AS GROWTH PROMOTERS

R.J. Heitzman

ARC Institute for Research on Animal Diseases,
Compton, Newbury, Berkshire, UK.

The use of anabolic agents for meat production is increasing. However, we do not yet understand how these substances produce increases in live weight gain and improvements in food conversion efficiency. The objective of this paper is to prevent some of the theoretical aspects behind the use, safety and mode of action of anabolic agents used in animal production.

The anabolic agents used in beef production are either related functionally to the sex steroids or are chemical substances whose mode of action is not yet understood (eg the resorcylic acid - β lactone, zeranol). The anabolic agents similar to the sex steroids may be divided into three classes -

Androgenic steroids

These are androgens, either natural or synthetic, which possess properties closely related to the male hormone, testosterone.

Oestrogenic steroids and synthetic oestrogens

These are compounds which possess oestrogenic properties similar to the female hormone, oestradiol-17β. The synthetic oestrogens, particularly diethylstilboestrol and hexoestrol, do not possess the same structural formula as the natural oestrogens.

Progestins

These are substances, both natural and synthetic, which possess progestational properties similar to the female steroid, progesterone.

Growth hormone

Growth hormone is not a sex steroid but has similar anabolic properties.

These anabolic agents have their primary action via alterations of intermediary metabolism in the animal and do not act via alterations in digestive efficiency. They are active by parenteral and oral administration.

Other classes of anabolic agents which affect the digestive system or are classed as feed additives are not discussed in this paper. They include antibiotics, mineral additives and vitamins and those compounds which alter rumen fermentation.

HYPOTHESIS FOR THE ROLE OF ANABOLIC AGENTS IN THE GROWTH OF CATTL

Heitzman (1976) suggested that androgens and oestrogens are both necessary to realise the maximum growth potential. In cattl the concentrations of steroids in blood which result in the fastest growth rates correspond approximately to a combination of the androgen level in a growing bull and the oestrogen level in a young cow. Thus, the optimal treatment should maintain this natural hormone status for as long as possible, preferably several months. If the hypothesis is valid, the greatest benefits would be seen in bulls treated with oestrogens, steers treated with androgen combined with an oestrogen and heifers and cows administered androgens. In practice, implants of oestrogens or combined preparations are used in steers, veal calves and bulls, and androgens in heifers and cows. The effect of anabolic agents on growth in pigs and sheep is similar to that in cattle.

RESIDUES OF ANABOLIC AGENTS IN EDIBLE TISSUES

Anabolic agents are administered either orally, or by intramuscular injection, or as subcutaneous implants. The latter method is the most widely used and the most common agents in use are preparations containing trenbolone acetate, testosterone oestradiol, hexoestrol, diethylstilboestrol or zeranol.

Investigations have concentrated on the determination of residues of anabolic agents in edible tissues. Although radiotracer studies are the methods of choice for absorption, excretic total residues and metabolite studies (Aschbacher et al., 1975; Pottier et al., 1975), radiommunoassay (RIA) techniques and immunofluorescence reactions are also used (Hoffman and Oettel, 1976; Hoffman and Karg, 1976; Heitzman and Harwood, 1977), especially to measure residues of anabolic steroids. It is now possible to measure residues of trenbolone, testosterone, oestradiol and progesterone by RIA but not DES, hexoestrol or zeranol.

The residues of trenbolone acetate, which include the active component trenbolone and as yet unknown metabolites, are present for several months in all tissues of cattle implanted with a commercial dose of 300 mg (Pottier et al., 1975; Heitzman and Harwood, 1977; Donaldson, 1977). It has been difficult to detect residues of testosterone, progesterone and oestradiol in edible tissues (Hoffman et al., 1975) but in treated animals it is possible to demonstrate increased levels of these steroids in blood samples collected from near the site of implantation (Heitzman, Donaldson and Pease, 1977).

MODE OF ACTION OF ANABOLIC AGENTS

The exact mode of action of anabolic agents still remains unclear, and it is difficult to describe a simple single mode of action. The common action of all anabolic agents is to increase N-retention and protein deposition (van der Wal, 1976). N-balance studies confirmed that anabolic agents with sex-steroid like activity and administered parenterally increased N-retention, but they did not alter absorption or metabolism in the alimentary tract (Chan et al., 1975).

THE HORMONAL REGULATION OF PROTEIN METABOLISM

In presenting current concepts on how hormones regulate protein metabolism, it should be emphasised that most of the investigations have been carried out in laboratory animals. We do not know how much of this work applies to farm animals.

356

Anabolic hormones may act directly at the muscle cell by
regulating protein synthesis and degradation, or they may act
indirectly by modification of a second growth promoting hormone.
Both are likely actions of the anabolic hormones.

An anabolic hormone arriving at the muscle cell has two
possible actions. Either it recognises a specific receptor on
the cell surface and attaches itself to the outer surface membran
or it enters the cell where it combines with a specific receptor
to form a complex. In both cases it is the attachment to a
specific receptor which initiates the events leading to increased protein accretion (for review seen Mainwaring, 1977).

It is thought that in cells lacking specific receptors
there is little direct activity of the hormone. Thus muscle
cells must have specific receptors for either the primary hormone
in the case of a direct action, or for the secondary hormone in
an indirect mode of action. An androgen receptor has been reported for rat muscle cells (Michel and Beaulieu, 1976).

The distribution of receptors may be determined by the sex
of the animal and this might explain the relative responses of
the sexes to different sex steroids.

PROTEIN ACCRETION

Proteins are continually synthesised and degraded in
muscle cells, and the rates of these processes determine the
protein turnover rate. When synthesis < degradation, there is
an increase in the net amount of protein, and this is called
protein accretion. However, protein accretion is a very inefficient process because of high turnover rates. Anabolic
hormones are known to increase protein accretion and they may
also reduce protein turnover rates (Vernon and Buttery, 1976).
This means that more protein is laid down at a lower energy cost
This action would explain why feed conversion efficiency is so
dramatically improved by anabolic agents.

MODE OF ACTION OF OESTROGENS

Considerable evidence supports the view that oestrogens (and progesterone) exert their primary effect at the level of chromatin transcription and gene expression in cell nuclei of particular target tissues. These include the uterus, liver and chick oviduct. In contrast, little is known of the action of oestrogens in skeletal muscle. Studies so far would indicate that oestrogens most likely act through the secondary hormones GH and insulin.

The combination of increased GH and insulin at the muscle cell is thought to increase protein accretion (Trenkle, 1976). Injections of GH increase LWG in pigs and probably in young cattle (Machlin, 1976).

The implantation of oestradiol-17β in wether sheep caused significant increases in the plasma concentration of GH and insulin (Donaldson, 1977). A similar effect was observed with DES in cattle (Trenkle, 1976). The primary action of the oestrogen is thought to be on factors controlling the production of GH from the pituitary, and an indirect effect of this causes an increased production of insulin. The effects on growth of exogenous oestrogen and GH are similar and it is concluded that one of the main actions of oestrogens, especially in males, is the increased production of GH and insulin.

MODE OF ACTION OF ANDROGENS

The evidence is not conclusive, but it is possible that androgens act directly at the muscle cell (Young and Pluskal, 1977; Mainwaring, 1977). Androgens regulate protein synthesis and degradation, with increased protein accretion and decreased protein turnover rate (Vernon and Buttery, 1976). Mainwaring (1977) has reviewed the possible mechanisms of action of androgens when they enter their target cells and form complexes with specific androgen receptors. These complexes can enter the nucleus of the cell and alter DNA replication and stimulate RNA synthesis, which in turn modifies protein synthesis.

A possible second mode of action is on the rate of protein degradation. Corticosteroids are potent catabolic agents in muscle tissue, and may serve a regulatory and suppressive role in normal growth. Androgens, but not oestrogens, are known to displace corticosteroids from their receptor sites (Meyer and Rosen, 1975). Therefore, it is not inconceivable that androgens may limit this role in animals by substitution at the receptor site.

A third possible mode of action of androgens is that they act indirectly by regulating the circulating levels of thyroxine. It is known that total circulating levels of thyroxine and free thyroxine are dramatically decreased in both cattle (Heitzman, Chan and Hart, 1977) and in sheep (Donaldson, 1977) and this may be a factor in regulating protein turnover in muscle cells.

CONCLUSIONS

The differing modes of action of androgens and oestrogens in the regulation of protein metabolism would offer biochemical support for the hypothesis that the best exogenous steroid treatment would be one that maintains or mimics maximum physiological levels of androgens and oestrogens in circulating body fluids.

Editor's note

Although not presented in the Seminar, this paper has been included in the proceedings as a completion to the general paper by D. Lister on a most important topic of actual interest. We feel indebted to Dr. R.J. Heitzman for the preparation of the text for this purpose.

REFERENCES

Aschbacher, P.W., Thacker, E.J. and Rumsey, T.S. 1975. J. Anim. Sci. 40,
    530-538.

Chan, K.H., Heitzman, R.J. and Kitchenham, B.A. 1975. Br. Vet. J., 131,
    170-174.

Donaldson, I.A. 1977. PhD Thesis, University of Reading.

Heitzman, R.J., 1976. In 'Anabolic agents in animal production' (Eds.
    F.C. Lu and J. Rendel). Environmental Quality and Safety, Suppl. V.,
    pp 54-59.

Heitzman, R.J., Chan, K.H. and Hart, I.C. 1977. Br.Vet. J. 133, 62-70.

Heitzman, R.J., Donaldson, I.A. and Pease, L. 1977. Proceedings of first
    meeting of Association for Veterinary Clin. Pharm. and Ther., Liverpool.
    In press.

Heitzman, R.J. and Harwood, D.A. 1977. Br. Vet. J. In press.

Hoffman, B. and Karg, H., 1976. In 'Anabolic agents in animal production'.
    (Eds. F.C. Lu and J. Rendel). Environmental Quality and Safety,
    Suppl. V, pp 181-191.

Hoffman, B., Karg, H., Vogt, K. and Kyrein, H.J. 1975. In 'Rückstände in
    Fleisch und Fleischzingnissen'. Ed. H. Boldt. K.G. Verlag.

Hoffman, B. and Oettel, G. 1976. Steroids. 27, 509-523.

Machlin, L.J. 1976. In 'Anabolic agents in animal production'. (Eds.
    F.C. Lu and J. Rendel). Environmental Quality and Safety, Suppl. V.
    pp 43-53.

Mainwaring, P. 1977. In 'The mechanisms of action of androgens'. Springer
    Verlag, New York, Inc.

Mayer, M. and Rosen, F. 1975. Am. J. Physiol. 229, 1381-1386.

Michel, G. and Beaulieu, E.E. 1976. In 'Anabolic agents in animal production'
    (Eds. F.C. Lu and J. Rendel). Environmental Quality and Safety.
    Suppl. V, 54-59.

Pottier, J., Busigny, M. and Grandadam, J.A. 1975. J. Anim. Sci, 41, 962-968.

Trenkle, A. 1976. In 'Anabolic agents in animal production'. (Eds. F.C. Lu
    and J. Rendel). Environmental Quality and Safety, Suppl. V. pp 79-88.

van der Wal, P. 1976. In 'Anabolic agents in animal production'. (Eds
    F.C. Lu and J. Rendel). Environmental Quality and Safety, Suppl. V.
    pp 60-78.

Vernon, B.G. and Buttery, P.J. 1976. Br. J. Nutr., 36, 575-579.

Young V.R. and Pluskal, M.G. 1977. Abstracts, 2nd Internat. Symp. on Protein
    Metabolism and Nutrition, Eur. Ass. Anim. Prod. Biddinghuizen. pp 13-26.

NUTRITIONAL EFFICIENCY OF PROTEIN AND FAT DEPOSITION

A.J.H. van Es

Institute for Livestock Feeding and Nutrition Research "Hoorn",
Lelystad, The Netherlands and

Department of Animal Physiology, Agricultural University,
Wageningen, The Netherlands.

ABSTRACT

Some aspects of intermediary metabolism of protein and fat deposition in growing cattle and other animals are treated first: energy supply and utilisation, protein synthesis and degradation (technical details and interpretation problems are discussed in an appendix) and fat synthesis and degradation at the tissue level. It is shown that due to insufficient quantitative information in this field - size of maintenance needs, of conversion of ME to fat and of change of rate of total protein synthesis with increasing rates of growth - energy requirements of growing cattle cannot yet be established reliably in a factorial manner.

The main lines of thought underlying the older and newer energy systems for feed evaluation and for requirement prediction are presented. These systems would be considerably more precise if a better prediction could be given of the size of the energy deposition from data on liveweight, daily liveweight gain and type of animal.

Finally the N requirements of growing cattle are discussed with special attention to N metabolism in the forestomachs, to new systems for their derivation in the factorial way and to estimates of these requirements resulting from feeding and balance trials.

INTRODUCTION

Protein and fat accretion during the growth period of cattle results from a great number of separate metabolic processes. To be able to grow the animal needs to maintain as well as to enlarge continuously its metabolic machinery. It does so in a well co-ordinated way showing a considerable interaction between maintenance and growth metabolism. Thus for giving information on the nutritional efficiency of protein and fat deposition during growth maintenance requirements have also to be considered. A precise understanding of all the processes involved is complicated further by the facts that during growth the quantities of protein and of fat deposited daily change in size and that the animal's physical activity and appetite decrease.

An attempt will be made to bring some order to this bewildering complexity of the growth phenomenon by first studying the separate processes of intermediary metabolism involved: energy supply and utilisation, protein synthesis and degradation and fat synthesis and degradation. Next the main lines of thought of the derivation of systems dealing with the requirements of growing cattle and with the evaluation of feeding stuffs with regard to energy and nitrogen will be considered. The order chosen is maintenance, protein deposition and fat deposition. Maintenance metabolism is a process of major importance in growing cattle, it increases from some 40 to 60% and more of total metabolism with age. Although daily protein deposition amounts only to 150 - 200 g in rapidly growing cattle in the weight range of 200 to 500 kg, its effect on liveweight gain is remarkable as its deposition is accompanied by about a threefold amount of water. In early maturing cattle fat deposition steadily increases with age from 200 to 600 g/d depending on animal type and feeding level. Fat deposition is not accompanied by a deposition of water, instead it appears to decrease slightly the water to protein ratio of the muscle tissues. In later maturing bulls it is much lower,

Geay et al. (1976) gave examples where both fat and protein
deposition averaged about 200 g per day in the weight range of
200 - 500 kg.

INTERMEDIARY METABOLISM

1. Energy supply and utilisation
     In growing cattle the main nutrients reaching the blood
are: volatile fatty acids (VFA), amino acids (AA), fats (F,
including higher fatty acids and glycerol) and some glucose.
When used for energy purposes - for muscle contraction, maintenance of ion concentrations, active uptake, synthesis of
enzymes, hormones, new tissues, etc. or for repair work, regulation, etc. - they are first converted into ATP. Next, hydrolysis of the terminal phosphate group of ATP supplies the required
free energy. The synthesis of one mole of ATP from ADP theoretically requires 80 - 100 kJ (19 - 24 Kcal) metabolisable energy
(ME) for the above mentioned nutrients assuming this ME to contain 10% fermentation heat. Glucose is most efficiently used,
fat some 5% less and VFA and AA 10 - 15% less (Armstrong, 1969).
Balance experiments with ruminants have indeed shown that the
utilisation of ME for maintenance, a process mainly requiring
ATP, does not vary much between ruminant rations since VFA and
AA are the major components of the ME. However, a slight
improvement was found in the ME utilisation with increasing
metabolisability (q, % ME in gross energy): + 0.5% per unit
increase of q (van Es, 1976). This might be due to a higher
percentage of propionic acid, fat and glucose in the absorbed
nutrient mixture, to less work for intake and digestion and/or
to less heat loss aue to fermentation at higher values of q.

     Mature non-lactating cattle need some 420 kJ (100 kcal)
ME per kg $^{3/4}$ for maintenance when fastened (van Es, 1972).
Figures of 450 - 600 kJ are given for growing cattle above 150
kg (Garrett, 1970; van Es, 1972; Webster et al., 1976; Vermorel
et al., 1976). Mature sheep require some 340 kJ. Pigs above
100 kg need, again per $W^{3/4}$, a similar quantity as sheep, at
lower weights the requirement appears to be higher (Thorbek,

1975). Technically it is difficult to measure the maintenance requirement of young animals. They are more easily stressed by the measuring procedures than fullgrowns and when fed only a maintenance ration they may feel hungry and show an abnormal behaviour.

Physical activity increases the requirements considerably (Wenk and van Es, 1976). Theoretically so does thermoregulatory heat production in a too cold environment; growing cattle, however, usually are above the lower limit of their thermoneutral zone. Near and slightly above this zone's upper limit the free energy needed for thermoregulation does increase but not very much, however voluntary intake, especially of roughages, decreases.

Free energy utilisation for synthesis of protein and fat will be discussed in the following sections.

2. Protein synthesis and degradation

Synthesis of protein requires linking the various AA in the correct order. The complicated process needs some 5 ATP hydrolyses per peptide linkage. The major quantity of this is due to processes of AA activation and of initiation, linking and translocation at the ribosomes. Small quantities are needed for mRNA and tRNA synthesis, moreover, part of the peptide chain synthesised is sometimes used to cross cell membranes and is lost during this transport (Campbell, 1977). The last - mentioned process lacks sufficient quantitative evidence; it seems greater for enzyme and hormone proteins than for muscle and milk proteins. Assuming a synthesis cost of 6 moles ATP for the former and 5 moles for the latter, a weight of 100 g (2 400 kJ; 574 Kcal) for a mole of 'peptide' and a cost of 80 kJ (19 kcal) ME for the synthesis of one mole of ATP from ADP + P results in a theoretical energetic efficiency of 2 400/(2 400 + (5 or 6) x 80) = 86% and 83% respectively for the synthesis of the two kinds of proteins from ME which contains the required AA.

The rate of synthesis of protein is many times greater than net protein accretion during growth. Much of the existing protein is degraded to AA in the course of time and has to be renewed. Most of the information on rates of synthesis comes from rats, some from rabbits, sheep, pigs and men. The techniques used are extremely complicated and their results are often difficult to interpret (see Appendix). Most experiments show that the rate of synthesis for whole-body protein is related to metabolic body weight, i.e. per kg liveweight the rate is highest for small species, that it decreases with age and that it is higher at higher rates of growth (Millward et al., 1975, 1976[3]; Munro, 1976[2]; Arnal, 1977). Visceral proteins have a much higher rate than proteins of muscles and brain. Synthesis rate of muscle slows down in case of protein or energy shortage, but the liver's rate is hardly influenced or increases slightly (Buttery et al., 1977; Garlick et al., 1976; Lobley and Harris, 1977; Munro, 1976[1]).

Degradation of protein to AA is a process which unfortunately does not result in ATP synthesis. The resulting AA are incorporated in the various AA pools and can either be used for protein synthesis again or for other purposes, e.g. energy supply. Schimke (1977) gives several reasons why there is a need for degradation of proteins: for regulation of metabolic rate via enzyme and hormone levels, for mobilisation in times of shortage, for removal of faulty proteins and for 'restructuring' cells. Due to this degradation AA pools are seldom exhausted with regard to essential AA's. Rates of degradation of proteins differ considerably, depending on the metabolic function of the protein. As in the case of synthesis they are high for enzyme and hormone protein, moderate for precursor proteins of structural tissue (e.g. procollagen, Robins, 1977) and low but certainly not negligible for structural proteins like muscle (Millward et al., 1975, 1976[1,2]; Lobley and Harris, 1977; Buttery et al., 1977).

The difference between synthesis and degradation gives the net protein accretion or loss. In mature animals the two

processes are nicely balanced and the synthesis process is
responsible for roughly 20% of the ME needed for maintenance
(rat 23%, rabbit 19%, sheep 21%, 11%, man 22%). During rapid
growth values near 30% can be computed (rat 36%, 29%, lamb 51%,
pig 28%). These values have been derived by the author from
rates of synthesis of whole-body protein given by Garlick et
al. (1976), Millward et al. (1976) and Nicholas et al. (1977)
assuming a ME requirement for ATP synthesis needed for peptide
linkage of 440 kJ per 100 g protein and maintenance requirements
per $kg^{3/4}$ of 420 kJ ME for rapidly growing rats, pigs and lambs
and the mature rabbit and of 340 kJ ME for mature rat, sheep
and man.

The figures show that the energy costs for protein synthesis are fairly low, even during rapid growth. Figures for
cattle are not yet available. In view of the small variation
among the animal species mentioned above it would be surprising
if those of cattle would be much higher. At energy equilibrium
these costs of protein synthesis are included in the energy
required for maintenance. Rapid growth appears to increase
the costs. Thus net protein accretion needs energy for peptide linkage not only of the net protein synthesised but also
of the increased protein synthesis for maintenance. It explains
why the efficiencies of the utilisation of ME for protein accretion, calculated by regression methods (see below), are always
lower than the theoretical 86 - 83% derived above. Lack of
information on the size of the increase of the total synthesis
rate due to growth in cattle prevents prediction of the net
accretion costs. It is probable that this increase will be
lower than in the case of eg the growing pig. Per metabolic
weight growth in cattle means a deposition of energy and
protein which is considerably lower than in pigs (Table 1),
ie metabolically seen growth in cattle is not very rapid.

An increase of the protein synthesis rate with rapid
growth does not occur always. Vernon and Buttery (1976) found
in rats a higher rate of gain after treatment with trienbolone
acetate while synthesis (and degradation) rates of protein

were decreased instead of increased. In veal calves and pigs
treated with anabolics van Weerden and van Es (see van Weerden
and Grandadam, 1975) using indirect calorimetry found no significant change in heat production despite a shift from fat accretion towards protein accretion. This appears to be in agreement with the results of Vernon and Buttery.

TABLE 1

ENERGY AND NITROGEN DEPOSITION OF RAPIDLY GROWING PIGS AND BULLS, EXPRESSED
PER METABOLIC BODY WEIGHT

	pig		bull	
	30 kg	80 kg	200 kg	500 kg
$RE/W^{3/4}$ [1]	0.33	0.67	0.23	0.28
$RN/W^{3/4}$	1.2	0.8	0.6	0.3

1) RE: energy deposition, MJ/d

   RN: N deposition, g/d

    W: body weight, kg

        (data from Schiemann et al., 1976 and Thorbek, 1975).

3. Fat synthesis and degradation

The precursors for fat deposition in cattle are VFA, fat
and longchain fatty acids, AA and small amounts of glucose.
The fat content of cattle rations usually is low (<5%) as ruminants cannot stand high levels. Absorbed long-chain fatty
acids or fats can be utilised with high efficiency for fat deposition as little biochemical conversion is needed. Also the
protein content is usually low (10 - 15%) except at pasture;
part of the protein is converted into VFA in the rumen. Absorbed AA from microbial origin and from feed protein which escaped degradation in the rumen are deaminated before use as precursors for fat synthesis. The excretion of the ammonia group
requires some free energy, this loss is compensated more or
less by the fact that the utilisation of the deaminated compounds for fat synthesis is slightly more efficient than that
of the VFA. Thus, the efficiencies of the energy utilisation
of absorbed VFA and AA will not differ much. Absorbed glucose

results from small amounts of starch which escaped degradation
by the rumen organisms.

Long-chain fatty acids are synthesised by serially linking
and modifying acetic-acid molecules, resulting from absorbed
VFA, AA and glucose. For this purpose free energy (ATP) is
needed to activate acetic acid, furthermore there should be an
adequate supply of reducing compounds in the form of NADPH
(Armstrong, 1969). If we assume that sufficient NADPH can be
produced via the pentose cycle from propionic acid and via NADP-
isocitrate dehydrogenase from acetate (Annison, 1976), the theoretical efficiency of the energy utilisation for the synthesis
of stearic acid from VFA is about 80%. There is little difference among the acids of the VFA in this respect. Above it was
found probable that the absorbed AA would have a similar efficiency of utilisation.

Balance and slaughter experiments, however, have shown that
the metabolisable energy (ME), mainly consisting of the energy
of absorbed VFA and AA (deaminated partially), is utilised for
fat synthesis far less efficiently than 80%. This is especially
the case with feeds of low q. A small part of the discrepancy
can be explained by the fact that ME includes some 10% fermentation heat which is useless to the animal so that an efficiency
of 72% might be expected for ME instead of the above 80%.

Measured efficiencies for feeds with a q of 70 are between
50 and 60% and for those with a q of 50 between 30 and 50% (van
Es, 1976). These values result from difference trials with
mature ruminants, fed at about the maintenance feeding level
and 1.5 - 2.5 times as much. All errors accumulate in the
differences in ME intakes and energy depositions of the two
trials, which is the reason why the efficiency figures have high
standard deviations. Errors are greatest for feeds with a low
q as animals on such feeds often do not even reach the 1.5 x
maintenance feeding level. Several reasons are given for the
discrepancy. At lower q there might be a shortage of NADPH,
losses with fermentation heat might be greater and more energy

might be needed for the whole process of eating, ruminating and digestion (Annison, 1976; van Es, 1976).

It is improbable that increased fat turnover during growth as seems to be the case with regard to protein turnover might be responsible for the discrepancy for the following reasons. Fat hydrolysis and subsequent resynthesis requires little energy. In mature and in growing monogastrics actual efficiencies of utilisation of ME for fat synthesis are close to theoretical values (see below). Moreover, lipolysis in well-fed ruminants seems to be low (Metz and van den Bergh, 1977) and the discrepancy is greatest at low q ie at low rates of fat deposition, and smaller at high rates.

Our knowledge of the basic aspects of fat deposition in ruminants clearly has serious gaps.

ENERGETIC EFFICIENCY IN GROWING CATTLE

In monogastric animals requirements of ME for growth have often been derived with regression methods using the following model:

$$ME = a \ W^p + b \ RF + c \ RP,$$

in which ME   = metabolisable energy, kJ/d

      W   = body weight, kg

      RF, RP= retained fat and retained protein, g/d

      a,b,c,p = constants

The term $a \ W^p$ stands for the animal's maintenance needs. Theoretically it should be possible when a sufficient number of balance data is available to calculate the values of the constants. Unfortunately this is only so provided that the data show wide variation in W, RF and RP but little correlation between these variables and provided that the constants are true constants, ie that they do not change to new values in the course of growth. Seldom are all these conditions fulfilled. In ad libitum fed growing animals, RF increases with W whereas RP does not change much, moreover the animals' activity usually declines so that coefficient a is not really constant. To obtain a greater variation in RF and RP data from the animals during feed restriction are also used for the calculations. Feed restriction depresses RF more than RP and sometimes changes the animals' behaviour (Wenk ana van Es, 1976), so that it does not solve all problems. Moreover, RF and RP have a rather high error which is most unsatisfactory for obtaining reliable estimates of the regression coefficients.

Up to now only Pullar and Webster's (1977) experiments were devised in such a way that most of the biases mentioned above were avoided. They measured ME, W, RF and RP of lean and fatty Zucker rats, both fed the same ration at two body weights, ad libitum as well as restrictedly. In their regression model maintenance requirements were only assumed equal for the same animal at about the same weight. In this manner they found energetic efficiencies for fat and protein deposition of 71 and 43%, respectively, values close to those obtained by several other research workers for growing monogastrics using experimental schemes which had more biases. The first value is slightly below the biochemically expected value, the second is considerably below the theoretical value, most probably a consequence of increased protein turnover at higher growth rate.

Results of similar regression computations with growing cattle are not available. This is mainly due to the lack of a sufficient number of results of balance trials. Even if they were available, it is doubtful whether the regression method would be successful for deriving reliable estimates of efficiencies of fat and protein depositions in cattle above 200 kg account for too small a part of total metabolism (Table 1).

Therefore it is not surprising that for cattle above 150 kg energy needs are simply derived by working with total energy deposition (RE) rather than fat and protein deposition separately. Cattle above that weight no longer show the increases in physical activity which appear linked with early age, so maintenance requirements can simply be related to metabolic weight. The model used, $RE = a(ME - bW3/4)$ can easily be converted to $RE/W^{3/4} = a \; ME/W^{3/4} - b^1$ and plotted graphically. Unfortunately in most experiments with growing cattle for statistical reasons high values of b^1, ie high maintenance needs, can easily be compensated by high values of a, ie a highly efficient conversion of ME in RE. The results obtained are not usually precise enough either with the balance or with the comparative slaughter technique and the experimental scheme used, to draw the conclusion that the computed values of a and b^1 give a correct picture of the physiological processes involved (Vermorel et al., 1976; Geay et al., 1976). Extensive series of balance experiments are needed for this purpose with early - and late - maturing breeds fed ad libitum, as well as restrictedly, ME and RE being measured several times during the growth period. Even then the question of the possibly higher maintenance requirement in the practical situation (free walking animals kept in groups and usually group-fed) has to be solved.

Most older energy systems for growing cattle are used to evaluate the various feed stuffs values derived for other purposes: TDN, Starch value, NEF_R. It is assumed that in this respect growing cattle do not differ from the animals used to derive such values. In the case of the latter two values it is argued that these values have been derived from fat deposition in mature cattle and that in growing cattle also the major part of the energy deposition is fat. In these systems requirement data were derived from feeding trials with growing cattle. Lofgreen and Garrett (see a.o. Garrett, 1970) were the first who worked out a system in which the differences between energy utilisation for maintenance and for production were included. They used the comparative slaughter technique, a not too precise and highly laborious technique, and restricted themselves to the Californian feed lot situation to which their requirement data also apply.

Two other systems (MAFF, 1975; van Es, in press), based partly on data obtained with growing ruminants but mainly on

those with mature ones, were presented recently. They use the
elegant solution of Mac Hardy (1966) and Harkins et al. (1974)
of the problem posed, with regard to correct feed evaluation in
a similar approach as Lofgreen and Garrett, when growth rates
vary: the ratio of maintenance and production energies is used
while predicting ME utilisation. These two systems differ only
in relation to the degree of simplification used. Their precision is still not very high due to insufficiently accurate
information for growing cattle on maintenance requirements and
on the efficiency of the utilisation of the ME for energy deposition as discussed above. Furthermore, both the requirements
and the utilisation depend on the size of the total energy deposition (RE, kJ). Thus, a precise estimate of this deposition
is needed. Under practical circumstances the estimate can only
be made from weight (W, kg), daily liveweight gain (ΔW, kg) and
breed, type and sex of the animal. The MAFF-system uses :

$$RE = (6\ 280 + 18.8\ W)\ \Delta W/(1 - 0.3\ \Delta W),$$

which was derived from results obtained with early-maturing
steers. While deriving the other system this equation gave
too high values for Friesian bulls, so it was adapted to

$$RE = (2\ 092 + 25.1\ W)\ \Delta W/(1 - 0.3\ \Delta W),$$

which appears better suited for Friesian bulls but needs further
improvement. Even this equation may give too high RE estimates
for bulls of the late-maturing breeds.

It will be clear that better information on maintenance
requirement, on utilisation of ME for energy deposition and on
the relationship between energy deposition, weight and daily
liveweight gain for the various types of animals is urgently
needed. When available by changing some constants or factors
both new systems can easily incorporate such better information.
The older systems too can easily change their requirement standards but their method of feed evaluation is very rigid.

NITROGEN UTILISATION BY GROWING CATTLE

At tissue level growing cattle need amino acids for the

synthesis of proteins for maintenance and production. The total amount is not very high: protein deposition in growing cattle seldom exceeds 200 g per day, an amount of protein present in only some 6 kg of cow's milk. Most of these amino acids result from microbial protein synthesised in the forestomachs which after hydrolysis is absorbed from the small intestine into the blood. This protein has an amino acid pattern which differs little from that of most of the animal's proteins, so its biological value is high ensuring an efficient utilisation. A minor amount of absorbed amino acids comes from undegraded feed-protein entering the small intestine. Its biological value depends on the kind of feed stuffs of the ration but due to its small size it hardly influences the amino acid pattern of the total quantity of absorbed amino acids.

The true absorption of microbial amino acids from the small intestine is high, although not extremely so, because proteins in the bacterial cell walls are not too easily hydrolysed. Some 20% of the microbial N is present in other forms than proteins or amino acids.

Rate of synthesis of microbial protein and of degradation of feed and feed-N (protein - and non - protein - N) in the forestomachs depends on the chemical and physical composition of the feed. The major part of the ingested N is broken down by the microbes to ammonia, degradability between dietary proteins, however, varies considerably. The microbes use the ammonia for their growth but need for this purpose and their maintenance metabolism free energy resulting from anaerobic degradation of organic matter. Sugar, starch and many proteins are good sources for their energy supply, cell wall organic matter has a much slower energy release. The end products of the degradation, the volatile fatty acids, are acid and tend to lower the pH of the rumen fluid. This is opposed by bicarbonate entering the forestomachs with the saliva. Insufficient salivation, eg due to a low percentage of forage in the long state, may lead to a lower pH of the rumen fluid. This is usually accompanied by a shift in the propionate/acetate ratio of the fluid to higher values and, if it lasts for the greater part of the day, it leads to slower breakdown of cell wall material of the forage, ie to lower fibre digestibility. Lower digestibility also occurs when the microbes have not enough ammonia for their growth. It is for this reason that the ration should contain more than 10% crude protein in the dry matter.

In several countries systems have been proposed for deriving the N requirements of ruminants in this factorial way (Roy et al., 1977; Kaufmann, 1977; Satter and Roffler, 1977). Lack of reliable quantitative data despite a good qualitative understanding of the processes involved mean that these systems still need further improvement.

Most N requirements for growing cattle have been derived, so far from feeding and balance trials in which rations with different crude protein contents were compared with regard to growth rate, feed-conversion and N balance. They show a good agreement for bulls of the Black and White type (Schulz et al., 1974; Poppe and Gabel, in press; de Boer and Hamm, in press). Rapidly growing late-maturing breeds might need slightly higher quantities.

APPENDIX

MEASUREMENTS OF PROTEIN SYNTHESIS AND DEGRADATION AND THE INTERPRETATION OF THEIR RESULTS

FLUX AND SYNTHESIS

The animal is given a continuous infusion of a labelled amino acid in the blood. The specific radioactivity (SR) of this amino acid (AA) of the blood plasma is measured at intervals.

(This survey made mainly from publications quoted in the list of references marked with *)

After some time (2 - 8 h), depending on the pool size of this
AA, the SR will reach a constant or plateau value and remain
at that level. The flux of the AA out of the free AA pool can
then be calculated from rate of infusion and SR. The flux is
equal to the total amount of the AA used for protein synthesis,
oxidised or converted to metabolites. Because of the short
length of the infusion period there is no risk that re-entrance
of the labelled AA in the free AA pool occurs; eg time is too
short and dilution too great for the labelled AA used for protein synthesis to return to the free AA pool after protein degradation to a measurable extent. Oxidation and conversion of the
chosen AA usually are low compared to synthesis, so flux can be
used as an estimate of the AA used for synthesis. Whole-body
protein synthesis is calculated from this estimate and the average content of the AA used in the protein of the body.

DISCUSSION

The method described can easily give misleading results.
Introducing the injection needle might upset the animal's
normal metabolism for hours, so infusion via a cannula applied
several days earlier seems inportant. It is the more important
if, as often is the case, from the data on protein synthesis via
flux, data on protein degradation are also derived by subtracting data on net accretion from those on synthesis: protein synthesis - protein degradation = protein accretion. Net protein
accretion is equal to 6.25 x the N-balance measured over a
sufficiently long time (at least a week) concurrent with the
infusion period or just before or thereafter.

Blood plasma is not the only large free AA pool, tissue
fluid is another one. It usually has a lower SR than blood
plasma, probably due to the inflow of unlabelled AA from protein
breakdown. Plasma and tissue pools show a considerable exchange,
thus the plasma pool receives labelled AA from the infusion as
well as AA (with a lower SR than the plasma AA) from the tissues.
The latter inflow lowers the SR of the plasma. If blood plasma

and tissue fluids would form one homogeneous free AA pool SR
would be lower than plasma SR, thus the calculation from AA flux
from infusion rate and plasma SR underestimates the true AA flux
and protein synthesis. On the other hand overestimation of protein synthesis occurs when the oxidation and conversion of the
infused AA is neglected. To reduce this overestimation essential AA are used for infusion which are less easily used for
other purposes than synthesis. The risk of oxidation, moreover,
might be kept low by feeding the animal ample energy and just
enough of the AA chosen for infusion.

Due to increased absorption of AA after each meal slight
deviations from plateau SR occurs, especially in monogastric
animals eating rapidly one meal a day. Unavoidable analytical
errors and sometimes pollution of the infused labelled AA with
its D-isomer are other sources of error. Finally the AA chosen
for infusion may not be present to the same extent in all proteins. As the various proteins differ considerably in synthesis
rate a considerable overestimation of whole-body protein synthesis would result if an AA was chosen of which a protein with
a high synthesis rate contained a high percentage.

It will be clear that estimates of body protein synthesis
rates from flux measurements require thorough planning, careful
handling of the animal and precise analytical and physical analysis. Moreover, their results should be checked using other
labelled AAs and other methods.

FRACTIONAL SYNTHETIC RATES OF PROTEIN SYNTHESIS IN DIFFERENT
TISSUES

A continuous or a single shot infusion of a labelled AA
is given for preference, an AA which is not used to a great
extent for other purposes than protein synthesis (if so, too
much label has to be infused and the use for other purposes
might vary in size with time). The increase of SR of the
chosen AA is measured at intervals in the free AA of blood
plasma and tissue fluid. After a precisely known number of

hours, often 6, the animal is killed and samples are rapidly taken from the various tissues and analysed for the SR of the chosen AA in the free and protein bound state. It is assumed that rates of protein synthesis and degradation remain constant from the start of infusion until slaughter. Thus, the SR of the protein bound AA is the result of uptake of label from the free AA pools, loss of label by degradation of protein usually being very small during the short period. Thus, from the final SR of the bound AA and the SR levels of plasma or tissue fluid free AA pools from start of infusion until slaughter the rates of synthesis of protein for the various tissues can be computed.

DISCUSSION

In this case also, the animal should not be aware that the isotope infusion is commenced, the more so as the effect of possible stress on rate of synthesis, appears to differ between tissues, muscle tissue being more sensitive. The precision obtained depends largely on the precision with which the average difference of SR between protein AA and free pool AA during the period from start of infusion to slaughter can be measured. It will be clear from this that the precision is higher for continuous infusion than single shot techniques, and for the first technique even more if plateau levels are reached sooner.

The choice of the free AA pool, blood plasma or tissue fluid, forms a problem. There is some evidence that free AA from extracellular fluid can be used for synthesis without complete equilibration with intracellular free AA. In the rabbit fractional synthesis rates using the plasma free AA pool were 0.014 and 0.15 for muscle and liver protein respectively versus 0.019 and 0.32 using tissue free AA pools. A further complication is that even for the same tissue the extent of this uptake of AA from the extracellular pool for protein synthesis seems to depend on the tissue's metabolic state. Finally there is doubt if synthesis and breakdown of myofibrillar protein is always random. There is some evidence that after great changes in AA supply at tissue level increase and decrease of the size of muscle myofibrils is not random. In that case infusion of label could result in heterogeneous labelling of muscle protein.

It is clear that also in the case of fractional synthetic rates the measurements require thorough planning whereas the results so far allow two ways of interpretation leading for some tissues to great difference in synthetic rate. Even so, whole-body protein synthesis rates, calculated by adding the products of the various estimates of fractional rates and the weights of these tissues, often agree fairly well with estimates obtained from flux techniques.

DEGRADATION

The animal is dosed for some time through infusion or with feed or drinking water with an isotope with the aim of labelling the amino acids of its proteins. Labelled AAs can be used but are expensive, thus use is often made of the fact that labelled bicarbonate-C easily exchanges with the C in the COOH-groups of several amino acids. After dosing the readily exchangeable AA-pools are allowed to lose their label. Next at intervals animals are killed and samples analysed for total protein-bound label and for SR of various AA. It is assumed that the labelled AA of degraded protein enters in such a large pool of unlabelled AA that resynthesis of labelled AA can be neglected. In that case loss of labelled protein-bound AA with time is a measure of rate of protein breakdown. Moreover rate of loss of SR of protein-bound AA gives information on the rate of protein synthesis.

A second technique makes use of a breakdown product of myofibrillar protein, 3-methylhistidine, which cannot be used for synthesis again and is excreted in many animal species with the urine. From knowledge of the 3-methylhistidine content of the muscle and the daily excretion degradation rate of myofibrillar protein can be computed.

DISCUSSION

Resynthesis of the AA resulting from protein degradation is the main problem of the first technique. For this reason attention should be paid particularly to those labels which turnover rapidly such as the carboxyl groups of glutamate and aspartate. The method cannot be used for proteins with a rapid turnover like those of liver, intestinal mucosa, enzymes, because of the interval needed after dosing.

Soon after the peptide synthesis a small part of the histidine of actin and myosin of the muscles, together with about half of the muscle's protein, is methylated. The content of 3-methylhistidine varies with animal species and age and has to be known. The method of excretion of the degraded 3-methylhistidine depends on animal species. In man and rabbit excretion with urine is rapid. In the rat part of it is acylated, so this part should be included in the analysis. In sheep its pool is large so that changes in degradation rate of myofibrillar protein do not give marked changes in urinary excretion. The pig retains most as a peptide and therefore excretes very little.

Even when the excretion of 3-methylhistidine and its metabolites is measured correctly, it provides information mainly on degradation of myofibrillar protein and not on that of sarcoplasmic protein, which is said to degrade more rapidly nor on other proteins. Furthermore, in the rat 10% of the total 3-methylhistidine is present in skin and gastrointestine. Its degradation and excretion with the urine was higher than that of myofibrillar muscle protein.

With both methods the risk of upsetting the animal's metabolism is considerably smaller than in flux-techniques. Unfortunately they are also not free from uncertainties.

REFERENCES

Annison, E.F., 1976. Energy utilisation in the body. In: Principles of
    cattle production, H. Swan and W.H. Broster, eds. Butterworths,
    London, p. 169-199.

Armstrong, D.G., 1969. Cell bioenergetics and energy metabolism. In:
    Handbuch der Tierernahrung I, W. Leinkeit et al., eds. Paul Parey,
    Hamburg/Berlin, p. 385-414.

Arnal, M., 1977. Muscle protein turnover in lambs throughout development.
    In: B[1], p.35-37.

Boer, F. de and Hamm, G.G.H., 1977. Protein requirements and NPN-supple-
    mentation in fattening bulls. Proc. EEC/FAO Symp. on recent devel-
    opments in the use of new sources of protein, essential amino acids
    and non protein nitrogen with special reference to ruminants, Geneva,
    in press.

Buttery, P.J., Beckerton, A. and Lubbock, M.H., 1977. Rates of protein
    metabolism in sheep. In: B[1], p. 32-34.

Campbell, P.N., 1977. Recent advances in eukaryotic protein synthesis.
    In: B[1], p. 12-14.

Es, A.J.H. van, 1972. Maintenance. In: Handbuch der Tierernährung, W. Len-
    keit et al., eds. Paul Parey, Hamburg, Vol. II, p. 1-54.

Es, A.J.H. van, 1976. Factors influencing the efficiency of energy utilis-
    ation by beef and dairy cattle. In: Principles of cattle production,
    H. Swan and W.H. Broster, eds. Butterworths, London, p. 237-253.

Es, A.J.H. van, Vermorel, M. and Bickel, H. Feed evaluation for ruminants.
    New energy systems in the Netherlands, France and Switzerland.
    Livest. Prod. Sci., in press.

Geay, Y., Robelin, J. and Jarrige, R., 1976. The influence of the metabol-
    isable energy content of the diet on the efficiency of energy utili-
    sation for young fattening bulls. In: A[1], p. 225-228.

Garlick, P.J., Burk, T.L. and Swick, R.W., 1976. Protein synthesis and
    RNA in tissues of the pig. Am. J. Phys. 230 (4) 1108-1112.

Garrett, W.N., 1970. The influence of sex on the energy requirements of
    cattle for maintenance and growth. In: Proc. 5th Symp. Energy meta-
    bolism of farm animals, A. Schürch and C. Wenk, eds. Juris, Zürich,
    p. 101-104.

Harkins, J., Edwards, R.A. and McDonald, P., 1974. A new net energy system
    for ruminants. Anim. Prod. 19, p. 141-148.

Kaufmann, W., 1977. Calculation of the protein requirements for dairy cows according to measurements of N metabolism. In: B[1)], p. 130-132.

Lobley, G.E., and Harris, C.I., 1977. Problems in estimated turnover. In: B[1)], p. 29-31.

MacHardy, F.V., 1966. Simplified ration formulation. Proc. 9th Int. Congr. Animal Production, Edinburgh, p. 25 (abstr.).

MAFF, 1975. Energy allowances and feeding systems for ruminants. Technical bull. 33, HMSO, London, 80 p.

Metz, S.H.M. and van den Bergh, S.G., 1977. Regulation of fat mobilisation in adipose tissue of dairy cows in the period around parturition. Neth. J. agric. Sci., 25, in press.

Millward, D.J., Bates, P. and Laurent, G., 1976_1. The relationship between the growth of the DNA-unit in muscle and protein turnover. Proc. Nutr. Soc. 36, 35A.

Millward, D.J., Garlick, P.J., James, W.P.T., Gender, P. and Waterlow, J.C., 1976. Protein turnover. In: Protein metabolism and nutrition, D.J.A. Cole et al., eds. Butterworths, London, p. 49-70.

Millward, D.J., Garlick, P.J., Nnanyelugo, D.O. and Waterlow, J.C., 1976_2. The relative importance of muscle protein synthesis and breakdown in the regulation of muscle mass. Biochem. J. 156, p. 185-188.

Millward, D.J., Garlick, P.J. and Reeds, P.J., 1976_3. The energy cost of growth. Proc. Nutr. Soc. 35(3) 339-350.

Millward, D.J., Garlick, P.J., Stewart, R.J.C., Nnanyelugo, D.O. and Waterlow, J.C., 1975. Skeletal-muscle growth and protein turnover. Biochem. J. 150, p. 235-243.

Munro, H.N., 1976_1. Eukaryote protein synthesis and its control. In: Protein metabolism and nutrition, D.J.A. Cole et al., eds. Butterworths, London, p. 3-18.

Munro, H.N., 1976_2. Regulation of body protein metabolism in relation to diet. Proc. Nutr. Soc. 35(3) 297-308.

Nicholas, G.A., Lobley, G.E. and Harris, C.I., 1977. Use of the constant infusion technique for measuring rates of protein synthesis in the New Zealand White rabbit. Brit. J. Nutr. 38 (1) 1-17.

Nishizawa, N., Noguchi, T., Hareyama, S. and Funabiki, R., 1977. Fractional flux rates of N^T-methylhistidine in skin and gastrointestine: the contribution of these tissues to urinary excretion of N^T-methylhist-idine in the rat. Brit. J. Nutr. 38 (1) 149-151.

Poppe, S. and Gabel, M., 1977. Views on the requirements of beef cattle

(including fattening cattle) for protein, essential amino acids and
non-protein nitrogen. Proc. EEC./FAO Symp. on recent developments in
the use of new sources of protein, essential amino acids and non protein nitrogen with special reference to ruminants, Geneva, in press.

Pullar, J.D. and Webster, A.J.F., 1977. The energy cost of fat and protein
deposition in the rat. Br. J. Nutr. 37 (3) 355-363.

Robins, S.P., 1977. Is collagen inert? In: B[1], p. 38-40.

Roy, J.H.B., Balch, C.C., Miller, E.L., Ørskov, E.R. and Smith, R.H., 1977.
Calculation of the N-requirement for ruminants from nitrogen metabolism studies. In: B[1], p. 126-129.

Satter, L.D. and Roffler, R.E., 1977. Calculating requirements for protein
and non protein nitrogen by ruminants. In: B[1], p. 133-135.

Schiemann, R., Jentsch, W., Wittenburg, H. and Hoffmann, L., 1976. Die
Verwerturng der Futterenergie durch wachsende Bullen. I. Mitt. Arch.
Tierern. 26, p. 491-517.

Schimke, R.T., 1977. Why is there protein turnover? In: B[1], p. 15-16.

Schulz, E., Oslage, H.J. and Daenicke, R., 1974. Untersuchungen über die
Zusammensetzung der Körpersubstanz sowie den Stoff-und Energieansatz
bei wachsenden Mastbullen. Fortschritte Tierphys. Tierern. 4. 70 pp.

Thorbek, G., 1975. Studies on energy metabolism in growing pigs. II. Beretnin
Stat. Husdyrbrugsforsøg 424, p. 93-126.

Vermorel, M., Bouvier, J.C. and Geay, Y., 1976. The effect of the genotype
(normal and double muscle Charolais and Friesian) on energy utilisation
by growing cattle at 2 and 16 months of age. In: A[1], p. 217-220.

Vernon, B.G. and Buttery, P.J., 1976. Protein turnover in rats treated
with trienbolone acetate. Brit. J. Nutr. 36 (3) 575-579.

Webster, A.J.F., Smith, J.S. and Mollison, G., 1976. On the prediction of
heat production in growing cattle. In: A[1], p. 221-224.

Weerden, E.J. van and Grandadam, J.A., 1975. The effect of an anabolic agent
on N deposition, growth and slaughter quality in growing castrated
male pigs. In: Proc. Symp. Environmental quality and safety, FAO/WHO,
Rome, suppl. vol. 5, p.15-122.

Wenk, C. and van Es, A.J.H., 1976. Energy metabolism of growing chickens
as related to their physical activity. In: A[1], p. 189-192.

1)A = Proc. 7th Symposium on Energy Metabolism of Farm Animals of the
EAAP. M. Vermorel, ed. De Bussac, Clermont-Ferrand, 1976.
B = Proc, 2nd International Symposium on Protein Metabolism and Nutrition
S. Tamminga et al., eds. Pudoc, Wageningen, 1977.

FEED EFFICIENCY AND GENOTYPE-NUTRITION INTERACTIONS
IN GROWING ANIMALS, PARTICULARLY IN CATTLE FOR BEEF PRODUCTION

Conclusions of CEE Colloquium held at Theix (France)
On 27-28th September 1976

C. Béranger

Institut National de la Recherche Agronomique
Laboratoire de la Production de Viande
Centre de Recherches Zootechniques et Vétérinaires
Theix - 63110 Beaumont - France.

The objective of this Colloquium was to bring together specialists in nutrition and genetics [1] to achieve a better comprehension between them on the problem of feed efficiency in growing animals. A report was written by J.C. Tayler [2] following this Colloquium. From the 22 communications [1] presented and from the discussion, some general ideas and conclusions can be summarised in this paper.

EXPRESSION OF EFFICIENCY

It is of importance to make clear and standardise means of expressing this efficiency.

Feed efficiency means efficiency of conversion of feed to animal tissue and should be expressed in terms in which an increase in value corresponds to an improvement : $E = \frac{output}{input}$ X 100. Its reciprocal $\frac{input}{output}$ X 100 is the feed conversion ratio (FCR) which is used more frequently in practice.

Input could be expressed in terms of dry matter, digestible organic matter, metabolisable energy, net energy, crude protein, digestible protein... each in its appropriate unit.

(1) 32 participants from 10 countries. See list of participants and communication in appendix

(2) CEE : coordination of Agricultural Research (non published).

Output could be expressed in terms of gain of liveweight, empty weight, carcass, muscle, fat, edible meat, energy, protein... in their respective units.

Energy as input and live weight gain as output will be the essential concern in the rest of this paper.

SOURCES OF VARIATION OF FEED EFFICIENCY

Efficiency depends, first of all, on the respective part of energy intake used for maintenance (heat losses related to basal or fasting metabolism) and for growth (protein and fat retained and heat losses related to their deposition).

Maintenance (400 to 500 KJME/kg $W^{0.75}$ above 150 kg) increase with live weight and varies with behaviour, sex (bulls > steers and heifers), genotype and environmental condition (humidity temperature...).

Energy available for growth depends on the difference between intake and maintenance requirement. So, voluntary intake influences growth rate and efficiency.

Efficiency for growth expressed as live weight, depends not only on the level of growth rate but also on the energy content of gain related to its composition (fat, protein, water). This composition depends on the protein production potential of the animal and on the available energy above maintenance ; the excessive energy above protein synthesis requirement induces fat deposition.

The energy efficiency of protein deposition is lower than that of fat. Therefore, in terms of energy, animals with high protein growth potential are less efficient than early maturing animals that fatten quickly. However, the energy cost per gram of organic matter is about the same for protein and fat because of the lower energy content of protein. Moreover, as protein in muscle is associated with three times its weight of water, in terms of live weight gain, efficiency increases as protein content of gain increases. Therefore, animals with high muscle growth potential are more efficient than animals with high fattening capacity, in terms of live weight, carcass, or muscle gain.

For these above reasons, for example, young bulls are less efficient than steers in terms of energy (higher maintenance) requirements, higher protein deposition, lower Kf [1] and more efficient in terms of live weight gain (less fat and more water in gain).

For these reasons, as well, the shape of the variation of feed efficiency with energy intake is a curve which comes to a peak corresponding to an optimum level of energy. Up to this maximum, as the energy intake increases feed efficiency increases, because the live weight gain increases and the part of energy used for maintenance decreases : beyond this maximum, feed efficiency decreases because fat deposition increases in the gain, removing the effect of live weight gain increment (Figure 1). However in terms of gain of energy, efficiency always increases with energy intake.

GENETIC VARIATIONS

Genetic variations in feed efficiency between animals, sex or breeds, are related to variations of the different factors influencing efficiency. Variations in mature weight induce variations in maintenance and in growth rate and therefore in feed efficiency. Variations in maturity induce variations in composition of gain. A comparison between genotype carried out at the same percentage of mature weight, and (or) at the same metabolic age reduces the variations but does not eliminate them.

(1) Kf : efficiency of utilisation of metabolisable energy above maintenance for fattening.

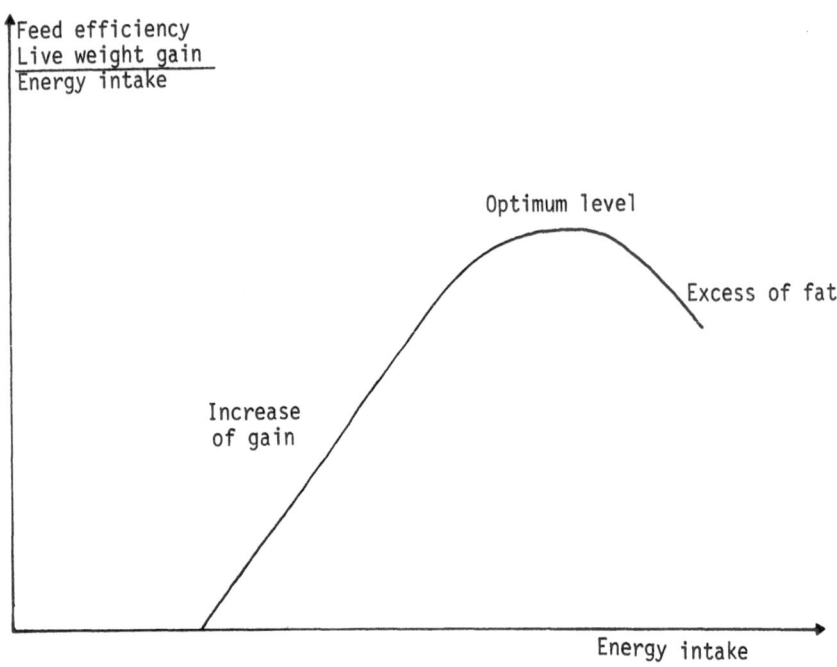

Fig.1. Variations of feed efficiency with energy intake.

Breeds and sex differences occur in maintenance requirement related to activity, response to stress... On the same
diet, large variations appear between breeds or animals in
voluntary intake (per metabolic weight) and therefore, in
available energy for growth and in growth rate. However, the
main differences appear in the composition of gain related to
variations in protein production potential and variations in
intake. The energy consumed in excess of maintenance and
protein production potential may be transformed into fatty
tissues or into heat (like in deer). Therefore, feed efficiency
varies largely with voluntary intake and muscle growth capacity
of the different genotypes.

For example, Friesian bulls have a high voluntary intake,
a low muscle growth rate, and a high fattening capacity : they
show a lower feed efficiency than Charolais or Limousin bulls,

which have low voluntary intake, high muscle growth capacity and therefore little fat deposition in gain. With Limousin and Friesian cattle having approximately the same mature weight, comparisons could be made simultaneously, at the same age and the same weight. However, it appears also that large breeds are more efficient than smaller breeds at the same weight as far as differences in muscle growth potential and maturity are concerned.

Therefore, if various genotypes are compared at the same weight and the same intake, differences in growth rate and in efficiency are essentially due to differences in composition of gain linked to protein production potential. Selection based on live weight gain selects also for protein deposition (18 - 20% of gain in bulls) and therefore for feed efficiency because this selection increases gain and modifies its composition in the right way to improve efficiency. Selection on relative growth rate could perhaps prevent an increase in mature weight.

To avoid variations between breeds resulting in differences in gain composition, efficiency should be compared to equal carcass fatness, or over the same period of metabolic age or at the same stage of body tissues development. Therefore, the serial slaughter technique should be used to provide this information.

GENOTYPE NUTRITION INTERACTIONS

From several experiments it is more or less clear that the differences between genotype in feed efficiency vary according to the level of feeding or the type of diet. If energy intake is reduced below the ad libitum level, feed efficiency decreases in animals with high protein growth potential because they reduced their growth rate without a large modification of the composition of gain. On the other hand energy level reduction induces an increase in feed efficiency in animals with low protein growth potential, high appetite and high fattening capacity. This increase is related to a decrease in their fat deposition and energy content of their gain.

So, large differences between animals can be observed when they are fed ad libitum by the fact that they thereby express their muscle growth potential and (or) their fattening capacity. However these differences could be largely reduced with restricted feeding because good energy efficiency in one type of animal would thus be reduced, while bad efficiency in another type, would be improved. These effects have been observed in pigs for example between Pietrain French, and Belgium Landrace, as well as in bulls for example between Friesian and Limousin or Salers and Charolais, or between various beef X dairy cattle crossbreds.

The effect of various types of diet frequently observed, is generally related to differences of net energy intake, if diet are well balanced in proteins. But protein deficiency could affect animal efficiency differently according to their protein growth potential. Environmental variations could also modify heat losses which have the same effects as variations in the energy supply.

It appears very important therefore to compare and to select animals at different levels of feeding, two at least, or with different types of diet, using in addition, the serial slaughter technique, in order to make comparisons at the same age, same weight, or fatness. Optimum level of feeding to maximise feed efficiency could be determined approximately with the above methods, for each type of animal, and a comparison could be made at this optimum level. Different dietary regimes could be chosen to maximise the difference between individuals for selection purposes, or to represent some different diets used in current practice.

On the other hand, it appears necessary to use different breeds in nutritional studies. Some well known and selected breeds could be chosen in relation to their differences in growth capacity, voluntary intake, biological type (dairy or beef) to provide "corner stones" in nutrition and genetic studies.

As a result of these studies, it is becoming possible to fit the diet correctly to the potential of the different types of animals or to choose the most efficient animals to utilise the available diets. Systems of production, more or less intensive, with entire or castrated animals, could also be adapted to the different genotypes in relation to their efficiency. Also, it should be possible to define carcass weight/fatness/efficiency/relationships for the different cattle populations of the Community to define the range of slaughter weight for optimum carcass composition and feed utilisation and the best way of feeding and selecting them. Meat quality assessment and market requirements were not discussed at this Colloquium, but have to be taken largely into account in this area.

In conclusion, having a more clear and common approach to feed efficiency problems, both nutritional and genetic research programmes should take into account the possibility of inter-actions and should include experiments to measure them with maximum accuracy.

APPENDIX

List of participants :

Australia : Frisch D.J., Graham, N. Mc. C. - Belgium : Buysse, F.
Denmark : Andersen, N.N., Just, A. - France : Beranger, C. Bonaiti, B.,
Colleau, J.J., Geay, Y., Henry, Y., Jarrige, R., Menissier, F., Robelin, J.
Rouvier, R., Sellier, P., Vermorel, M., Vissac, B. - Germany : Langholz, H.J.
Oslage, H.J. - Ireland : Hanrahan, J.P., Mc Carthy, J.P., More O'Ferrall,
G.J. Italy : Malossini, F. - The Netherlands : Bakker, H., Van Es, A.J.H.
United Kingdom : Fawsett, H.R., Tayler, J.C., Taylor St. C.S., Thomson, D.J.,
Webster, A.J.F. - United States of America : Garrett, W.N.

List of communications :

Van Es, A.J.H. : Maintenance requirements.

Webster, A.J.F. : Variation in the maintenance requirements of young
        cattle when the energy value of food is defined according to the
        M.E. System.

Kielanowski, J. (Poland) : Comparative efficiency of energy utilisation
        for protein and fat deposition.

Oslage, H.J. : Protein and energy utilisation by fattening bulls.

Webster, A.J.F. : The energy cost of fat and protein deposition in the
        Zucker rat.

Taylor, St. C.S. : Models of expressing the genetic variability in the
        food utilisation efficiency.

McGraham, N. : A growth model.

Just, A. : Differences in energy and nitrogen utilisation between litters
        and sexes in growing pigs of Danish Landrace.

Colleau, J.J. and Bonaiti, B. : Analysis of differences in feed efficiency
        between genotypes in growing cattle.

Rouvier, R. : Influence of adult size on growth, body composition and food
        efficiency in the Rabbit.

More O"Ferrall : Selection for postweaning gain and feed efficiency in mice.

Langholz, H.J. : Aspects of genotype nutritional interactions in
        cattle.

Geay, Y. : Fitting the diet to the potential of the animal.

Andersen, B.B. and Andersen, H.R. : Danish results on nutrition/genotype
    interactions.

Oslage, H.J. : Growth and body composition of fattening bulls of two
    breeds related to different levels of nutrition.

Frisch, J.E. : Genotype-nutrition interactions in cattle in varying
    environment.

Fawcett, R.H. : Some aspects of the appropriate nutrition of pigs with
    differing inherent ability for protein growth.

Sellier, P. and Henry, Y. : Interactions between feeding level (energy
    and protein) and genotype in pigs.

Bakker, H. : Interaction between feeding level and genotype within a
    population of mice.

Mc Carthy, J. : Evidence from selection work with small animals and
    poultry regarding the genetical components of gross efficiency in
    feed utilisation.

EFFECT OF ENERGY LEVEL ON GROWTH AND EFFICIENCY

H. Refsgaard Andersen

Department of Cattle Experiments
National Institute of Animal Science, Copenhagen V, Denmark

ABSTRACT

The effect of energy level on growth and efficiency is examined in a 7 x 4 factorial experiment including 168 RDM-bulls fed on four energy levels (100%, 85%, 70% and 55%) and slaughtered at seven weights from 180 to 540 kg live weight. The main results from this and other trials are reported. Energy level by sex interaction is discussed by references to reported studies.

By reducing the energy level, the daily carcass gain is decreased relatively more than daily live weight gain, especially at a high live weight. Daily gain of fat is reduced relatively more than lean and bone and the differences increase with increasing weight. Lean : bone ratio decreases by reducing the feeding level. Compared at a low carcass weight the carcass composition is only slightly influenced by feeding level, but at a higher carcass weight the relative weight of lean and bone increases, and fat decreases when the energy level is reduced. The relative weight of the fattest cuts in the carcass decreases with decreasing energy level. Compared at the same total muscle weight the muscle distribution is only slightly affected by feeding level.

Compared at a low live weight the feed conversion ratio (fu : kg gain) is practically unaffected by increasing energy level from moderate to very high. At high weights the animals are most efficient at moderate feeding. It is, however, noted that this conclusion is in disagreement with more energy standards.

Most studies suggest that differences between sexes in growth rate as well as feed conversion ratio are reduced by decreasing feeding level.

INTRODUCTION

The level of energy supply is undoubtedly the main factor affecting growth in beef cattle, and it is well known that increasing energy intake increases daily gain and carcass gain. In contrast opinions vary about the influence of feeding level on relative growth of lean, fat and bone. The disagreement, however, may to some extent be caused by difficulties in interpreting growth data from different experiments due to:

a) Feeding level x slaughter weight interactions. In most experiments animals are only slaughtered at one stage of development, either a fixed age, a fixed weight or an expected fat content.

b) Differences from one experiment to another according to slaughter age or slaughter weight and feeding levels

c) Differences between breeds and between sex and interactions with feeding levels.

d) Different dissection methods.

Because of the points mentioned it is also difficult to interpret results concerning the influence of feeding level on feed efficiency. Feed efficiency can be expressed as input (dry matter, digestible organic matter, metabolisable energy, net energy) per kg output (live weight gain, carcass gain etc.). Therefore differences in concentrates:roughages ratio will also affect the efficiency depending on how the inputs are expressed. Even in using the same energy unit the value of the feed is not calculated in the same way in different countries.

For the reasons mentioned above, the influence of feeding level on growth and efficiency can best be examined from experiments with different slaughter weights and feeding levels, and where energy concentration in the ration has been nearly the same for feeding levels. Therefore growth and feed efficiency in this paper will mainly be discussed in relation to a Danish experiment with young bulls fed on four feeding

levels (100% ad libitum, 85%, 70% and 55%) and slaughtered at
seven live weights (from 180 to 540 kg), and where ration energy
concentration on all intensities was nearly the same (Andersen,
1975a, 1975b). Also feeding level x sex interactions will be
discussed by reference to reported studies.

EFFECT OF ENERGY LEVEL ON GROWTH AND CARCASS COMPOSITION

Live weight and carcass growth

The effect of different levels of feeding on daily gain
has been examined in many experiments. Results from a number
of them are shown in Table 1. They indicate that the decrease
in daily gain is relatively lower than the reduction in feeding
level. The daily carcass gain is more influenced by feeding
level than daily live weight gain, as carcass weight is reduced
relatively more than live weight. Many experiments show that
dressing percentage decreases with decreasing feeding level,
especially at a high live weight (Broadbent, 1967; Hiner and
Bond, 1971; Andersen, 1975a). In our experiment the dressing
percentage at ad libitum feeding was increased by about 1% unit
for every 100 kg increase in live weight in the range from 200
- 550 kg, while at a very low feeding level (55% of ad libitum)
the dressing percentage was unaffected by weights.

Using the same net energy ratio between intensities in all
weight intervals, maximum daily gain occurred at the same live
weight independent of feeding level (Andersen, 1975a). The
same was observed in experiment with monozygous heifer twins
(Sejrsen et al., 1976), but the maximum daily gain for the
heifers occurred at a lower weight than for the bulls. However,
changing the ratio between energy levels in different weight
intervals will affect the above mentioned gain pattern. It was
also shown by Sejrsen and Larsen (1977) feeding a different
silage : concentrate ratio ad libitum to heifers. The maximum
daily gain occurred at higher weights with increasing amount of
roughages in the ration, because the differences between feeding
levels were reduced with increasing weight.

TABLE 1

DAILY GAIN AND FEED CONVERSION RATIO AT DIFFERENT ENERGY LEVELS

Sex	Breed	Number of Animals	Weight Interval kg	Relative Feed Intake per Day *	Daily Gain g	Daily Gain relative *	Relative Feed Conversion Ratio *	Energy Unit	Reference
Bulls	SDM	3	230–520	100	1302	100	100	Scand. f.u.	Sørensen et al., 1972
		3	221–517	84	1182	91	94		
		3	235–518	66	957	74	88		
Bulls	MRY	21	251–454	100	1211	100	100	SE	De Boer et al., 1971
		20	251–443	93	1147	95	99		
		21	249–427	79	1017	84	95		
		20	251–414	64	942	78	83		
Bulls	SDM	11	149–456	100	1304	100	100	Scand. f.u.	Andersen and Sørensen 1975.
		11	149–456	93	1184	91	101		
		11	152–457	77	1104	85	90		
		12	151–455	60	928	71	84		
Bulls	Friesian	7	240–489	100	1016	100	100	ME	Levy et al., 1974
		7	238–482	85	938	92	88		
Bulls	Friesian	6	239–477	100	748	100	100	ME	Levy et al., 1974
		6	240–472	86	676	90	95		
Bulls	Friesian	13	221–443	100	1131	100	100	ME	Levy et al., 1976
		13	217–450	85	955	84	101		
Bulls	Salers	6	338–561	100	1186	100	100	ME	Geay et al., 1976
		6	340–553	88	1132	95	92		
		6	339–562	90	1183	100	90		

cont./.

TABLE 1 continued

Sex	Breed	Number of Animals	Weight Interval kg	Relative Feed Intake per Day *	Daily Gain g	Relative*	Relative Feed Conversion Ratio *	Energy Unit	Reference
Bulls	Charolais x Salers	6	343–591	100	1322	100	100	ME	Geay et al., 1976
		6	340–565	77	1198	91	84		
		6	342–555	78	1134	86	90		
Bulls	Charolais	5	339–564	100	1224	100	100	ME	Geay et al., 1976
		5	341–539	85	1075	88	97		
		5	343–533	84	1035	85	99		
Steers	SDM	3	218–506	100	936	100	100	Scand. f.u.	Sørensen et al., 1972
		3	226–514	92	907	97	91		
		3	218–517	71	844	90	76		
Steers	Hereford		220–240	100	854	100	100	Net energy	Henrickson
			223–406	73	699	82	88		
Heifers	RDM	28	150–400	100	872	100	100	Scand. f.u.	Sejrsen et al., 1976
		28	150–400	66	669	77	87		

* Feed intake, daily gain and feed conversion ratio at the highest energy level = 100.

Growth of lean, fat and bone

Figure 1 shows that lean, fat and bone are all influenced by feeding level, but to a different extent depending on the weight of the animal. On all feeding levels the maximum daily gain of bone occurred before 200 kg live weight, and the daily gain of lean at 270 kg. The maximum daily gain of fat occurred at lower weights, when feeding levels were decreased. When the feeding level was reduced, the daily gain of fat decreased more than lean and bone, and the differences increased with increasing live weights.

Lean : bone ratio is often discussed in relation to the influence of feeding level on growth. It is often mentioned that the ratio is not influenced by feeding level, when comparisons are made at same live weight (Callow, 1961; Elsley et al., 1964). It means that lean and bone should be reduced relatively equal. However, our results show that the lean : bone ratio decreased with decreasing feeding level (Figure 2). The decrease was less pronounced with increasing weight except at extreme low intensity. Also Henrickson et al. (1965), Broadbent (1967), Waldman et al. (1971) and Andersen and Sørensen (1975) found a decreasing lean : bone ratio, when feeding levels were reduced.

Carcass composition

Because of the influence of feeding level on growth mentioned above, carcass composition is affected as shown in Figure 3. At a low weight the difference in carcass composition was small in spite of a large difference in feeding level. This is in agreement with results from many Danish feeding experiments with veal calves (120 - 140 kg carcass weight). However, at higher slaughter weights fat percentage increased strongly with increasing feeding level, while the percentage of bone and lean decreased. This is in agreement with the results of Callow (1964) Broadbent (1967), Waldman et al. (1971) and Sørensen et al. (1972)

Also the relative weight of the individual cuts as well as the composition of the cuts is affected by feeding level, when

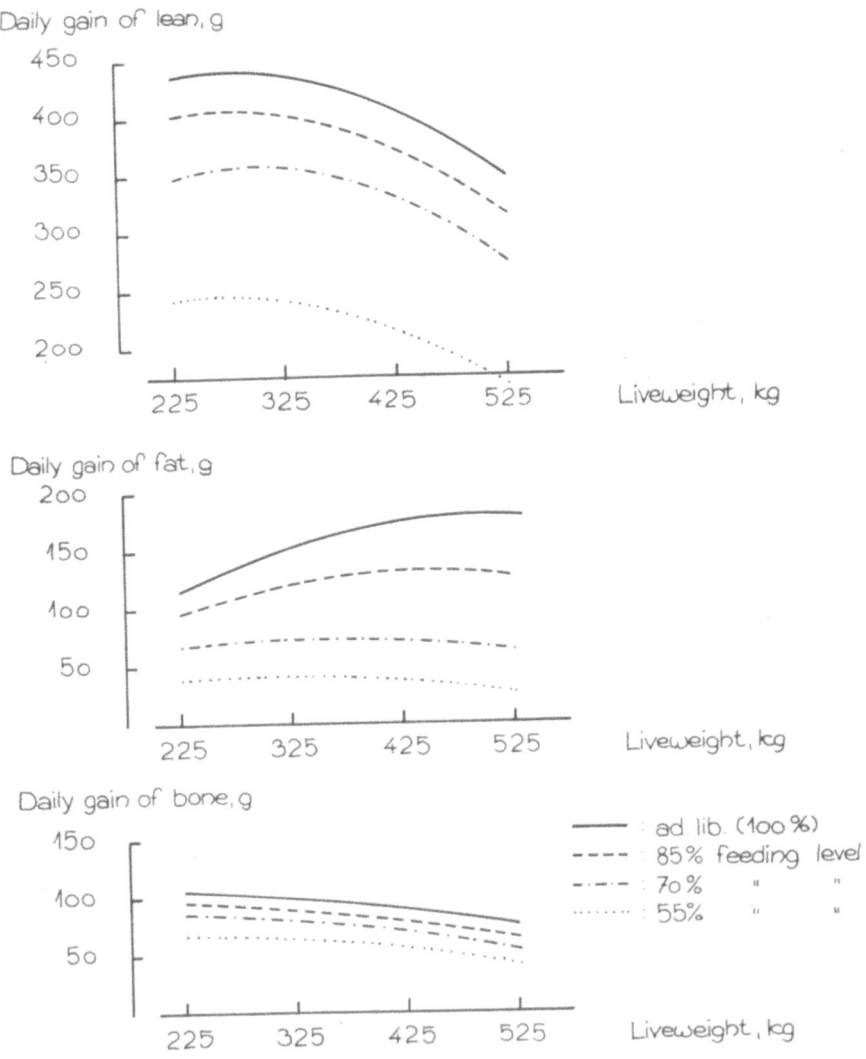

Fig. 1. Marginal daily gain of lean, fat and bone at different slaughter
weights and feeding levels (Andersen, 1975a).

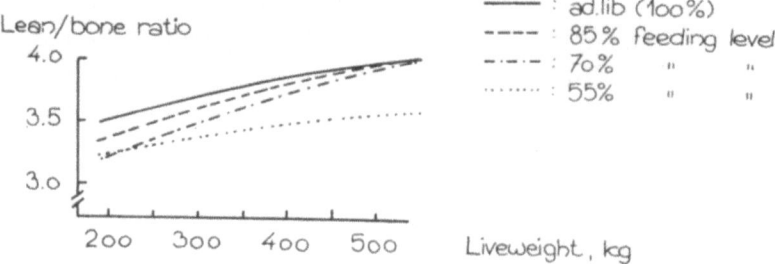

Fig. 2. Lean/bone ratio of different slaughter weights and feeding levels.

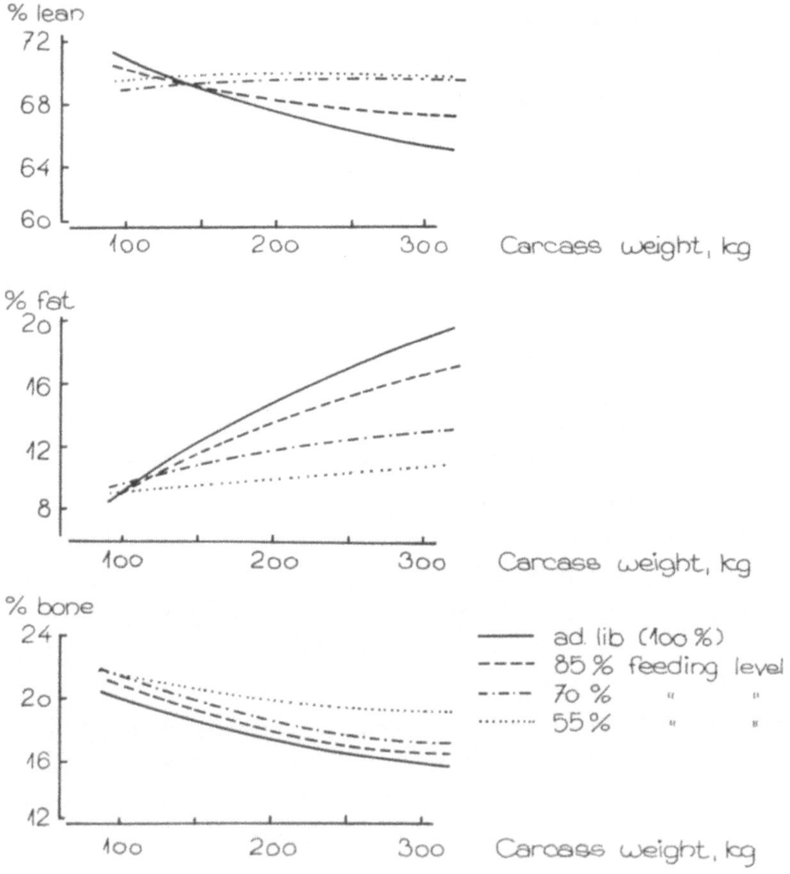

Fig. 3. Per cent lean, fat and bone at different carcass weights and feeding levels (Andersen, 1975a).

comparisons are made at the same carcass weight. The results from our experiments with different slaughter weights and feeding levels are shown in Figure 4.

The relative weight of the fattest cut (flank + brisket + flat ribs + fore shank) was reduced by restricting feeding level, while the weight of the pistol cut was nearly independent of feeding level. The estimated composition of the individual cuts at different carcass weights is shown in Figure 5. Compared at the same total muscle weight the muscle distribution was slightly influenced by feeding level (Andersen, 1975b). By restricted feeding the muscles in the flank and brisket (*Pectoralis profundus, Transversus, Obliquus externus abdominis, Obliquus internus abdominis*) and loin (*Longissimus dorsi*) were reduced and some muscles in the neck were increased.

ENERGY LEVEL BY SEX INTERACTION ON GROWTH AND COMPOSITION

That sex affects growth pattern is well known and needs no documentation. On the other hand it has not been clearly demonstrated whether bulls, heifers and steers, in different weight intervals, react to the same extent on different feeding levels. However, it seems that the differences in growth rate between sexes depend on the plane of nutrition available. Harte (1969) summarised the results of a series of experiments with bulls and steers at pasture and found the greatest difference in growth rate in favour of bulls on good quality pastures. This is in agreement with conclusions drawn by Price and Yeates (1969).

In our own experiments, comparing bulls and heifers fed silage ad libitum (low energy intake), the energy intake and daily gain were nearly the same, but with high concentrate : silage ratio the difference in growth rate between sexes was strongly in favour of bulls even though the energy intake was the same.

Also preliminary results from a Danish experiment comparing bulls and steers fed at three different feeding levels and

percent of carcass

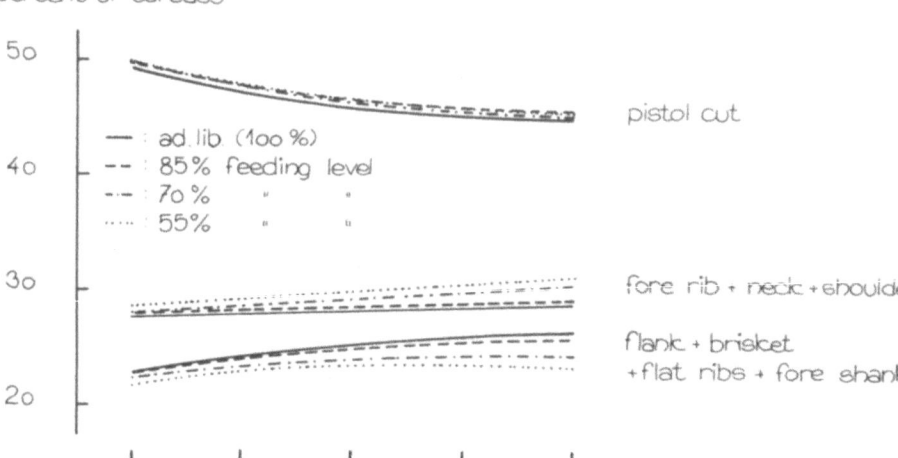

Fig. 4.   Relative weight of the three main cuts at different carcass weights
and feeding levels (Andersen, 1975a).

LEGENDS TO FIGURES 5 AND 6 ON THE FOLLOWING PAGES

Fig. 5.   Per cent of lean, fat and bone in the three main cuts at different
carcass weights and feeding levels. Feeding levels: A = 100%,
B = 85%, C = 70% and D = 55%. (Andersen, 1975a).

Fig. 6.   Relative feed conversion ratio at different daily gain for bulls
according to Schiemann et al., 1972 (-), Schulz et al., 1974 (...),
van Es et al., 1976 (---) and Andersen, 1975a (—).

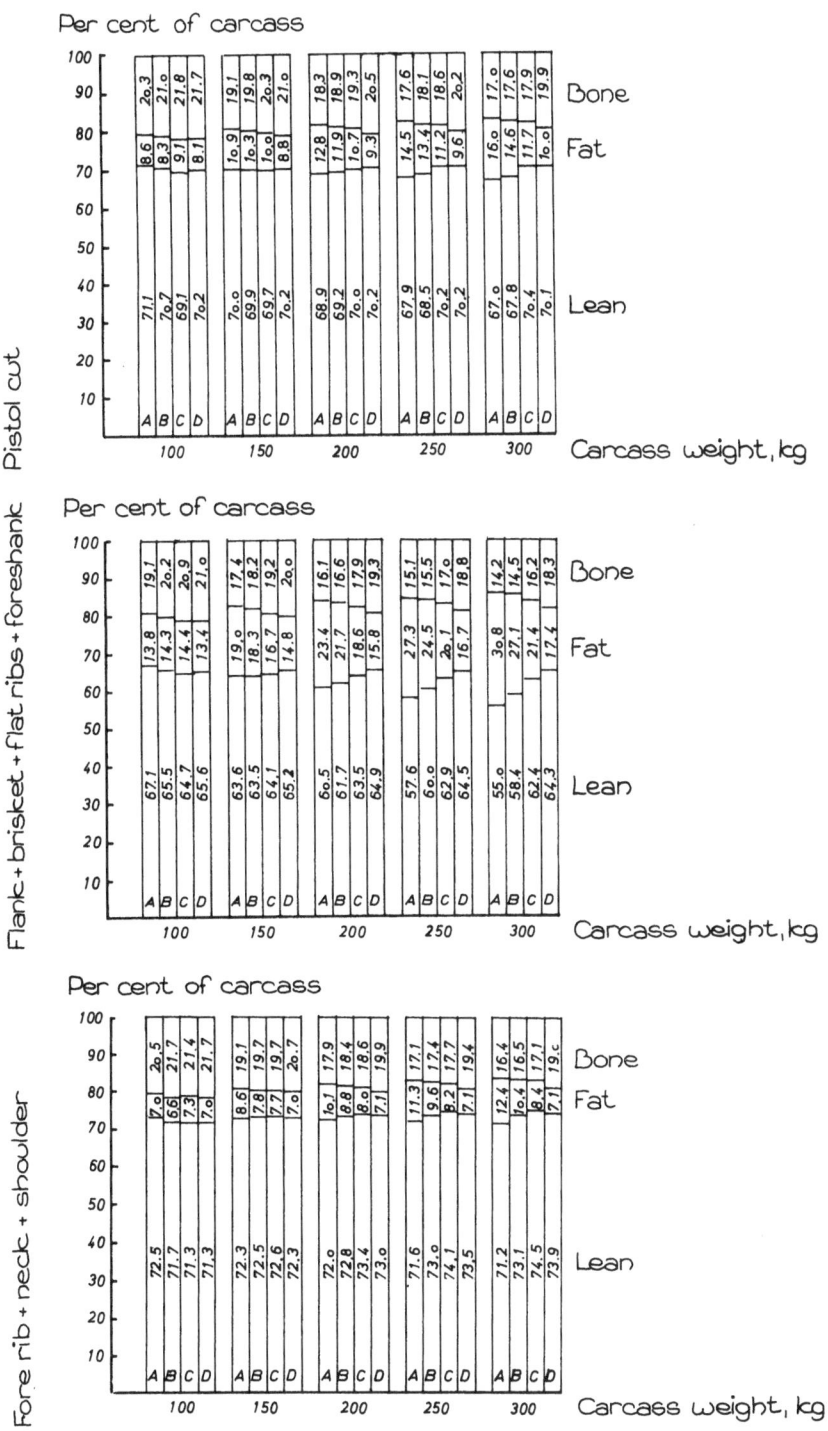

Fig. 5.

404

Relative net energy/kg gain at 200 kg liveweight

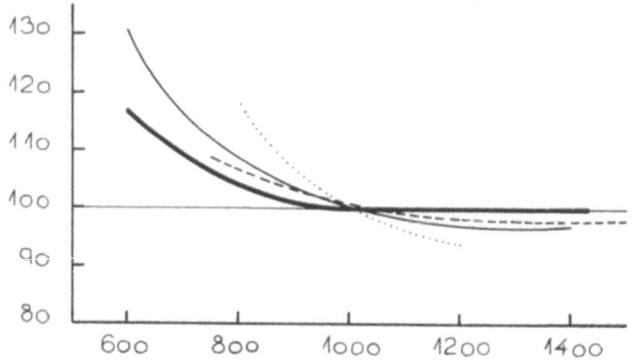

Relative net energy/kg gain at 350 kg liveweight

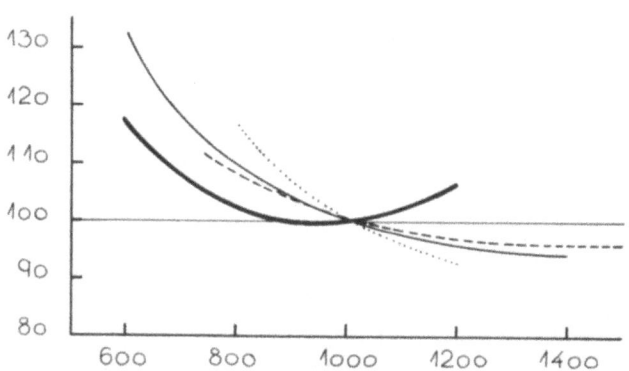

Relative net energy/kg gain at 500 kg liveweight.

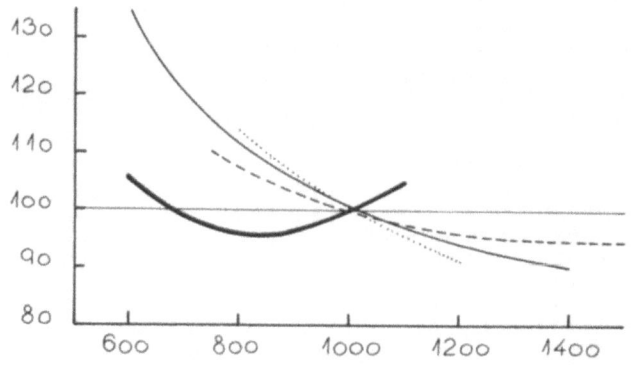

Fig. 6.

slaughtered at different weights indicate that the relative difference in growth rate between sex increases with increasing feeding level. Similar results were obtained by Sørensen et al. (1972).

There is an absence of information showing whether restricted feeding delays growth of lean, fat and bone to the same extent in bulls, heifers and steers.

EFFECT OF ENERGY LEVEL ON EFFICIENCY

Results from metabolic studies

The influence of feeding level on feed efficiency has been examined in short term metabolism experiment as well as in long term feeding experiments. However, the determination of energy requirements in beef cattle has nearly exclusively been calculated from metabolism experiments, although discrepancies exist with the results achieved in practical feeding experiments.

The relationship between energy level (daily gain) and feed conversion ratio estimated from different energy standards for bulls at 200 kg, 350 kg and 500 kg live weight are shown in Figure 6. According to Schiemann et al. (1972) Schulz et al. (1974) and van Es et al. (1976) the requirement : kg gain decrease with increasing daily gain and at the same rate irrespective of live weights. Also standards for steers (ARC, 1965; Ministry of Agriculture, Fisheries and Food, Department of Agriculture and Fisheries for Scotland and Department of Agriculture for Northern Ireland, 1976; NRC, 1976), show a better feed utilisation with increasing daily gain.

Results from practical feeding experiments

Contrary to the energy standards mentioned most feeding experiments show the best feed conversion ratio at moderate feeding (Table 1). The same was also shown by Langholz (1976) in more experiments using animals of different growth capacity fattened at different nutritional levels.

Corresponding results are also found under practical feeding conditions (Henneberg and Østergaard, 1976). On the basis of the production of 2898 young bulls on the Danish pilot cattle farms the herd results were divided in two groups according to daily feed intake. The difference in daily intake of Scandinavian feed units between the two groups was 14%, but the difference in daily gain was only 9%. Therefore the difference in feed conversion ratio was about 6% in favour of the lowest intake.

Most of the feeding experiments are carried out in the weight interval between 200 and 500 kg live weight, and the feed conversion ratio can only be calculated for the whole fattening period. However, according to our energy standard (Andersen 1975a) which is based mainly on the above mentioned feeding experiment, feeding level x live weight interaction exists. The advantage in feed conversion ratio in favour of moderate feeding increased with increasing weight (Figure 2) Also for heifers and steers the advantage in feed conversion ratio in favour of moderate feeding compared with high feeding level was increased with increasing weight (Sejrsen and Larsen, 1977; Almquist et al., 1970). The probable reasons for the lower efficiency with increasing daily gain (increasing feeding level) at the highest weights in spite of a shorter feeding period and with that a theoretical relatively lower total requirement for maintenance are :

(a) The deposit of energy per kg gain increases with increasing feeding level. Because of increasing fat content, especially at a high live weight, and nearly the same amount of lean per kg gain (Andersen, 1975) the deposited energy increases. The efficiency of protein and fat deposition is discussed by van Es at this seminar.

(b) The efficiency of the utilisation of the metabolisable energy for maintenance (k_m) and growth (k_f) differs, maintenance being the more efficient. Therefore, when a single term (net energy) is used to describe the total requirement (growth + maintenance) a ration has a relatively higher value at a low feeding level because a relatively higher proportion of the energy is used for maintenance.

(c) The digestibility of the feed decreases with increasing feeding level. This is due to an increased rate of passage through the digestive tract and decreased rate of microbial degradation in the reticulorumen (Kromann, 1973) and the decrease in digestibility with intake is higher the more grain the ration contains. On the other hand the methane production per unit of food is decreased with increasing feeding level (Kromann, 1973).

(d) The energy requirement for maintenance depends on feeding level. In most feeding recommendations the calculated net energy for maintenance is regulated in relation to metabolic body weight, but kept constant irrespective of the production level (van Es et al., 1976; NRC, 1976; Schulz et al., 1974; Schiemann et al.,1972). However, the question whether production affects maintenance metabolism can be raised.

More experiments suggest that maintenance metabolism decreases with decreasing feeding level. We fed 55% of ad libitum intake (which nearly corresponds to the maintenance requirement of van Es et al., 1976) and the bulls grew nearly 600 g per day during the whole fattening period (Andersen, 1975a). Also in an experiment with monozygous twin heifers, the animals grew although the energy intake was lower than the recommended maintenance requirement. Furthermore it was found that with decreasing feeding level the respiration and heart rates decreased relatively strongly, indicating a decreasing maintenance metabolism (Hansson et al., 1953). There are also a few metabolic studies which indicate decreasing maintenance requirement with decreasing feeding level (Thorbek and Henkel, 1976). Decreasing maintenance requirement with decreasing feeding level could explain some of the discrepancies between results from recognised energy standards and from practical feeding experiments.

EFFECT OF ENERGY LEVEL BY SEX·INTERACTION ON EFFICIENCY

It is well known that on a relatively high feed intake
bulls are more efficient in converting feed into live weight
gain than steers, and steers are more efficient than heifers.
On the other hand only a few experiments in the literature allow
comparison of sex at different feeding levels, but more experiments suggest that the relative difference between sexes is
reduced with decreasing feeding level (Johnsson, 1973; Sørensen
et al., 1972). This can be explained by the fact that the
fat content is reduced relatively more by restricted feeding in
the fattest animals. However, comparing ad lib. fed bulls
with restricted fed steers in the same weight interval, Brännäng
et al. (1970) found about the same feed conversion ratio for
bulls and steers in spite of a higher fat content in the steers.
On this background the author concluded that steers convert
feed energy into body energy more efficiently than bulls. The
relatively better feed conversion ratio for steers compared
with bulls at a low feeding level could also be explained by
difference in maintenance metabolism between sexes. Comparing
the sexes Webster et al. (1977) found 20% higher maintenance
requirement for bulls than for steers. Differences in maintenance metabolism could also explain why steers on poor quality
pasture (low feeding level) are just as efficient as bulls.

Table 2 shows that at a feeding level leading to about
700 g gain per day the feed conversion ratio for heifers and
bulls was nearly the same. In contrast at a high feed intake the
differences between sexes was strongly in favour of bulls,
especially at a high live weight.

TABLE 2

FEED CONVERSION RATIO FOR RDM-HEIFERS AT TWO FEEDING LEVELS (SERJSEN ET AL., 1976) COMPARED WITH THE REQUIREMENT FOR RDM-BULLS AT THE SAME DAILY GAINS (ANDERSEN, 1975)

Weight interval kg	Low Feed Intake for Heifers			High Feed Intake for Heifers		
	Daily gain g	F.U./kg gain		Daily gain g	F.U./kg gain	
	Heifers and Bulls	Heifers	Bulls	Heifers and Bulls	Heifers	Bulls
150-200	894	3.4	3.2	1075	3.9	3.0
200-250	701	4.6	4.2	948	5.0	3.9
250-300	662	5.1	4.8	872	6.1	4.2
300-350	707	5.6	5.2	838	7.2	5.1

CONCLUSIONS

The influence of feeding level on growth and efficiency depends on the stage of development. Because of these feeding by weight interactions the influence of feeding level is best examined in experiment with at least three feeding levels and three slaughterweights or slaughterages.

Feeding level by sex interaction also exists. Therefore, comparing sexes the results are best examined in experiments where the sexes are fed on different feeding levels and slaughtered at different weights or ages.

REFERENCES

Agricultural Research Council (A.R.C.). 1965. The nutrient requirements
    of farm livestock No. 2. Ruminants. Agricultural Research Council,
    London, 264 pp.

Andersen, H.R., 1975a. Slagtevægtens og foderstyrkens indflydelse på
    vækst, foderudnyttelse og slagtekvalitet hos ungtyre. (The influence
    of slaughter weight and feeding level on the growth, feed conversion,
    carcass composition and conformation in bulls). (In Danish with
    English summary and subtitles.) 430. Beretn. Statens Husdrybrugsforsøg,
    København, 124 pp.

Andersen, H.R., 1975b. The influence of slaughter weight and level of feed-
    ing on growth rate, feed conversion and carcass composition of bulls.
    Livestock Production Science 2, 341-355.

Andersen, H.R. and Sørensen, M. 1975. Formalet halm i foderblandinger til
    ungtyre. Meddelelse nr. 19 fra Statens Husdyrbrugsforsøg.

Broadbent, P.J., 1967. The growth and carcass characteristics of pure
    Ayrshire steers. Report from Ayrshire Cattle Society of Great Brit-
    ain and Ireland. 44 pp.

Brännäng, E., Henningsson, T. Liljedahl, L.E., and Lindhé B, 1970. Studies
    on monozygous cattle twins. XXI. The effect of castration and
    intensity of feeding on the growth rate, feed conversion and carcass
    traits of Swedish red and white cattle. Landtbr. högsk. Annlr., 36,
    91-113.

Callow, E.H., 1961. Comparative studies of meat. VII, A Comparison between
    Hereford, dairy Shorthorn and Friesian steers on four levels of
    nutrition. J. Agric. Sci., 56, 265-282.

De Boer, F., Smiths, B. and Dijkstra, K.T.J. 1971. Voederhoeveelheid,
    groei en slachtkwaliteit bij jonge vleesstieren. (Feeding level,
    daily gain and carcass-quality in fattening young bulls). Landbouw-
    kundig Tijdschrift, Dijkstra-nummer, 354-359.

Elsley, F.W.H. McDonald, I. and Fowler, V.R. 1964. The effect of plane of
    nutrition on the carcasses of pigs and lambs when variations in fat
    content are excluded. Anim. Prod. 6. 141-154.

Geay, Y., Robelin, J., and Béranger, C. 1976. Influence du niveau alimentaire
    sur le gain de poids vif et la composition de la carcasse de taurillons
    de differentes races. Ann. Zootech, 25, 287-298.

Hansson, A., Brännäng, E. and Claesson, O. 1953. Studies on monozygous cattle twins XIII. Body development in relation to heredity and intensity of rearing. Acta. Agric. Scand. III, 61-95.

Harte, F.J., 1969. Six years of bull beef production research in Ireland. In: Meat Production from Entire Male Animals, 153-172. J. and A. Churchill Ltd., London.

Henneberg, U., and Østergaard, V. Slagtekalveproduktionens styring og økonomi. Helårsforsøgsdata 1969-75. (Management and economy of the production of young bulls). 445. Beretn. Statens Husdyrbrugsforsøg, København, 40 pp.

Henrickson, R.L., Pope, L.S. and Hendrickson, R.F. 1965. Effect of rate of gain of fattening beef calves on carcass composition. J. Anim. Sci., 24, 507-513.

Hiner, R.L. and Bond, J. 1971. Growth of muscle and fat in beef steers from 6 to 36 months of age. J. Anim. Sci., 32, 225-232.

Johnsson, S., 1973. Input-output relationships in beef production. The effects of different levels of feeding, substitutions of concentrate, hay and silage for artificially dried grass, and slaughtertime. Doktorafhandling, Uppsala.

Kromann, R.P., 1973. Evaluation of net energy systems. J. Anim. Sci. 37, 200-212.

Langholz, H.J., 1976. Aspects of interactions between beefing capacity and feeding level in beef production. Contribution to the EEC Colloquium on feeding efficiency and genotype environmental interaction in growing animals, Theix/Clermont Ferrand.

Levy, D., Holzer, Z., Neumark, H and Amir, S. 1974. The effects of dietary energy content and level of feeding on the growth of Israeli Friesian intact male cattle. Anim. Prod. 18, 67-73.

Levy, D,, Holzer, Z and Folman, Y. 1976. Effects of plane of nutrition, diethylstilboestrol implantation and slaughter weight on the performance of Israeli-Friesian intact male cattle. Anim. Prod. 22, 55-59.

Ministry of Agriculture, Fisheries and Food, Department of Agriculture and Fisheries for Scotland and Department of Agriculture for Northern Ireland, 1976. Energy allowances and feeding systems for ruminants. Her Majesty's Stationery Office, London 79 pp.

National Research Council (N.R.C.), 1976. Nutrient Requirements of Beef Cattle, 5 ed. National Academy of Sciences. Washington, D.C. 56 pp.

Price, M.A. and Yeates, N.T.M. 1969. Growth rate and carcass characteristics in steers and partial castrates. In: Meat Production from Entire

Male animals, 69-77. J. and A. Churchill Ltd., London

Schiemann, R., Nehring, K., Hoffmann, L., Jentsch, A. and Chudy, A. 1971
Energetische Fütterung und Energienormen. VEB Deutscher Landwirt-
schaftsverlag Berlin, 344 pp.

Schulz, E., Oslage, H.J. and Daenicke, R. 1974. Untersuchungen über die
Zusammensetzung der Körpersubstanz sowie den Stoff - und Energieansatz
bei wachsenden Mastbullen. Zs. Tierphys., Tierern. Futtermk. Beiheft 4.
70 pp.

Sejrsen, K., Larsen, J.B., Foldager, J., Agergaard, E. and Klausen, S. 1976.
Kæ lvningsalderens indflydelse på foderforbrug, frugtbarhed, kæ lvnings-
forløb og mæ lkeydelse - 2.5 contra 1.5 års kæ lvningsalder hos RDM.
Meddelelse nr. 102 fra Statens Husdyrbrugsforsøg.

Sejrsen, K. and Larsen, J.B. 1977. Effect of silage:concentrate ratio on
feed intake, growth rate and subsequent milk yield on early calving
heifers. Livest. Prod. Sci, 4, 313-325.

Sørensen, M., Lykkeaa, J. and Andersen, H.R. 1972. Fodringsintensitetens
indflydelse på tilvæ kst og kødkvalitet hos ungtyre og stude. Årbog
Landøk. Forsøgslab., København. 370-376.

Thorbek, G. and Henckel, S. 1976. Studies on energy requirements for maintenance in Farm Animals. Proc. 7th Symp. on Energy Metabolism, Vichy,
France. EAAP Publ. No. 19.

Van Es, A.J.H., Vermorel, M. and Bickel, H. 1978. Feed evaluation for
ruminants. New energy system in Netherlands, France and Switzerland.
Livest. Prod. Sci. (in press).

Waldman, R.C., Tyler, W.J. and Brungardt, V.H. 1971 Changes in the carcass
composition of Holstein steers associated with ration energy levels
and growth. J. Anim. Sci., 32, 611-619.

Webster, A.J.F., Smith, J.S. and Mollison, G.S. 1977. Prediction of the energ
requirements for growth in beef cattle. 3. Body weight and heat production in Hereford x British Friesian bulls and steers. Anim. Prod.
24, 237-244.

INFLUENCE OF NUTRITION ON THE GROWTH PATTERN OF FATTENING BULLS OF TWO DIFFERENT BREEDS (FRIESIAN AND SIMMENTAL)

K. Rohr and R. Daenicke

Institute of Animal Nutrition, Federal Research Centre
for Agriculture, Braunschweig-Völkenrode, Federal Republic of
Germany.

ABSTRACT

Fifty six bulls of the Friesian and Simmental breeds were fattened from 160 kg to 565 and 615 kg live weight respectively, with two different levels of energy intake. Body gain composition and energy utilisation were measured by means of the comparative slaughter technique (whole body analysis).

Simmental bulls showed higher daily gains (1098 vs. 1050 g) in spite of 3% lower dry matter and energy intakes. Consequently, feed conversion ratio was better than in Friesians. Daily gains increased with higher feeding level but were unchanged with elongation of the fattening period. Feed conversion was neither influenced by feeding level nor by final body weight.

On an average, 27 g more protein and 63 g less fat were daily deposited in Simmental bulls when compared with Friesians. Protein deposition was constant in Friesians but increased with higher feed supply in Simmental bulls. Body fat gain increased as well with feed supply as with age of the animals.

Maintenance requirements for energy were not different between breeds.

INTRODUCTION

Within the Federal Republic of Germany, fattening bulls account for nearly 50% of total beef production. The contributio of steers is negligible. Females (cows and heifers) normally do not get special fattening rations. Therefore, systematic beef production in our country is synonymous with formulating rations for young growing bulls. The calves to be fattened are born almost entirely in dairy herds ie they descend from dual-purpose breeds. The most important breeds are Friesians and Simmentals, each contributing about 35%.

In comparative feeding trials, Simmental bulls have sur-passed Friesians with respect to daily gains (Bogner, 1967; Schwark and Ebendorff, 1970; Huth, 1977). The question arises whether this superiority is due to differences in feed intake, protein deposition, or any other factor (eg maintenance requirements). In order to contribute to this question, we studied the growth pattern and body composition of Friesian and Simmental bulls at two feeding levels.

MATERIAL AND METHODS

The experimental design is given in Table 1. Fifty six bulls (28 of each breed) with an initial live weight of 160 kg were kept on two different feeding levels until they reached a live weight of 565 kg respectively 615 kg (corresponding to 500 kg respectively 550 kg empty body weight). All animals got 2 kg concentrates daily. Differences in feed supply were achieved by feeding maize silage either in restricted amounts or unlimited. With 'restricted' feeding daily live weight gains of 1 kg were aimed at. Concentrates consisted of soyabean meal, wheat, barley, and minerals. The chemical composition of all feed-stuffs was analysed weekly. Sheep were used to measure the digestibility of maize silage. The empty bodies of all slaughtered animals were dissected into several fractions (muscles, adipose tissue, skin, bones, internal organs and

offals, blood) and analysed for their chemical composition.
Gains in protein and fat were calculated by subtracting the
respective amounts analysed in 5 calves of each breed (160 kg
live weight). Results were analysed statistically using the
'DUNCAN'-Test.

RESULTS AND DISCUSSION

Data on intake of dry matter, digestible crude protein,
and energy are given in Table 2. Friesians consumed 3% more
dry matter and energy than Simmental bulls. Ad libitum against
restricted feeding of maize silage resulted in 7% higher intakes. Average daily food consumption increased by some 5%
with elongation of the fattening period (550 kg vs. 500 kg empty
body weight).

TABLE 1

DESIGN OF BULL FATTENING EXPERIMENT (NUMBER OF ANIMALS IN BRACKETS)

Breed	Feed supply	Final live weight (kg)
German Friesian (F) [28]	Restricted (R)[1] [14]	565 [7] 615 [7]
	High (H)[2] [14]	565 [7] 615 [7]
German Simmental (S) [28]	Restricted (R) [14]	565 [7] 615 [7]
	High (H) [14]	565 [7] 615 [7]

1: restricted maize silage plus 2 kg concentrates

2: maize silage ad libitum plus 2 kg concentrates.

TABLE 2

INTAKE OF DRY MATTER, PROTEIN, AND ENERGY

Daily intake	Factor				Final weight		Statistical significance[1]		
	Breed		Feed supply						
	F	S	R	H	565	615	F:S	R:H	565: 615
Dry matter (kg)	6.28	6.11	5.98	6.41	6.03	6.36	**	***	***
Digestible crude protein (g)	734	726	722	738	717	733	ns	***	***
Metabolisable energy (Mcal)	16.66	16.16	15.87	16.95	16.02	16.80	**	***	***
Net energy (SE)	4326	4185	4113	4398	4157	4354	**	***	***

1: ns = not significant, * = P<0.05, ** = P<0.01, *** = P<0.001

In spite of the lower energy intake, daily live weight gains were significantly higher in Simmental than in Friesian bulls (Table 3). Differences in empty body weight gain showed the same trend but were not significant. As one might expect, daily gains increased with feed supply. On the other hand, the length of the fattening period had no influence. Feed conversion ratio - ie SE/kg live weight respectively empty body gain - was significantly better in Simmental than in Friesian bulls. Feeding level and length of fattening period were not effective in this respect.

An explanation for higher gains and better feed conversion in Simmental bulls is given by the data in Table 4. Daily protein deposition was 27 g or 17% more in these animals with fat deposition being 63g lower. From the mean values in Table 4, feed supply and final weight appear to have only a small influence on protein deposition, while body fat is increased significantly by both factors. However, it must be emphasised that Friesians and Simmentals responded differently to feed supply. The former had a constant daily protein deposition of c. 160 g with all treatments. the latter deposited 12 g respectively 6.6% more with the higher feeding level. Nevertheless, this difference was not significant. As feed protein supply was fairly high in this experiment, it may be suggested that energy intake was a limiting factor with respect to maximum protein deposition in Simmental bulls. Thus it cannot be excluded that with a higher portion of concentrates protein synthesis still may be increased.

Maintenance requirements for net energy may be estimated by subtracting energy gain from NE-intake. Such a calculation shows no difference between Friesian and Simmental bulls in this experiment (Table 5). Maintenance requirements were 80 kcal respectively 81 kcal NE/kg $^{0.75}$. These figures are similar to those found by Schulz et al. (1974) with Friesian bulls.

418

TABLE 3

DAILY GAINS AND FEED CONVERSION RATIO (FATTENING PERIOD: 160 - 565 KG LW RESP. 160 - 615 KG LW)

Daily gains resp. feed conversion	Factor						Statistical Significance [1]		
	Breed		Feed supply		Final weight		F:S	R:H	565: 615
	F	S	R	H	565	615			
Live weight gain (g)	1050	1098	1042	1106	1066	1083	*	**	ns
Empty body weight gain (g)	960	991	943	1008	970	981	ns	**	ns
SE/Kg live weight gain	4138	3837	3975	4001	3940	4035	**	ns	ns
SE/Kg empty body weight gain	4527	4259	4396	4390	4334	4452	*	ns	ns

[1] : for symbols see Table 2

TABLE 4

AMOUNTS OF PROTEIN, FAT AND ENERGY DEPOSITED IN FATTENING BULLS

Daily deposition	Factor						Statistical Significance [1]		
	Breed		Feed supply		Final weight		F:S	R:H	565:615
	F	S	R	H	565	615			
Protein (g)	161	188	171	178	177	172	***	ns	ns
Fat (g)	270	207	220	257	221	256	***	**	**
Energy (Mcal) [2]	3.39	2.94	2.97	3.35	3.01	3.31	***	**	**

1 : for symbols see Table 2

2 : calculated as 5.32 kcal/g protein and 9.37 kcal/g fat (after Blaxter and Rook 1953)

TABLE 5

UTILISATION OF ENERGY FOR MAINTENANCE AND GROWTH

Breed	NE intake (Mcal)	Energy deposited (Mcal)	NE for maintenance (Mcal)	Energy portion used for		Maintenance requirement (Kcal NE/kg$^{0.75}$)
				growth (%)	maintenance (%)	
Friesian	10.21	3.39	6.82	33	67	80
Simmental	9.88	2.94	6.94	30	70	81

CONCLUSIONS

Our overall results lead to the following conclusions:
Simmental bulls are superior to Friesians with respect to daily
gains and feed conversion ratio. This superiority is due to a
higher capacity of protein synthesis with a correspondingly
lower fat deposition. As energy above maintenance is utilised
in a different manner, breed-specific recommendations for
nutrient supply may be necessary in the future. This last
conclusion is supported by results of French workers
(Geay, 1977).

REFERENCES

Blaxter, K.L., Rook, J.A.F. 1953. The heat of combustion of the tissues of cattle in relation to their chemical composition. Br. J. Nutr. 7, 83-91.

Bogner, H. 1967. Das Rind. DLG-Verlag, Frankfurt (Main).

Geay, Y. 1977. Interactions entre genotype et nutrition chez les bovins destinés à la production de viande. 28th Annual Meeting of EAAP, Brussels, 22 - 25.8.1977.

Huth, F.W. 1977. Influence of age, nutrient intake, and body type on weight gain and body composition in young fattening bulls of the breeds German Schwarzbunte and German Fleckvieh. Paper presented at this Symposium.

Schulz, E., Oslage, H.J., Daenicke, R. 1974. Untersuchungen über die Zusammensetzung der Körpersubstanz sowie den Stoff- und Energieansatz bei wachsenden Mastbüllen. Fortschritte in der Tierphysiologie und Tierernährung, 4.

Schwark, H.J., Ebendorff, W. 1970. Untersuchungen über die Mastleistung und den Schlachtwert von Jungbullen verschiedener Rassen und Kreuzungen Arch. Tierzucht 13, 3-17.

INFLUENCE OF NUTRITION ON BODY COMPOSITION AND CARCASS QUALITY OF FATTENING BULLS OF DIFFERENT BREEDS (GERMAN FRIESIAN AND SIMMENTAL)

R. Daenicke and K. Rohr

Institute of Animal Nutrition, Federal Research Centre
for Agriculture, Braunschweig-Völkenrode, Federal Republic of
Germany.

ABSTRACT

Body composition and carcass quality as related to breed, energy intake, and final live weight was studied in a total of 56 fattening bulls (28 Friesian and 28 Simmental bulls). Body composition (protein, fat, ash) was determined by chemical analysis. Carcass quality was measured by dissection into primal cuts and by partitioning primal cuts into muscular tissue, adipose tissue and bones. The results can be summarised as follows:

1) Empty bodies of Simmental bulls had 10% more protein, 23% less fat, and 9% less ash than Friesians. They had a higher dressing percentage, better meatiness but lower portions of fat and bones in the carcass than Friesian bulls.

2) Higher energy supply (maize silage fed restrictive versus maize silage fed ad lib.) resulted in higher fat deposition in whole bodies as well as in carcasses.

3) Higher final weights (565 versus 615 kg) resulted in higher proportions of fat and lower proportions of protein. With respect to carcass quality there was a decrease in valuable cuts. Proportions of muscle and bone tissues decreased, proportion of fat tissue increased.

INTRODUCTION

The fattening of young bulls in the Federal Republic of Germany aims at the production of slaughter animals with a high dressing percentage. Carcasses with good muscle development, especially at the round, back, and shoulder and with a low proportion of bones are desired. Fatness is tolerated up to that point which guarantees a sufficient marbling of the meat.

Apart from these carcass characteristics there is an increasing interest in the chemical composition of beef animals. The reason for this is that systematic recommendations for nutrient supply have to be based on the amounts of protein and fat deposited.

This experiment was designed to examine the influence of breed, energy supply, and final weight on chemical composition and carcass quality of fattening bulls.

MATERIAL AND METHODS

Details on the experimental design, animals and methods used have been given in a previous paper (Rohr and Daenicke, 1977, this symposium).

RESULTS AND DISCUSSION

Dissection data of empty bodies are summarised in Table 1. Empty body weight (EBW) was determined after removing the contents of the digestive tract and of the urinary and gall bladders.

On an average, EBW amounted to 89% of final live weight. This figure was neither influenced by breed and energy supply nor by final weight.

TABLE 1

FINAL WEIGHTS AND DISSECTION DATA OF EMPTY BODIES

| | | Factor | | | | | | Statistical significance[1] | | |
| | Breed | | Feed Supply | | Final weight | | | F:S | R:H | 565:615 |
	F	S	R	H	565	615				
Final weight (kg)	588.1	590.3	589.6	588.9	563.6	614.9	—	—	—
Empty body weight (kg)	522.6	523.1	521.9	523.7	500.6	545.1	—	—	—
% of final weight (%)	88.9	88.6	88.5	88.9	88.8	88.6	n.s.	n.s.	n.s.
Empty body composition (%)									
Carcass	64.1	65.5	65.2	64.4	64.6	65.0	**	n.s.	n.s.
Edible organs + tongue	3.1	3.0	3.1	3.1	3.1	3.0	***	n.s.	*
Skin	9.6	11.7	10.6	10.7	10.8	10.5	***	n.s.	n.s.
Head + 4 feet	4.5	4.3	4.4	4.4	4.5	4.3	**	n.s.	**
Belly cavity fat	8.1	5.4	6.3	7.3	6.1	7.4	***	*	***
Alimentary tract + offals	5.7	5.7	5.6	5.7	5.8	5.5	n.s.	n.s.	*
Blood[2]	5.0	4.5	4.8	4.6	5.1	4.5	n.s.	n.s.	*

1. n.s. = not significant, * = p< 0.05, ** = p< 0.01, *** = p< 0.001
2. calculated by difference

425

With respect to the fractional composition of empty bodies, Simmental bulls showed higher proportions of carcass and skin. On the other hand, proportions of belly cavity fat (ie kidney, stomach and intestine fat), organs, head and feet were lower than in Friesians. These results are in line with those of Huth (1977). The main effect of higher inergy intake resp. higher final weight was an increase in belly cavity fat.

Data on chemical composition of empty bodies are given in Table 2.

Large differences between breeds can be stated. Protein content was 10% higher and fat content 23% lower in Simmental bulls when compared with Friesians. A 9% lower ash content in Simmental bulls already indicates a correspondingly lower portion of bones (see below). Data on the chemical composition of Friesian bulls are in agreement with the results of a previous study concerning animals of 510 kg EBW (Schulz et al., 1974).

The fat content was increased by higher energy intake resp. higher final weight while the protein content was only little affected. Nevertheless, breeds responded differently to the level of energy intake in the live weight range 565 - 615 kg (Table 3).

With restricted feed supply similar amounts of protein and fat were deposited by animals of both breeds. At high feed supply definitely less protein and more fat was synthesised by Friesians than by Simmental bulls. From this result the conclusion can be drawn that with Friesian bulls at a high feeding level final live weight should not exceed 560 kg.

Simmental bulls may well be fattened to higher final weights.

Cutting data according to German cut, (Schon, 1961) are given in Table 4.

TABLE 2

CHEMICAL COMPOSITION OF EMPTY BODIES

| | | Breed | | Factor Feed supply | | Final weight | | Statistical Significance[1] | | |
		F	S	R	H	565	615	F:S	R:H	565:615
Protein	(%)	17.2	19.0	18.3	18.0	18.4	17.9	***	n.s.	*
Fat	(%)	23.0	17.7	19.6	21.2	19.1	21.7	***	*	**
Ash[2].	(%)	4.4	4.0	4.3	4.1	4.3	4.1	**	n.s.	n.s.
Water	(%)	55.4	59.3	57.8	56.8	58.3	56.4	***	n.s.	*

1. for symbols see Table 1.   2. calculated by difference.

TABLE 3

PROTEIN - AND FAT DEPOSITION IN THE LIVE WEIGHT RANGE 565 - 615 KG

| | | Restricted feed supply | | High feed supply | |
		F	S	F	S
Protein	(kg)	7.1	7.5	4.3	7.3
% of live weight gain		14.2	15.0	8.6	14.6
Fat	(kg)	19.5	21.4	30.6	23.7
% of live weight gain		39.0	42.8	61.2	47.4

TABLE 4

PRIMAL CUTS OF CARCASSÈS

	Factor						Statistical Significance[1]		
	Breed		Feed supply		Final weight				
	F	S	R	H	565	615	F:S	R:H	565:615
Carcass weight[2] (kg)	334.8	342.8	340.5	337.1	323.5	354.1	*	n.s.	***
Portions of primal cuts (%)									
Round of beef including rump	27.7	28.4	28.1	28.0	28.3	27.7	**	n.s.	*
Sirloin, best ribs, fillet	10.2	10.5	10.3	10.4	10.4	10.3	**	n.s.	n.s.
Neck, fore ribs	18.5	19.5	19.2	18.7	18.8	19.1	***	n.s.	n.s.
Shoulder	12.9	12.7	12.8	12.8	13.0	12.7	n.s.	n.s.	*
Brisket, flat ribs	13.5	12.7	13.1	13.1	12.9	13.4	***	n.s.	**
Flank	10.7	9.8	10.0	10.5	10.0	10.5	***	*	*
Shank, shin, tail	6.5	6.5	6.5	6.4	6.6	6.3	n.s.	n.s.	**

1. For symbols see Table 1.
2. Without kidney fat and kidneys.

The proportions of round of beef, sirloin, best ribs, fillet, neck, fore ribs were higher in Simmental bulls with the proportions of brisket, flat ribs, and flank being significantly lower.

The proportion of primal cuts was hardly influenced by energy supply. Increasing the final live weight from 565 to 615 kg resulted in lower portions of round of beef and shoulder plus shank, skin, and tail, while brisket, flat ribs, and flank were increased.

Data on tissue composition are listed in Table 5.

Fat-free muscle tissue was 7% more and adipose tissue was 20% less when comparing Simmental bulls with Friesians. Nevertheless, the muscle adipose tissue ratio of 1 : 0.28 in Friesians was still much better than that found in pigs of 100 kg live weight (1 : 0.6 in females, 1 : 0.8 in males) by Schon et al. (1977). Percentage of bones was remarkably lower in Simmental bulls than in Friesians.

The tissue composition of carcasses was altered as well by feed supply as by final weight. With higher energy intake resp. higher weight more fat and less bones were found. The proportion of muscle tissue decreased with the higher final live weight.

Table 6 gives a survey of the chemical composition of carcasses.

These values essentially reflect what has already been said about the influences on empty body composition. The results differ insofar as fat content is lower and ash is higher in carcasses than in empty bodies. With both breeds about 65% of total protein is bound to the carcass. The respective values for fat are 59% in Friesians and 63% in Simmental bulls. 81% (Friesian) resp. 83% (Simmental) of total ash was found in the carcass.

TABLE 5

TISSUE COMPOSITION OF CARCASSES[1]

		Factor						Statistical Significance[2]		
		Breed		Feed supply		Final weight				
		F	S	R	H	565	615	F:S	R:H	565:615
Muscular tissue (without fat)	(%)	67.2	71.9	69.9	69.2	70.4	68.7	***	n.s.	*
Adipose tissue[3]	(%)	18.2	14.6	15.8	17.0	15.3	17.5	***	*	**
Bones	(%)	14.7	13.5	14.3	13.8	14.4	13.8	***	*	*
Muscles: adipose tissue 1:		0.28	0.21	0.23	0.25	0.22	0.26	***	n.s.	*

1. Without kidney fat and kidneys.

2. For symbols see Table 1.

3. Determined by analysis

TABLE 6

CHEMICAL COMPOSITION OF CARCASSES 1

		Factor						Statistical Significance[2]		
		Breed		Feed supply		Final weight		R:S	R:H	565:615
		F	S	R	H	565	615			
Protein	(%)	17.7	18.8	18.3	18.1	18.4	18.1	***	n.s.	n.s.
Fat	(%)	21.0	17.0	18.5	19.6	18.0	20.1	***	*	**
Ash[3]	(%)	5.5	5.0	5.4	5.2	5.4	5.2	**	n.s.	n.s.
Water	(%)	55.8	59.2	57.8	57.1	58.3	56.7	***	n.s.	*

1. Without kidney fat and kidneys

2. For symbols see Table 1

3. Calculated by difference.

CONCLUSIONS

Our results show Simmental bulls to be superior to
Friesians with respect to chemical composition and carcass
quality. The most important difference is a higher dressing
percentage, a higher proportion of muscle tissue with simultaneously lower proportions of fat and bones. Because of the
higher capacity of protein synthesis - respectively lower fat
synthesis - Simmental bulls may be kept on a high feed intake
level even at the end of the fattening period. The optimum
slaughter weight of Simmental bulls is essentially higher than
that of Friesians.

REFERENCES

Huth, F.W. 1977. Influence of age, nutrient intake and body type on weight gain and body composition in young fattening bulls of the breeds German Schwarzbunte and German Fleckvieh. Paper presented at this symposium.

Schön, L. 1961. Schlachttierbeurteilung - Schlachtkörperbewertung. Arbeiten der Deutschen Landwirtschaftsgesselschaft. Bd. 25, DLG-Verlag.

Schön, L., Niebel, E., Fewson, D., Scholz, W. 1977. Die Wirtschaftlichkeitskoeffizienten beim Schwein und deren Bedeutung für die Zuchtarbeit. 3. Mitteilung: Abschätzung des Fleischanteils beim Schwein auf Grund von Teilstücken und deren grobgeweblicher Zusammensetzung. Zuchtungskunde, 49, 253 - 269.

Schulz, Oslage, Daenicke, 1974. Untersuchungen über die Zusammensetzung der Körpersubstanz sowie den Stoff - und Energieansatz bei wachsenden Mastbullen. Fortschritte in der Tierphysiologie und Tierernährung, 4.

Rohr, Daenicke, 1977. Influence of nutrition on the growth pattern of fattening bulls of two different breeds (Friesian and Simmental). Paper presented at this symposium.

NITROGEN UTILISATION IN YOUNG FATTENING BULLS KEPT ON TWO DIFFERENT ENERGY LEVELS

E. Farries

Institute of Animal Husbandry and Animal Behaviour,
Federal Agricultural Research Centre, Braunschweig-Völkenrode,
Mariensee, 3057 Neustadt 1, Federal Republic of Germany

ABSTRACT

Twenty six young fattening bulls of the breed German Schwarzbunte and 42 of German Fleckvieh were distributed at random in two groups, which were fed at two different levels of energy supply during the total fattening period from week 14 to 78 of life.

The daily energy supply was determined in relation to the body weight:

1% of body weight + 600 SE = group H
1% of body weight + 300 SE = group M.

The protein supply was kept at the same level for all animals in relation to body weight ranges. It amounted to 15 - 18% of the energy supply.

From the total number of animals 4 Schwarzbunte and 8 Fleckvieh bulls per group were selected randomly for a 2 weeks survey on metabolism, beginning with week 39, 47, 55, 63 and 71 of life.

The following criteria were investigated with regard to breed, age and energy supply:

1) digestibility of crude protein

2) nitrogen retention

3) utilisation of digestible nitrogen.

The results showed, that the digestibility of crude protein was influenced neither by breed, age or body weight nor by the energy supply. For the utilisation of digestible nitrogen and nitrogen retention with regard *to age, weight, and energy supply no difference could be observed in bulls of the German Schwartzbunte breed.*

In contrast, bulls of the German Fleckvieh breed showed a decrease in nitrogen retention and nitrogen utilisation with reduced energy supply.

INTRODUCTION

Good economical results of the fattening of young bulls can only be obtained by taking full advantage of the growth capacity for obtaining a product that can be successfully marketed. It is essential to adjust the nutrient supply in the various growth periods in correlation to the optimal conversion rates for a maximal gain in muscle protein and the lowest possible deposition of fat. Protein synthesis requires considerable amounts of energy, the ratio of feed protein to energy is therefore of decisive importance for carcass composition. Since differences between breeds must be expected, young bulls of the German Schwarzbunte and the German Fleckvieh were investigated for their capacity in muscle growth and energy utilisation.

The experiments were designed to check whether a variation in energy supply influences the digestibility of the nitrogen offered, its utilisation and the resulting protein gain. Possible influences of age and breed were also to be investigated.

The plane of energy supply was adjusted to the body weight and thus remained relatively unchanged throughout the experiment, moreover, for each group and for all growth periods investigated, the protein/energy ratio fed was constant. The difference in energy supply between the two intensity levels of feeding should be constant at 300 SE per day.

MATERIAL AND METHODS

Twenty six young fattening bulls of the breed German Schwarzbunte resp. 42 of German Fleckvieh were distributed at random in two groups each with different levels of energy supply.

The nitrogen utilisation of 4 Schwarzbunte and 8 Fleckvieh bulls per group was determined in five experiments - each lasting 14 days - intermittently over the entire fattening period as outlined in Table 1.

TABLE 1

EXPERIMENTAL DESIGN

Breed	n	Energy supply level (SE/day)	Nitrogen balance experiment
G.Sbt.	4	High (H) (1% of body weight + 600)	
	4	Medium (M) (1% of body weight + 300)	
G.Fl.	8	High (H) (1% of body weight + 600)	39. 47. 55. 63. 71. week of life
	8	Medium (M) (1% of body weight + 300)	

Only dry feed was given:

Grass hay   Concentrate

        HF I   HF IIa

    60% extracted   49% oat meal
       soyabean meal

    35% oat meal   49% barley meal

    5% mineral mixture   2% mineral mixture

The nutrient value of the components in 1 kg:

	g DCP	in SE
Grass hay	65	370
HF I	285	648
HF IIa	84	668

The nutrient supply is given in Table 2.

TABLE 2

AVERAGE NUTRIENT IN SUPPLY (g DCP IN SE/DAY)

Breed	n	Energy supply level (SE)		Nutrient intake in ....week of life				
				39	47	55	63	71
G. Sbt.	4	H	g DCP	612	623	682	776	873
			in SE	3610	4030	4510	4920	5380
	4	M	g DCP	606	669	672	790	807
			in SE	3820	3700	4190	4720	5180
G. Fl.	8	H	g DCP	592	645	670	770	846
			in SE	3820	4320	4740	5280	5800
	8	M	g DCP	570	633	658	747	786
			in SE	3520	3850	4290	4830	5330

The deviations of the mean energy supply from the calculated value of 300 SE/day are due to the weight differences within the same age groups. Therefore, the ratio DCP : SE fluctuates only slightly. The digestible crude protein was calculated from the experimentally obtained digestion co-efficients.

RESULTS

The results of the nitrogen utilisation experiment are summarised in Table 3.

1) Nitrogen intake

In all groups, the nitrogen intake increases linearly with age, respective weight, amounting to approximately 890 g of crude protein/day in the 39th and to 1210 g/day in the 71st week of life.

2) Nitrogen digestibility

It was found that age, body weight, breed, nitrogen intake and energy supply had no influence on nitrogen digestibility.

The slight fluctuations within and between groups were not indicative for any tendency. This ascertains that the digestibility for a crude nutrient is primarily dependent on its biochemical composition and that it is hardly influenced by individual variations within a species. Since the rations fed contained the same components in only slightly varied proportions, the quality of the feed stuff - the quality of the protein in this context - accordingly, was quite constant.

TABLE 3

NITROGEN METABOLISM

Breed	n	Energy supply level (SE)		Week of life				
				39	47	55	63	71
G. Sbt.	4	H	a	144.8	147.0	158.3	183.6	194.6
			b	67.6	67.8	68.9	67.6	71.8
			c	25.4	25.2	21.6	21.0	28.0
			d	+24.6	+24.9	+23.6	+20.7	+23.8
	4	M	a	139.0	147.5	155.7	179.6	187.7
			b	69.8	72.6	69.1	70.4	68.8
			c	24.6	26.7	24.8	18.9	20.4
			d	+23.8	+28.6	+23.3	+24.3	+26.8
G.Fl.	8	H	a	147.2	154.4	162.4	186.0	199.0
			b	64.3	66.9	66.0	66.2	68.0
			c	30.6	31.9	31.4	30.2	26.5
			d	+28.6	+33.4	+33.6	+37.4	+36.1
	8	M	a	140.3	150.3	156.0	181.4	191.9
			b	65.0	67.4	67.5	65.9	65.5
			c	23.4	24.6	32.2	23.6	22.1
			d	+20.7	+25.1	+33.8	+28.6	+27.8

a = N-intake g/day; b = N-digestibility %; c = utilisation of digestible N %
d = N-balance g/day.

3) Deposition and utilisation of nitrogen

In this study, the amount of digestible nitrogen absorbed from the small intestines was about the same for all animals when referred to body weight. Therefore, utilisation depends on the individual capacity for protein synthesis. For these processes sufficient energy must be available in the inter-mediate metabolism which in turn depends on the extent of energy supplied in the feed.

Under the experimental conditions used, an effect of the energy supply on the nitrogen deposition and the utilisation of digestible nitrogen was not observed in Schwarzbunte bulls. On average, the daily N-retention and -utilisation were for the H-group 23.5 g and 24.2% and for the M-group 25.3 g and 23.1% respectively. An influence of age could not be observed.

4) Variation of body weight

The body weight data of the experiment are shown in Table 4, although the variation of body weight without knowing the composition of the gain hardly allows for a comparison to nitrogen metabolism.

TABLE 4

VARIATION IN BODY WEIGHT

Breed	n	Energy supply level (SE)	Average body weight kg, daily gain g in .... week of life				
			39	47	55	63	71
G. Sbt.	4	H	301	343	391	432	478
				+750	+857	+732	+821
	4	M	298	340	389	442	488
				+750	+875	+946	+821
G.Fl.	8	H	322	372	414	468	520
				+893	+750	+946	+929
	8	M	321	355	399	353	503
				+607	+786	+964	+893

The difference in body weight at the beginning of the experiment of about 20 kg between Schwarzbunte and Fleckvieh bulls of the same age (39th week) increased until the 71st week of life to 42 kg in the H-groups and decreased to 15 kg in the M groups.

The average daily weight gains were

Schwarzbunte	H	790 g
"	M	848 g
Fleckvieh	H	884 g
"	M	812 g

These results also demonstrate that there is a different response of the two breeds to plane of supply.

CONCLUSION

The energy supply provided in this experiment had no effect on the digestibility of feed nitrogen. Nitrogen digestibility is neither age nor breed dependent, but only related to the composition of the crude protein offered.

In contrast the utilisation of digestible nitrogen for the synthesis of body protein seems to be clearly energy dependen in meat accentuated breeds like Fleckvieh. The growing capacity cannot be fully utilised when the intensity of protein synthesis is reduced by a lack of energy. A higher energy supply in Fleckvieh is probably transformed to body fat to a lesser degree than is the case with Schwarzbunte bulls.

Evaluated on the basis of identical body weights, no change in tendency is apparent for the data collected.

GENETIC VARIATIONS IN GROWTH AND BODY COMPOSITION
OF MALE CATTLE

J. Robelin*, Y. Geay*, B. Bonaiti**
Institut National de la Recherche Agronomique
*Laboratoire de la Production de Viande
Centre de Recherches Zootechniques et Vétérinaires
Theix - 63110 Beaumont - France
** Département de Génétique Animale
Centre National de Recherches Zootechniques
78350 Jouy-en-Josas - France

ABSTRACT

The relationship of growth potential, relative growth of body components, and chemical composition of the body weight gain to genotype was studied in bulls of some French breeds. It was evident that the animals of large beef breeds such as Charolais and Maine Anjou have a birth weight and growth potential superior to those of smaller beef breeds or dairy breeds. All the beef breeds had carcass, muscle and protein yields greater, but an energy content of gain lower than the dairy breeds. They were less efficient in converting the energy intake into net energy but were more efficient in converting the energy intake into meat.

INTRODUCTION

The first studies of body composition in cattle have
allowed some general principles to be made clear about the
composition of body weight gain (Trowbridge et al., 1919;
Callow, 1948). Later studies carried out on castrated animals
have shown a genetic variability (Callow, 1961; Anon., 1966).
Since then, many studies have been undertaken around the world
on meat producing capacity of different genotypes and more
particularly, those breeds used for that purpose. Generally,
these beef breeds are used in cross-breeding for young bull
production, in Europe (Colleau, 1974; Langholz and Kanning, 1975;
Daenicke and Oslage, 1976; Andersen et al., 1976; Bibe et al.,
1976) or steer production in Great Britain (Baker, 1975) or in
the United States (Smith et al., 1976). The purpose of this work
is to determine the important differences that exist between larg
size and/or late-maturing breeds and dairy breeds, as far as
weight gain and composition of gain are concerned, as well as
feed-efficiency.

VARIATIONS IN GROWTH RATE AND CARCASS COMPOSITION AT SLAUGHTER

Two experiments were carried out to study growth rate and
carcass composition in beef breeds (Charolais and Limousin) or
dual-purpose breeds (Maine Anjou), as well as, the range of
dairy-beef variability among Holstein Friesian, Normand and
Charolais. The results obtained were joined together by
expressing the data graphically in relation to Charolais. The
first experiment, carried out at the Experimental Station of
Bourges, involved a reciprocal crossing scheme between Charolais
(CH), Maine Anjou (MA) and Limousin (LI) with a purebred
Hereford control (HE). In the second experiment, at the
experimental farm of Le Pin au Haras, three purebreeds,
Charolais, Holstein-Friesian (H) and Normand (NO) and two
crossbreds (Charolais x Normand (CH x NO) and Holstein x
Normand (HO x NO)) were compared (J. Colleau, 1974, 1976). The
numbers of slaughtered animals are given in Table 1. In the

TABLE 1

EXPERIMENTAL SCHEME

Sire breed Dam breed	LI	CH	MA	HE
LI	9/9	13/4	12/4	
CH	8/7	12/11	15/10	
MA	10/10	13/8	10/6	
HE				12/11

Experiment undertaken at the experimental station of Bourges
showing number of animals slaughtered at 15 and 18 months
respectively

Sire breed Dam breed	CH	NO	HO
CH	6	X	X
NO	15	59	30
HO	X	X	32

Experiment undertaken at the experimental station of Le
Pin showing number of animals slaughtered at 18 months

first experiment, the calves were suckled by their own dams up
to six months and then fed with hay and 1 - 2 kg concentrate
up to nine months. In the second experiment, they were
artificially reared up to three months and then fed with hay or
grass and concentrate (2 kg) up to ten months. At the age of
250 days, the weights of Charolais bulls were 294 kg and 260
kg respectively in each experiment. However, this difference
was perhaps the result of an inadaptation of Charolais to
early weaning. From the ages of 9 and 10 months respectively,
the bulls were fattened in the same experimental unit, near
the National Research Centre of Jouy-en-Josas. They received
concentrate *ad libitum* : 1/3 dehydrated sugar beet pulp, 2/3
dehydrated lucerne. The animals were controlled up to 15 or
18 months for the first experiment and 16 months for the
second. The results, in this paper, only concern males, except
for birth weight. In the first experiment, we doubled the
observed deviations for the sire and dam breed effects. This
resulted in a purebreed estimation comparable to that observed
between purebreeds on the second experiment.

a) Growth

Three criteria were used to compare the growth potential
of the genotypes: weight at birth, 9 months and 14.5 months
(Figure 1).

Birth weight results from the conjoint action of maternal
(related to the dam) and direct (related to the calf) effects.
Among purebreeds, Maine Anjou and Holstein showed extreme
values with a difference of 9.5 kg (24% of the mean). The
birth weight of Charolais was 2.5 kg below that of Maine Anjou,
whereas Normand and Limousin showed approximately the same
birth weights as Holstein. Only a comparison between sire
breeds gives an estimation of genetic growth potential. These
differences were smaller than those observed between purebreeds
or maternal breed effects. Maine Anjou and Charolais showed
the same birth weights. In Normand dams, Holstein reduced the
birth weight very little; Charolais increased it by 6 kg. As
the difference between Limousin and Charolais represented only

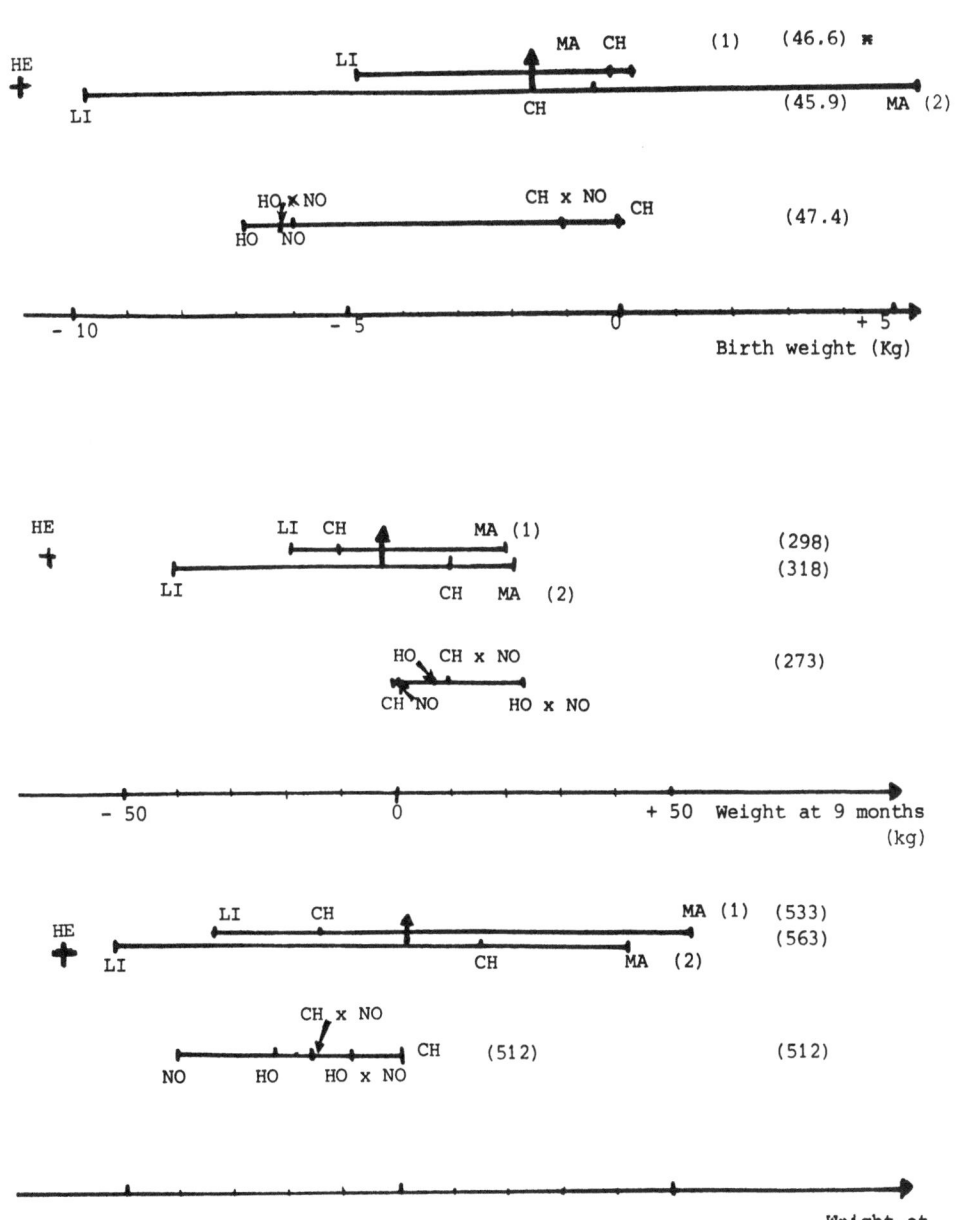

(1) Deviations of sire breeds (x 2)　　(2) Deviations of dam breeds (x 2)

ӿ Figures between brackets indicate real values for Charolais in the two experiments

Fig.1　Weight at birth, 9 and 14.5 months

5 kg or 2.5 kg on a crossbred basis, we might conclude that
Limousin leads to a birth weight exceeding dairy genotypes by
a few kilograms. These differences account for the increase in
calving difficulties caused by commercial beef crossing in the
dairy herds and therefore the small interest of dairy producers
for cross-breeding. However, Limousin might compensate for the
disadvantages of large sized breeds such as Charolais.

There were only small differences between Charolais,
Normand and Holstein at 9.5 months, whereas a difference of
around 40 kg was observed at 14.5 months between Charolais and
Normand. A deviation of the same magnitude was observed
between Charolais and Limousin. Limousin and dairy breeds may
therefore have the same growth potential. Contrary to that,
Maine Anjou leads to a 50 kg weight increase as compared to
Charolais. Therefore, there is a range of variability of 90
kg between the two extreme breeds, ie 15 - 20% of the mean.
This variability did not occur without interaction with the
environment which was due to the inadaptation to artificial
suckling and too early weaning of Charolais (and perhaps more
generally of suckling breeds).

All these genotypes showed a higher growth potential than
Hereford: 12 kg of difference at birth and 99 kg at 14.5
months between Hereford and Charolais. The difference was
relatively larger at birth.

b) Body composition
A definite hierarchy seemed to exist between the dressing
percentage (carcass weight/empty body weight) in dairy and beef
breeds (Figure 2). The ratio was only 61% for Holstein, whereas
it was between 66 and 68% for Charolais and Limousin. Limousin,
which had a lower growth potential, gave the best dressing
percentage (cf. communication of Geay, 1977). The dual-purpose
breeds (Normand and Maine-Anjou) fell between these two
extremes, Maine Anjou being closer to beef breeds and Normand
to dairy breeds.

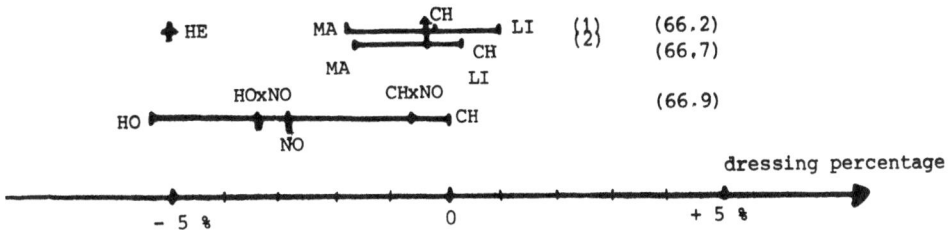

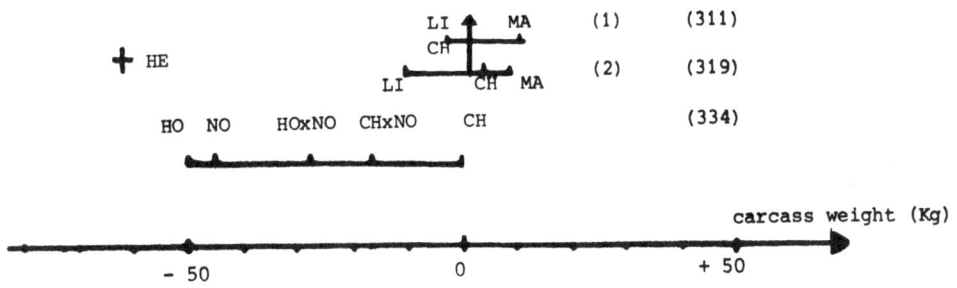

(1) Deviations of sire breeds effects (x 2)
(2) Deviations of dam breeds effects   (x 2)
ж Figures between brackets indicate real values for Charolais in the two
    experiments

Fig.2   Dressing percentage and carcass weight at 15 and 16 months
        respectively in the two experiments

The differences between breeds in dressing percentage coupled with the differences in growth potential, resulted in higher differences in carcass weight (50 kg between Holstein and Charolais). Opposite to that, Limousin might compensate for its handicap towards genotypes with a higher growth potential such as Charolais and especially Maine Anjou, by a higher dressing percentage.

The carcass composition was estimated from the 11th rib cut, kidney fat, cannon bone (Robelin et al., 1975; Figure 3). A deviation of 5% was observed for fat percentage between Holstein and Charolais, Normand showing intermediate values. Among beef breeds, the difference observed at 15 months was much smaller. Maine Anjou showed a 1 : 2 higher percentage of fat than Charolais and Limousin. For muscle percentages, we observed the same classification (beef breed showing higher muscle percentage) but the magnitude of variation was larger. In addition Limousin showed a higher muscle percentage than Charolais because of a lower bone percentage. In comparison to these genotypes, Hereford showed the same dressing percentage as Holstein, but a higher fat content.

VARIATIONS IN RELATIVE GROWTH OF ANATOMICAL AND CHEMICAL BODY COMPONENTS

Relative growth of carcass tissues, fatty tissues of the fifth quarter, chemical components of the carcass (lipids, proteins, energy) was measured by the slaughter-technique on 155 entire male cattle (84 Friesian, 29 Charolais, 42 Limousin) between 120 to 650 kg body weight (see communication of Robelin, 1977 ; Table 1). The data have been analysed by covariance analysis after logarithmic transformation using Seebeck's (1973) model and computer programme. The presentation of results mainly lays stress on the differences observed previously between the two types of cattle 1) early maturing Friesian (FF); 2) late maturing Charolais (CH) and Limousin (LI).

The allometric coefficient of carcass fat differed

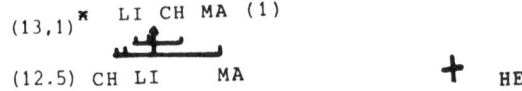

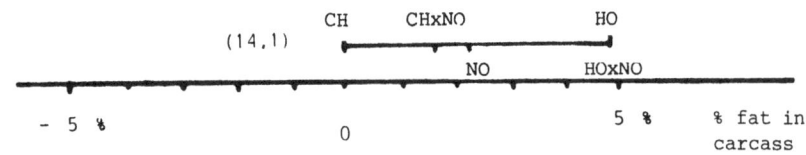

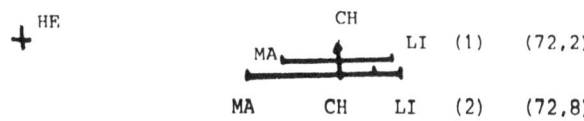

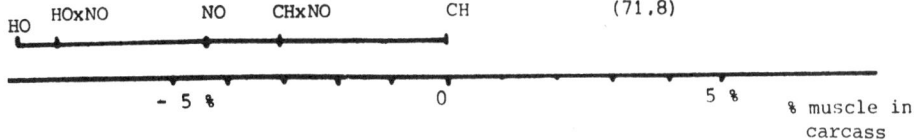

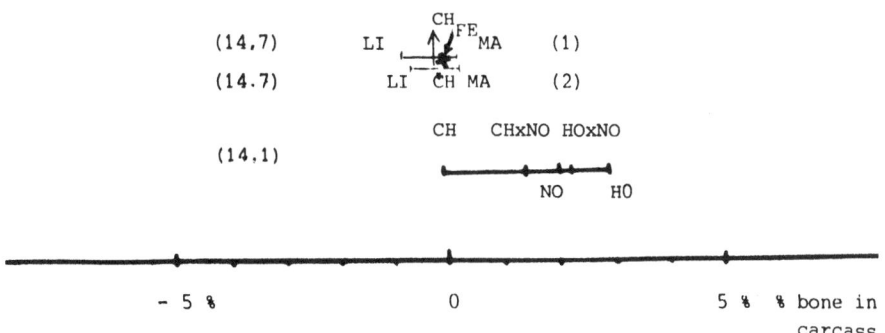

(1) Deviations of sire breeds effects (x 2)

(2) Deviations of dam breeds effects (x 2)

 ✗ Figures between brackets indicate real values for Charolais in the two
   experiments

Fig.3 Carcass composition at 15 and 16 months respectively in the two
       experiments

significantly between the two types of bulls (1.57 - 1.61 and
1.80 respectively for LI, CH and FF) (Table 2). The slopes of
muscle and bone growth were also significantly different
between FF and CH or LI, but the magnitude of variation was
much lower : 0.63 to 0.79 for bone and 0.92 to 1.00 for muscle.
Compared with the same carcass weight (220 kg ; Table 3) the
percentage of carcass fatty tissues varied from 9 to 10% for LI
and CH to 15% for FF. This difference is very similar to that
observed earlier between dairy and beef bulls at the same age.
The fat content of Friesian was close to the results obtained by
Schulz et al. (1974) and H.R. Andersen (1975) on Friesian and
Red Danish bulls, but it was much lower than the values obtained
with Friesian, Hereford or Angus steers (23 to 29%) by Anon.
(1966) and Seebeck and Tulloh (1968), (Table 3).

The magnitude of variation was increased when the
comparison was made with the same weight of muscle + bone
(Table 3). The same remarks that apply to the carcass fatty
tissues may also be made for the whole body fatty tissues.

The relative growth of various fatty tissues in proportion
to whole body fatty tissues did not differ widely between
breeds : the slope of carcass fat was slightly lower than 1,
that of subcutaneous fat (SCF) varied from 1.41 (FF) to 1.49
(CH), that of intermuscular fat (IMF) from 0.90 (FF) to 0.98
(LI). At the same weight of whole body fatty tissues (WBF =
38.3 kg), the weight of carcass fat and intermuscular fat did
not vary with the breed. On the other hand, there was
significantly less subcutaneous fat in CH (3.7 kg) and LI
(4.1 kg) than in FF (4.7 kg). The ratio of subcutaneous fat
over subcutaneous + intermuscular fat (SCF/SCF + IMF), at the
same weight of SCF + IMF followed approximately the same
hierarchy between breeds (Table 2) as the ratio carcass fat
over carcass weight. The earlier the animals matured, the
more developed was the subcutaneous fat. The results obtaned
by Charles and Johnson (1976) on crossbred steers indicated
the same tendency.

TABLE 2

VARIATION OF RELATIVE GROWTH OF BODY COMPONENTS OF BULLS WITH BREED

Independent variate (x)	Dependent variate (y)	Allometric coefficient (b)			Adjusted mean of Y (kg or Mcal)		
		LI	CH	FF	LI	CH	FF
	Carcass Tissues:						
Carcass weight	Fatty tissues	1.57^a	1.61^a	1.80^b	21.9^a	25.1^b	36.4^c
	Muscle	0.98^a	1.00^a	0.92^b	176.7^a	170.3^b	157.1^c
($\bar{x}$ = 233 kg)	Bone	0.70^a	0.63^a	0.79^b	32.1	35.4	37.1^c
Empty body weight	Whole body fatty tissues	1.64^a	1.69^b	1.84^c	30.4^a	33.3^b	44.7^c
($\bar{x}$ = 355 kg)							
Whole body fatty	Carcass fatty tissues	0.98^a	0.98^a	0.98^a	29.9^a	30.1^a	29.0^b
tissues	Subcutaneous f.t.	1.43^a	1.49^a	1.41^a	4.1^b	3.7^b	4.7^a
($\bar{x}$ = 38.3 kg)	Intermuscular f.t.	0.98^a	0.92^a	0.90^a	20.7^a	22.2^a	20.2^a
	Muscle groups:						
	Neck	1.33^a	1.17^a	1.07^b	26.0^a	25.2^a	25.2^a
	Thorax-Thoracic limb	1.21^a	1.12^{ab}	1.09^b	17.6^a	17.7^a	18.0^b
Muscle weight	Abdominal wall	1.02^a	1.14^b	1.04^a	11.1^a	11.4^a	12.5^a
($\bar{x}$ = 170 kg)	Surrounding spinal column	0.88^a	0.96^b	1.03^c	21.1^a	21.0^a	21.4^a
	Pelvic limb	0.83^a	0.87^{ab}	0.92^b	64.0^a	63.4^a	60.5^b
	Thoracic limb	1.00^a	0.99^a	0.99^a	22.2^a	22.8^{ab}	23.8^b
	Carcass chemical components:						
Carcass weight	Lipids	1.41^a		1.64^b	23.8^a		36.5^b
	Water	0.96^a		0.91^b	166.0^a		153.0^b
($\bar{x}$ = 254 kg)	Proteins	1.00^a		0.98^a	50.1^a		47.5^b
	Energy	1.20^a		1.34^b	499.0^a		603.0^b

Friesian (FF : n = 83), Charolais (CH : n = 29) and Limousin (LI : n = 42)
(Values in the same group with different subscripts are significantly different at the level 0.05)

TABLE 3

VARIATION WITH BREED OF CARCASS FATTY TISSUES, MUSCLE AND SUBCUTANEOUS FATTY TISSUE

Sex	Breed	Carcass weight (CW = 220 kg)		Muscle + Bone (M+B = 180 kg)	Subcutaneous + Intermuscular fat (SC + IM = 40 kg)			Authors
		Carcass fat (% CW)	Muscle (% CW)	Carcass fat (% M + B)	Carcass weight (kg)	Carcass fat (% CW)	Subcutaneous fat (% SC + IM)	
Castrated male	Hereford	29	54	50	190	25	36	anon (1966)
	Angus	24	58	41	204	23	41	Seebeck & Tulloh (1968)
	Friesian	23	60	38	211	22	28	anon (1966)
Entire male	Red Danish	15	68	20				H.R. Andersen (1975)
	Friesian	13	70	14				Schultz et al. (1974)
	Friesian	15	68	17	263	17	22	Results presented in this paper
	Charolais	10	73	11	333	14	17	
	Limousin	9	76	9	383	12	20	

As previously shown by Butterfield (1963) on steers, wide variations did not exist between breeds in the relative growth rate of the different parts of the musculature (Table 2); late maturing bulls had a slightly higher relative growth rate in the neck and thorax to thoracic limb muscles and a slightly lower relative growth rate of the muscles surrounding the spinal column. At the same time, late maturing bulls had, at the same weight of musculature (170 kg), more muscle in the pelvic limb (64.0 and 63.4 kg respectively for LI and CH) than early maturing bulls (60.5 kg for FF). These differences could be explained by differences in the physiological age of animals with the same muscle weight.

Variations between breeds, in the chemical composition of the carcass, were very similar to those observed on an anatomical basis (Table 2). The relative growth of lipids and energy were greater in FF (respectively 1.64 and 1.20) than in LI (respectively 1.41 and 1.20). On the contrary, the relative growth rate of water was lower in FF (0.91) than in LI (0.96). At the same carcass weight (254 kg) the FF bulls had 53% more lipids, 21% more energy, 8% less water and 5% less protein than LI bulls.

CONSEQUENCES ON COMPOSITION OF GAIN AND FEED EFFICIENCY

The differences between breed in the evolution of body composition are related to differences in the composition of body weight gain. Table 4 reports the results on body weight gain composition measured by the slaughter technique on 226 bulls (77 Friesian, 69 Charolais and 80 Limousins). For the same range of body weight (350 to 550 kg) and the same body weight gain (BWG = 1.2 kg/day), the percentage of muscle and protein in BWG was greater by approximately 15 to 20% in late maturing animals(CH or LI) in proportion with early maturing ones (FF). On the other hand, the percentage of fatty tissues and lipids in the whole body weight gain were lower by approximately 20% in late maturing bulls (Table 3).

TABLE 4

COMPOSITION OF BODY WEIGHT GAIN OF BULLS : VARIATION WITH BREED

Item	Breed		
	Friesian	Charolais	Limousin
Number of animals	77	69	80
Initial weight (kg)	341 ± 48	366 ± 66	350 ± 43
Final weight (kg)	515 ± 21	602 ± 26	559 ± 21
Daily gain (g/d) :			
Full body weight	1194 ± 170	1258 ± 183	1179 ± 152
Empty body weight	1080 ± 140	1156 ± 161	1141 ± 153
Carcass	698 ± 102	790 ± 135	788 ± 101
Muscles	452 ± 81	581 ± 122	581 ± 81
Lipids	231 ± 69	190 ± 63	185 ± 49
Proteins	203 ± 31	235 ± 38	232 ± 32
Water	607 ± 90	688 ± 107	681 ± 93
Energy (Kcal/d)	3.27 ± 0.65	3.07 ± 0.61	3.00 ± 0.54
Composition of body weight gain (g/kg) :			
Lipids	193	151	157
Proteins	170	187	197
Energy (Kcal/kg)	2.74	2.44	2.54
FU intake (*) per day	7.7 ± 0.6	7.8 ± 0.6	7.0 ± 0.6
body weight gain (g per FU)	155	161	168
carcass gain (g per FU)	91	101	112
muscle gain (g per FU)	59	74	83
protein gain (g per FU)	26	30	33

(*) 1 FU = 1 885 Kcal of net energy for growth and maintenance

These variations explain the differences in the calorific value of body weight gain 2.74, 2.54 and 2.44 Mcal/kg gain respectively for FF, LI and CH bulls.

Thus for the same energy intake (Table 4), the late maturing and/or large beef breeds deposited more protein and associated water than Friesian and had consequently a higher body weight gain. Besides, the maintenance requirement of Friesian seemed to be higher than that of Charolais bulls (Vermorel et al., 1976). Lastly, the voluntary food intake of late maturing animals was lower, as was attested by the results presented at the Theix Seminar in France (Béranger, 1977). These animals thus have a better feed efficiency. Jarrige et al. (1970) showed, in fact, that the Friesian bulls ate *ad libitum* almost 30% energy more than the Charolais, for the same weight gain (1400 g/d) at the same weight (400 kg). This better efficiency of beef breed animals appeared, too, in crossbreeding comparisons (Smith et al., 1976; Liboriussen et al., 1976).

CONCLUSION

The results presented in this study revealed clear differences in the capacity of bulls of different French breeds, for meat production. The large, late-maturing, breeds (Charolais) have higher birth weight and higher growth rates, than dairy breeds and the smaller beef breeds (Limousins). All the beef breed animals produced more carcass, muscle and protein for the same energy intake. They were less efficient in converting the energy intake into net energy but were more efficient in converting the energy intake to meat.

458

REFERENCES

Andersen, H.R. 1975. The influence of slaughter weight and level of
feeding on growth rate, feed conversion and carcass composition of
bulls. Liv. Prod. Sci., 2, 341-355.

Andersen, B.B., Liboriussen, T., Thysen, I., Kousgaard, K. and Butcher, L.
1976. Crossbreeding experiment with beef and dual-purpose sire
breeds on Danish dairy cows. Liv. Prod. Sci. 3, 227-238.

Anonymous. 1966. A comparison of the growth of different types of cattle
for beef production. Report of Major Beef Research Project. The
Royal Smithfield Club, London.

Baker, H.K. 1976. Beef breeds for beef herds in the United Kingdom Seminar
on "Optimisation of cattle breeding schemes". Dublin Eur. J. 490 e,
173-185.

Béranger, C., 1977. Feed efficiency and genotype-nutrition interactions
in growing animals. Conclusions of CEE Colloquium. Theix 1976.
Seminar on "Patterns of growth and development in cattle", Ghent.

Bibe, B., Frebling, J., Ménissier, F. and Perreau, B., 1977. French crossbreeding experiment between beef breeds. CEC Seminar on "Crossbreeding
experiments and strategy of breed utilisation to increase beef
production". Verden. EUR 5429e, 174-195.

Butterfield, R.M., 1963. Relative growth of the musculature of the ox.
In "Carcass composition and appraisal of meat animals" : 7, 1-14.
(Ed. D.E. Tribe). East Melbourne, CSIRO.

Callow, E.H., 1948. Comparative studies of meat. II. The changes in
the carcass during growth and fattening and their relation to the
chemical composition of the fatty and muscular tissues. J. Agric.
Sci. 38, 174-199.

Callow, E.H., 1961. Comparative studies of meat. VII. A comparison
between Hereford, Dairy Shorthorn and Friesian Steers on four levels
of nutrition. J. Agric. Sci., 56, 265-282.

Charles, D.D. and Johnson, E.R., 1976. Breed differences in amount and
distribution of bovine carcass dissectible fat. J. Anim. Sci., 42
332-341.

Colleau, J.J., 1974. Comparaison entre la race mixte Normande, les races
spécialisées Holstein, Canadienne et Charolaise et leurs croisements.
I. Performances de croissance des mâles et des femelles. Ann. Gén.
Sel. Anim. 6, 445-462.

II. Performances d'engraissement et de carcasse des mâles. Ann. Gén. Sel. Anim. 7, 35-48.

Colleau J.J., 1976. Troisième bilan de l'expérimentation bovine du Pin au Haras. I. Résultats analytiques. Station de Génétique Quantitative et Appliquée, CNRZ.

Daenicke, R. and Oslage, H.J., 1976. Nährstoffbedarf und Rationsgestaltung in der Rindermast. Landbauforschung Völkenrode, 26, 85-90.

Geay, Y., 1977. Dressing percentage in relation to weight and breed. Seminar on "Patterns of growth and development in cattle". Ghent.

Jarrige, R., Béranger, C., Geay, Y., Grenet, N., Malterre, C. and Robelin, J. 1970. Besoins énergétiques des jeunes bovins. In "La Production de Viande par les Jeunes Bovins". p. 167, SEI-INRA Etude n° 46.

Langholz, H.J. and Kanning, K.; 1975. Bestimmung der günstigesten Verwertung von Kälbern ans der Fleischrinderhaltung. Research report, 12. pp. Mimeograph.

Liboriussen, T., Niemann-Sørensen, A. and Andersen, B.B. 1977. Genotype environment interaction in beef production. CEC Seminar on "Crossbreeding experiments and strategy of breed utilisation to increase beef production". Verden. EUR 549e, 427-441.

Robelin, J. and Geay, Y. 1975. Estimation de la composition des carcasses de jeunes bovins à partir de la composition d'un morceau monocostal prélevé au niveau de la 11eme côte. I. Composition anatomique de la carcasse. Ann. Zootech. 24, 391-402.

Robelin, J. 1977. Development with age of the anotomical composition of the carcass of bulls. CEC Seminar on "Patterns of growth and development in cattle". Ghent.

Schulz, E., Oslage, H.J., and Daenicke, R. 1974. Untersuchungen über die Zusammensetzung der Körpersubstanz sowie den Stoff und Energieansatz bei wachsenden Mastbullen. Fortschritte in der Tierphysiologie und, Tierernährung. 4.

Seebeck, R.M., and Tulloh, N.M. 1968. Developmental growth and body weight loss of cattle. III. Dissected components of the commercially dressed carcass, following anatomical boundaries. Aust. J. Agric. Res., 19, 673-88.

Seebeck, R.M. 1973. The effect of body weight loss on the composition of Brahmin cross and Afrikaaner cross steers. I. Empty body weight, dressed carcass weight, and offal components. J. Agric. Sci., 80, 201-210.

Smith, G.M., Laster, D.B., Cundiff, L.V., and Gregory, K.E. 1976. Characterisation of biological types of cattle. II. Postweaning growth and feed efficiency of steers. J. Anim. Sci. 43, 37-47.

Trowbridge, P.F., Moulton, C.R. and Haigh, L.D. 1919. Composition of the beef animal and energy cost of fattening. Min. Agr. Expt. Station, Research Bulletin N$^{\circ}$ 30.

EFFECT OF BREED AND INTERACTION WITH NUTRITION

H.-J. Langholz,
University of Göttingen, GFR.

ABSTRACT

Despite growing intensity in cattle production much of the diversity in EEC cattle production environment is likely to remain. This will be especially valid for the beef production environment, for which the overall level and the seasonal distribution of the metabolisable energy will be the main controlling factors.

There is much evidence from breed comparison experiments on breed differences in growth and slaughter performance at a given feeding intensity. These differences tend to increase with feeding intensity and production period increasing. They are more distinct for bulls than for steers and heifers. This, in fact, indicates the existence of breed (genotype) – environmental interaction which has been experimentally proved when a wide range of feeding intensities is applied and/or an extreme range of breeds with regard to growth capacity is involved. Detailed information is given from the stratified commercial crossbreeding experiment on German Friesians. From the numerous comparative data available on carcass evaluation of different breeds or breed groups there is a strong indication for the hypothesis that breed differences and interactions with nutrition are mainly due to the different capacity of the daily protein retention. Breeds should be evaluated with regard to their lean production capacity both in the total amount and in relation to fat deposition resulting from a range of production intensities representative for the field conditions. On the basis of such evaluations a more diversified feeding and breeding strategy for beef production could be developed.

THE EEC BEEF PRODUCTION ENVIRONMENT

For the EEC cattle production Cunningham (1975) documented
more than 40 cattle populations differing widely in their genetic
potential and their production milieu. Despite the fact that
growing production intensity especially through increased use
of concentrates will level out the existing breed differences
to some extent or even remove some of the breeds (Lauvergne,
1976) much of their diversity is likely to remain. This applies
especially to the beef production potential, which also, in the
future, will have to cope with a wide range of ecological and
farm economical conditions.

This wide range in beef production conditions is mainly
caused by the following factors:

1) Differing ecological potential with special regard both
to the amount of nutritive energy available throughout the
year and the concentration of nutrients in the basic diet;
varying from the unfavourable hill and upland conditions
to the very intense systems of beef production on maize
silage.

2) Climatic differences especially with regard to the need
of overwintering cattle indoors ranging from 0 - 200
feeding days.

3) Differences in the economic environment especially
concerning the price ratios beef : milk and beef :
concentrates, the liveweight: milk price ratio being with
9.0 : 1 most favourable for beef production in France and
with 6.3 : 1 least favourable in Denmark.

4) Varying production conditions due to different farm
structures and land utilisation with a tendency of beef
production assorting with larger farm units.

5) Varying marketing and consumption habits indicated eg
by a traditional preference of steer beef in Britain and
Ireland, of light carcasses in Italy, of heavy mature
carcasses in France and Belgium and of lean beef in Germany.

Most important for the beef production environment resulting from this range of production conditions are differences:

- in the overall level of metabolisable energy supply

- in the variation (seasonal) of metabolisable energy supply during the production period and

- in the length of production period terminated either by weight or by time.

In order to produce a carcass of a determined quality - in terms of weight, dressing out percentage and degree of fatness - at maximum economic efficiency - the production systems have to be optimally adapted to these basic environmental factors.

This also means optimal adaptation of the genetic potential to the production environment. In this context of main importance seem to be the differences in the genetic control of growth intensity and growth capacity. These genetic differences may be explained predominantly by a different capacity of the daily protein retention and subsequently by a different use of the free metabolisable energy either for protein or for fat retention at a given weight/age stage of the growing animals.

EEFECT OF BREED ON GROWTH AND CARCASS PERFORMANCE

There is much evidence on breed differences in growth and carcass-performance at a given level of feeding intensity. Given a sufficient energy supply breeds or crossbred groups with high growth capacity tend to grow faster. The superiority of concentrate fed commercial crossbred bulls out of the German-Friesian dairy herd sired by large beef type breeds eg amounts to about 10% compared with straightbred Friesian bulls. A similar superiority was found in Great Britain for crossbred steers sired by large breeds on British Friesians under commercial production conditions (Baker, 1976$_1$). Direct comparisons of the straightbred sire breeds and German Friesians in bull beef production show differences in growth which are about twice the observed differences between the crossbred and purebred Friesian bulls (Witt et al., 1971; Huth, 1977). Thus we may conclude that breed differences in growth are closely related to the mature size of the parental breeds provided a suffecent energy supply. There is little evidence for special combining effects (heterosis for growth rate of breed crosses on a high plane of nutrition (Langholz, 1971). On the other hand breed differences tend to get smaller with feeding intensity and/or fattening period decreasing.

Breed differences in feed conversion, eg in the amount of nutrients used per kg liveweight gain, are more or less of the same order as the differences in the daily liveweight gain being in favour of the fast growing breeds. This applies particularly for the situation where different breeds are compared on a total weight gain basis.

From the numerous experiments on breed comparison there is a strong indication for the hypothesis that breed differences in growth are mainly due to differences in the capacity of daily protein retention. This means that at a given feeding level, differences are to be expected between breeds concerning the proportional use of the free metabolisable energy for protein and fat synthesis which in turn lead to the observed breed differences in dressing out percentage degree of fatness and to a lesser extent of differences in the meat : bone ratio.

It is to be expected that breed differences in this respect will increase with feeding intensity and lengthening of fattening period. However, according to our own observations, breed differences turn out to be already very distinct at a medium level of indoor fattening on farm feeds (Table 1).

From total dissections we have, meanwhile, direct proof of remarkable variability between breeds with respect to the total amount of lean produced in a given weight/age fattening period. These differences tend to be higher when high feeding intensities relative to the growth capacity are applied, guaranteeing the full expression of protein retention, and allowing for exceeding

TABLE 1

SLAUGHTER PERFORMANCE OF COMMERCIAL CROSSBREDS ON GERMAN FRIESIANS FED ON A SILAGE-CONCENTRATE DIET (IN % OF PUREBRED GERMAN FRIESIANS - PABST, 1977)*

| Breed ** | Bulls | | | | Heifers | | | |
| | 518/588 days - 532/584 kg | | | | 484/539 days - 420/449 kg | | | |
	No.	Dressing %	Pistol weight	% Kidney and pelvic fat	No.	Dressing %	Pistol weight	% Kidney and pelvic fat
CH x GF	73	+ 4.3	+ 13.1	- 22.6	70	+ 3.4	+ 13.0	- 11.6
GS x GF	62	+ 2.3	+ 9.5	- 20.3	66	+ 1.3	+ 11.9	- 12.5
GY x GF	39	+ 2.5	+ 9.3	- 29.0	49	+ 1.9	+ 8.3	- 12.9
GRW x GF	48	+ 1.9	+ 7.4	- 14.5	57	+ 1.4	+ 3.1	- .9
GF x GF	63	55.9%	59.6 kg	3.6%	66	54.0%	46.3 kg	5.4%

* figures derived from least squares constants

** CH = Charolais: GS = German Simmental: GY = German Yellow; GRW = German Red and White; GF = German Friesian

fat deposition. Thus differences are more distinct for heifers and steers and for comparison where extreme breeds regarding size are involved (Andersen and Liboriussen, 1977; Andersen, 1977; Bergström, 1976; Kempster et al., 1976). On the other hand, there is little evidence for breed differences in the relative distribution of lean in the carcass. Recent studies on Danish, Dutch and British data (Andersen, 1977; Andersen and Liboriussen, 1976; Bergström, 1976; Kempster et al., 1976) confirm the limited variability of lean distribution in the carcass first documented in the comprehensive studies of Berg and Butterfield (1976). Thus breed differences simply can be reduced to the different capacity of total lean production.

Regarding fat deposition some breed characteristics have been isolated especially for the kidney-knob and channel fat but still this fat depot in connection with the subcutaneous fat was found to be the most reliable indicator for predicting the total fat deposition (Kempster et al., 1976).

Exact information on breed differences in lean/bone ratio is limited. From the experimental results available up to now between breed variation is of minor order (Andersen and Liboriussen, 1976; Andersen, 1977; Bergström, 1976; Truscott et al., 1976). Breed differences tend to be more distinct for bulls than for steers. Regarding the high percentage of fattening stock coming from dairy herds we should however recognise that lean/bone ratio in the European dairy type breeds is lowest being 4.0 : 1 or less, and is showing a negative trend. This trend will be accelerated by the increasing influence of American dairy breeds predominantly US and Canadian Holsteins on the European dairy herd. A slight negative genetic correlation between milk yield and bone percentage was also recently stated by Rutzmoser (1977) for the German Simmental population.

Finally, we have to point out that the breed differences in growth and slaughter performance mainly recognised for entire males on the continent, also apply for steers and heifers both with regard to the traits affected and with regard to the magnitude of differences. The maximum values appear however on a much lower feeding level especially regarding the heifers.

SIGNIFICANCE OF INTERACTIONS BETWEEN BREED AND NUTRITION

As was already pointed out breed differences tend to gain significance with increasing feeding intensity and vice versa, thus in fact indicating the existence of breed-environmental interactions.

However, there is up to now little experimental proof of genotype-environmental interaction. Most of the few studies related to this problem and also those on sire group-environmental interactions have shown insignificant results (Skjervold and Gravir, 1961; Haring et al., 1963; Rave, 1973). However, we have to recongise that the investigated differences in the environmental situation were not very extreme. From more recent studies we learn that genotype environmental interaction becomes significant if changes in enviroment are very distinct especially if at the same time extreme breed groups regarding size are involved. (Andersen and Andersen, 1974; Liboriussen et al., 1976; Baker, 1976$_2$, Langholz, 1976). A very clear picture of the interactions betweeen genetic potential and nutritional level is given by the German stratified cross-breeding trial on commercial crossbreeding of German Friesian with sires of beef type breeds. As indicated in Table 2, the superiority of bulls sired by the large framed breeds, Charolais and Fleckvieh, is distinct when feeding intensity is increasing from pasture regimes over to indoor feeding predominantly with maize silage, to fattening on a concentrate diet only. It has to be mentioned that the pasture regimes also include to some extent concentrate feeding on pasture and finishing indoors. There are indications that finishing on pasture diet only would further equalise the differences between the breed goups, but the restricted amount of data unfortunately does not permit a reliable quantification.

468

TABLE 2

DAILY GAIN OF COMMERCIAL CROSSBRED BULLS AND HEIFERS VERSUS STRAIGHTBRED GERMAN FRIESIANS BY FEEDING SYSTEMS
(PRELIMINARY RESULTS ACCORDING TO ERNST, 1977; PABST, 1977 AND HUTH, 1976).

Breed	Bulls Feeding system			Heifers Feeding system	
	Pasture *	Maize silage	Concentrates	Pasture	Maize silage
	723 days/ 560 kg	554 days/ 558 kg	500 days/ 560 kg	701 days/ 447 kg	511 days/ 435 kg
CH x GF	103	105	113	104	108
GS x GF	104	105	118	107	108
GY x GF	–	105	–	–	106
GRW x GF	101	106	106	102	102
GF x GF	711 g	904 g	966 g	560 g	749 g

* partly with concentrate feeding on pasture, partly with finishing indoors.

TABLE 3

CARCASS YIELD OF COMMERCIAL CROSSBRED BULLS AND HEIFERS VERSUS STRAIGHTBRED GERMAN FRIESIANS RELATED TO FEEDING SYSTEMS (PRELIMINARY RESULTS ACCORDING TO ERNST, 1977; PABST, 1977 AND HUTH, 1976).

Breed	Bulls			Heifers	
	Feeding systems			Feeding systems	
	Pasture *	Maize silage	Concentrates	Pasture	Maize silage
	723 days/ 560 kg	554 days/ 558 kg	500 days/ 560 kg	701 days/ 447 kg	511 days/ 435 kg
CH x GF	107	111	116	108	111
GS x GF	108	108	118	110	109
GY x GF	-	108	-	-	106
GRW x GF	103	108	106	105	103
GF x GF	318 kg	287 kg	296 kg	223 kg	216 kg

* partly with concentrate feeding on pasture, partly with finishing indoors

470

TABLE 4

PERCENT KIDNEY AND PELVIC FAT (IN RELATION TO HOT CARCASS WEIGHT) OF COMMERCIAL CROSSBRED BULLS AND HEIFERS VERSUS STRAIGHTBRED GERMAN FRIESIANS BY FEEDING SYSTEMS (PRELIMINARY RESULTS ACCORDING TO ERNST, 1977, PABST, 1977 AND HUTH, 1976).

Breed	Bulls Feeding system			Heifers Feeding system	
	Pasture * 723 days/ 560 kg	Maize silage 554 days/ 558 kg	Concentrates 500 days/ 560 kg	Pasture 701 days/ 445 kg	Maize silage 511 days/ 435 kg
CH x GF	(2.0%)	2.8%	3.5%	3.1%	4.7%
GS x GF	(2.3%)	2.9%	3.7%	3.8%	4.7%
GY x GF	–	2.6%	–	–	4.7%
GRW x GF	(1.9%)	3.1%	3.9%	3.8%	5.3%
GF x GF	(1.9%)	3.6%	3.4%	4.5%	5.4%

* bulls from two farms only out of three

As indicated by Tables 3 and 4 the superiority of the crossbred groups arises mainly from higher gain in lean. Due to a significant higher dressing out percentage the breed group differences as well as the degree of interaction with feed intensity turn out to be much more distinct for the carcass yield. As indicated by the reduced fat deposition (% kidney and pelvic fat) in the case of medium-high feeding intensity the superiority in lean deposition is expected to be even higher.

Concerning fat deposition the interdependences obviously do not behave in a linear way. On a lower plane of nutrition in the pasture regimes, we observe a small and insufficient, but, in relation to carcass weight, equal fat deposition. On a medium to high plane of nutrition eg with maize silage fed bulls or heifers from pasture, a significant lower fat deposition can be observed for the crossbred groups. On the very high plane of nutrition represented by heavy concentrate feeding to bulls and by maize silage feeding to heifers breed group differences in fat depostion are levelling out again.

Coming back to the hypothesis presented earlier that genetic differences in growth are mainly due to the different capacity of protein retention, we have to conclude that at a low nutritional level, growth is limited by the amount of free metabolisable energy independent of the genetic capacity for protein retention. The proportion of the free energy for fat retention for all breed groups approaches a minimum and an equal percentage in relation to their body weight. Assuming that energy demand for maintenance is mainly controlled by body weight already at a very early stage (van Es, 1976) it is unlikely that 'true' interactions will occur changing the rank of different breeds in their growth rate. However, cattle breeding experiments in the tropics (Langholz, 1977) indicate that severe limitations in energy supply reduce the adaptability of breeds with high growth potential leading to a high mortality rate in early stages. At a very high level of energy supply on the other hand the maximum fat retention rate for all breed

groups might be reached, which again seeems to be similar in
relation to body weight for all the breeds. Breed differences
in growth at this stage will have to be referred to as
differences in feed intake only. The Danish results on ad lib.
concentrate feeding by Andersen and Andersen (1974) and
Andersen (1977) are confirming this statement.

However, under the normal range of fattening intensities
in the field, this maximum energy situation will not occur,
especially not for fattening of bulls. This means that we
have to expect breed nutritional interaction with regard to the
percentage of free energy used for protein respectively fat
retention. From all experiments on bull beef production with
straightbred German Friesians and with commercial Charolais-
Friesian-crosses this statement clearly is to be proved (Figure
1). Taking daily gain in a given fattening period as an indicator for feed intensity, no difference in fat deposition can
be observed at a low level of 700 g/day and below. Above that
level, the difference in fat deposition between Charolais x
Friesian-crosses and straightbred Friesian bulls steadily
increases up to about 1.5% at 1000 g daily gain according to
the regression derived from all available experiments. If the
final weight for straightbred Friesians is about 100 kg lower
than that from the Charolais the equilibrium of fat deposition
turns out at the much higher intensity level of 900 g gain per
day.

Comparing Charolais x Friesian-crosses and Friesian heifers
under predominantly high feeding intensities no sign of an
interaction can be observed (Figure 2). Feeding levels
guaranteeing 500 g daily gain and more, apparently lead to a
proportional fat retention for both breed groups. The
difference in the relative fat deposition refers solely to the
higher protein retention capacity of the Charolais crossbred
heifers. We might expect interactions if the nutritional
level is falling below that investigated range of intensities.

Finally, we have to realise that the significance of the

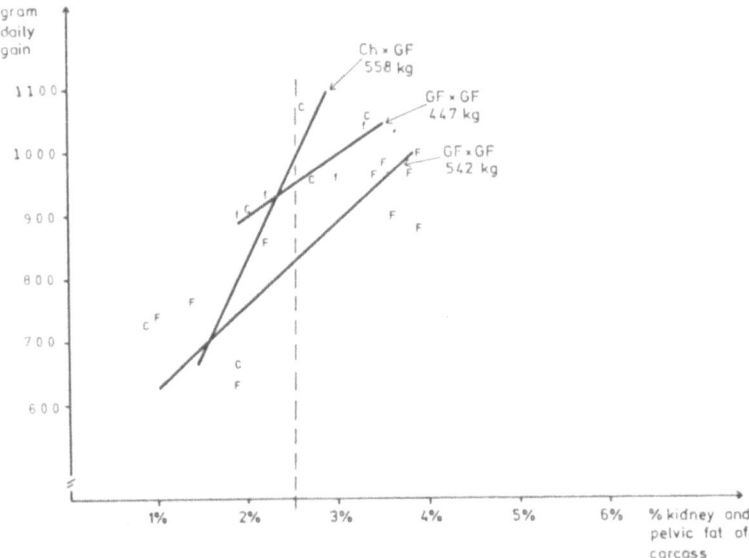

Sources: 9, 10, 11, 18, 19, 20, 26, 27, 28, 33, 35, 36

Fig. 1. Deposition of kidney and pelvic fat at increasing feeding intensities for Charolais crossbred and purebred German Friesian bulls

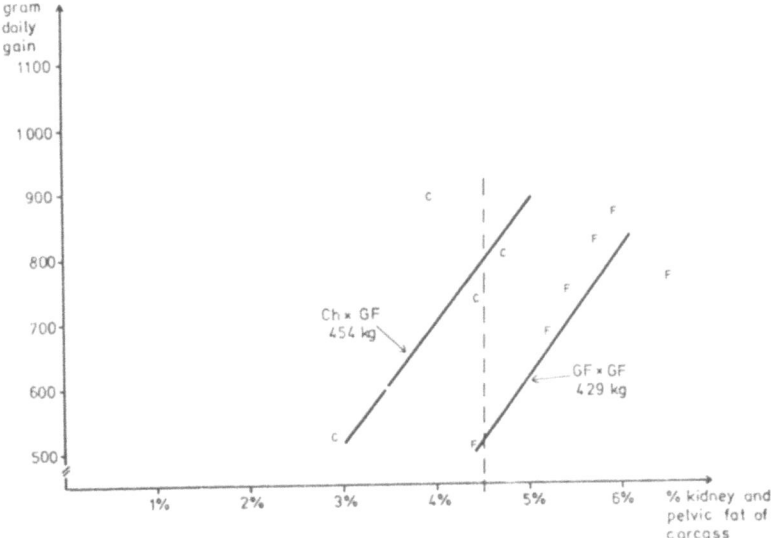

Sources: 9, 10, 27, 36

Fig. 2. Deposition of kidney and pelvic fat at increasing feeding intensities for Charolais cross bred and purebred German Friesian heifers

interdependences between feeding intensity and the genetic
potential for lean production respecting the genetic tendency for
fat deposition is quite different from experiment to experiment.
These differences may be due partly to individual variation but
to a greater extent seem to be due to seasonal effects in the
feeding regime.

CONSEQUENCES OF BREED DIFFERENCES AND INTERACTIONS WITH
NUTRITION FOR BEEF PRODUCTION - AND BREEDING - STRATEGY

On account of the range of production environments and the
differences in breed response of varying feeding intensities, we
have to consider both the environmental and the genetic aspects
simultaneously when attempting further improvement. This means
that on one hand:
    - the breed potential, in particular for producing lean,
    has to be included as a variable factor when setting up
    feeding standards and feeding regimes
and on the other hand:
    - the nutritional situation with regard to level,
    variability and concentration of energy supply has to be
    considered when assessing the beefing potential of breeds
    for relevant production systems (Figure 3).

The essential problem in designing an optimal breeding
strategy is to match the right feeding regime with the right
genetic potential, aiming at the production of a carcass of a
certain weight and quality, especially with regard to the lean/
fat ratio.

If the environmental conditions can be changed only to a
limited extent because of economical or ecological reasons, the
breed has to be adapted to the production systems. For beef
production under range conditions, eg in hill and upland areas,
a breed of limited growth potential has to be chosen in order to
ensure the production of a sufficiently mature carcass in a
given time. Conversely, to make optimum use of an intensive
fodder production, eg of maize silage, high quality fattening

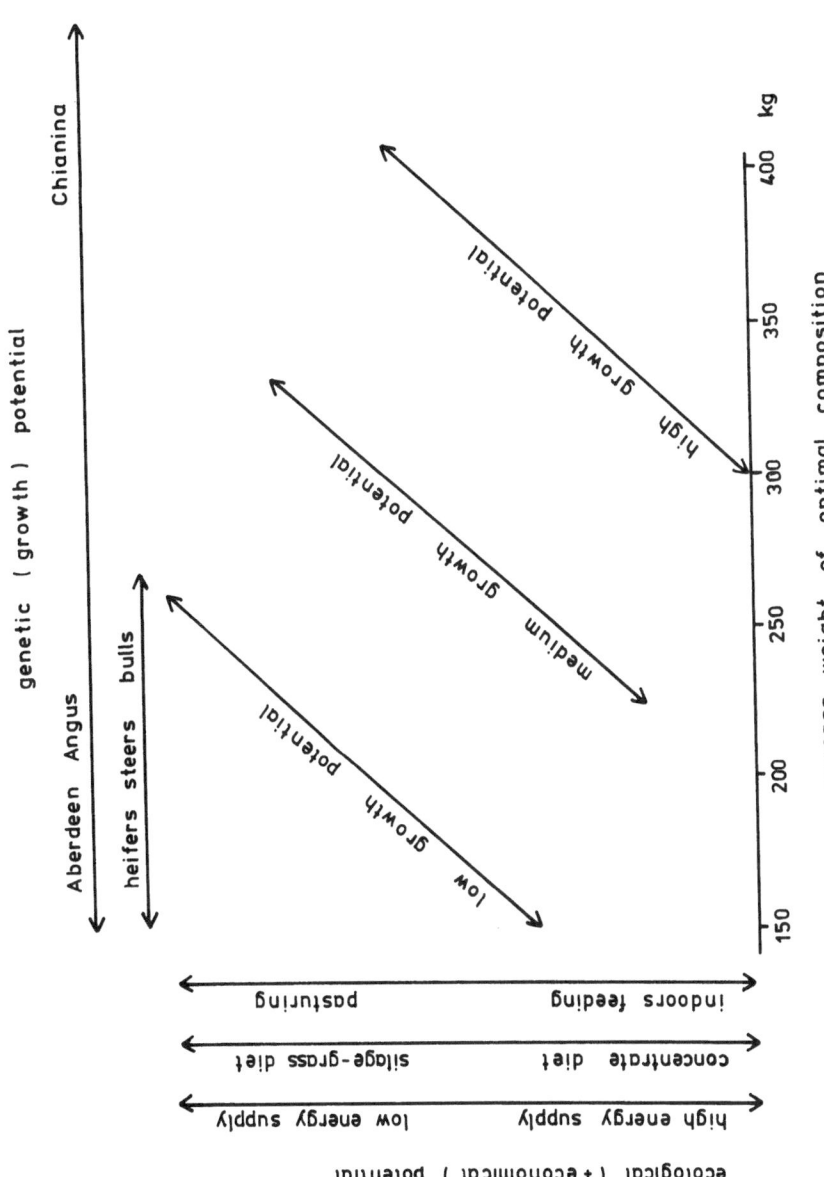

Fig. 3. Illustration of the interaction between growth potential and feeding level.

stock of high growth potential is needed to produce carcasses of sufficient weight and without excessive fat deposition.

If, on the other hand, the breed is fixed eg where the fattening stock is recruited from dairy herds, the feeding regime and intensity has to be adapted to the genetic material. For example if we intend to produce a very lean carcass of about 300 kg at a degree of fatness indicated by 2.5% kidney and pelvic fat the overall daily gain should not exceed 825 g for the German Friesians (Figure 1). Working with Charolais x Friesian crossbred bulls a diet allowing for a daily gain of close up to 1000 g is needed. If we have to work with Friesians on a higher plane of nutrition we have to reduce the final weight if a higher degree of fatness is unacceptable. If we have to work with Friesians on a lower feeding level, the only way out will be to produce a heavier carcass or to reduce genetic potential by castration in order to obtain sufficient fatness in the carcass. What is urgently needed in order to develop optimal production strategies in the ways under discussion, is a mapping out of the breed's performance with regard to a relevant range of nutritional environments. Of particular interest is the ex-pected lean production capacity both in the total amount and in relation to fat deposition resulting from a defined feeding intensity in a defined weight/age fattening period. Plant breeders, for example, for a long time, have introduced a 'variety passport' in which the ecological requirements of each variety is recorded.

In order to overcome the difficulties resulting from the large number of breeds available the evaluation should be started for a range of biological types of breeds. On the basis of such a breed type evaluation, more diversified feeding standards can be derived as outlined in Table 5, which would enable the stockman to fix the feeding strategy optimally to the breeds potential or vice versa.

TABLE 5

OUTLINE OF A DIVERSIFIED FEEDING STANDARD FOR BEEF PRODUCTION WITH BULLS

Growth capacity of breed	Designed carcass weight	Length of fattening period	Daily energy requirement		
			Starting phase	Medium phase	Finishing phase
Small	170 kg 200 kg 230 kg				
Medium	230 kg	12 months 15 months 18 months	X X X	X X X	X X X
	260 kg	15 months 18 months 21 months	X X X	X X X	X X X
	290 kg	18 months 21 months 24 months	X X X	X X X	X X X
Tall	290 kg 320 kg 350 kg				

Consequently, any breed or breeding animal evaluation should be carried out with respect to the level, the variability and the concentration of nutrient supply representative for the field conditions. This means that the normally applied station tests should be performed as a simulation of the feeding conditions in the field as successfully proved by MLC testing procedures (Southgate, 1976).

REFERENCES

1. Andersen, B.B. and Andersen, H.R. 1974. Genotype x environment interaction for beef production traits in dual purpose cattle breeds. Acta
   Agric. Scand. 24, 335-338.

2. Andersen, B.B. and Liboriussen, T. 1976. Danish crossbreeding experiments
   with beef breeds on dairy and dual purpose breeds. Proc. EEC Seminar
   Verden, EUR 5492e, 230-239.

3. Andersen, B.B. 1977. Genetiske undersøgelser vedrørende kvaegets tilvaekst,
   kropsudvikling og for udnyttelse. Diss. Copenhagen 1977.

4. Baker, H.K. 1976_1. Experiments on crossing for beef in dairy and beef
   cows in UK, Proc. EEC Seminar in Verden. EUR 5492e, 258-276.

5. Baker, H.K. 1976_2. Notes on genotype-nutrition interactions. Proc.
   EEC Seminar Verden. EUR 5492e, 414-424.

6. Berg, R.T. and Butterfield, R.M. 1976. New concepts of cattle growth.
   Sydney Univ. Press, 240 pp.

7. Bergström, P.L. 1976. Crossbreeding for beef production in the
   Netherlands. Proc. EEC Seminar in Verden (in press).

8. Cunningham, E.P. 1975. The structure of cattle populations in the EEC.
   Proc. EEC Seminar Dublin, EUR 5490e, 13-27.

9. Ernst, E. 1969. Die Effektivität der Einkreuzung von Charolaisbullen
   in deutsche Schwarzbunte Rinder aus züchterischer und ökonomischer Sicht.
   Diss. Kiel 1969.

10. Ernst, E., Langlet, J.-F. and Martin, H. 1973. Vergleichende
    Untersuchungen an Deutschen Schwarzbunten, Holstein-Friesian sowie den
    Kreuzungen aus beiden Populationen. Schriftenreihe Agrarwiss. Fak.
    Kiel, Heft 50.

11. Ernst, E. 1977. Weidemast von männlichen und weiblichen Einfach-
    gebrauchskreuzungen auf der Grundlage von Deutschen Schwarzbunten in
    Schleswig-Holstein. Proc. 3. Göttinger Tagung 'Rindfleischerzeugung'
    (in press).

12. Van Es, A.J.H. 1976. The utilisation of energy (and nitrogen) by growing
    animals. Paper presented at EEC Colloqium, Theix mimeograph, 8 pp.

13. Haring, F., Weniger, J.-H., Gruhan, R. and Langholz, H.-J. 1963.
    Einsatz von Kraftfutter und Trockenfutter in der Mast von Jungbullen.
    Zdke. 35, 98-113.

14. Huth, F.W. 1976. Pers comm.

15. Huth, F.W. 1977. Wachstumsverlauf von Rassen und Kreuzungen mit
    unterschiedlicher Wachstumskapazität. Proc. 3. Göttinger Tagung
    'Rindfleischerzeugung' (EEC Colloquium) (in press).

16. Kempster, A.J., Cuthbertson, A. and Smith, R.J. 1976. Variation in lean
    distribution among steer carcass of different breeds and crosses. J.
    Agric. Sci. Camb. 87, 583-542.

17. Kempster, A.J., Cuthbertson, A. and Harrington, G. 1976. Fat distribution
    in steer carcass of different breeds and crosses. Anim. Prod. 23,
    25-34.

18. Krüger, L. and Meyer, F. 1966. Untersuchungen zur Frage der Erzeugung
    und der Wertbestimmung von Rindfleisch. Z. f. Tierz. u. Zübiol. 83,
    139 u. 224.

19. Küther, K.-H. 1972. Optimaler Kraftfuttereinsatz in der Stallendmast
    von Bullen bei verschiedenen Wirtschaftsfuttermittein. Diss. Kiel, 1972.

20. Langholz, H.-J. 1964. Die Machkommenprüfung auf Station als
    züchterischer Weg zur Verbesserung der Rindfleischerzeugung. Diss.
    Göttingen, 1964, 133 pp.

21. Langholz, H.-J. 1971. Die Kreuzung als züchterische Massnahme zur
    Erhöhung der Rindfleischerzeugung. Zkde. 45, 307-317.

22. Langholz, H.-J. 1976. Aspects of interaction between beefing capacity and
    feeding level in beef production. Paper presented at EEC Colloquium,
    Theix, mimeograph, 24 pp.

23. Langholz, H.-J. and Claus, J. 1977. Ansatzpunkte zur Förderung der
    Tierproduktion am tropischen und subtropischen Standort - Probleme
    des Technologietransfers - Verbesserung des Tiermaterials. Paper
    presented at conference of Dachverband Agrarforschung Berlin, 19 pp,
    mimeograph.

24. Lauvergne, J.J. 1976. Was spricht fürdie Erhaltung Kleinerer
    Rassengruppen in der europäischen Rinderzucht. Der Tierzüchter 28,
    442-446.

25. Liboriussen, T., Neimann-Sørensen, A. and Bech Andersen, B. 1976.
    Genotype x environment interaction in beef production. Proc. EEC
    Seminar Verden. EUR 5492e, 427-441.

26. Mähl, H. 1977. Pers. comm.

27. Pabst, W. 1977. Praxisübliche Wirtschaftsmast von männlichen und
    weiblichen Einfachgebrauchskreuzungen auf der Grundlage Deutsches
    Braunvieh und Deutsche Schwarzbunte. Proc. 3. Göttinger Tagung
    'Rindfleischerzeugung' (EEC Colloquium) (in press).

480

28. Raue, F. 1975. Ein Beitrag zur Frage des Wachstumsausgleichs in der Rindermast. Diss. Kiel, 1975.

29. Rave, G. 1973. Genotyp-Umwelt-Interaktion für zwei Prüfmethoden der Nachkommenprüfung auf Fleischleistung beim Rind. Zkde. 45, 22-30.

30. Rutzmoser, K. 1977. Neuere Ergebnisse zur Beziehung zwischen Milch- und Fleischleistung beim Rind. Der Tierzüchter 29, 292-294.

31. Skjervold, H. and Gravir, K. 1961. Interaction between genotype and environment in selection for rate of gain in cattle. Meldinger fra NLH Nr. 150.

32. Southgate, J.R. 1976. Stationsprüfungen auf Fleischleistung beim Rind unter simultanen Praxisbedingungen in Grossbritannien. Der Tierzüchter 28, 350-352.

33. Trappmann, W. 1972. Schätzung phänotypischer und genetischer Parameter der Fleischleistung von Jungbullen bei Stations- und Feldprüfung. Zkde. 44, 17-28.

34. Truscott, T.G., Lang, C.P. and Tullok, N.M. 1976. A comparison of body composition and tissue distribution of Friesian and Angus steers. J. agric. Sci. Camb. 87, 1-14.

35. Witt, M., Huth, F.W. and Selhausen, D. 1967. Jungrindermastversuche mit den Rassen Deutsche Schwarzbunte und deren Kreuzungen mit Aberdeen Angus. Zkde. 39, 159-169 and 239-252.

36. Witt, M., Andreae, U., Kallweit, E., Huth, F.W., Wehrhahn, E. Selhausen, D. and Röseler, W. 1971. Mastversuche mit der Charolais-Rasse und deren Kreuzungen. Schriftenreihe MPI Tierz. u. Tierern. 53, 174 pp.

EFFECT OF SIREGROUP WITHIN BREEDS ON GROWTH AND EFFICIENCY AND INTERACTION WITH NUTRITION

B. Bech Andersen

Department of Cattle and Sheep Experiments
National Institute of Animal Science
Rolighedsvej 25, 1958 Copenhagen V Denmark

ABSTRACT

An important genetic within breed variation in growth capacity, feed utilisation and carcass composition makes it possible to improve these traits in beef and dual purpose cattle breeds by selection.

Performance test selection based on daily gain leads to an improvement in feed utilisation, but also to an indirect increase in birth weight and mature weight and to a reduction of the muscle bone ratio. However, by incorporation of the traits daily gain, gestation length and ultrasonic muscle area in a selection index it is possible to regulate the correlated effect of a performance test selection on birth weight, mature weight, feed utilisation and carcass composition.

INTRODUCTION

Several scientific papers have documented an important genetic within breed variation in growth capacity, feed utilisation and carcass composition in animals of dual purpose and beef breeds. It is, therefore, possible to improve the efficiency and quality of the beef producing animals by testing and selection. Breeding plans with that goal have been established in many European countries.

However, the elaboration of a selecting and breeding strategy optimum in all respects, aiming at an improvement of the economic efficiency of the breeds will require that not only the genetic variation in the various traits is known but also the interactions between them and their relationships to gestation length, birth weight, feed utilisation and mature weight.

DAILY GAIN OF LIVE WEIGHT, CARCASS WEIGHT AND WEIGHT OF LEAN, FAT AND BONE

For young bulls tested at experimental stations with controlled feeding and environment, the phenotypic coefficient of variation for total daily gain ranges from 6 to 9 and the corresponding coefficient of heritability from 0.3 to 0.6 (Rittmannsperger, 1966; Flatnitzer et al., 1969; Lindström and Maijala, 1970; Dietert et al., 1970; Gravert et al., 1971; Langholz and Jongeling, 1972; Trappmann, 1972; Fimland, 1973; Linner, 1973 and Andersen, 1977). The coefficient of heritability for daily carcass gain is found to be a little higher than for total live weight gain (Bränning, 1968; Trappmann, 1972 and Langholz and Jongeling, 1972).

In an analysis of dissection data on 2330 young bulls distributed on 136 progeny groups, Andersen (1977) found a relatively great genetic variation in the composition of the gain (Table 1). The results demonstrate that when the final live weight increases from 250 kg to 450 kg the total average daily gain increases, which is due to an important increase in fat gain and some increase in lean gain. The demonstrated breed differences in total gain are almost exclusively caused by breed differences in lean gain. The coefficients of heritability are highest for lean and bone gain, but the daily gain of fat shows the highest genetic coefficient of variation.

TABLE 1

EFFECT OF FINAL WEIGHT, BREED, SIREGROUP WITHIN BREED (h^2 VALUE) AND PHENOTYPIC AND GENETIC VARIATION IN TOTAL GAIN AND COMPOSITION OF GAIN (ANDERSEN, 1977).

	Average Daily Gain (grams) of:				
	live wt.	carcass wt.	lean wt.	fat wt.	bone wt.
Final Weight					
250 kg	1097	599	410	55	108
450 kg	1151	649	444	94	102
Breed (450 kg bulls)					
Red Danish	1127	623	421	92	100
Black and White Danish	1154	650	441	95	103
Red and White Danish	1172	675	470	94	103
Phenotypic S.D.	67.3	36.2	32.4	11.9	7.8
h^2 - value	0.42	0.45	0.54	0.34	0.53
S.E. (h^2)	0.11	0.08	0.08	0.07	0.08
Genetic Coeff. of Var. (%)	3.8	3.8	5.3	7.4	5.6

CARCASS COMPOSITION AND TISSUE WEIGHT DISTRIBUTION

The number of scientific reports describing the within breed variation in anatomic carcass composition is relatively small owing to the high costs of dissection of beef carcasses. For young bulls fed in controlled environments the h^2 - value in carcass composition ranges from 0.2 to 0.9 (Hinks and Andersen, 1969; Averdunk, 1969; Torreele and Slawinski, 1970;

Andersen, 1970 and Langholz and Jongeling, 1972).

Andersen (1977) found in the analysis described on the preceding page and in Table 2 that the genetic differences in the composition of gain lead to a corresponding genetic effect on carcass composition. The genetic within breed variance is highest for lean : bone ratio and lean : fat ratio. The h^2 - values for weight distribution of lean, fat and bone are very low, which is in agreement with findings of Lindhé and Henn- ingsson (1968) and Berg and Butterfield (1976) showing that the effect of breed on muscle distribution is low and unimportant.

FEED UTILISATION

Feed efficiency expressed as the input : output ratio (feed conversion ratio) has a coefficient of heritability ran- ging from 0.2 to 0.7 (Langlet et al., 1967; Preston and Willies, 1970; Langholz and Jongeling, 1972 and Trappmann, 1972). This genetic variation in feed efficiency can be due to genetic differences in daily feed intake, composition of the gain, maintenance requirement or digestion capacity. Taylor and Young (1968) showed in an experiment with twins that the gene- tic coefficient of variance in maintenance requirement is 4 to 6%.

Andersen (1977) calculated the feed utilisation as Scand. f.u. per kg total gain, carcass gain and lean gain. As shown in Table 3 the genetic coefficients of variation for these traits are 4.7, 5.1 and 6.7% respectively. The percentage of the total feed consumed available for growth is calculated as:

$$\% \text{ f.u. for growth} = \frac{(\text{total f.u. consumed} \div \frac{70.5 \times w^{0.73}}{1650} \times \text{days in test}) \times 100}{\text{total f.u. consumed during test}}$$

where w = average weight in the test period. This trait shows a h^2 - value of 0.46. Feed utilisation expressed as input : output is strongly correlated to daily gain capacity with gen-

TABLE 2

EFFECT OF FINAL WEIGHT, BREED, SIREGROUP WITHIN BREED (h^2 - VALUE) AND PHENOTYPIC AND GENETIC VARIATION IN CARCASS COMPOSITION AND TISSUE WEIGHT DISTRIBUTION (ANDERSEN, 1977).

	Dressing percent.	Lean bone	Lean fat	% pistollean	% lean	LD area	LP/LT	FP/FT	BP/BT
Final Weight									
250 kg	53.8	3.6	7.5	34.2	70.6	47.2	48.5	41.3	48.0
450 kg	55.8	4.2	5.0	31.5	69.1	64.8	45.7	39.8	46.8
Breed (450 kg bulls)									
Red Danish	54.8	4.1	4.8	30.8	68.5	64.3	45.0	39.2	46.7
Black and White Danish	55.7	4.1	4.9	31.8	68.8	63.8	46.2	39.9	47.0
Red and White Danish	56.9	4.3	5.2	32.0	70.0	66.2	45.7	40.3	46.8
Phenotypic S.D.	1.41	0.26	0.79	1.09	1.88	5.73	0.98	2.44	1.22
h^2 - value	0.58	0.52	0.50	0.54	0.52	0.58	0.29	0.16	0.08
S.E. (h^2)	0.09	0.08	0.08	0.08	0.08	0.09	0.06	0.05	0.04
Genetic Coeff. of Var. (%)	1.92	4.52	11.2	2.57	1.95	6.71	1.16	2.46	0.75

LD - area = M. long. dorsi area in cm^2

LP/LT = 100 x lean in pistol/lean in rest of carcass

FP/FT = 100 x fat in pistol/fat in rest of carcass

BP/BT = 100 x bone in pistol/bone in rest of carcass

etic correlations ranging from - 0.6 to - 0.9 (Lindhé and
Henningsson, 1968; Averdunk, 1969; Dietert et al., 1970; Langholz and Jongeling, 1972; Trappman, 1972 and Andersen, 1977).

TABLE 3

EFFECT OF FINAL WEIGHT, BREED, SIREGROUP WITHIN BREED (h^2-VALUE) AND
PHENOTYPIC AND GENETIC VARIATION IN FEED UTILISATION (ANDERSEN, 1977).

	Scand. f.u. per kg			
	total gain	carcass gain	lean gain	% of feed to production
Final Weight				
250 kg	3.04	5.58	8.18	49.6
450 kg	4.13	7.32	10.76	48.7
Breed (450 kg bulls)				
Red Danish	4.25	7.69	11.42	49.2
Black and White Danish	4.14	7.35	10.86	49.0
Red and White Danish	3.99	6.93	9.99	47.8
Phenotypic S.D.	0.32	0.54	1.00	1.60
h^2 - value	0.36	0.47	0.51	0.46
S.E. (h^2)	0.07	0.08	0.08	0.08
Genetic Coeff. of Var. (%)	4.65	5.10	6.67	2.23

On the data described previously, Andersen (1977) has
by use of partial regression analysis calculated the relative
importance of daily gain, growth curve form index, dressing
percentage and lean percentage on Scand. fu/kg gain. The results show that growth capacity and shape of growth curve explain 85% of the variation in the feed conversion ratio, while
dressing percentage and lean percentage have no effect. This
forms a contrast to the general opinion, but the explanation
can be that bulls with a tendency to relatively large fat
deposition also have a lower maintenance requirement at a
given weight.

GENOTYPE x NUTRITION INTERACTION

Few papers have dealt with the problem about siregroup x feeding level interaction. However, Andersen and Andersen (1974) have shown that genetic differences in appetite can lead to genotype x feeding system (restricted or ad lib. feeding) inter-action (example in Table 4). Testing and selection of breeds and/or animals within breeds must therefore be based on feeding systems which are not too different from common practice.

TABLE 4

GENOTYPE x FEEDING SYSTEM INTERACTION FOR DAILY GAIN AND FEED
CONVERSION RATIO (TEST FROM 28 DAYS TO 310 KG) (Andersen and Andersen, 1974).

| | | Feeding on Concentrate | |
		ad lib. by automat	restricted by age
RDM	Number of animals	25	27
	Feed intake, Scand. fu/day	4.6	4.1
	Av. daily gain, g.	1267	1241
	Scand. fu/kg gain	3.61	3.30
SRB x RDM	Number of animals	30	28
	Feed intake, Scand. fu/day	4.7	4.1
	Av. daily gain, g.	1322	1254
	Scand. fu/kg gain	3.55	3.29
FA x RDM	Number of animals	29	26
	Feed intake, Scand, fu/day	4.5	4.1
	Av. daily gain, g.	1277	1226
	Scand. fu/kg gain	3.53	3.34
MRI x RDM	Number of animals	30	27
	Feed intake, Scand. fu/day	4.6	4.1
	Av. daily gain, g.	1317	1270
	Scand. fu/kg gain	3.51	3.23

RDM = Red Danish Cattle   FA = Finnish Ayrshire

SRB = Swedish Red and White   MRI= Dutch Red and White

CORRELATED RESPONSE TO SELECTION

In a Canadian selection experiment with Shorthorns (Andersen et al., 1974), eight years of selection for high one -
year weight resulted in an increase in this weight of approximately 40 kg. At the same time, however, the birth weight
rose by approximately 3.5 kg. Calculations showed that if the
selection was based on daily gain during the test period instead of annual weight the correlated increase in birth weight
would almost be halved. Flock et al. (1962); Martin et al.
(1962); Shelby et al. (1963); Brinks et al. (1964); Witt et al.
(1964); Lindström and Maijala (1970) and Lindström (1974) have
also demonstrated a positive correlation between rate of growth
and birth weight.

The genetic determined growth capacity can also influence
the composition of the growth. In experiments, where the
animals were slaughtered at a constant age, increasing growth
capacity and consequent increasing weight at slaughter caused
a small, indirect increase in the degree of fatness and a
slight fall in the percentage of valuable cuts (Shelby et al.,
1963; Averdunk, 1969; Cundiff et al., 1971 and Andersen et al.,
1974. At a constant weight at slaughter, however, increased
growth capacity will result in a reduction of the degree of
fatness and a corresponding increase in the relative content
of lean (Gallagher, 1963; Cunningham and Broderick, 1969;
Dietert et al., 1970; Trapmann, 1972 and Langholz and Jongeling,
1972.).

A genetic improvement of the growth capacity will, as
already discussed lead to a better overall feed efficiency
provided that the animals are fed on the same amount of feed
per day. Koch et al. (1963) have shown that selection for
greater growth capacity will indirectly increase both feed
utilisation and appetite. Selection for better feed utilisation will indirectly increase the growth capacity but not the
appetite. Selection for better appetite will increase the
growth capacity, but according to Koch et al. (1963) it will

not influence the feed utilisation either in a positive or negative direction.

Andersen (1977) has analysed the expected long-term effect of a performance test selection based on different indices. The results are presented in Table 5. In a series of selection indices the traits daily gain, muscle area (ultrasonic) and gestation length are combined in different ways, and it is demonstrated that by the strategy of selection (composition of index) it is possible to regulate the correlated effect on birth weight, mature weight, feed utilisation and carcass composition.

A performance test selection for daily gain carried out to such an extent that the average daily gain will increase by 100 g (intensity of i. = 3.45) causes an indirect increase in birth weight of 3.3 kg, and in mature cow weight of 48 kg. However, it is important to notice that these increases do not affect the ratio between birth weight and cow weight. Dressing percentage and lean : bone ratio will decrease by such a one-sided selection for daily gain.

Selection with the same intensity for ultrasonic muscle area causes a strong improvement in dressing percentage and muscularity, a slight increase in birth weight and a slight decrease in mature cow weight.

As will be apparent from other examples in Table 5 it is also possible to include the length of gestation period for the tested bulls in the index. By restriction on the indices it is possible to keep birth weight and cow weight constant together with a genetic improvement of daily gain, but such indices cause an important decrease in carcass quality. The restricted index including daily gain and muscle area with lean : bone ratio constant demonstrates that it is possible to improve the growth capacity of the breeds without an adverse effect on carcass quality.

TABLE 5

EXPECTED GENETIC EFFECT ON PERFORMANCE TEST SELECTION BASED ON DIFFERENT INDICES

| Index | Daily gain in g of: | | | | | Response in: Scand.fu per: | | L D | D % | L B | L F | LP LT | G L | B W | Mat cow wt. |
	live wt.	carc. wt.	lean	fat	bone	kg gain	kg lean								
DG	100	36	36	-12	10	-0.4	-1.2	-1	-1.5	-0.15	0.7	0.4	1.2	3.3	48
LD	-5	23	30	-2	-4	-0.0	-0.8	11	1.7	0.35	0.4	0.6	-0.3	0.9	-8
DG + LD	52	36	37	-5	3	-0.3	-1.3	9	0.5	0.20	0.7	0.7	0.5	2.7	19
DG + (LD) [1]	90	41	37	-6	9	-0.5	-1.3	4	-0.8	0	0.8	0.6	1.0	3.5	40
DG + GL	90	37	31	-5	10	-0.5	-1.2	-1	-1.0	-0.09	0.7	0.5	5.6	5.3	82
DG + (GL) [2]	75	22	17	-5	9	-0.3	-0.6	0	-1.5	-0.17	0.5	0.1	-4.1	0	-7
DG + LD + (GL) [2]	9	20	24	-3	-2	-0.1	-0.7	10	1.1	0.28	0.4	0.6	-2.7	0	-25
DG + LD + (GL) [3]	77	17	11	-4	10	-0.3	-0.4	-3	-2.0	-0.26	0.4	0	-3.5	0	0

1) Restricted index with lean/bone ratio constant.

2) Restricted index with birth weight constant.

3) Restricted index with birth weight and mature cow weight constant.

s.f.u. = Scand. feed units

LD   = Longissimus dorsi muscle area

D % = Dressing percentage

L/B = Lean/Bone ratio

L/F = Lean/Fat ratio

LP/LT = 100 x lean in pistol/total lean

GL = Gestation length

BW = Birth weight

REFERENCES

Andersen, B.B., 1970. Individprøve for slagtekvalitet. Licientiatafhand-
ling, K.V.L., København. 110 pp.

Andersen, B.B., Fredeen, H.T. and Weiss, G.M. 1974. Correlated response
in birth weight, growth rate and carcass merit under single-trait
selection for yearling weight in Beef Shorthorn Cattle. Can. J. An.
Sci. 54, 117-125.

Andersen, B.B., and Andersen, H.R. 1974. Genotype x environment interaction for beef production traits in dual purpose cattle breeds. Acta
Agric. Scand. 24, 335-338.

Andersen, B.B., 1977. Genetic investigations on growth, body development
and feed utilisation in dual purpose cattle. Diss. Copenhagen. 137 pp.

Averdunk, G., 1969. Ergebnisse und Problematik der Eigenleistungs - und
Nachkommenprüfung auf Fleischleistung beim Rind. Züchtungskunde 41,
152-161.

Berg, R.T. and Butterfield, R.M. 1976. New concepts of cattle growth. Sydney University Press. 240 pp.

Brinks, J.S., Clark, R.T., Keiffer, N.M. and Urich, J.J. 1964. Estimates
of genetic, environmental and phonotypic parameters in range Hereford
females. J.An.Sci. 23, 711-716.

Brānnāng, E. 1968. Growth records of cattle: Methods, expression and
interpretation of results. 19th Annual Meeting EAAP, Dublin.
Mimeographed.

Cundiff, L.V., Gregory, K.E., Koch, R.M. and Dickerson, G.E., 1971. Genetic
relationships among growth and carcass traits of beef cattle. J. An.
Sci. 33, 550-555.

Cunningham, E.P. and Broderick, T. 1969. Genetic and environmental parameters of growth and carcass traits in dual-purpose cattle. Ir. J.
Agric. Res. 8, 397-416.

Dietert, W., Weniger, J.H. and Pfleiderer, U. 1970. Untersuchungen über
verschiedene prüfungsverfahren auf mastleistung und schlachtkörperwert
beim Rind. Züchtungskunde 42, 349-361.

Fimland, E.A., 1973. Estimates of phenotypic and genetic parameters for
growth characteristics of young potential AI bulls. Acta Agr. Scand.
23, 209-216.

Flatnitzer, F., Averdunk, G. and Bogner, H. 1969. Gewichtsermittlung weib-

licher Tiere im Feld im Vergleich zur Stationsprüfung über mannliche
Tiere zur Zuchtwertschätzung von Vätern auf Zuwachsleistung. Bayeri-
sches Landwirtschaftliches Jahrbuch 46, 828-853.

Flock, D.K., Carter, R.C. and Priode, B.M. 1962. Linear body measurements
and other birth observations on beef calves as predictors of pre-
weaning growth rate and weaning type score. J.An.Sci. 21, 651-655.

Gallagher, R.M., 1963. The influence of growth rate on some carcass
characteristics of beef cattle. In Symposium on carcass composition
and appraisal of meat animals (Ed. Tribe, D.E.). CSIRO. Melbourne,
Australia, 9-1- 9-9.

Gravert, H.O., Rosenhahn, E.,Ernst, E. and Feddersen, P. 1971. Entwicklung
und Bedeutung der Nachkommenschaftsprüfungen auf Fleischleistung beim
Rind in Schleswig-Holstein. Züchtungskunde 43, 155-161.

Hinks, C.J.M. and Andersen, B.B. 1969. Genetic aspects of growth and carcass quality in veal calves, Anim. Prod. 11, 43-45.

Koch, R.M., Swiger, L.A.,Chambers, D. and Gregory, K.E. 1963. Efficiency
of feed used in beef cattle. J.An. Sci. 22, 486-494.

Langholz, H.J. and Jongeling, C. 1972. Untersuchungen zum genetischen
Aussagewert der stationären Nachkommenprüfung auf Mastleistung und
Schlachtkörperwert beim Rind. Züchtungskunde 44, 368-384.

Langlet, J.F., Gravert, H.O. and Rosenhahn, E. 1967. Untersuchungen über
die Erblichkeit der Fleischleistung bei schwarzbunten Rindern. Z.
Tierz. Züchtungsbiol. 83, 358-370.

Lindhe, B. and Henningsson, T. 1968. Crossbreeding for beef in Swedish
Red and White cattle. Part II. Growth and efficiency under standardised conditions together with detailed carcass evaluation. Lantbr.
högsk. An. 34, 517-550.

Lindström, U. and Maijala, K. 1970. Evaluation of performance test results
for AI bulls. Acta Agric. Scand. 20, 207-218.

Lindström, U.B., 1974. Points of view on performance testing dual purpose
bulls. Z. Tierz. Züchtungsbiol. 91, 11-21.

Linner, J.L.M., 1973. Beziehung zwischen Merkmalen der Milchleistung und
der Mast- und Schlachtleistung beim deutschen Fleckvieh. Diss.
München 1973. 79 pp.

Martin, T.G., Jacobsen, N.L. and McGilliard, L.D. 1962. Factors related
to weight gain in dairy calves. J. Diary Sci. 45, 886-892.

Preston, T.R. and Willis, M.B. 1970. Intensive beef production. Pergamon
Press, New York, 544 pp.

Rittmannsperger, F., 1966. Schätzung phänotypischer und genetischer Parameter von Masteigenschaften bei Jungbullen des österreichischen Fleck
und Braunviehs. Züchtungskunde 38, 346-353.

Shelby, C.E., Harvey, W.R., Clark, R.T., Queensberry, J.R. and Woodward, R.R.
1963. Estimates of phenotypic and genetic parameters in ten years
of Miles City R.O.P. Steer data. J. An. Sci. 22, 346-353.

Taylor, St. C.S. and Young, G.B. 1968. Equilibrium weight in relation to
food intake and genotype in twin cattle. Anim. Prod. 10, 393-412.

Torreele, G. and Slawinski, T. 1970. A study about the relationship
between growth rate, dressing percentage and carcass composition for
intensively fattened young bulls. Med. Facult. Landbouwwetensch.
Rijksuniversiteit 35, Gent, Belgium. 401-408.

Trappmann, W., 1972. Schätzung phänotypischer und genetischer Parameter
der Fleischleistung von Jungbullen bei Stations-und Feldprüfung.
Züchtungskunde 44, 17-27.

Witt, M., Walter, E. and Rappen, W.H. 1964. Untersuchungen über den Ein-
fluss verschiedener Faktoren auf das Geburtsgewicht und die Beziehung
zwischen dem Geburtsgewicht und dem ½ und 1-Jahresgewicht bei schwartz-
bunten Rindern. Z. Tierz. Züchtungsbiol. 80, 3-24.

EFFICIENCY OF LEAN MEAT PRODUCTION BY DAIRY STEERS

E. Hind

ARC Animal Breeding Research Organisation, Edinburgh, UK.

ABSTRACT

Friesian and Jersey steers were slaughtered at 1, 12, 24, 48 and 72 weeks of age and their carcasses were dissected. They had been fed, ad libitum a standard complete diet and individual food intakes were recorded.

Results are presented for growth of body weight and lean tissue, and total food intake. Gross efficiences of lean meat production are derived and are contrasted with efficiences obtained when allowance is made for the food consumed by the dam.

INTRODUCTION

A large experiment described by Monteiro (1974) involved the slaughter of Friesian and Jersey steers at 24, 48 and 72 weeks of age. The ARC Meat Research Institute (MRI) carried out the steers' slaughter and full carcass dissections. The steers had been fed milk substitute until weaning and from birth were offered, ad libitum, a complete cobbed diet, AA6, described by Wainman et al (1975). Body weights and food intakes were recorded regularly until slaughter. The MRI also slaughtered and dissected the carcasses of several Friesian and Jersey male calves aged one or twelve weeks. The two combined sets of data gave a description of growth from birth to 18 months of age. The numbers of steers in each age group are given in Table 1.

TABLE 1

NUMBERS OF FRIESIAN AND JERSEY STEERS SLAUGHTERED AT FIVE AGES

Age at slaughter (weeks)	Breed	
	Friesian	Jersey
1	10	2
12	9	4
24	17	14
48	24	12
72	21	12
Total	81	44

METHODS

Four periods of growth could be studied. Weights of total lean tissue, transformed by taking logarithms, were subjected to a least squares analysis. Regular body weight and food intake records enabled growth of these variables over each period to be evaluated for all steers alive during that period. Intake of milk substitute and complete diet from birth to weaning were combined on a metabolisable energy basis and expressed as weight of complete diet.

RESULTS

Growth of body weight and total lean tissue, and total food consumed during the growth periods are given in Table 2. Growth of body weight accelerated over the first twelve months of life. In both breeds the average rate of growth from weaning (at 3 months) to 6 months was more than 1½ times that from birth to weaning. Growth rate over the next six months was almost 1½ times that over the previous six months. From 12 to 18 months the increase in weight in the Jerseys was almost the same as that from 6 to 12 months and, in the Friesians, was slightly lower (Table 2).

TABLE 2

GROWTH OF BODY WEIGHT AND LEAN MEAT AND TOTAL FOOD INTAKE OF FRIESIAN AND JERSEY STEERS OVER FIVE AGE INTERVALS

Breed	Age interval (weeks)	Growth of body weight (kg)		Growth of lean (kg)		Food intake (kg)	
		Mean	SE	Mean	SE	Mean	SE
Friesian	Conception to 1	40	0.5	12	0.7	-	-
	1-12	47	1.0	7	1.4	161	2.6
	12-24	72	1.7	25	2.6	372	7.0
	24-48	172	2.2	54	4.4	1353	25.5
	48-72	151	6.5	39	6.9	1724	58.5
Jersey	Conception to 1	25	0.5	6	0.7	-	-
	1-12	27	1.3	7	1.5	101	2.9
	12-24	45	2.1	11	2.0	243	8.7
	24-48	108	4.3	31	3.3	921	36.2
	48-72	104	6.3	29	5.2	1200	49.5

The two breeds were less alike in their accumulation of lean tissue. The Friesians' most rapid acceleration was between weaning and 6 months of age and during this period their rate of growth was more than three times that over the previous three months. Over the next six months the Friesians' average rate of growth was similar to that over the previous three months; their rate of accumulation of lean tissue was considerably reduced over the final six months. The Jerseys seemed to accumulate lean tissue more steadily; their rate of growth increased gradually to a maximum at around 12 months and then gradually declined (Table 2).

The increase in rate of food consumption in the first year of life was similar to that for body weight. In both Friesians and Jerseys intake over the three months after weaning was double that over the first three months of life; from 6 months to 12 months intake was over 2½ times that in the previous six months. Thereafter the rate of growth declined and the increase in food intake also declined (Table 2).

EFFICIENCY OF LEAN MEAT PRODUCTION

High efficiency of lean meat production is always desirable. Efficiency is sometimes defined as the ratio of total lean tissue to total food consumed after birth. For example, at 12 weeks Friesian calves had consumed 152 kg food and accumulated 19 kg lean (Table 2), so that their efficiency was 19/152 = 0.13 kg lean/kg food. As the steers grew their overall efficiency declined rapidly from an infinitely high value at birth (Figures 1a and 1b). This decline in efficiency with age can be explained by the increase in the proportion of food required to maintain body size and the consequent decrease in the proportion used to form lean tissue.

A more meaningful measure of efficiency would be obtained by including the input, in food units, required to produce the new-born calf. For example, the current market cost of a male calf was approximately equivalent to 500 kg complete diet for

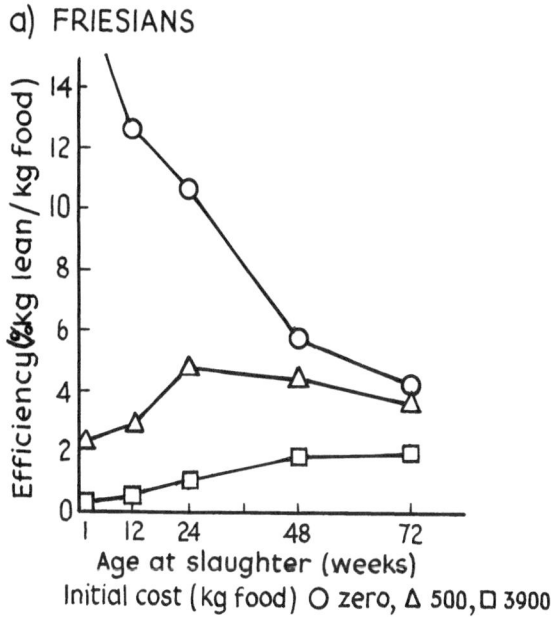

a) FRIESIANS

Efficiency (kg lean / kg food)

Age at slaughter (weeks)

Initial cost (kg food) ○ zero, △ 500, □ 3900

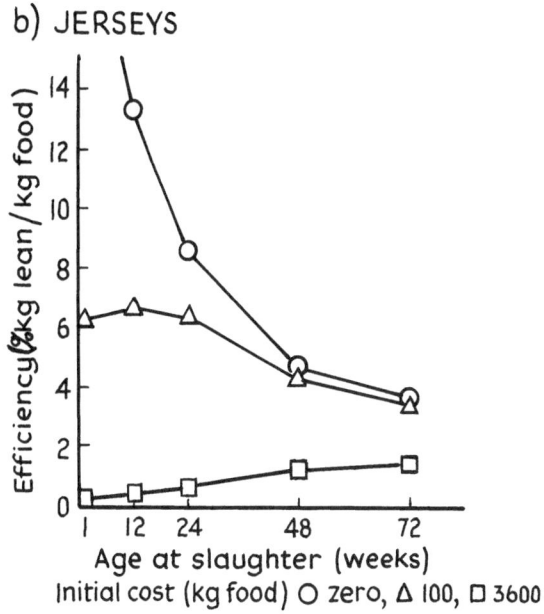

b) JERSEYS

Efficiency (kg lean / kg food)

Age at slaughter (weeks)

Initial cost (kg food) ○ zero, △ 100, □ 3600

Fig. 1. Efficiency of lean meat production of Friesian and Jersey steers
for three levels of initial cost and at five ages.

a Friesian and 100 kg complete diet for a Jersey. Overall efficiency of a 12 week old Friesian calf was then, from Table 2, 19/(500+152) = 0.029 kg lean/kg food. The finite efficiency at birth increased to a maximum at about 6 months after birth for Friesians and at about 3 months for Jerseys (Figures 1a and 1b). Thereafter efficiency fell as the food required for maintenance became increasingly important.

The cost at birth of the calf of a beef-only cow would include the food intake of the cow herself. The closest analogous estimate of dam intake in this experiment was the food consumed by a cow between one calving and the next, after allowing for her milk production. This was approximately 3 900 kg for Friesians and 3 600 kg for Jerseys. At birth the overall efficiency of the calf's lean production was very low and increased until beyond 18 months of age in both breeds (Figures 1a and 1b).

Figures 1a and 1b demonstrate the effect that inclusion of the dam's intake, or some other cost at birth, has on the efficiency of the calf. The initial cost determines not only the average level of overall efficiency but also the way in which the level of efficiency will change throughout life.

REFERENCES

Monteiro, L.S. 1974. Food efficiency in cattle. ARC Anim. Breed. Res. Org.
Ann. Rep. 40-46.

Wainman, F.W., Smith, J.S. and Dewey, P.J.S. 1975. The nutritive value
for sheep of ruminant diet AA6, a complete cobbed diet containing
30% barley straw. J. Agric. Sci., 84, 109-111.

MULTIBREED COMPARISONS OF BODY WEIGHT AND FOOD INTAKE IN CATTLE

R.B. Thiessen

ARC Animal Breeding Research Organisation,
West Mains Road, Edinburgh, EH9 3JQ, UK

SUMMARY

A multibreed cattle experiment has been established by the Animal Research Breeding Organisation with the objective of estimating degrees of genetic variation and genetic inter-relationships between body weight, food intake, milk yield and food efficiency. The design is based on about 30 British breeds with 10 to 12 animals per breed from 2 to 4 sires. The animals were housed indoors and fed a standard complete pelleted diet ad libitum through a system of Calan-Broadbent electronic gates.

Body weight and food intake were measured on 114 animals from 22 breeds over the age range of 12 to 72 weeks. The curves for body weight were approximately linear in form and were quite distinct for the individual breeds. The food intake curves were curvilinear and more variable than those for body weight. There was a strong correspondence in the ranking of the breeds for body weight and food intake.

The proportion of variation between breeds for 12 week averages of body weight and food intake was estimated by the intraclass correlation over 5 age intervals from 12 to 72 weeks of age. For body weight the proportion of variation between breeds increased from 59% at 12 - 24 weeks of age to 72% at 60 - 72 weeks of age. For cumulated food intake the between-breed variation increased from 29% at 12 - 24 weeks to 63% at 60 - 72 weeks.

The relationships between food efficiency, absolute growth rate and relative growth rate (which is absolute growth rate as a percentage of body weight) were estimated over the age range from 26 to 52 weeks. The correlations between relative growth rate and food efficiency, both between and within breeds (0.54 and 0.66) were always higher than those between absolute growth rate and food efficiency (0.47 and 0.50).

REFERENCE

Thiessen, R.B., Hnizdo, Eva, Maxwell, D.A.G., Gibson, D. and Taylor, St.C.S.
    Multibreed comparisons of body weight and food intake in cattle.
    Anim. Prod. (submitted).

INFLUENCE OF AGE, NUTRIENT INTAKE AND BODY TYPE ON WEIGHT GAIN AND BODY COMPOSITION IN YOUNG FATTENING BULLS OF THE BREEDS GERMAN SCHWARZBUNTE AND GERMAN FLECKVIEH

F.W. Huth

Institute of Animal Husbandry and Animal Behaviour Mariensee,
Agricultural Research Centre, Braunschweig-Volkenrode,
3057 Neustadt 1, Federal Republic of Germany

ABSTRACT

The fattening performance of intensively fed young bulls of the breeds German Schwarzbunte and German Fleckvieh was compared. The animals were slaughtered and dissected at 65 and 78 weeks of age.

At both ages, the Fleckvieh bulls weighed on average 100 kg more than the Schwarzbunte bulls, they had a better meat: fat ratio, a significantly higher proportion of hide and a slightly higher bone proportion.

The proportions of the wholesale cuts did not differ significantly.

When slaughtered at the same weight of about 560 kg, the Fleckvieh bulls were on average 13 weeks younger than the Schwarzbunte bulls. Since the body composition of the animals changes with increasing age, differences between the two breeds with respect to the meat : fat ratio and the hide and bone proportions, accordingly became more pronounced.

The deposition of fat, muscle and bone increased in both breeds – when maintained at identical nutrient supply – with increasing growth capacity (daily weight gain). Since all parts of the body grow at about the same rate, no significant differences in relative body composition were observed.

INTRODUCTION

The aim of this study was to compare the fattening performance and the body composition of young bulls of dairy-accentuatec
German Schwarzbunte and beef-accentuated German Fleckvieh cattle
which are the major dual-purpose breeds stocked in the Federal
Republic of Germany. In order to allow for a differentiation
between the influences exerted by breed and age, emphasis was
placed on the results obtained from slaughtering by weight or
by age. The experimental procedure chosen, slaughtering after
various feeding regimens, either by weight or by age, aimed at
the differentiation between the possible influences exerted by
the breed and by the age.

MATERIAL AND METHODS

In two series of experiments, young fattening bulls of the
breeds German Schwarzbunte and German Fleckvieh, fed individually
during the stable fattening period, were maintained on different
nutrient supply levels and slaughtered at different ages.
Besides milk, the animals received concentrate, dry sugar beet
pulp and hay. The experiment was commenced at week 14, after
an identical feeding for all animals lasting from the 3rd to
the 13th week of life.

In the first set of experiments, 443 animals of the breed
German Schwarzbunte were kept at three levels of nutrient supply

Level a) Body weight (kg) x 10 $\pm$ 0 = SE supply
(e.g. body weight 300 kg = 3 000 SE/day)
- restricted feeding

Level b) Body weight (kg) x 10 + 300 = SE supply
(3 300 SE/day) - medium feeding

Level c) Body weight (kg) x 10 + 600 = SE supply
(3 600 SE/day) - intensive feeding.

Of each group, one half of the animals was slaughtered at
week 65 and the second half at week 78 of life.

In the second set of experiments, 144 bulls of the German Schwarzbunte and 68 bulls of the German Fleckvieh breed were fed intensively (level c) from week 14 until slaughter at either the age of 65 or of 78 weeks.

The carcass sides were dissected as outlined in Figure 1.

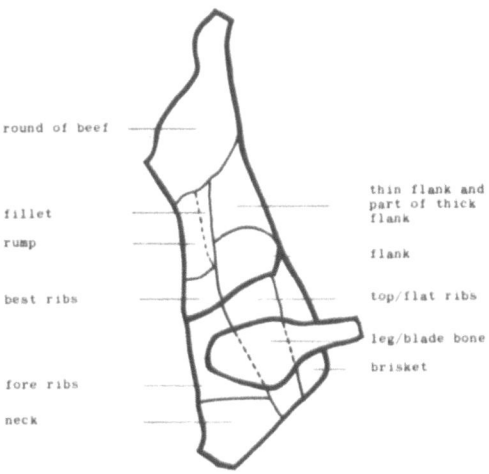

Fig. 1: Wholesale cuts of beef

The extent of fat deposition was evaluated from the fat content (%) of the three-rib-piece and the weights of the abdominal fat deposits (kidney, pelvis, rumen and small intestines).

The data obtained from both breeds on characteristics of fattening performance and carcass composition are listed together in Tables 1a (absolute values) and 1b (relative values) according to :

(A) The difference in age at slaughter (65 resp. 78 weeks) after intensive feeding;

(B) The differences in nutrient intake and subsequent slaughter at the same age;

TABLE 1a

FATTENING PERFORMANCE AND BODY COMPOSITION OF YOUNG FATTENING BULLS (GERM.
SCHWARZBUNTE AND GERM. FLECKVIEH) WITH DIFFERENCES IN AGE AT SLAUGHTER (A)
NUTRIENT INTAKE (B) AND BODY TYPE (C)

		A Age (weeks)		B* Nutrient intake restr. med. inten-sive			C body type (daily gain) low medium high		
		65	78	restr.	med.	intensive	low	medium	high
n		109	103	149	150	144	50	109	53
Total kSE-intake		1416	1969	1123	1373	1552	1559	1697	1776
Total DCP	kg	319	423	265	316	355	347	371	388
Weight: end of fattening period		497	595	443	485	512	486	546	599
Daily gain from 3rd week of life	g	1013	1028	821	904	965	912	1020	1123
Total carcass	kg	292	357	254	283	302	287	324	358
Carcass dressing	%	61.6	62.8	59.9	61.2	61.8	61.8	62.2	62.3
Head with tongue	kg	16.8	19.0	16.2	17.1	17.4	16.8	17.7	19.3
Hide	kg	44.3	50.8	38.6	40.9	41.7	42.9	47.6	51.3
4-feet-weight	kg	9.0	10.1	8.0	8.4	8.6	8.9	9.5	10.1
Fat in 3-rib piece	%	22.3	25.5	18.3	22.3	25.3	24.0	24.2	24.6
Abdominal cavity fat	kg	22.0	36.4	14.9	21.8	28.9	25.2	29.2	32.4
Heart	kg	2.0	2.3	1.7	1.9	2.0	2.0	2.1	2.3
Liver	kg	5.9	6.8	5.3	5.9	6.2	5.7	6.3	6.9
Fillet and roast beef	kg	10.3	12.5	8.9	9.9	10.6	10.2	11.3	12.3
Best ribs	kg	18.3	22.7	16.0	18.0	19.1	18.0	20.6	22.6
Neck	kg	11.4	14.1	10.2	11.3	11.8	11.0	12.6	14.0
Brisket and top ribs	kg	17.2	21.9	14.4	16.5	18.0	17.2	19.5	21.6
Leg	kg	22.8	27.0	20.4	22.4	23.4	22.5	24.9	27.3
Thin flank	kg	14.2	18.7	10.8	13.1	14.8	14.3	16.3	18.4
Round of beef	kg	44.9	52.1	40.3	43.4	45.2	43.4	48.5	53.1
Total carcass side	kg	139.1	169.0	121.0	134.6	142.9	136.6	153.7	169.3

* only German Schwarzbunte

(C) The difference in body type, derived from the daily weight gain of intensively fed animals, slaughtered at the same age.

RESULTS

Table 1a shows that the characteristics: weight at slaughter, dressing, fat deposition, weight of organs and wholesale cuts increase - as expected - both with age (A) and nutrient supply (B). Growth capacity (C) remarkably precipitates an overall effect, meaning that the weight gain is not restricted to the muscles, but is also evident in the skeleton, the total body fat, the weight of the organs and all other parts of the body, some of which increased at a linear rate.

TABLE 1b

THE RELATIVE VALUE OF BODY COMPOSITION OF YOUNG FATTENING BULLS (GERM. SCHWARZBUNTE AND GERM. FLECKVIEH) WITH DIFFERENCES IN AGE AT SLAUGHTER (A), NUTRIENT INTAKE (B) AND BODY TYPE (C)

	A age (weeks) 65 78	B * nutrient intake restr. med. inten- sive	C body type (daily gain) low med. high
	n 109 103	149 150 144	50 109 53
% of total carcass:			
Head with tongue	5.8 5.3	6.4 6.0 5.8	5.9 5.5 5.4
Hide	15.2 14.2	15.2 14.5 13.8	14.9 14.7 14.3
4-feet-weight	3.1 2.8	3.1 3.0 2.8	3.1 2.9 2.8
Abdom. cavity fat	7.5 10.2	5.9 7.7 9.6	8.8 9.0 9.1
Carcass side = 100:			
Fillet and roastb.	7.4 7.4	7.4 7.4 7.4	7.5 7.4 7.3
Best ribs	13.2 13.4	13.2 13.4 13.4	13.2 13.4 13.3
Neck	8.2 8.3	8.4 8.4 8.3	8.1 8.2 8.3
Brisket and top ribs	12.4 13.0	11.9 12.3 12.6	12.6 12.7 12.8
Leg	16.4 16.0	16.9 16.6 16.4	16.5 16.2 16.1
Thin flank	10.2 11.1	8.9 9.7 10.4	10.4 10.6 10.9
Round of beef	32.3 30.8	33.3 32.2 31.6	31.8 31.6 31.4

* only German Schwarzbunte

The comparison of the relative carcass composition (Table 1b) with reference to age at slaughter (A), nutrient intake (B) and daily weight gain (C) shows, that in each case the proportion of dressing including head with tongue, hide and 4-feet decreased while that of abdominal fat increased.

In the older and in the more intensively fed animals, the proportion of the round of beef (thigh) and to a lesser extent that of the legs (blade bones) decrease, while the relative values for the brisket, the top ribs and the thin flanks increase The changes in carcass composition of animals slaughtered at the same weight become more prominent with age.

A similar tendency, but to a much lesser degree was observed in the animals with enhanced growth capacity (C). The changes were too small to allow for an unequivocal assessment, but it can be said, that enhanced growth capacity does not result in proportional weight increase of the round of beef.

In Tables 2a (absolute values) and 2b (relative values), the fattening performance and the carcass composition of young fattening bulls of the breeds German Schwarzbunte and German Fleckvieh are compared. The differences between the two breeds are not age-dependent.

Fleckvieh bulls either fed 80 kg and slaughtered at week 65 or fed 88 kg and slaughtered at week 78, weighed on average 100 kg more than Schwarzbunte bulls fed and slaughtered correspondingly (Table 2a).

The differences in hide weight are remarkable. Although the weights of abdominal cavity fat were virtually identical in both breeds, the German Fleckvieh bulls had on average a 3.6 % lower fat content in the three-rib-piece. Due to the enhanced growth capacity of Fleckvieh bulls, organ and wholesale cut weights markedly surpassed those of the Schwarzbunte bulls.

After intensive feeding, Fleckvieh bulls weighed on average

TABLE 2a

A COMPARISON OF FATTENING PERFORMANCE AND BODY COMPOSITION OF YOUNG
FATTENING BULLS OF THE BREEDS GERMAN SCHWARZBUNTE AND GERMAN FLECKVIEH AT
THE SAME AGE AT SLAUGHTER AND THE SAME WEIGHT AT SLAUGHTER

		German Schwarzb.	German Fleckv.	German Schwarzb.	German Fleckv.
	n	76	33	68	35
Age (weeks)		65		78	
Total kSE-intake		1314	1651	1817	2265
Total DCP	kg	308	345	408	453
Gain	kg	423	503	517	605
Weight : end of fattening period	kg	468	563	561	662
Daily gain from 3rd week of life	g	959	1139	972	1137
Total carcass	kg	273	336	334	402
Carcass dressing	%	61.2	62.4	62.5	63.4
Head with tongue	kg	16.3	17.9	18.6	19.9
Hide	kg	39.2	56.0	44.5	62.9
4-feet-weight	kg	8.2	10.8	9.1	12.0
Fat in 3-rib piece	%	23.7	20.4	27.1	23.2
Abdominal cavity fat	kg	22.2	21.6	36.3	36.5
Heart	kg	1.9	2.2	2.1	2.6
Liver	kg	5.8	6.1	6.6	7.1
Fillet and roast beef	kg	9.7	11.6	11.7	14.0
Best ribs	kg	17.2	20.9	21.2	25.6
Neck	kg	10.6	13.1	13.2	15.9
Brisket and top ribs	kg	15.8	20.4	20.4	24.9
Leg	kg	21.6	25.5	25.4	30.0
Thin flank	kg	12.9	17.1	17.0	21.9
Round of beef	kg	42.1	51.5	48.6	59.0
Total carcass side	kg	129.9	160.1	157.5	191.3

563 kg at 65 weeks of age, while Schwarzbunte needed 78 weeks to attain a 561 kg weight average. In the 13 weeks older Schwarzbunte bulls, the fat content of the three-rib-piece was 3.1 % higher and they contained 15.3 kg more abdominal cavity fat than the Fleckvieh bulls. The wholesale cuts, with exception of the round of beef, showed no marked difference.

TABLE 2b

A COMPARISON OF THE RELATIVE VALUE OF BODY COMPOSITION OF YOUNG FATTENING BULLS OF THE BREEDS GERMAN SCHWARZBUNTE AND GERMAN FLECKVIEH AT THE SAME AGE AT SLAUGHTER AND THE SAME WEIGHT AT SLAUGHTER

	Germ. Schwarzb.	German Fleckv.	German Schwarzb.	German Fleckv.
n	76	33	68	35
Age (weeks)	65		78	
% of total carcass:				
Head with tongue	6.0	5.3	5.6	5.0
Hide	14.4	16.7	13.3	15.6
4-feet-weight	3.0	3.2	2.7	3.0
Abdominal cavity fat	8.1	6.4	10.9	9.1
Carcass side = 100:				
Fillet and roastbeef	7.5	7.2	7.4	7.3
Best ribs	13.2	13.1	13.5	13.4
Neck	8.2	8.2	8.4	8.3
Brisket and top ribs	12.2	12.7	13.0	13.0
Leg	16.6	15.9	16.1	15.7
Thin flank	9.9	10.7	10.8	11.5
Round of beef	32.4	32.2	30.9	30.8

Similar results were obtained when both breeds were fed with a total of 1 800 kSE /animal. Fleckvieh bulls gained in 68 weeks 12 kg more weight than did Schwarzbunte bulls in 78 weeks. This also indicates that the differences in body composition are age-dependent.

Fat content and wholesale cut weights are the most interesting parameters for a performance comparison of the two breeds, with respect to age and nutrient supply.

Table 2b shows that in both age groups, the Fleckvieh bulls surpassed the Schwarzbunte bulls in the meat : fat ratio to some extent in the 4-feet-weight and significantly so in the weight of the hide.   In contrast the wholesale cut weights were not markedly different.

When slaughtered at the same body weight, differences between the two breeds in proportional fat deposition, 4-feet- and hide-weight are observed, the relative values of carcass side wholesale cuts differ little with exception of the round of beef value, which is increased in Fleckvieh.

The differences in body composition between the final fattening weight and the age at slaughter are primarily age-related and do not depend on the breed.

CARCASS COMPOSITION OF DIFFERENT BREEDS

S. Kögel and H. Alps

Bayerische Landesanstalt für Tierzucht,
8011 Grub, Federal Republic of Germany

ABSTRACT

Characteristics of 1028 dissected carcasses of Simmental, German Braunvieh and Gelbvieh bulls as well as crossbred Simmental and Braunvieh bulls are shown. Relations and ratios of muscle, fat and bone tissues are calculated.

The more beefy typed dual purpose breeds and crossbreds with Limousin have a higher percentage of muscle and less fat and bone compared with Friesian x Simmental and Brown Swiss x Braunvieh. Simmental bulls with heavier carcasses show lower proportions of high priced cuts (loin, round) than bulls with lower carcass weights.

INTRODUCTION

Since the main aim of beef production is to produce red
meat with a certain amount of fat, the development of muscle
and fat tissues is most important for the classification of
cattle or a breed according to their suitability for beef production. From the results of Berg and Butterfield (1976) it is
evident that there is little chance of changing muscle distribution by selection. Therefore we should concentrate our
attention on improving the total amount of muscle on the carcass without producing an extra amount of fat. In addition to
sex and nutritional factors the genetic potential has an important effect on carcass composition.

MATERIAL

The material available for this evaluation originates from
two State experimental farms. All bulls were randomly sampled
out of progeny test groups or from different trials where the
crossbred bulls have been compared with Simmental bulls, with
the exception of the special trial with German Braunvieh and
Brown Swiss crossbreds (Table 4). All bulls were fed corn silage and a small amount of concentrate with soyabean meal as
protein supplement. The animals were slaughtered and dissected
at the Bayerische Landesanstalt für Tierzucht Grub.

RESULTS

Comparison of different genetic groups

In Table 1 the number of dissected bulls, the averages of
age at slaughter, carcass weight and net gain are summarised.
The results are calculated on non-corrected data, since the
differences in age and carcass weight are not very large. The
purebred bulls were slaughtered over a period of years, the
crossbred bulls over shorter periods. The percentage of muscle,
fat and bone tissues of the different breeds is shown in Table
2. With 72.3% muscle Braunvieh and Limousin x Simmental cross-

TABLE 1

AGE, CARCASS WEIGHT AND WEIGHT OF CARCASS PER DAY OF AGE OF THE
DISSECTED BULL CARCASSES OF DIFFERENT BREEDS

Breed	No. of bulls	Age at slaughter (days)	Hot carcass weight (kg)	Hot carcass weight per day of age (g)
Simmental	827	502	333.5	664
Braunvieh	69	503	302.7	602
Gelbvieh	27	522	334.8	641
Simmental x Pinzgauer	35	501	347.8	694
Simmental x G. Friesians	14	549	327.2	596
Limousin x Simmental	18	501	363.5	726
Red Holst.Fries. x Simmental	38	501	328.8	656

TABLE 2

MUSCLE, FAT AND BONE TISSUES AS A PERCENTAGE OF THE BULL CARCASSES
OF DIFFERENT BREEDS

Breed*	Muscle %	SD	Fat %	SD	Bone %	SD
Simmental	71.4	2.1	10.3	2.4	15.3	1.0
Braunvieh	72.3	1.9	8.7	1.9	15.6	0.8
Gelbvieh	71.5	2.4	9.6	2.7	15.8	1.1
Simmental x Pinzgauer	70.6	2.5	12.0	3.1	14.9	1.2
Simmental x G. Friesians	70.2	1.2	11.4	1.8	15.9	1.0
Limousin x Simmental	72.3	1.5	11.5	1.8	14.2	0.9
Red Holst.Fries. x Simmental	68.5	2.6	12.9	3.0	16.2	1.2

* No. of bulls see Table 1

TABLE 3

MUSCLE : BONE, MUSCLE : FAT AND FAT : BONE RATIOS OF BULL CARCASSES
OF DIFFERENT BREEDS

Breed*	Muscle : bone ratio	Muscle : fat ratio	Fat : bone ratio
Simmental	4.65	6.90	0.63
Braunvieh	4.63	8.26	0.54
Gelbvieh	4.52	7.35	0.57
Simmental x Pinzgauer	4.72	5.85	0.74
Simmental x G. Friesians	4.41	6.14	0.69
Limousin x Simmental	5.10	6.29	0.78
Red Holst.Fries. x Simmental	4.22	5.26	0.75

* No. of bulls see Table 1

TABLE 4

COMPARISON OF CARCASS CHARACTERISTICS OF BRAUNVIEH AND CROSSBRED
BULLS (3/4 BROWN SWISS, 1/4 BRAUNVIEH)

	Braunvieh (BV)	3/3 Brown Swiss (BS)	Difference BV-BS
Number	19	19	–
Age at slaughter (days)	459	463	-4
Hot carcass weight (kg)	310.0	298.5	+11.5
Hot carcass weight per day of age (g)	675	645	+30
% muscle (LSQ-Mean)	70.6	69.3	+1.3*
% fat (LSQ-Mean)	10.9	11.1	-0.2
% bone (LSQ-Mean)	15.9	16.9	-1.0**
Muscle : bone ratio	4.44	4.10	+0.34
Muscle : fat ratio	6.48	6.24	+0.24
Fat : bone ratio	0.69	0.66	+0.03

* $p \leq 5\%$

** $p \leq 1\%$

breds are the best breeds in this respect. Braunvieh also
shows the lowest amount of fat tissue (8.7%). The highest
percentage of fat and bone was found in Red Holstein Friesians
x Simmental crossbreds. These animals have a more flat shape
of the muscular system due to the different type of the Holstein Friesian breed. Limousin crossbreds have some advantages
in muscle development, at the same time they also have the
lowest percentage of bones. Thus the muscle : bone ratio of
the Limousin x Simmental bulls is the highest of all the genetic groups as indicated in Table 3. The muscle : fat ratio
varies more than the muscle : bone ratio due to the greater
differences in the percentage of fat tissue. The highest
ratio of muscle : fat was found with Braunvieh followed by
Gelbvieh and Simmental. The fat : bone ratio varies mainly
according to the differences of the percentage of fat tissue.

In Table 4 carcass characteristics of Braunvieh bulls are
compared with crossbred bulls of the combination 3/4 Brown
Swiss and 1/4 Braunvieh (Alps et al., 1976). The differences
show that the carcass of the more beefy typed German Braunvieh
has a slightly better weight gain and produces more muscle and
less bone tissues than the more dairy typed crossbred with
Brown Swiss.

The correlations between muscle and fat (Table 5) could
be an indicator of the rate of maturity, because the development of fat in relation to the other tissues increases faster
after a certain stage of maturity than the growth of muscles
(Berg and Butterfield, 1976). The negative correlation between
muscle and bone with Gelbvieh and Simmental x Pinzgauer bulls
indicate that they could be fattened to a higher weight without producing carcasses with too much fat. There are relatively
close correlations between muscle and bone for all breeds,
therefore the relations fat : bone cannot be very different
from those of muscle : fat.

Comparison between different slaughter weights

Comparisons of carcass characteristics of Simmental bulls

520

TABLE 5

CORRELATIONS BETWEEN MUSCLE : FAT, MUSCLE : BONE AND FAT : BONE TISSUES
OF BULL CARCASSES OF DIFFERENT BREEDS BASED ON NON-CORRECTED DATA

Breed*	Muscle : fat r	Muscle : bone r	fat : bone r
Simmental	0.38	0.73	0.28
Braunvieh	0.16	0.72	0.20
Gelbvieh	-0.11	0.53	-0.15
Simmental x Pinzgauer	-0.07	0.86	-0.17
Simmental x German Friesians	0.61	0.73	0.15
Limousin x Simmental	0.58	0.81	0.27
Red Holst. Fries. x Simmental	0.32	0.69	0.16

* No. of bulls see Table 1

TABLE 6

COMPARISON OF CARCASS CHARACTERISTICS OF SIMMENTAL BULLS AT DIFFERENT
WEIGHTS

Weight at slaughter (kg) Number	400-499 149	500-599 338	600-699 39
Weight of half carcass (kg)	143.4	166.0	193.9
% muscle	71.2	71.2	71.2
% fat	9.8	10.4	11.4
% bone	15.7	15.3	14.8
Muscle : bone ratio	4.54	4.65	4.80
Muscle : fat ratio	7.27	6.85	6.24
Fat : bone ratio	0.62	0.68	0.77
Percentage of carcass:			
Neck and fore rib	16.8	17.4	18.1
Shoulder and fore shank	16.0	15.8	15.6
Flat rib and brisket	12.1	12.0	11.9
Loin	7.9	7.8	7.7
Flank	9.5	9.8	10.3
Round and hind shank	32.1	31.4	30.5
Fillet	2.2	2.2	2.2
Kidney and pelvic fat	2.4	2.5	2.8

at different carcass weights show that the heavier bulls have
a higher percentage of fat and a lower one of bone (Table 6).
The development of different parts of the carcass (Schon, 1976)
shows an increase in the relative growth of the neck and fore
rib, the flank, and the kidney and pelvic fat, whereas the high
priced cuts (round, loin) decrease relatively. This confirms
the results of Berg and Butterfield (1976) and Alps et al. (1977)
who found a negative genetic correlation (r = -0.36) between
percentage of pistol and weight at 500 days with Simmental
progeny test groups.

CONCLUSIONS

Bulls of the more beefy typed dual purpose breeds (Simmental, German Braunvieh, Gelbvieh) and Limousin x Simmental crossbreds show better results concerning the percentage of muscle
and have a lower amount of fat than Simmental crossbreds with
more dairy typed breeds. This should be considered in breeding
programmes with single or continued crossing of these different
breeds. Therefore the improvement of the percentage of muscle
in beef and dual purpose breeds should be part of the breeding
aims; but the experiences with some specialised beef breeds
show that there is a negative correlation between muscling and
calving difficulties if a too high degree of muscling is reached.

REFERENCES

Alps, H., Kögel, S., Mittelstädt, W. and Gottschalk, A. 1976. Mast - und
    Schlachtleistung von Bullen mit 75% Brown Swiss - Genanteil im Vergleich
    zu Deutschem Braunvieh. Allg. Bauernbl. <u>44</u>, (48), 1917-1920.

Alps, H., Averdunk, G., Matzke, P. and Kögel, S. 1977. Zur Frage der
    Schlachtkörperbewertung in der Nachkommenprüfung auf Station beim Rind.
    Referat: Tagung der DGfZ und der Gesellschaft für Tierzuchtwissenschaft,
    Bonn.

Berg, R.T. and Butterfield, R.M. 1976. New Concepts of Cattle Growth.
    Sydney University Press.

Schön, I. 1976. Wholesale and retail jointing, techniques and standard-
    isation for experimental purposes. In 'Criteria and Methods for
    Assessment of Carcass and Meat Characteristics in Beef Production
    Experiments'. EEC Seminar, Zeist, 1975. Luxembourg, EUR 5489, 171-178.

GROWTH RATE AND CARCASS COMPOSITION OF WATER BUFFALO CALVES AND BOVINE CALVES SLAUGHTERED AT 20, 28 AND 36 WEEKS

A. Romita, A. Borghese, S. Gigli, A.Di Giacomo.
Istituto Sperimentale per la Zootecnia. Rome - Italy.

ABSTRACT

The trial was carried out on 30 bovine male calves and 30 water buffalo male calves fed milk replacers and concentrate, subdivided into three groups and slaughtered at 20, 28 and 36 weeks.

Live weight and carcass components are higher in calves and the difference increases with age. Meat percentage of carcass is higher in water buffaloes, while fat and bone are higher in calves. The relative growth coefficients for meat, bone and fat on carcass respectively for calves and water buffaloes are as follows: 0.97 $\pm$ 0.03, 0.73 $\pm$ 0.04, 2.11 $\pm$ 0.20; 0.96 $\pm$ 0.05, 0.70 $\pm$ 0.07, 2.36 $\pm$ 0.25.

The relative growth of some muscles on total meat varies between species.

INTRODUCTION

In Italy the commercial value of water buffalo (*Bubalus bubalis*) meat has a negative influence on breeding of this species that otherwise could compete with cattle for its higher rusticity and lower quantity of feeding requirements. In addition the disease resistance and the lower replacement of the herd (10%) due to its higher longevity allow more animals to be reared for meat production. To gain more information about this species, it seemed to us necessary to compare the growth rate and carcass characteristics of bovine calves and water buffalo calves reared under identical condition of feeding and environment and slaughter ed at different ages.

Comparison of live-weight gain and carcass composition were made by Charles and Johnson (1972) between water buffaloes slaughtered at 14 - 18 months of age and Brahman, Shorthorn and Hereford; by Johnson and Charles (1975) between water buffalo steers and Aberdeen Angus, Friesian and Hereford steers follow-ing similar periods of lot feeding; by Joksimovic and Ognjanovic (1977) between ten water buffalos and six young Busha bulls (14 - 18 months old); and by Romita et al., (1976, 1977_1, 1977_2) between calves and water buffaloes slaughtered at 20, 28 and 36 weeks.

The relative growth patterns of different muscles and organs has already been studied by different authors on rabbits (Cantier et al., 1969; Vezinhet et al., 1972); on sheep (Benevent 1971, Lohse et al., 1971, Boccard and Dumont, 1973); on sheep, cattle and pigs (Tulloh, 1963), on English cattle (Butterfield and Berg, 1966_2; Seebeck and Tulloh, 1968_2, Mukhoty and Berg, 1971), on French cattle (Robelin et al., 1974; Robelin and Geay, 1976); on Italian cattle (Lanari, 1972), but we have not found data regarding comparison among young buffaloes and cattle reared under identical environmental conditions.

MATERIAL AND METHODS

The trial was carried out on 30 bovine male calves and 30
water buffalo male. calves individually fed the same quantity
of milk replacers and concentrates (Romita et al., 1976; 1977_1,
1977_2) and reared in the same barn. Each couple water buffalo -
bovine were of the same age and were located in contiguous pens
and slaughtered on the same day; 10 animals per species were
slaughtered at 20 - 28 and 36 weeks.

After slaughtering, carcasses were chilled for 9 days at
$3-4^\circ$ C, then half carcass was divided in forequarter, loin,
flank and round as shown in Figure 1. All the carcasses were
dissected into meat, bone and fat and then some muscles: pesce
(supraspinatus), filetto (psoas major), girello (semitendinosus),
controgirello (biceps femoris+gluteo biceps); muscle groups:
noce (semimembranosus, adductor, gracilis, pectineus), rosa
(vastus lateralis, rectus femoris, vastus medialis, vastus
intermedius), pezza (gluteus medius, gluteus accessorius); bones:
scapula, humerus, os femoris, tibia-fibula, were weighed. The
Huxley equation (1932) was used to analyse relative growth of
tissues: in fact describing relative growth in terms of allometric relationship, where proportionate increase is the criterion, is probably more satisfactory (Berg and Butterfield,
1968). The independent variables used were; net live weight,
carcass weight, total muscle weight and total bone weight.

RESULTS

In Table 1 the characteristics of the live animals and of
the carcasses are reported. All the differences in the reported
parameters, with some exceptions in the first age, are significant
and increase with the age. Calves have more meat, fat and bone
due to their higher live weight, but the meat percentage in water
buffaloes is higher, while the fat and bone percentage is lower.
(Table 1a).

In Table 2 the coefficients of relative growth of some

TABLE 1.

NET LIVE WEIGHT, CARCASS COMPOSITION

		20 weeks		28 weeks		36 weeks	
		Calves Mean	Buffaloes Mean	Calves Mean	Buffaloes Mean	Calves Mean	Buffaloes Mean
Net live weight	kg	145.77	141.00	201.70	185.33*	259.91	210.06***
Cold carcass	kg	87.40	78.00**	118.41	101.46***	151.28	114.59****
Total meat	kg	56.80	52.80	75.94	67.36**	95.70	74.94***
Total bone	kg	23.00	19.40***	27.45	22.33****	34.26	25.16****
Total fat	kg	7.60	5.80**	15.03	11.77*	21.32	14.49**

TABLE 1a

DRESSING PERCENTAGE AND TISSUES PERCENTAGES ON CARCASS

Dressing	%	63.70	59.60**	62.20	58.25**	61.24	57.6***
Meat	%	65.00	67.60**	64.13	66.39**	63.26	65.40*
Bone	%	26.40	24.90**	23.18	22.01	22.65	21.96
Fat	%	8.60	7.50	12.69	11.60	14.09	12.64

body components on net live weight are reported. Both in calves
and in water buffaloes, higher impetus of growth is shown by
fat tissue followed by meat and bone. For the components of the
fifth quarter, the highest coefficient of relative growth is
the one of the skin, especially in water buffaloes, but the only
significant difference is found in liver. The main tissues,
if we consider as independent variable the carcass (Table 3),
show the same trend.

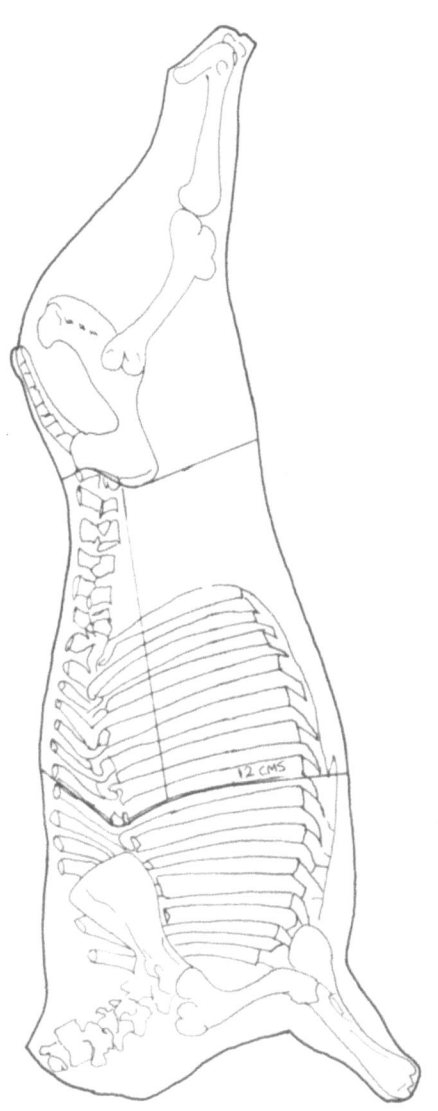

Fig. 1.

TABLE 2

RELATIVE GROWTH RATE OF SOME COMPONENTS (Y) OF THE BODY ON NET LIVE WEIGHT (X)

	Calves		Buffaloes	
	b ± sb	R^2	b ± sb	R^2
Carcass	0.95 ± 0.02	0.99	0.91 ± 0.05	0.92
Total meat	0.93 ± 0.03	0.97	0.91 ± 0.05	0.93
Total bone	0.70 ± 0.04	0.90	0.67 ± 0.06	0.82
Total fat	2.04 ± 0.18	0.82	2.37 ± 0.18	0.87
Kidney	0.83 ± 0.07	0.84	0.84 ± 0.17	0.76
Heart, spleen lungs	0.70 ± 0.06	0.82	0.65 ± 0.08	0.70
Liver	0.81 ± 0.09	0.75	0.49 ± 0.09	0.50*
Skin	1.03 ± 0.04	0.96	1.44 ± 0.29	0.68
Head	0.81 ± 0.03	0.97	0.90 ± 0.04	0.95
Ant.Shin bone	0.70 ± 0.06	0.83	0.59 ± 0.07	0.74
Post. Shin bone	0.63 ± 0.05	0.87	0.68 ± 0.07	0.76

TABLE 3

RELATIVE GROWTH RATE OF MEAT, BONE AND FAT ON CARCASS

	Calves		Buffaloes	
	b ± sb	R^2	b ± sb	R^2
Meat	0.97 ± 0.02	0.99	0.96 ± 0.05	0.92
Bone	0.73 ± 0.04	0.91	0.70 ± 0.07	0.79
Fat	2.11 ± 0.20	0.81	2.36 ± 0.25	0.77

Loin has the highest rate of growth, round the lowest, while flank and forequarter meat have intermediate values both in calves and water buffaloes (Table 4).

The relative growth of muscle groups is different between species (Table 5) except for 'filetto' but significant differences are found only in 'noce' and 'rosa' and 'pesce'.

The relative growth of bones is higher for scapula without significant differences between species (Table 6).

TABLE 4

RELATIVE GROWTH RATE OF MEAT OF DIFFERENT REGIONS ON TOTAL LEAN MEAT

	Calves		Buffaloes	
	b $\pm$ sb	R^2	b $\pm$ sb	R^2
Forequarter	1.08 $\pm$ 0.03	0.98	1.04 $\pm$ 0.04	0.96
Loin	1.45 $\pm$ 0.09	0.90	1.73 $\pm$ 0.16	0.81
Flank	1.06 $\pm$ 0.15	0.77	0.93 $\pm$ 0.12	0.74
Round	0.90 $\pm$ 0.02	0.99	0.92 $\pm$ 0.03	0.97

TABLE 5

RELATIVE GROWTH RATE OF SOME MUSCLES ON TOTAL LEAN MEAT

	Calves		Buffaloes	
	b $\pm$ sb	R^2	b $\pm$ sb	R^2
Pesce	1.03 $\pm$ 0.06	0.92	0.70 $\pm$ 0.12	0.77**
Filetto	1.05 $\pm$ 0.07	0.89	1.01 $\pm$ 0.08	0.86
Noce	0.82 $\pm$ 0.03	0.96	0.95 $\pm$ 0.05	0.92*
Rosa	0.87 $\pm$ 0.03	0.97	0.74 $\pm$ 0.06	0.86*
Pezza	0.89 $\pm$ 0.03	0.96	0.79 $\pm$ 0.05	0.89
Controgirello	0.92 $\pm$ 0.03	0.97	0.86 $\pm$ 0.03	0.96
Girello	0.95 $\pm$ 0.06	0.90	0.82 $\pm$ 0.08	0.82

TABLE 6

RELATIVE GROWTH RATE OF SOME BONES ON TOTAL BONE.

	Calves		Buffaloes	
	b $\pm$ sb	R^2	b $\pm$ sb	R^2
Scapula	1.31 $\pm$ 0.08	0.91	1.18 $\pm$ 0.10	0.85
Humerus	1.17 $\pm$ 0.07	0.92	1.16 $\pm$ 0.09	0.87
Os femoris	1.03 $\pm$ 0.06	0.93	1.06 $\pm$ 0.07	0.91
Tibia-fibula	1.07 $\pm$ 0.05	0.94	1.11 $\pm$ 0.06	0.92

CONCLUSION

Over the growth period represented in the present study
net weight gain is higher in calves but we think that much can
be done for buffaloes in improving the feeding by a better
knowledge of their specific requirements, eg by changing the
fat/protein ratio in milk replacers (Romita and Dias Da Silva).
Meat characteristics of water buffalo carcasses are better than
calves as shown by Charles and Johnson (1972).

As this trial is a short term one it is impossible to give
a clear picture of the developmental patterns of muscle, fat
and bone; for the same reason, the comparison between species
appears to be incomplete. Thus, we intend to go on with this study
through to mature age of the animals.

The coefficient of relative growth of carcass on net live
weight is slightly lower than the value found by Robelin et
al. (1974), probably because of the young age of the animals
considered and of the importance that they still have some compo
nents of the fifth quarter as is also shown in the data of
dressing percentage reported in Table 1a.

The order of the relative growth coefficients of the main
tissues both on net live weight and on carcass weight is the
same as found by the authors   cited in the Introduction.

The allometric coefficient of fat is very high and we
think it is due to the high plane of nutrition that has often
been shown to increase the proportion of fat in the carcass
(Callow, 1961, Henrickson et al., 1965, Meyer et al., 1965);
Elsley (1963$_1$, 1963$_2$) found that higher levels of feeding led
to precocious fat development.

The relative growth coefficient of meat and bone are
similar between species, but fat deposition is higher in water
buffaloes.

No significant differences were found between species in
the meat of four carcass regions considered. The round has the
lower relative growth while the same values have forequarter
and flank as found by Robelin et al. (1974) but our higher
coefficient for loin is probably due to the inclusion of some
muscles of the abdomen that have high growth (Butterfield and
Berg, 1966$_1$, 1966$_2$, Seebeck and Tulloh, 1968$_1$).

The different organisation of muscle groups between species,
is interesting;all muscles considered of round have higher growth
in calves, except 'noce' which is found in the internal side
of the round.

In future further analysis could be undertaken when
more forequarter muscles and a serial slaughter up to maturity
become available.

532

REFERENCES

Benevent, M., 1971. Croissance relative ponderale post-natale dans les deux sexes des principaux tissus et organes de l'Agneaux Mérinos d'Arles. Ann Biol. Anim. Biochim. Biophysic, 11, 5-39.

Berg,R.T., and Butterfield, R.M. 1968. Growth patterns of bovine muscle fat and bone. J.Anim. Sci., 27, 611-619.

Boccard, R., and Dumont, B.L., 1973. Etude de la production de viande chez les ovins. IX Variations de l'organisation de la musculature de l'Agneau en fonction de la vitesse de croissance. Ann. Zootech, 22, 423-431.

Butterfield, R.M. and Berg, R.T. 1966_1. A classification of bovine muscles, based on their relative growth patterns. Res. Vet.Sci. 7, 326-332.

Butterfield, R.M. and Berg, R.T. 1966_2. Relative growth patterns of commercially important muscle groups of cattle. Res.Vet.Sci. 7 389-393.

Callow, E.H. 1961. Comparative studies of meat. VII. A comparison between Hereford, Dairy Shorthorn and Friesian steers on four levels of nutrition. J.Agric. Sci. 56, 265-282.

Cantier, J.,Vezinhet, A., Rouvier, R. and Dauzier, L. 1969. Allométrie de croissance chez les lapins. 1. Principaux organes et tissus. Ann. Biol. Anim. Biochim. Biophysic. 9. 5-39.

Charles, D.D. and Johnson, E.R. 1972. Carcass components of the water buffalo (*Bubalus bubalis*). Aust. J.Agric. Res. 23, 905-911.

Elsley, F.W.H. 1963_1. Studies of growth and development in the young pig. Part 1. The carcass composition at 56 days of age of pigs reared along different growth curves. J.Agric. Sci. 61, 233-241.

Elsley, F.W.H. 1963_2. Studies of growth and development in the young pig. Part II. A comparison of the performance to 200lb. of pigs reared along different growth curves to 56 days of age. J.Agric.Sci. 61, 243-251.

Henrickson, R.L.,Pope, L.S. and Hendrickson, R.F. 1965. Effect of rate of gain of fattening beef calves on carcass composition. J.Anim.Sci. 24, 507-513.

Huxley, J.S. 1932. Problems of relative growth. Methuen London.

Johnson,E.R. and Charles, D.D. 1975. Comparison of liveweight gain and changes in carcass composition between buffalo (*Bubalus bubalis*) and bos taurus steers. Austr.J.Agric. Res. 26, 415-422.

Joksimovic, J. and Ognjanovic, A. 1977. Comparison of carcass yield, carcass composition and quality characteristics of buffalo meat and beef. Meat Sci. 1, 105-110.

Lanari, D. 1972. The choice of parameters for the evaluation of veal carcasses and the calculation of growth coefficients for muscle, fat and bone. Riv. Zootech, 45. 77-88.

Lohse, C.L., Moss, F.P. and Butterfield, R.M. 1971. Growth patterns of muscles of merino sheep from birth to 517 days. Anim.Prod. 13, 117-126.

Meyer, J.H., Hull, J.L., Weithamp, W.H. and Bonilla, S. 1965. Compensatory growth responses of fattening steers following various low energy intake regimes on hay or irrigated pasture. J.Anim. Sci. 24, 29-37.

Mukhoty, H. and Berg, R.T. 1971. Influence of breed and sex on the allometric growth patterns of major bovine tissues. Anim. Prod. 13, 219-227.

Robelin, J., Geay, Y. and Beranger, C. 1974. Croissance relative des différents tissus organes et régions corporelles des taurillons Frisons, durant la phase d'engraissement de 9 à 15 mois. Ann. Zootech. 23, 313-323.

Robelin, J. and Geay, Y. 1976. Répartition des masses musculaires chez le jeune bovin mâle entier et son evolution au cours de la période d'engraissement entre 8 - 9 et 16 - 17 mois. Ann. Zootech. 25, 273-279.

Romita, A., Borghese, A. and De Maria Carmela. 1976. Prove comparative fra vitelli bovini e bufalini allevati fino a 20 settimane di età. I. Accrescimento, indice di conversione, resa al macello, caratteristiche della carcassa. Ann. Ist. Sper. Zoot. 9, 79-92.

Romita, A., Borghese, A. and Gigli, S. 1977_1. Prove comparative fra vitelli bovini e bufalini allevati fino a 28 settimane di età. I. Accrescimento, indice di conversione resa al macello e caratteristiche della carcassa. Ann. Ist.Sper. Zoot, (in press).

Romita, A., Borghese, A. and Gigli, S. 1977_2. Prove comparative fra vitelli bovini e bufalini allevati fino a 36 settimane di età. I. Accrescimento, indice di conversione, resa al macello e caratteristiche della carcassa. Ann. Ist. Sper. Zoot. (in press).

Romita, A. and Dias Da Silva, A.V. Accrescimenti, indici di conversione e rese al macello di bufalotti allevati con due differenti tipi di latte ricostituito (unpublished data).

Seebeck, R.M. and Tulloh, N.M. 1968_1. Developmental growth and body weight loss of cattle. II. Dissected components of the commercially dressed and jointed carcass. Aust. J. Agric. Res. 19, 477-495.

Seebeck, R.M. and Tulloh, N.M. 1968$_2$. Developmental growth and body weight loss of cattle. III. Dissected components of the commercially dressed carcass, following anatomical boundaries. Aust. J. Agric. Res. 19, 673-688.

Tulloh, N.M. 1963. The carcass composition of sheep cattle and pigs as function of body weight. In carcass composition and appraisal of meat animals, 5, 1-16. (Ed.D.E. Tribe). East Melbourne CSIBO.

Vezinhet, A., Rouvier, R., Dulor, J.P. and Cantier, J. 1972. Allométrie de croissance chez le lapin. III. Principales régions du système musculaire. Ann. Biol. Anim. Biochim. Biophysic. 12, 33-45.

DISCUSSION ON SECTION 2

H.J.H. MacFie *(UK)*

Regarding the production you have just mentioned, Dr. Fowler, at the end do you become more efficient by reducing the age of slaughter? Do you decrease the overall cost of the meat production? Is that simply the same thing as reducing the lipid by 10% to give the same effect?

V.R. Fowler *(UK)*

It is very much greater in terms of efficiency if you reduce the slaughter weight, because you have not supplied maintenance for that period. If you slaughter an animal at 81 kg instead of 90 kg you do not have to feed it for ten or twelve days, and that is a tremendous saving, and the amount of lean tissue which you have lost is comparatively small.

H.J.H. MacFie

So that the other one is compounded in it.

V.R. Fowler

Yes, it is.

R.G. Kauffman *(USA)*

You have talked about the individual animal unit. A couple of times you mentioned the number of pigs per year and so on. Have you ever extrapolated your information back to a 'per sow' unit, or 'per acre of land' unit, rather than on an individual animal basis, for the total energy efficiency?

V.R. Fowler

Yes, but I have not got all the data here.

R.G. Kauffman

It is difficult, I know, but is that not the ultimate answer we want?

V.R. Fowler

I do not know what the limiting resource in animal production is. It is not acreage: you could have ten decks of pigs
on one acre. It is not really useful to do it on an area basis.
The limiting factor is capital.

J. Kempster *(UK)*

I find the answer of 60 kg a little low when you compare
it to some crude estimates that we, and others, have made. Coulc
you give us some definition in your methods of calculating the
dam overheads, and also explain what you meant by lean. Was thi₤
lean carcass - or total protein accretion in the whole body?

V.R. Fowler

It was dissectable skeletal muscle. That is the first
thing. I ignored the cost of raising the gilt because she can
always be sold for meat herself; she can always be replaced by
selling her for meat and using the money to buy a new gilt. You
can ignore the energy costs of rearing her.

J. Kempster

Not in real terms because the price of the gilt is much
more than the price of the sold animal.

V.R. Fowler

Not if you are keeping one of your own bacon pigs; the
difference is trivial and is taken into account in a biological
situation. It is much easier to ignore it. And you create difficulties because, if you put in extra factors like the extra
cost of housing for a baby pig per unit of body tissue, the cost
of housing its mother, etc., then the sum progressively pushes
the economic optimal point up to perhaps 70 to 75 kg. I would
agree with you: in economic terms it will be higher, but it is
interesting to see where the biological balance point is, and to
think of all the reasons why we tend to go higher. Part of it
is because the risk of keeping an older animal is much less than
that of keeping a younger animal. It is less likely to die of
natural causes.

R. Fawcett *(UK)*

    I would like Dr. Fowler to tell me whether or not this was simply taking ad libitum intake of conventional European-type diet for a pig, because the result is critically dependent upon the amount of fat that this pig is laying down in its growth.

V.R. Fowler

    These are absolutely wonderful pigs, being fed ad libitum, and I have made the costing as favourable to the pig as I possibly can. The conversion is about 2.8 : 1 on a diet of 12.5 KJ per kg.

R.G. Kauffman

    Is not the weight of 60 kg simply arbitrary at this point because you could have a pig which would be of different composition at 80 kg, and which might be even better than your 60 kg pig depending on the genetics. Is this not possible?

V.R. Fowler

    The same argument could apply to 40 kg. I am just saying that, taking very representative examples of the top 5% of British pigs, I think the biological optimum in these terms would be very close to 60 kg.

D.N. Rhodes *(UK)*

    Would that picture change much for the boar?

V.R. Fowler

    Yes, it might well push the weight up, because he would not be putting down so much fat with each increment of growth after 60 kg. It might be a little better for the boar, but not that much. I do not think it would change much more than to 70 kg.

V.R. Fowler *(UK)*

I would like to ask Dr. Lister a question.

Are the animals with the high thyroid activity really converting fat that they do not want into heat which we cannot see?

D. Lister *(UK)*

No, I do not think they are. Some of it goes as heat that we cannot see but I do not think this is a very serious issue. I think they are certainly mobilising fat which they may or may not use. If they use it for heat production, then obviously we lose it as heat, but perhaps they are just recycling, and the energy cost of this is not particularly great.

V.R. Fowler

It is clear that the Piétrain is not that much more efficient at converting energy into protein than is the Large White and yet they are extreme examples of what you have been describing. Surely it is that the Piétrain is merely expending as heat energy which the Large White is just depositing as fat.

D. Lister

I think that is right. We are talking in extremes between Piétrains and Large Whites, and undoubtedly the major reason why a Piétrain is lean is because it does not deposit fat. It does burn it off as heat, but in the normal range I think that differences in maintenance are not particularly crucial.

R.G. Kauffman *(USA)*

Have you ever had an opportunity to study double muscled cattle in relation to Friesians or others?

D. Lister

No, but I would expect the same thing as in Piétrains.

M.J. Clancy *(Ireland)*

If I got the essence of the difference correct, it seems that the Piétrain depends more on fat metabolism in its muscle during life than it does on carbohydrate.

D. Lister

I think it is likely.

J.H. Oslage *(West Germany)*

If you state that the Piétrain does not deposit fat as much as other breeds do, I would say that it undoubtedly is because they have a higher protein synthesis capacity. On the other hand the intake of the Piétrain is lower, and the lower deposit of fat might be a result of the lower intake of feed.

D. Lister

There are two issues here. However I am a little unsure about protein deposition because, in the experiments which Dr. Wood and Dr. Perry did, there was really no difference at all between the Piétrains and the Large Whites which we had, in the amounts of lean which they deposited. The substantial difference lay in the amount of fat which they deposited, even on the same intake of food. I am sure that these are two separate issues: lower intake and higher heat production.

A.J.H. van Es *(Netherlands)*

I would like to ask a question on Dr. Refsgaard Andersen's paper: what was the nitrogen content of the ration - was it changed? Let us say the animals received 55%: did they get the same amount of protein, or the same percentage of protein, as this is important for protein supply.

T. Liboriussen *(Denmark)*

We used the same percentage of protein.

A.J.H. van Es

So these 55% animals might have been low in protein?

T. Liboriussen

That is possible.

R.G. Kauffman

A question to Miss Hind: at the total food consumed at 2 000 kg, how would you compare Friesians against Jerseys?

E. Hind *(UK)*

The level of efficiency depends upon what you are going to take as the maternal cost. If they had the same efficiency then their maternal costs would be in a given ratio - and that is true at any point, in fact.

R.G. Kauffman

You are saying that therefore there perhaps would be no difference in the two breeds?

E. Hind

If their maternal inputs were in a given ratio there would be no difference.

R. Fawcett *(UK)*

I am rather intrigued by the exposition that has been presented here. I think it is a very interesting one but I am a little at a loss to understand the definition of maternal input.

E. Hind

I have tried to skate over that because I think it really depends what you are talking about. If you are a dairy farmer you may say that your maternal input is zero, because you must get the calf to be able to get the milk, when your optimum point at slaughter would be at birth where efficiency is infinite. If you are a beef farmer, then your maternal input is everything which the mother consumes, so it would mean the food consumed from one calving to the next. Or it could be a market cost if you bought the calf to rear it.

R. Fawcett

Would it not be appropriate to consider that there is some value attached to the mother when she is slaughtered? You have got a payment some time later in her life for the value of her carcass, so you would have to adjust it downwards for the sale value.

E. Hind

Yes, but you would have to replace her as well so, if you add that, you would have even more cost - not less.

J. Kempster (UK)

If you put in a realistic figure for the dam overhead and draw your lines up, is the optimum efficiency of the Friesian always above the optimum efficiency of the Jersey?

E. Hind

They have the same optimum efficiency, in the ratio of about 1 000 to 330, so you can get it so that they are the same.

J. Kempster

And the maternal input is dependent upon the size of Jersey versus Friesian dams?

E. Hind

Yes - because of this particular relationship, it is always in a ratio of about 2.8 : 1, if they are to have the same optimum efficiency.

D.N. Rhodes (UK)

I would like to make a short comment. Dr. Fowler made a point on putting nutrients from plants straight into a human being. This is fine. There would be no reason to have meat production at all - except that human beings like to eat meat, and I think they will always want to eat meat because we have two or three thousand generations of tradition behind us. That is one reason. The second reason is that the energy inputs into the production of meat come from two main sources: oil, and the sun.

The sun produces, in different parts of the world, sources of feedstuff which are not necessarily assimilable by human beings so this route is not always available. Of course, for most of our foods it is available, and this makes up the rest of food science.

The second point I want to make is that many of the paper are - in fact the meeting itself is - about the transformation of nutrients from plants into meat, and here we come back to this word, 'efficiency'. The point I tried to make earlier was that the important measure, it seems to me, of efficiency is the amoun of lean plus fat which is required by the consumer, in the form in which it is required by the consumer, divided by the amount of feedstuff not assimilable by human beings which was put in. In Miss Hind's paper there were various references to the transfer of feed into total gain on the hoof. These were very interesting but, what I think is more important is the ratio of the produced meat to the amount of feed which goes in.

A.J.H. van Es

Dr. Lister mentioned the measurements of hormones. I would like to ask whether he measured them once a day, or let us say every two minutes? It is very important with regard to the pattern of hormone level.

D. Lister

A major point in my paper was that it is difficult to conclude anything from a single assay of a hormone. What you have to do is to look at the interrelationships of hormone systems. If you rely on one measure of something to do with thyroid hormones, it is liable to lead you into a wrong con-clusion, and so we now attempt to integrate all our hormone assays and so forth, to get a broad picture. It may well mean that we need to sample animals continuously during and after feeding. We need to play a number of tricks in order to establis what the baseline criteria are. We would accept that this is a very serious limitation to the approach which has been practised so far.

A.J.H. van Es
       I agree completely with that.

B.L. Dumont *(France)*
       You have said that the Piétrain, as compared to Large
White Males, is typified by reduced fat deposition, which arises
from increased glycolysis. One can also say that Piétrain males
compared to Large White males are typified by a quite different
muscle distribution, or by a different muscle morphology and
thickness, or by different muscular fibre metabolic typing. My
question is, is there any relationship to be expected from a
physiological point of view between these various aspects of
Piétrain males as compared to Large White males' typification?

D. Lister
       I can only talk about the Piétrains which we have used
in our work. It just so happens that the Piétrains and the Large
Whites which we used grew protein at the same rate when fed the
same amount of food, and that was a very lucky finding because
it helped us to interpret a number of other things about the use
of food. But when we have dissected Piétrains and Large Whites
we have found it very difficult to establish any differences in
the distribution of muscles and I think this has also been true
in the work of Davies and others in Edinburgh. The distribution
is not changed very much but I agree that the geometry of the
carcass is different in a Piétrain, and I think this is associated with differences in the skeletal size. We always suspect
that we have relatively short and small bones in Piétrain pigs,
and if you have the same amount of muscle then the muscle must
appear compact rather than stretched out along long bones, so
I think that there are very obvious differences in the geometry
of the carcass which affects the appearance. But certainly in
our own work - and I think that this is perhaps true of other
people who have worked with Piétrains of a similar kind in our
country - we have never been able to find substantial differences
in muscle distribution. There are differences, we know, in fibre
types between Piétrains and Large Whites, and I think this is a
reflection of the nervous and hormonal factors that are involved

in the overall control of body composition.

The relation between all these factors and the typification according to the glycolysis is a very complicated issue, and perhaps I can suggest to you what we think is an overall mechanism. It seems that animals which have differing abilities to mobilise fat have different settings of sensitivity threshold perhaps in the hypothalamus, so that a stress for one animal is not a stress for another; and the animal which normally has been in a 'stressed' condition will be one that is constantly mobilising fat, and probably has a high metabolic rate also. We think that this also leads to differences in the distribution of fat, and also to the amount of fat which is being deposited, but it is a very difficult subject and I really do not know.

R. Boccard *(France)*

Just a word about this which might explain some differences between Piétrains and Large Whites. I think the Piétrain is close to being a double muscled animal, and in their muscle they have less collagen than the Large White. When we cross the Piétrain and Large White we obtain an intermediate value for collagen content in specific muscles. Perhaps the low fat content in Piétrains, and their low collagen content, can be considered as the same phenomenon in double-muscled cattle.

D. Lister

I think that, physiologically, Piétrains and double-muscled cattle are very similar, and this especially relates to things such as the way in which the connective tissue framework is built up. I am pretty sure that what we find in double-muscled cattle is also there in Piétrain pigs. What I am not sure about is the extent to which Piétrains are double-muscled animals; it would be nice to show that they had muscles which were hypertrophied, but in our experience we have not yet seen any.

R.W. Pomeroy *(UK)*

I wonder if our French friends would like to comment on

the Limousin cattle, because, from what little I have seen of
the literature, the Limousin seems to have at least some of the
characteristics we are dealing with. It has a rather small
appetite, it has a very high lean content and, from the figures
which I have seen, it does not seem to grow all that rapidly.
Is the Limousin stress-susceptible, for example, like the Piétrain,
or what? Are there any parallels between the Limousin cattle
and the Piétrain pig?

R. Boccard

The Limousin certainly has a pronounced muscularity, and
its muscle to bone ratio is very high as compared to other breeds.
I do not know about the food intake - it may be lower than in
other animals. Its growth is also a little less. I think that
its stress susceptibility is similar to that in other breeds:
young bulls can be stressed by transportation, resulting in dark
cutting. There is no conclusive evidence, however, as to whether
this occurs more or less frequently than in other breeds.

J. Kempster

Could I ask how the Texel fits into this because here
we have a breed of sheep which again differs from the others,
certainly, compared with our population. It does have wide
differences in lean distribution. Is it a matter of double-
muscling in the Texel, or is it just lean distribution? I am
confused between double-muscling and lean distribution: the
Piétrain definitely has a larger leg; the Limousin does not have
a particularly large leg.

R. Boccard

The main problem is to characterise exactly what is meant
by double muscling. In this respect different conformation or
a different aspect of shape are misleading, and we have to find
specific characteristics. Double-muscling is essentially a
phenotypic aspect of some metabolic or biochemical change inside
the muscle. In my opinion, the hydroxyproline to protein ratio
is characteristic. Piétrains answer to this criterion, but I
do not know about the Texel sheep.

R. Jarrige *(France)*   .

The EEC Commission for beef production research has decided to hold a seminar on double-muscled cattle in 1978, and I
am sure that we will find answers there to many of the questions
that have been raised at this meeting.

V.R. Fowler

Could I address a question to Dr. Lister, because I think
there need to be some balancing thoughts on the Piétrain. Dr.
Fuller at the Rowett Institute measured the heat production from
Piétrain pigs and three-quarter Large White pigs, and he found
that the only difference on a range of feeding levels between
the two breeds was that the Piétrain did not eat as much as the
Large White did. The heat productions were the same, their
protein deposition, on the same feed intake, was identical and
similarly their fat deposition. Now, the argument may be that
these were not typical Piétrains, but they looked like Piétrains
and I would like to know how far one should generalise when the
Piétrain is supposed to be the extreme example in the pig of
this particular characteristic as described by Dr. Lister. Would
you like to comment on that?

D. Lister

This is exactly the point which I was trying to make
earlier. You have to define your phenotype very accurately indeed and we know very well from our experience that animals can
masquerade as any breed you care to mention. Physiologically
they may not represent that breed at all, and I know that in
our case the only difference which we could find was in terms of
their fat mobilisation and deposition.

We now know that there are animals masquerading as Large
Whites which have selectively become leaner, and when they do
this, they become Piétrain in terms of their physiology.
Exactly the same thing has happened in the Norwegian Landrace
selected by Dr. Standal. As they became leaner they became more
Piétrain, but they still look like Norwegian Landrace. In the
case of the Rowett experience, as in our own experience too with

animals from the same source, we know that those particular
Piétrains were Large Whites, so one cannot expect them to be
different physiologically.

J. Kempster

You cannot argue that way and then put a graph up which
shows the breeds ranking in nice order in terms of either being
beef or dairy type.

D. Lister

I am asking people to be more precise in their definition.
The animals which I referred to have been characterised in a
number of ways to show that they have the Piétrain characteristics. The Rowett Piétrains were characterised only in terms
of their ability to deposit muscle and fat, and of their energy
balance generally, and they were not different so why should we
expect them to be different in other characteristics as well?
I used breeds simply as examples of genotypes which varied in an
apparently consistent way if you also have all the other information. But taking a breed, any example of a breed from anywhere,
and saying that this is characteristic of Hereford, or Piétrain,
or Limousin is just nonsensical. You must have defined the
thing which you are dealing with.

R. Fawcett

I am at a loss now to understand what is a Piétrain and
what is a Large White!

J.H. Oslage

I would like to raise two short questions for Dr. van Es
but first of all I have a question on terminology. In the past
we have heard the expressions, 'nitrogen retention' or 'deposition' used, and today some speakers used the term 'accretion'. I
would like to know whether this has any different meaning, or
whether it is just a synonym.

Dr. van Es mentioned that the coefficient of efficiency
for fat deposition or accretion is very close to the expectation

we can calculate according to the theoretical values. In the
case of protein accretion, the measured values are lower than
the theoretically expected ones and you tried to explain this by
the degradation of protein. This would mean that we should have
the greatest differences between the theoretical values and the
actual measured values in animals which have a high synthesis
and at the same time a high degradation. However, we actually
measure the higher efficiencies for protein deposition in young
animals which have a high rate of synthesis and a high turnover.

My second question is concerned with protein turnover.
You mentioned that this is part of maintenance requirements and
would explain why young animals have a larger maintenance requirement per kg of body weight. I think you calculated the
amount of this energy as 20% of maintenance energy. My question
is, to what intensity of growth, or intensity of protein synthesis does this figure of 20% refer?

A.J.H. van Es

'Deposition' and 'accretion' and 'net accretion' all have
the same meaning. Formerly the expression 'protein synthesis'
was very often used, which in fact is something different,
because the accretion or deposition is the difference between
synthesis and degradation. A lot of people working on protein
turnover prefer the term 'accretion' to indicate something completely different from synthesis or degradation. I think that
covers your first question.

You asked how I computed the 20% energy cost of maintenance compared with the 30% in a rapidly growing animal and how
rapidly that animal was growing. The information so far about
the 20% and the 30% relies on the amount of ATP needed for total
protein synthesis, and the few figures available are only from
small animals. I have, as you can see in my paper, started from
the data for adult animals, and I have assumed that these mature
animals will hardly have any protein deposition, compared to
rapidly growing animals or animals which were said to be growing,
and will have deposited protein. Then I found for the mature

animals the 20% and for the other animals a figure between 25
and 50% with an average of about 30%. We need more data. I
think the difference of the 10% is due to this increased protein
synthesis, which is the reason why the nett accretion is much
more energy requiring.

Your other question dealt with the values for the KF,
and the value of K for protein deposition, but I did not understand you properly.

J.H. Oslage

My question was, would you explain the differences between the measured values of K protein against the theoretical
values which are about 82%, due to the degradation of protein.
If that is the right explanation we should expect the greatest
differences between the actual measured values and the theoretical
values with animals which have a high rate of protein synthesis.

A.J.H. van Es

I think that the difficultiy with older animals is that
you have so many other metabolisms going on and your error in
knowing the amount of energy which is needed additionally for
protein accretion is larger. I think you will agree with that.

In the mature animal we have synthesis and a degradation
which is about the same. We need ATP for the synthesis, but we
do not get any return for breaking down the protein - it is just
not giving any new ATP. If you now move from that mature animal
to a young animal with a high percentage of protein synthesis in
its total metabolism then we have also a 50% higher degradation.
The difference is the accretion. The problem, of course, is that
we have insufficient data for establishing the right percentage,
and we have no data at all for cattle.

K. Rohr (West Germany)

I have a question for Dr. van Es. You have stressed the
weakness of the regression models. Nevertheless, we have to
rely on these models, for instance in the dairy cow. Of course

there is little data on growing and fattening cattle, and I
think the greatest bulk of data is coming from Rostock. In a
publication last year, efficiencies of 50% and about 100% were
given for protein and deposition respectively, and this in turn
influences the maintenance requirement, which is pretty high in
this equation. But if the maintenance requirement is fixed to
110 k/cal metabolisable energy per kg of metabolic body weight,
which is the commonly adopted value, the efficiency of fat deposition is still twice that of the efficiency of protein deposition. I would like you to comment on this if you would.

A.J.H. van Es

I think again we have a problem with regression models.
In fact, what the Rostock people have done is to use the complete equation, and you know also that this retained part of
the protein in cattle is a very small part of the total metabolism. However, they did not feed the animals at very different
levels of nutrition as our Danish colleagues did. There was
some variation in feeding level but not very much, so the variation in retained fat and the variation in retained protein was
very small. Therefore you can never expect the values found in
this regression to have any scientific or physiological significance.

R.M. Seebeck *(Australia)*

I would just like to make a few comments touching on
the three first speakers this morning.

As part of my work, I am now looking at components of
maintenance requirement. Originally, this came from body composition work where I have tried to relate body composition to
fasting metabolism measurements and for some reason, in my data
and other data which I have looked at, I just cannot get this
relationship, although you would expect it because, as Dr. van Es
has said, energy cost of fat turnover is small, and there is
not much fat turnover; therefore a fat animal should have a
lower maintenance requirement. We get breed differences in
maintenance requirement in the sort of cattle we are looking at,

say, Brahman crossbred which have a lower maintenance requirement in terms of both maintenance or in fasting metabolism, but we cannot explain it in terms of body composition. I have now started to look at physiological functions in relation to maintenance requirement, or what I prefer to call the non-productive use of energy, maintenance requirement or fasting metabolism being artificial measurements of that. Protein turnover, as Dr. van Es has said, is obviously one of these, and in Australia we are running a large experiment now with various people working on it, especially my colleague John Black in Sydney. We are looking at in vivo incorporation of fatty acids and at numerous other things, looking at maintenance and trying to work out partition of nutrients.

Getting back to maintenance, a particular thing which I have been looking at is the sodium-potassium pump and its contribution and at the moment we are still trying to see whether we can estimate sodium-potassium pump activity. If so, and if it is related to maintenance requirement, we may be able to use this as an indirect measurement of maintenance requirements either for genetic purposes or to follow changes in maintenance with fluctuation in the feed supply. We need some sort of indication of maintenance that we can follow in the dynamic state rather than having either to put them on to a maintenance diet or to fast them, which completely destroys many experiments. As soon as you do that you want to be able to do it with changes in food intake particularly.

Just as an aside to Dr. Lister's work on adrenalin, I have actually injected adrenalin into cattle. I expected, because of the marked increase in volatile fatty acids, that the respiratory quotient of the animals would drop, whereas in fact it went up, suggesting that in that situation we may have utilisation of glucose as the main energy contribution rather than of fatty acids. Of course glucose goes up as well.

A.J.H. van Es

I think that measuring maintenance requirements is an extremely difficult thing. If you use fasting you can, just by using this fasting technique, change the behaviour and the physical activity of your animal. I have much literature and evidence that fasting metabolism is a rather poor means of getting at maintenance requirements. I think also that you are referring to maintenance under rather specific conditions - let us say, out of doors, with animals walking around freely in an environment which may be somewhat hot and dry. The figures which I have given all apply to tied animals, kept indoors in a favourable climate, with hardly any water supply problems.

You mentioned the sodium/potassium pump. This is only part of total maintenance metabolism, maybe an important part, but is it not also influenced by the amount of water available? I think this is only part of the total, as is the case in measuring heart rate. Did you measure physical activity? We have done some measurements on some chickens with 'burglar' devices with radar just to have a measure of activity. What is activity? It is certainly a requirement of ATP for muscle contraction, and if you have your animals freely walking around, it is a very difficult thing to measure. Also, what about the disease level of your free animals - worm levels and such things?

R.M. Seebeck

At this stage we are not particularly worried about movement although we have stalled animals which I am working with. Nevertheless, the estimates of the sodium-potassium pump for other species is actually a higher proportion than protein turnover of this non-productive use of energy. It is in the order of 30%, while protein turnover may be 10 to 20% of non-productive use. At the moment it is a technique which is worth looking at, and of course it is related to thyroid turnover, towards thyroid activity, in relation to the calorigenic effect of thyroid hormone. Some people argue with that, but I think the evidence is pretty much in favour of it, at least that is, part of the calorigenic effect of thyroid hormones which is in general related to

basal metabolism. There is a great complex of things. I want
to do two things with maintenance requirements. Maintenance
requirements in our conditions are probably more important than
in Europe, because we have fluctuating feed supplies throughout
the year, and for part of the year we are in a sub-maintenance
diet. If we have animals such as the purebred Brahman, its
efficiencies under extreme conditions will probably be greater
than those of, say, a Hereford, due to its low maintenance requirements, which will get it through a period of low nutrition;
and, furthermore, it will be resistant to diseases and so on in
the hot time of the year. The main emphasis is on trying to
increase efficiency when, for some reason we have in general a
high correlation, at least between breeds, of feed intake on
the one hand and fasting metabolism or maintenance on the other.
What we want to do is to break that correlation with animals
with high food intake and low maintenance requirements, or vice
versa, to keep the feed intake the same and decrease maintenance
requirement. Depending upon just how bad the conditions are,
and what the nutritional and other components involved are, is
what the absolute level of either food intake or maintenance is
for the most efficient production.

B.L. Dumont

Have any studies been made on the variation between strains
of animals, and especially of laboratory animals, on the amount
of protein degradation? In other words, do we know if it is
possible to select animals on their low protein degradation
ability - for instance by lengthening the turnover period: is
this possible, or not?

A.J.H. van Es

Unfortunately, these measurements of the rate of protein
synthesis and the rate of protein degradation are rather difficult.
You can see it in the Appendix to my paper, which I have also
given some people working in the field - Dr. Lobley and the
Rowett, and Dr. Butterey in Nottingham, who agreed. Interpretation of the results obtained is very difficult and, if you are
asking for such small differences, selecting an animal which

may have a degradation rate 5 or 10% lower than another animal,
I think it would be impossible to measure it at the moment. That
is the reason why I think it is so necessary to do that work
because, if you work on it then you get better methods. At the
moment there really is no information.

J.H. Oslage

I would like to give a short comment to Dr. Seebeck. I feel
that the experiment to break the correlation between food intake
and maintenance requirement is rather difficult. Animals with a
high protein synthesis capacity could be expected to have a highe:
maintenance requirement. So if we try to breed animals with a
high food intake and low maintenance requirement, these animals
should be those with a high fat synthesis, and that we do not
want.

R.B. Thiessen

I have a question for Dr. van Es. Earlier in his talk he
quoted that the adult sheep had a much lower maintenance requirement than adult cattle. Could he give a reason as to why
this is so, and if you have a lower maintenance cost in growing
the sheep relative to growing cattle. If sheep were fed under the
same feeding regime as cattle and slaughtered at the same degree
of maturity or body composition, would you expect that sheep
would be more efficient than cattle?

A.J.H. van Es

It is a very difficult question. We do not know why there
are lower figures for mature sheep per kg metabolic body weight -
about 20% lower than in the case of cattle. The same is true
for the other question about growing lambs compared with growing
cattle - we do not know anything at all about that.

R.M. Seebeck

Just a short reply to Dr. Oslage. You may not want animals
which are fat, but of course we do not really know whether animal:
which are fat have an advantage under fluctuating nutritional
conditions. When we run into drought, fat may be a very useful

fat store, or it may not; there is no real definitive experimental
evidence.

There is another way in which we may actually change it.
Going back to the sodium-potassium pump, this utilises a vast
amount of ATP to keep the potassium inside the cell and the
sodium outside. If we can have animals with less permeable membranes we can still maintain that same sodium-potassium gradient
with less expenditure on the sodium-potassium pump, so there may
be a way of actually changing that. Whether than, in turn, is
linked with food intake I do not know. It may be that that same
potassium gradient is there solely to get nutrients into the
cell, because of the sodium carrier system which brings some
nutrients into the cell. At the moment we just do not know
whether we can in fact get animals of different membrane per-
meabilities, but I would think that would be a target to aim for
if we can get them, and see what their efficiencies are.

D. Lister

I wonder if I can say a little about our experience with
sodium-potassium clearance, because, like Dr. Seebeck, I reckoned
once upon a time that it was quite an important phenomenon and,
as we normally do, we tested out all potentially important
phenomena in Large Whites and Piétrains. We found that the
clearance of sodium and potassium was identical in both breeds
but of course Piétrains excrete very much less urine; they put
out a very concentrated urine, but their daily clearance of
sodium and potassium was exactly the same in both breeds.

V.R. Fowler

I think that Dr. van Es would agree that all we are trying
to do is to partition heat production, and when we say "maintenance" it is something which is statistical and which we cannot
measure except under very defined conditions. When we talk
about protein turnover, all we are interested in is how much heat
is associated with it, and the same with the lipid. We think
that $W^{.75}$ makes everything all right, but it does not. It could
be a different exponent, and Webster, for example, would propose

an exponent of 1 for growing bulls, and Byron would propose 0.56
for pigs, and anywhere between those two may be right in differen
situations. I think it is important that people who are not
engaged in the field do not get the idea that we measure these
things with tremendous precision. All we are doing is dividing
the cake three ways, often only two ways. I wondered if you
might comment on whether some of the discrepancies in the literature arise from forcing everything into this exponent 0.75.
For example, whether some of the differences between small and
large animals are not due to different ways of measuring nitrogen
retention, which is done by comparative dissection in some cases,
and analysis, and in other cases by nitrogen balance which, in
some circumstances, could have a serious error, the younger the
animal, the more serious the error possibly becoming.

A.J.H. van Es

I think that in ruminating cattle we are in the weight range
of 150 to 500 kg, and there the influence of using 0.75 or 0.65
or 1.0 is not so serious. If you are very close to a live weight
of 1 to 1.5 kg, then you have the real trouble because there it
is very curvilinear, but in the range mentioned we have hardly
any curvilinearity, so you could use 1.0 with the same results
and with about the same error.

With regard to your other question, if you do measurements
and extrapolation of regression work with chickens which are
very small animals, and then you do the same with mature cattle,
you end up with figures very nicely in the region of 0.75 - with
the chickens also. I would prefer not to use 0.75, but ¾ power.
This avoids giving the impression that 0.75 is so very precise.

I agree with you completely that neither the energy balance
nor the nitrogen balance technique is completely free from error.
It has considerable errors - errors of measurement, and also
errors from keeping the animals in an environment which is not
what it would be in practice. However, the slaughter trial also
gives problems because you have to keep that animal, you have to
feed it, you have to know how much feed is going into that animal

After that you slaughter it, serially, but then you have another
animal to be slaughtered, so also there we are running into
errors. We will have to extend the experimental basis.

J.D. Wood *(UK)*

I have a short question for Dr. Fowler on Table 4 in his
paper. Here he examines the effects of changes in several of
the parameters on the efficiency of lean tissue growth on the
mean cost of lean tissue. He concluded that there were two ways
of making a significant reduction in this cost: one was to change
the weight at slaughter; and the other was to increase by selection the rate of lean tissue growth. May I ask him what a difference of 10% in an increase in the rate of lean tissue growth
represents in terms of body composition at 90 kg - is it, for
instance, the difference between a castrate and a boar, and is
it possible to achieve this simply by selection, as he suggests?

V.R. Fowler

Yes, I think it is possible to do it by selection, because
the commercial product evaluation tests, the first two of which
have been published by the Meat and Livestock Commission, show
that the difference between the best and the worst herds in the
growth rate of lean tissue is about of that order - 10%.
Similarly, most experiments on boars versus castrates indicate
differences ranging between 10 and 15%, so I think that is
totally accessible by breeding.

R. Jarrige

As co-ordinator, up to now I have said nothing. I would now
like to try to summarise some aspects of the discussion and of
the exposés presented this morning.

The principal consideration is muscle growth potential. In
biochemical terms, this is protein growth potential, which is
the difference between synthesis and degradation. As regards
fat deposition, we can distinguish roughly two parts. One is
necessarily correlated to the protein deposition. Another part
is the storage fat when the intake of energy is in excess of

the muscle growth potential. The genes act on the muscle growth
potential, and the hormones are some of their tools. When the
male animal is castrated, the muscle growth potential is decreased.

We can assume that the genes act on the maintenance requirements as they act on the synthesis or degradation, and they act
on the ingestion capacity of the animal. From a nutrition point
of view, we have to supply the muscle with a quantity of energy,
and the quantity of amino acids necessary to express the full
potential in our intensive systems of production. In some animals
the capacity of ingestion is too high according to their muscle
growth potential when they are given a high energy diet. It is
frequently the case with the Friesian and the excess of energy
is then stored as fat. We can limit this amount of storage fat
by decreasing the energy concentration of the diet or by restricting the quantity fed to the animal. At the opposite end,
other animals, such as the double muscled, probably have an
ingestion capacity lower than their muscle growth potential.
We have to give them higher energy diets, for example. I think
that many problems raised during this meeting could be synthesised
within this very rough and inadequate scheme.

Then we have the different points where selection is able
to affect the rate of synthesis, perhaps the rate of degradation,
though this is entirely hypothetical, in the maintenance requirements and in the ingestion capacity; but many of these parameters are cross-linked together and it is difficult to know
what is cause and what is effect. Another point is the use of
anabolic agents, and Dr van Es says that they decrease the rate
of degradation. I am not sure whether Dr. Lister agrees with
this.

A.J.H. van Es

There is some preliminary evidence for it. There is also
some evidence for a decrease in synthesis but the greater degree
seems to be in degradation.

R. Jarrige

Finally, taking into account the 'short comment' by Dr.
Rhodes, we have to bear in mind the object of producing something
palatable for the consumer, and this requires the inclusion of
some fat. The meat industry also requires some fat as well, so
I wonder whether the 60 kg pigs are really fitted to the meat
industry and to the consumers' preferences - but I am not a pig
expert.

R.G. Kauffman *(USA)*

I would like to ask Professor Rohr whether he treated his
data with analysis of covariance to rule out the effect of
fatness per se in comparing the two breeds.

K. Rohr *(West Germany)*

The data was treated by analysis of variance with a factorial
design, and the differences were checked by the Duncan tests.

R.G. Kauffman

But you did not attempt to hold the effects of fat itself
constant. It is my view that that is a difference you have in
all your data.

A.J.H. van Es *(Netherlands)*

Could you convert your maintenance estimate expressed in
net energy into metabolisable energy? You have a figure of
about 8 k cal net energy.

K. Rohr

This is in line with the 110 k cal metabolisable energy $kg^{\frac{3}{4}}$.

A.J.H. van Es

Yes, but these were rather rapidly growing animals and so
you are not in agreement with the latest Rowett Institute work,
because they are higher.

K. Rohr

No, this is right.

J. Harte *(Ireland)*

Is it reasonable to put down the protein deposition per day? How do you get that figure?

K. Rohr

We made a whole body analysis of all these animals and we subtracted from the total amounts of protein and fat analysed in these animals, the amount of protein and fat we found in calves of 160 kg liveweight; we analysed 5 calves of each breed in order to subtract. I think the error should be low. We only used these two steps - calves of 160 kg and these heavy animals.

J. Harte *(Ireland)*

I would submit that in fact if you killed the two breeds at the lighter weight (which is in line with what Dr. Kauffman is saying about the confusion of fat) that you might have obtained a different result. I do not think it is fair to list down the daily deposition of protein since you have not done serial slaughter to prove it.

K. Rohr

We have not got such figures with Simmental bulls. So far we only have serial figures on serial slaughterings in Friesian bulls from the first experiment by Schultz and Daenicke, with intervals of 100 kg beginning with 150 kg and then adding every step with 100 kg. On average we obtained in this experiment similar values to this 160 g per day which we calculated from the previous experiment. Of course, we had a maximum in the range of 250 to 350 kg in this experiment with Friesians.

J. Kempster *(UK)*

I see nothing wrong in calculating rate of protein deposition as has been done here; it is a critical measurement, and you improve the position in no way by doing a serial slaughter experiment. This is an experiment where they compared breeds with fixed liveweight which seems perfectly fair.

The point I would like to ask is, because your sample is

quite small, how were the animals selected? Can we take these as being representative of the population?

K. Rohr

You can always doubt whether a representative sample of the population has been taken. We got these calves from the market and the Simmental calves from the southern part of Germany. But surely the results may shift, especially as we are now using in our dual purpose breeds sires from Canada and the United States. However, this was a tremendous difference of 17% in protein synthesising capacity between these two dual purpose breeds.

R. Jarrige *(France)*

How do you explain this lack of variation in feed efficiency between the two final weights and is there any interaction with the breeds on this aspect between Friesians and Simmentals?

K. Rohr

This high feed intake we had was certainly not the highest we could get in this case. We did not use, as you have seen, a very high amount of concentrates. The difference in intake was only 7% between the two treatments, but you could see from the picture which Dr. Daenicke has given, that within this range of 565 to 615 kg the Friesians and the Simmentals responded differently.

C. Béranger *(France)*

Thank you Dr. Rohr. I think we can move on now to a more general discussion.

We have seen a large variation between breeds or different types of animals, and interactions with the level of feeding and environmental conditions. I think we will discuss first the magnitude of the differences between breeds.

From the different papers it becomes evident that the variation between breeds is quite different, depending upon the breeds involved and also on the interaction between the different

conditions in which breeds are compared. The methods of
expressing these differences may play a role as well.

It might be interesting to discuss first the magnitude of
the rate of variation of differences we have observed between
breeds and genotypes of animals, and how to express this difference under the best conditions. Are there any questions or comments?

R.G. Kauffman (USA)

I am concerned about the fact that, when we make comparisons, in this case between different animals or breeds of animal
we put a handicap on one breed and not on the other. When we
compare two breeds at different levels of fatness, then we have
to say: all right, we are interested in a weight at a given
end point, and if that is the reason, then we have to turn our
economists loose, not our dissectors, to establish this. I
think that in most of the data which I have seen here (and I
think that the climaxing point was Miss Hind's work this morning
very short and very appropriate I thought) there is very little,
or no, difference in breed or in method of feeding, or anything
when we compare them on the basis of a fat standardised method.
There are obviously differences in weight, the live animal or
its carcass, or its parts, and perhaps even differences in
quality, because we do know that there are differences in fat
distribution; but other than that it seems to me that it depends
on whether we want to raise one 6 000 kg elephant or thousands
of 300 gm rats. Maybe this point is not quite clear to some of
you. What I am saying is that maybe there is a size factor here,
but we have not established that yet. So far I have not heard
what the minimum amount of fatness is. I think we need to establish that. We also need to establish whether we raise bulls
or steers - and I am not sure whether that is certain or not -
and thirdly we need to know whether we are going to feed them
grain or grass.

H.J. Langholz *(West Germany)*

I do not fully agree with your view that we have to make
the comparison mainly by standardising the fat content. I see
it in a range, as there is quite a difference in the preferences
of fat between the countries. Also I think we should compare
breeds under a range of environmental conditions, and study the
effect by means of serial slaughtering.

There are two aspects to be borne in mind. The first is
the total gain, or the average daily gain of lean; and the
second is the proportion of fat and lean at the point at which
you slaughter the cattle. When we have this diagram filled up
for all our breeds, then we can make up our mind what we are
going to do under certain situations. I was very pleased to
hear from the nutritionists that they are going to bring out
feeding standards for the different breeds. I think that, as
long as you do not follow this pattern, breed comparisons do not
make much sense. This I fully agree with. We need to define
our environmental conditions very carefully and then we have to
define the outcome of our experiments in terms of the gain of
lean and the percentage of fat and lean in the carcass. This
would be enough for me and the breeders to make a decision on
the policy of breeding companies.

K. Rohr *(West Germany)*

I would like to support what Dr. Langholz has said. I
believe that the difficulty is that this group is not very
homogeneous. When Dr. Kauffman stresses the importance of making an analysis of co-variance, I myself, as a nutritionist, do
not know what to do with this analysis. This is because we all
have a different orientation. What we are trying to do is to
define certain requirements or to give recommendations for
nutrient supply and we start off with calves of different breeds,
and of course, we have different final weights.

G. Harrington *(UK)*

I think for the first time we have isolated a point of fundamental disagreement. This is anticipating tomorrow's papers to a degree, but I think we disagree with one of Dr. Langholz's points. Of course we agree about environmental specificity, but on the objective on the carcass side we would not support the two criteria he mentioned but alternatively the one implied by Miss Hind's paper - the optimal efficiency of producing lean meat from feed, within the particular environmental situation. Accepting variations in carcass weight and in carcass fatness, accepting that we cannot produce lean meat at its maximum efficiency without producing some fat in the carcass, it is most unlikely, since we have different breeds under different systems, that the variations we get in fatness if we achieve slaughter at maximal efficiency in feeding to lean meat terms, will not be adequate to meet a varying market requirement. Perhaps we can discuss these further tomorrow.

C. Béranger *(France)*

Perhaps the problem is more complicated with regard to the optimum composition of carcass, depending on the requirements of the market, and on the meat quality requirements of the consumer. Moreover, the value of the calves interferes, and their belonging to a variety of breeds. It is possible to produce a certain amount of meat to an equal fatness during a longer or shorter period with the different breeds. If we compare them at equal carcass fatness, there are large differences in carcass weight, and large differences in the time taken to reach this carcass weight. If we adapt these different genotypes to the different conditions of feeding and environment, you have certainly optimum types of animals adapted not only to the requirements with regard to carcass weight, meat quality and carcass composition, but also in terms of feed efficiency which is related to this amount of meat produced by one calf.

R. Boccard *(France)*

There is the same problem with fatness. Are you rich
enough to trim the fat and get the meat as you like it because
I think that the consumer has never demanded exactly the level
of fat he wants inside the meat. In France, the butcher was for
a long time obliged to cut off the unwanted fat and throw it
away. On the other hand, the consumer would be very happy to
have a high level of fat inside the muscle.

H.J. Langholz

Just a few points in this connection. We have to take
account of several factors, in particular what weight should be
put on maintenance requirements *(on blackboard)*. Many of our
studies show that we have a high correlation between gain and
feed efficiency, and, as this says indirectly, we have a little
variation in the genotype for maintenance. Dr. Rohr has shown
in his small experiment that he did not find differences, and
we may perhaps extrapolate to some extent under our situation.

R.G. Kauffman

I would like to ask a question of our nutritionists with
regard to the overall efficiency of production if we talk about
milk production and maintenance of the dam from which the calves
come. I want to look at the total programme as I am a little
concerned sometimes by looking at just the calf. It seems to
me when we are talking about beef production, we are talking
about milk production, the maintenance of the cow and obviously
the size must come into this, and of course a certain level of
fat production along with the lean in the meat product of the
calf. That is why I am very concerned when we start comparing
breeds especially, that we maintain some sort of level of fatness
(say 10%). I would like to ask Dr. van Es if he would comment
on whether large size or small size animals are the kinds we
ought to raise under European conditions.

A.J.H. van Es

This is difficult to answer without a calculator. If you
start with the dairy cow, this means that you are usually selecting

your animal for milk production, and you will end up with a
rather wide variation in male offspring with regard to lean potential. That is one uncertainty that you have to calculate
and which will work out in a negative way. Now assume we have
a dairy type: what is the conformation of that calf? Here again
a lot of things are coming in, so your calculation is a very
difficult one because it is not only the 10% fat which you will
have in your final product of a lower or of a higher weight, it
is also what is the value of the other parts. When this is about
the same for all lean tissue, then I do not think it matters so
much. It is important how soon a given maturity is reached. It
makes an enormous difference if you need 200 days to reach a body
weight of 300 kg or if you can do it (10% fat content of course)
in 150 days. Here your growth rate, your rate of deposition of
lean tissue, I think, comes in again. Then it is just a matter
of maintenance; the more rapidly-growing animal of course will
need less maintenance, so that is a more economic animal. However, it is rather late in the day to give a good answer!

R.M. Seebeck *(Australia)*

I would just like to make a point on Dr. Thiessen's paper.
As I understand it he is advocating that we should use relative
growth rate to select for food efficiency. If you have animals
of the same mature size, and we want to select animals without
increasing the mature size for most purposes, differences in
efficiency really mean differences in the growth curve. The animals with a high relative growth rate, they are more efficient
and they reach a certain proportion of mature size more quickly.
Unfortunately, after having reached this higher proportion of
their mature size they show a low relative growth rate, compared
to the animals of the same age with a lower fraction of their
mature size. This puts a practical difficulty onto this situation. Theoretically the concept is there, but in practice for
selection purposes I cannot see how we can get around it at the
moment. Either Dr. Thiessen or Dr. Taylor might like to comment
on my difficulty in that situation.

R.B. Thiessen *(UK)*

I would just like to say one or two words. I would agree with you that if you select on relative growth rate that you would be changing the growth curve, and, as you point out, the point at which you select would have an effect. The age range which I was speaking about was, in fact, 6 to 12 months, which would correspond to what a cattle breeder might choose his replacement cattle at over that age range, the yearling growth period, so that would take take in a certain proportion of this curve. Also at that age range there is the zero correlation with mature size. What we do not know is how much variation there is in this characteristic of relative growth rate, whether there is enough to work on. Also we do not really know how it might affect body composition, if we did select on it, so there are still a lot of unknowns about relative growth rate which we would like to look at a lot further.

J. Kempster *(UK)*

I would like to ask Dr. Thiessen a question. How sensitive are your intraclass correlations to the sample of breeds involved? What would happen if you broke it down and only used the breeds that are of commercial importance? Is it sensitive on the small variation which you probably obtained in breeds which are numerically small? How sensitive is it to the breeds involved? I know it is a clear objective.

R.B. Thiessen

I think Dr. Taylor has worked out how to calculate the standard errors on these intraclass correlations. If I recall correctly it is based on the number of animals per breed plus the number of breeds, and if you reduce the number of breeds markedly, I think this would increase the standard error if I am correct in this. So the whole theory of developing these estimates between breed variants and within breed variants or intraclass correlations so expressed, in fact depend upon having a large range of breed types which span this distribution, and I would think if you do bring them down into a small group of breeds which you might like to look at which are very similar in type, that the intraclass correlation, the error on it will increase a great deal.

The idea of this multibreed approach is, in fact, to be
able to establish general relationships such as those between
body size and growth rate, body size and food intake, body size
and lactation, lactation and food efficiency. To get this regression relationship you have to get this whole span - that is
how you most accurately estimate it - and you then start to look
for deviations around this regression relationship. But if you
have a whole clump in the middle you are going to be far less
effective in establishing this relationship.

B.L. Dumont (France)

I would like to put a question, not on the magnitude of
the variation between breeds but on the possible origin of this
variation. The question is directed to Dr. Robelin. You have
found a significant difference in allometric coefficient of neck
muscles between breeds. On the other hand you have found significant differences in fatty tissue between breeds. If we consider that the growth of the neck muscles and the fatty tissue
depositions are both sexual characters, could we conclude that
the Friesian bull is not a complete male, or that the Charolais
bull is a superb male? In other words: the variation we find
between bulls, has it or has it not its origin in the amount of
hormone which could be found in the different breeds?

C.Béranger

We observe in the results, especially from Dr. Langholz,
that there is not the same interaction in heifers with the level
of feeding as in bulls. It is striking that in the past, before
we produced a lot of bull beef, we did not observe so clearly
the differences between breeds in respect of growth potential,
fatness and so on. Nowadays we see more precisely differences
between breeds, genotypes and interaction between genotype and
environment, all of which may be of importance. Are there the
differences you observe between genotypes more or less related
to their hormone status? If we compare Charolais or Limousin
castrates, is it possible to obtain the same results as if we

compare entire males of these types of cattle? Is there any
response to this aspect?

J. Robelin *(France)*

I think that part of the answer was in the text of Dr. Bergström's paper, which states that the differences which can be
found between breeds may be due to the fact that even inside
males of the same muscle weight there could be differences of
maturity. Certainly at the same muscle weight the Friesians are
older than the Charolais, for example, and that could be a reason
why the neck is more developed in these animals.

B.L. Dumont

I do not know. I look at your reasons and I find first
that the allometric coefficient for neck muscle is higher in
Charolais cattle than in Friesian. In my opinion, that means
different hormonal status because the development of neck muscles
is a sexual character.

J. Robelin

That the Charolais and the Friesian are not at the same
stage of maturity for the same muscle weight is also an explanation and I think that it would be best to have the opinion of
an endocrinologist.

D. Lister *(UK)*

I do not profess to be anything, but I am inclined to go
along with your view that it is a question of the stage of maturity that the carcass is at, at the time it is dissected,
because, if you measure circulating testosterone, for instance,
in a range of animals, it is more likely to be those animals
which are of a beefy type in which the circulating levels are
lowest. In Friesians the levels are higher, and this is a well-
known characteristic of animals with high thyroid activity; thus
I do not think that it is associated with maleness. I suspect
it is due to the relative maturity of the carcass.

C. Béranger

Are there questions or comments about variation of intake or appetite between genotypes? It is a somewhat new point that there is large variation; and also that it is perhaps correlated with the muscle growth potential. We observe in many of these data that animals with high growth potential also have a lower appetite. Perhaps it is a correlation which does not mean anything, such as the colour of animals and their appetite, but perhaps also, there is a relationship, coming from selection of these different types of animals, between the possibility of a large intake and the muscle growth potential.

Y. Geay (France)

I cannot give a complete answer but I can fill in some parts of the puzzle. We have obtained a great difference in intake capacity between Friesian and Charolais on the same diet, for example, we have found that it is about 15% higher in Friesians compared to Charolais. On the other hand, the level obtained by my colleague Vermorel in respiratory chambers shows that Charolais have a lower maintenance requirement than is the case with Friesians. The difference is about 13%; the maintenance requirement for Charolais bulls was about 100 k/cal/metabolic weight, and for Friesian, 113. Perhaps the difference in quantity of intake explains the difference in maintenance requirements.

R.W. Pomeroy (UK)

I am a little bit uneasy about a good many of these breed comparisons, unless the conditions are well defined and explained and what I mean is this. Unless you are breeding the animals yourself, the beef breed and the dairy breed, if you are taking them from breeders, the first thing is that it is very much easier in my experience, to buy representative dairy breed calves than it is to buy representative beef breed calves. But even if you have a lot of money and are prepared to pay for the beef bred calves the chances are that they will perhaps be 3 or 4 weeks old before you get them, whereas you can buy the dairy bred calves virtually at birth. The beef bred calves come from herds where suckling the calves is the general rule, so that by 3 or 4

weeks these calves have got a very strong suckling reflex established, and you have to break that if you are going to rear
them parallel with the dairy calves. In other words, beef bred
calves are subjected to a complete upheaval at about 3 weeks,
and they have a very severe set-back. I know that Hereford
calves are very difficult to rear under these conditions, and I
am willing to bet that the Charolais will suffer in the same
way. As I say, I am a little bit uneasy about the results of
these breed comparisons if these sort of conditions have been
imposed on the calves.

I think this leads on to a proposition that we do not really
know enough about the effect of growth in the perinatal and prenatal and immediately postnatal period on the growth over the
period in which we are apt to do our experiments.

C. Béranger

I can say that in France we are comparing Normandy cattle
with Charolais cattle which are reared in the same conditions
as suckling animals, suckling their mothers for 6 to 7 months,
and thereafter they are weaned under the same conditions and
thus they are reared under exactly the same feeding and management systems before they are turned out onto the feeding barn
for the fattening period. We always observe the same difference
in intake as we have observed when we compare dairy animals
coming from dairy herds, and animals coming from beef herds.

R.B. Thiessen

I just happen to have two slides here! Slide 1 gives a
graph of body weight against cumulative food intake of those
initial 9 breeds which I mentioned in our experiment in which
we have the large breeds and the small breeds, including some
breeds that are almost twice the size of others. If you are
going to compare the same body weight, 250 or 300 kg, you see
that a given large breed has only eaten about 1 000 kg of food
to get to that point, whereas the Dexter has eaten about 3 000
kg to get to that point, and so the Dexter would appear to be
far less efficient than the first animal. The dots give the

point at which the animals are one year of age, and, if you compare them at the same age which is what I was doing previously,
you find that the comparison is much fairer in that the large
breeds, although they are much bigger, have eaten quite a bit
more food, so that at one year of age it may be similar in
efficiency compared to the small Dexter at one year of age.

However, to bring up Dr. Kauffman's point as to how to compare breeds at the same degree of finish, that you should not
compare breeds of different fat levels and so forth, we are also
faced with the same problem. So again, at Dr. Taylor's suggestion,
we compare breeds at the same degree of maturity; we divide the
body weight of the different breeds by their estimated mature
size and similarly divide the food intake by their mature size.
The results are given in my second slide, which is the 'pears'
slide, and you see that all the breeds come together when compared at the same degree of maturity. At 60% maturity, which
might be a common slaughter point for most breeds, they have
eaten between 4 to 5 times their estimated mature weight of the
complete pelleted diet and at the point of calving which is up
here on my slide (where the curves are starting to break up and
where animals are going into lactation) they have eaten about 7
times their estimated mature weight in food intake, Thus, although this is not exactly the same degree of finish as Dr. Kauffman was talking about, I think it is trying to bring large and
small breeds together and you see that there does not seem to be
a great deal of variation in efficiency at that point of maturity.

C. Béranger

It seems to me that there is perhaps some problem where we
are speaking about this breed comparison. Most of the comparisons
did not include the breeds which we have studied, with a high
muscle growth potential such as Charolais and Limousin. We have
some breeds of the same mature point. For example, Limousin and
Friesians have the same mature weight and have approximately
the same growth rate, and, under these conditions, you have large
variations in appetite. If you compare animals of the same fatness, this variation becomes wider because the Limousin appetite

is decreasing with fatness, whilst the Friesian, at the same degree of fatness, remains at a high level. This is perhaps a problem because if we have this type of animal with a very low appetite, and a very high feed efficiency, we have the reverse system of what you have when you compare animals of different mature weight and which are certainly closer together if we eliminate this mature weight effect or if we eliminate the metabolic physiolgical age by comparing fatness percentage.

The question is, is there a biological effect, or is it due to very different types of animal as Dr. Taylor says, such as the colour of these animals? Are there any relationships between this intake and this protein growth potential, apart from the aspect of maturity where, at the same mature weight for different cattle, you have different growth potentialities for protein deposition? Is there any relationship with the appetite? This is certainly an important question to have answered.

J. Martin *(Belgium)*

I also have a question. For a given animal, do we admit there is an upper limit for protein deposition and for fat deposition?

A.J.H. van Es *(Netherlands)*

I think there is the example of the sheep at Rowett Research Institute, where they are feeding sheep on and on at very high feeding levels and these sheep are still putting on weight, and putting on especially, fat. So I wonder whether there is a ceiling. With protein the top is reached earlier.

C. Béranger

There is a similar effect with pigs. If the appetite centre is destroyed the pigs have a very high intake, they do not stop eating and they put on a large amount of fat.

Are there any more questions or comments about the interaction between genotypes and the levels of feeding or environmental conditions?

H.J. Langholz *(West Germany)*

I would just like to make a final remark and support the
idea of the multiple breed experiment in our work. I think we
have heard and learned a lot from this discussion and we will
have to look for general conclusions on the relation between protein deposition and fat deposition in relation to the levels of
the food and energy supply we have, and including the problem
of the feed intake. I think there are very critical and important questions for the breeder, since one generation in dual purpose breeding is 6 - 7 years, so in order to change anything we
need at least 30 years. It is somewhat more favourable on the
beef cattle breeding side. However, in view of the large invest-
ments, we have to consider the problems over and over again in
general terms and in general patterns, not forgetting what may
happen to the general policy. Therefore, I would very much
emphasise the necessity the establish the general relations and
interpretations, utilising the cornerstones of breeds as long
as we have them, as long as they are not pushed away by political
or economical stringencies which are very intensive right now.

This is what I would like to say at the end of this session.

D.N. Rhodes *(UK)*

Could I make another point, Mr. Chairman, which I think is
probably even more important than Dr. Langholz's and absolutely
supports his general thesis, and that is about meat technology
and the consumer. Unfortunately this programme in my opinion
has been put quite the wrong way round: it should start with
the consumer, and then go back to meat technology and then provide the basis for production. I have said this many times
before; it does not seem to have much effect; however, I will
say it again. Technology could completely change your thoughts
about efficiency in production. The constraints under which you
are working at the moment about size of animals and the like,
going right through all the species, are linked directly with the
retail trade with the ideas which the retail trade have about
what its consumers want in terms of a piece of meat for a particular process of cooking, and these ideas are rooted in tradition.

That retail industry must, I suppose, be the most backward of
all retail industries which we have. It does not seem to be able
to think that it is possible even to change the ideas which consumers have about the way to eat meat. In addition to that,
technology in the abattoir, the wholesale and retail trades is
also very traditional and very difficult to change or influence,
and yet it must change, it will change, and already we have many
new methods of handling meat, taking it out of carcass form into
other forms in which the identity of the muscle disappears, and
a new form of eating meat appears. Then all these size constraints
for production disappear, and you will be asked, not to produce
animals of any size but to do what I said originally - turn the
feed which we cannot use directly for human nutrition into as
much lean as you possibly can do, with the minimum amount of fat,
as long as that minimum of fat is not less than the minimum that
the consumer will ask for - and leave the rest to the technologist.
He will take it off the carcass, the muscular identity will disappear, the worst parts of the connective tissues will be removed
and a series of products of different degrees of comminution to
eliminate the disadvantages of what is left will be applied;
television advertising will make sure that the consumer will want
it; and it will be the most efficient meat - cheapest - in better
competition with other convenience foods - sold and packaged in
a way that the consumer can get it straight out of the super-
market, take it home and deal with it in a rational way.

Of course, your 30 years is about right, I should think, for
any of these sorts of ideas to get in, but how to utilise then
all of these research data, doesn't really matter I would have
thought, so long as you include all sources of animal feed and
all conditions of growing and record them accurately. In the
end they will be needed.

A.J.H. van Es
I think in your first remarks you said, "Let us look at the
end, because the consumer wishes so and so". If the consumer
likes it another way, however, you have to know how the animal
machinery works, so let us start at the beginning, let us see

how it works, and then it will be much easier to come to that final conclusion. If the market changes or if the consumer is influenced by advertising to think that something is better to eat than something else, then we know how the machinery works and then it is easier to come to any particular requirement.

This conference was called "Growth and Development" - it starts at the beginning and then it adds up.

C. Béranger

Perhaps this is the last comment because time is flying.

R. Boccard

I would like to stress the importance of good raw material, which could enable the technology to produce a final product of good quality. Therefore animal husbandry has definitely a great importance in the maintenance and improvement of quality.

SECTION 3

POSSIBILITIES FOR THE IMPROVEMENT OF BEEF PRODUCTION
IN RELATION TO THE CUSTOMER'S REQUIREMENTS

Co-ordinator:

H. de Boer

POSSIBILITIES FOR FULFILLING TRADE AND CONSUMER REQUIREMENTS FOR MEAT QUALITY IN PRESENT AND FUTURE BEEF PRODUCTION SYSTEMS

Lis Buchter

Danish Meat Research Institute
Roskilde, Denmark

ABSTRACT

The paper discusses future developments in the production and distribution of beef meat and relates these to consumer requirements and meat
quality research. Results obtained in Danish experiments with young bulls
are compiled to show the relative influence of factors connected with genotype,
fattening, and slaughter, on the colour, flavour, and tenderness of the meat.
On the basis of the results the paper concludes that consumer requirements
of uniformity and quality can be met, but these necessitate a close co-operation betweeen research, advisory systems, beef producers, and meat industry.

INTRODUCTION

The prognosis for beef foresees that the total amount to
be consumed in the EEC will increase. In order to keep prices
down more emphasis will have to be put on production systems
which, through controlled breeding and feeding, produce animals
with effective conversion rates and high carcass quality. Within the beef trade the amount of wholesale distribution will
decrease and the meat will often be delivered directly from the
slaughter plant to the retailer. At the same time both retailer
and consumer will be interested in replacing the traditional
customer dependence on the butchers' recommendations with a
system where prepacked meat can be sold with declarations not
only of weight, keeping quality and nutritive value, but also
with a declaration on eating quality characteristics such as,
for example, tenderness and flavour.

In the light of these developments the paper intends to
deal with :

How meat quality can be influenced by different beef production systems.

How research on meat quality can be a factor in ensuring
that future beef production systems not only aim at producing fast-growing animals with excellent carcass quality,
but also at incorporating consumers' requirements of good
and uniform meat quality.

CONSUMER REQUIREMENTS AND THE ROLE OF MEAT RESEARCH

What is meat quality?

As a consumer first of all I expect to get wholesome,
hygienically safe and fresh looking meat. Then I want the
price of the meat products to be low enough that I can afford
to include meat in my family's daily diet. When these requirements are fulfilled I start considering the meat quality aspects
dealt with in this paper ie colour, flavour and tenderness.

If the price depends on the quality then the most favoured cut on a beef carcass is the psoas major muscle. This muscle does not differ much in colour and flavour from several other cuts, but it is outstanding in respect of tenderness and is also appropriately called tenderloin.

Apart from the requirements for quality in such internationally accepted luxury cuts as the psoas major, it is difficult to specify consumer requirements. Traditions and opinions vary widely and it is not possible to point to a type of colour, flavour and tenderness that in general indicate the best quality.

In common for most consumers, however, is the desire to be able to get a known quality each time a specific product is bought. The high price paid for the psoas major muscle might thus not only reflect that it is a tender cut but might also cover the fact that the customer, who buys this cut, very seldom risks getting a tough piece of meat and therefore has a certain guarantee.

In relation to consumer requirements an aim for meat quality research can, therefore, be to investigate how meat quality is influenced by different factors connected with the breeding, fattening, and slaughtering of the animal. This knowledge can then be used to inform and advise beef producers and the meat industry of the possibilities of producing

1) Meat qualities to suit different customers' requirements.

2) More uniform qualities of meat.

EXPERIMENTS

Examples are compiled to show the relative magnitude with which different factors such as for example, breed, sire, weight at slaughter, feeding level, chilling, and ageing can influence meat quality characteristics. The examples are all based on results from experiments which were carried out in co-operation between the National Institute of Animal Science in Copenhagen and the Danish Meat Research Institute. To understand the background of this work an explanation of the beef production traditions in Denmark might be useful.

The majority of the beef animals are dual purpose. Each year 1.1 million heads of beef are slaughtered. Of this number young bulls constitute 47%, bulls 2%, steers 1%, heifers 12%, cows 35%, and veal 3%. As can be seen the production of steers is very low. The reason for this is that under normal conditions a steer will cost approximately 20% more to produce than a young bull of the same weight and furthermore the price for the steer will be approximately 5% lower due to higher fat and lower muscle content in the carcass.

Young bulls can be used either for production of light animals with pale, very tender veal-type meat or for production of heavier animals with red beef-type meat.

To estimate the relative influence of different factors on the meat quality, results from different series of experiments are collected. All results are obtained on the *Longissimus dorsi* muscle, which according to our experience shows some correlation with the meat quality in other muscles of the carcass. References are given to detailed descriptions of the experimental design and the methods used in the fattening period. All animals have been treated according to the standardised procedure for transport, slaughter, and ageing as described by Bech Andersen et al. (1977). Methods used for meat quality analyses are also given by Bech Andersen et al. (1977).

SIGNIFICANCE OF MEAT QUALITY CHARACTERISTICS

The colour of the carcass or the meat is often seen by the customer as an indication of the age of the animal. In several markets light meat colour is a desirable characteristic. The meat colour is measured by two methods ie, by reflectance measurements of the lightness of the meat at 535 nm, and by chemical determination of the total pigment content expressed as ppm hematin. To give some values of reference pork loins have pigment contents of approximately 25 ppm and cows have approximately 200 ppm. The change between veal and beef type colour takes place in the range 100 - 150 ppm.

The flavour in experimental beef animals is normally 'clean', but it varies of course with the maturity and age of the animal. Part of this age related development in flavour comes from an increased amount of fat in the meat. According to our experience there is a relationship between the intra-muscular fat content and the organoleptic quality. Correlation coefficients are often low ($r \sim 0.2 - 0.6$), but plots show that there seems to be a threshold value at about 2.5% intramuscular fat. Thus when the meat has a content of 2.5% or higher, it is on the average more likely to get high scores for flavour, tenderness, and juiciness in an organoleptic evaluation, while meat with fat contents lower than 2.5% might get both high and low organoleptic scores. An intramuscular fat content of approximately 2.5% corresponds to a 'just noticeable' marbling in the meat.

A characteristic of entire males as opposed to steers and heifers is that the tenderness in the loin muscle can vary considerably from one animal to another (Buchter, 1971). This variation seems to be of limited interest to the consumers of veal-type meat, but it will be one of the main difficulties for heavier young bulls sold as beef-type meat. In order to make meat from young bulls able to compete in quality with meat from steers and heifers, it is essential to decrease the variation between animals and to increase the tenderness level in general. Shear force values in very tender meat are close to 6kg, while shear force values of above 15 kg in young bulls indicate tough meat.

RESULTS

In Table 1 the influence of the genotype on .meat quality

is illustrated with results obtained on 300 kg young bulls from
a progeny test and a crossbreeding experiment. On the average
no difference is found between Red Danish (RDM) and Black Pied
Danish (SDM) cattle for colour or organoleptic qualities. By
crossbreeding RDM and SDM with beef or dual purpose sires
significant differences in colour and fat content can be found
between sire-breeds. In the results differences between sire-
breeds were also found for shear force and organoleptic qualitie
These differences might, however, depend more on the sires
than on the sire-breeds.

According to our experience the most influential genotype
factor is the sire. Results from the progeny tests for beef
production show that within both RDM and SDM, sires can be
found, which give very tender meat or which give very tough
meat. Heritability coefficients for shear force and colour are
both calculated to be approximately 0.4 with phenotypic coefficients and variation on 36% and 7% for shear force and
colour respectively (Bech Andersen et al., 1977).

During the fattening period the meat quality is influenced
by the weight. With increasing weight the colour darkens, the
fat content rises, and the meat becomes tougher especially in
the muscles with a higher content of connective tissue. The
results in Table 1 came from an experiment where both slaughter
weight and feeding level were investigated. The results for
the influence of the feeding level showed rather surprisingly
that although the age of the animals was increased considerably
by keeping the young bulls at a feeding level of 55% and 70%,
both the colour and the shear force were only slightly different
from that found in animals kept at 100% (ad libitum). The
feeding level had, of course, a considerable influence on % fat.

In experiments young bulls are usually tied individually
in stalls during fattening. In ordinary farming they are often
kept loose in boxes. In a recent experiment the two systems of
housing were compared. The results in Table 1 show that the
animals which were kept in boxes and thus were allowed to

TABLE 1

MEAT QUALITY RESULTS FROM DIFFERENT EXPERIMENTS ON YOUNG BULLS

	Factor and (reference)	Material	No of anim.	Reflectance value	Total haem ppm	% fat	Shear force kg	Organoleptic diff.
G E N O T Y P E	Breed	RDM (300 kg) SDM (300 kg)	85 92	15.7 15.6	78 72	1.05 0.97	7.7 9.2	No
	Cross-breed	DRK ⎰RDM+ Charolais x⎱SDM, Hereford (300 kg)	15 14 14	16.8 17.1 16.7	69 63 76	0.90 0.82 1.22	6.6 8.3 9.7	yes
	Sire	Highest ⎰results from SDM-progeny Lowest ⎱groups	92	16.3 14.9	78 67	1.12 0.81	13.0 6.3	yes
F A T T E N I N G	Weight at slaughter	180 kg (RDM) 360 kg (RDM) 540 kg (RDM)	24 24 24	19.0 15.6 13.3	54 86 124	0.76 1.36 1.98	6.8 7.9 9.8	yes
	Feeding level	100% ≤180-540 kg 70% ≤180-540 kg	42 42	16.4 15.6	81 90	1.72 1.25	7.6 7.9	no
	Housing	Tied (SDM, 500 kg) Loose (SDM, 500 kg)	14 28	14.4 12.8	117 146	2.9 2.4	6.8 7.7	no
S L A U G H T E R I N G	Transport	Final pH = 5.5 Final pH = 6.3	3 3	15.0 12.1	90 90	X	X	X
	Chilling	6°C ⎰RDM + SDM, 0°C ⎱450 kg	32 12	X	X	X	8.1 12.2	yes
	Ageing at 4°C	3 days ⎰RDM + 7 days ⎱SDM, 14 days ⎱450 kg	10 10 10	X	X	X	18.0 11.4 8.6	yes

exercise had considerably higher pigment content and darker meat colour than the animals which had been tied.

All the previously described results have been obtained on meat treated according to a standarised procedure which includes careful transport of the animal, relatively slow chilling of the carcass (6°), and full ageing of the meat. The results at the bottom of Table 1 illustrate how the meat quality can be damaged during slaughter. Long transport time, rough handling, or overnight stays in the slaughter house might all result in meat with high final pH (dark cutting meat), which besides having an unpleasant dark colour also has a low keeping quality. Rapid chilling results in tough loin muscles and finally all meat intended for steaks or roasts should be aged at least one and preferably two weeks post mortem.

DISCUSSION AND CONCLUSION

When discussing the possibilities of improving meat quality in beef it has to be realised that the economic incentive is low. The reason for this is that only a small part of the carcass can be sold as cuts for steaks and roasts. Thus, for each additional unit of cost a certain programme for improvement of the meat quality put on the live weight of the animal the price for roasts and steaks will have to be increased by as much as 5 - 10 units.

Beef producers will always aim at producing a maximum amount of meat for a minimum of cost, and the improvements in meat quality will, therefore, have to be included in the programmes for producing fast growing young bulls. The results from the beef production progeny tests (Bech Andersen et al., 1977) show that it is possible to improve the quantity and the quality of the meat at the same time, because no correlation has been found between production traits and tenderness.

Tender, flavourful meat, which in quality and uniformity is comparable to meat from steers and heifers, can be produced

on young bulls. The basis for such production is, however, that
farmers and the meat industry through the advisory systems or
through production contracts can be shown the significance of :

1) The use of progeny tested sires, which have shown
their ability to give young bulls with good tenderness
under the present production and slaughtering systems.

2) Proper finishing of the animals before they are
slaughtered, so that the carcass and meat contains a reasonable amount of fat to improve flavour and tenderness.

3) Direct transport from farm to slaughter, slow chilling,
and sufficient ageing.

If these instructions can be followed it will be possible
for the producers to meet consumer requirements regarding
quality and uniformity at reasonably low production costs.

588

REFERENCES

Andersen, H. Refsgaard. 1975. The influence of slaughter weight and feeding
level on the growth, feed conversion, carcass composition and conformation of bulls. 430. Beretning, National Institute of Animal
Science, Copenhagen.

Bech Andersen, B., Liboriussen, T., Thysen, I., Kousgaard, K., and Buchter,
L. 1976. Crossbreeding experiment with beef and dual-purpose sirebreeds
on Danish dairy cows. Livest. Prod. Sci., 3, 227-238.

Bech Andersen, B., Lykke, Th., Kousgaard, K., Buchter, L., and Wismer
Pedersen, J., 1977. Growth, feed utilisation, carcass quality and meat
quality in Danish dual-purpose cattle. 453. Beretning, National
Institute of Animal Science, Copenhagen.

Buchter, L. 1970. Development of a standardised procedure for the slaughter
of experimental beef animals from the Danish progeny station 'Egtved'.
16th European Meeting of Meat Research Workers, Varna, Bulgaria, Vol.1
45-56.

Buchter, L. 1971. Eating quality in commercial Danish Friesian Cattle as
related to sex and age. 17th European Meeting of Meat Research Workers,
Bristol, England. p.450-454.

Liboriussen, T., Bech Andersen, B., Buchter, L., Kousgaard, K. and Juel Møller
A. 1977. Crossbreeding experiment with beef and dual-purpose sire
breeds on Danish dairy cows. IV. Physical, chemical and palatability
characteristics of longissimus dorsi and semitendinosus muscles from
crossbreed young bulls. Livest. Prod. Sci., 4, 31-43.

Lykke, Th., Nielsen, A., Bech Andersen, B., Kousgaard, K., and Buchter, L.
1974. Progeny tests for meat production 1972-1973. 414. Beretning,
National Institute of Animal Science, Copenhagen.

TARGETS FOR BEEF PRODUCTION IN RELATION
TO MARKET AND CONSUMER REQUIREMENTS

A. Cuthbertson and G. Harrington
Meat and Livestock Commission
P.O. Box 44 Queensway House, Bletchley,
Milton Keynes, UK

ABSTRACT

This paper is not concerned with growth and development as such, but with the implications for future beef production of trends in market and consumer requirements.

To maintain the demand for meat against an increasing range of consistent, convenient and sometimes cheaper products, there appears to be an increasing need to market consumer cuts in a form which maximises convenience, minimises waste and ensures consistent textures, as well as providing value for money.

The meat industry now has the technology to make the most of the available meat to satisfy these requirements and to do so much more efficiently. This involves removing constraints which inhibit change and adopting a new and integrated approach to beef slaughter, dressing and meat preparation.

The producers task must be primarily to produce lean tender meat at the lowest possible cost. Growth studies suggest that efficient lean meat production will involve a variety of breeds and production systems and this will lead to variations in the weight, fatness and shape of cattle at slaughter. However, given a more flexible and innovative approach to processing and marketing, such variations should not present the industry with a serious problem.

INTRODUCTION

The papers presented so far in this Seminar have considered the effect of genotype and environment on body growth and
development in beef and on the type of carcass produced. The
aim of this paper is to try to put this work into the perspective of likely trends in meat marketing and consumption and to
assess the extent to which an appreciation of these may permit
realistic beef production targets to be drawn up.

The structure of the meat industry, at least in Britain,
has led to the development of a series of constraints which have
inhibited innovation along the marketing chain. The traditional
pattern of slaughtering, distribution and retailing is only
slowly changing, although the particular economic pressures of
the last year or two are causing all sections of the meat industry to examine more closely the efficiency of their operation
and the type of product they market. In industry terms, there
is the beginning of a close re-examination of resource use. In
view of the long cycle of beef production it is particularly
important to anticipate consumer and meat marketing methods some
years ahead. Also, are market considerations in conflict with
efficiency considerations?

It seems logical to start by assessing the future needs
of the consumer and then to work back along the meat marketing
chain to see how these are likely to be met by retailers, whole-
salers and, ultimately, by producers.

TRENDS IN MEAT BUYING AND EATING HABITS

Of the major beef consuming countries of the EEC, Germany
appears to be showing the most marked increase in per capita
consumption; by 1980 beef consumption per head is estimated to
be nearly double the level of the early 1950's (OECD, 1977$_1$).
By contrast, in spite of increased disposable income (after adjusting for inflation), beef consumption per head in Britain was
similar in 1976 to that of 1955 and, by the mid 1980's is pred-

icted to be 13% less than in 1955 (Bansback, 1977). Processed meat products may be less affected by any decline in consumption than fresh beef sales. This is partly because under economic pressures the cheaper end of the market range is favoured and partly because of the continuing switch towards convenience foods, with some 45% of housewives in Britain and every third housewife in Germany (Liebler, 1977) now engaged in some form of paid activity outside the home and, therefore, having less time to spend on cooking. There is also a change of attitude, with many young people having neither the skills nor the desire to spend time on cooking requiring significant preparation time. Admittedly, promotional agencies are reacting to the social changes by emphasising to school children the merits of good nutrition, meat as a food and simple methods of preparation. A further important factor is that there is considerable potential for economies in processing beef by the use of extenders, etc., so that not only are processed meats generally less variable and more reliable than fresh meat in eating qualities, but they can be reduced in cost to a degree by formula manipulation.

The Germans already eat much of their beef in processed form (some 60% quoted by Grünewald, 1972), reflecting the fact that traditionally German households serve cold meats for the evening meal and no doubt influenced by the traditional need to improve and extend as far as possible the available meat.

In France, on the other hand, there is dominance of grilling with some roasting so only about 20% of beef is consumed in processed form (Grünewald, 1972). The French butcher is noted for the effort which he applies in preparing meat which involves separating out many individual muscles or muscle blocks and trimming off connective tissue in order to make more of the meat suitable for quick cooking methods. Even so, the French use some 1½ times more hindquarter than they do forequarter meat. Britain occupies an intermediate position with about a third of the beef consumed in processed form - although much of the raw material for processed beef products is, of course, imported manufacturing quality.

Also relevant is the growth in freezer ownership (ie deep freezers separate from refrigerators). It is forecast that by 1985 about half the households in Britain will own deep freezers compared with about one third at present (Bansback, 1977). This growth could have the important effect of persuading the consumer who in Britain has traditionally believed (with some justificatio frozen beef to be an inferior imported product, that the highest quality home produced beef can retain all its qualities if frozen and stored properly. If this were to happen, it opens the door to the merchandising of frozen retail cuts which could have a dramatic effect. The development of the frozen meat market in France has been slow reflecting a consumer preference for fresh meat (OECD, 1977_2). In Britain, the growth of self-service of pre-packaged meat has been slower than in the USA and certain European countries, but it is clear that even in Britain the purchase of meat is likely to become a less frequent event.

It is probable that in all member countries more meat will be bought through supermarkets and hypermarkets in the course of buying other goods, and that the decline in the number of specialist service butchers will continue. In Britain, sales of meat in stores selling groceries and other items now total some 36% of the total market for beef, lamb, pork and offal, leaving 64% in service butchers' shops whether owned by multiples or individual traders, and even within the last year there has been a 2% swing against the latter. However, beef of all the meats has the highest proportion still being sold through service shops (67%) indicating that the consumers' need for advice, service and reassurance is greatest for beef. Where meat is sold in a counter service situation consumer choice tends to be more influenced by the practices, preferences and advice of individual retailers than in a self-service situation. In Britain, at least, this has an important bearing on, for example, the fatness of beef offered to the housewife.

The relatively high level of fatness of British beef compared to most other EEC countries often puzzles European observers. The percentage of fat in British beef carcasses

(excluding cow beef) is higher on average than in other EEC
countries (except Eire) probably by some 5% units of dissectable
fat in the carcass, while Britain, in turn, has leaner carcasses
than those in the USA by, perhaps, another 5% units of dissectable fat. The difference between Britain and continental Europe
is due in part to the greater contribution of early maturing
cattle to beef production and to the small percentage of males
reared as bulls (under 5%). This situation has been slow to
change because there is still a strong belief among many in the
meat trade, and a limited number of their customers, that fat
is essential to achieve good eating quality in the meat, although
extensive testing has failed to show that fatness contributes
to the eating quality of the lean meat from young animals. However, there is evidence that consumer resistance to fat in
Britain is increasing. Rhodes (1977) gives survey results which
show that approximately half the consumers questioned rejected
fat on the plate, and multiple grocers confirm that housewives,
and the younger ones in particular, are rejecting fat in self-
service shops. Not unnaturally, given the fatness of existing
supplies, butchers want in the short term to sell as much fat
on their beef as possible. While some of the trimmed fat can
be used in processed products it would be more efficient, with
greater fat trimming being forced upon butchers, if less fat
were produced initially. We expect, therefore, that under consumer pressure the trend towards leaner beef in Britain will
continue.

It is appropriate to recognise at this stage that lean
meat production in cattle will always be associated with some
fat deposition. The fat carried by the lean meat of the animal
slaughtered at its economic optimum (that is when the feed cost
per unit of lean meat produced is lowest) may be considered a
'free' by-product. Recent studies have suggested that about a
fifth of the fat produced on cattle in Britain arises from
their being carried beyond the point of optimum efficiency
(Harrington and Kempster, 1977), and that this factor is much
more important in overall productive efficiency than the precise

amount accepted by consumers. However, selection for greater
effiency may well lead to cattle which reach optimum lean meat
production at lower levels of fatness.

Other important factors at the time of purchase to emerge
from consumer attitude studies, at least in Britain, are the
convenience aspect of the food, referred to earlier, the fresh-
ness and likely eating characteristics of the meat reflected by
its colour, leanness and to some extent apparent texture, and
its value for money. At the point of consumption, tenderness
appears to be a key factor and to maintain the demand for meat
against an increasing range of consistent, convenient, nutritiou
and sometimes cheaper products, there appears in our view an
increasing need to market consumer cuts in a form that maximises
convenience, minimises waste and ensures consistent textures.

TRENDS IN MEAT PROCESSING AND MARKETING

How is the industry going to satisfy these consumer requir
ments in future? The short answer is that the producers task
must be primarily to produce lean tender meat at the lowest
possible cost. Given this, the meat industry now has the
technology to make the most of the meat to satisfy the requirements noted above and to do so much more efficiently.

Quite an impressive case is being built up in Britain and
America for a whole new technological approach to beef slaughter
and dressing. It appears that improvements in handling cost,
yield, tenderness, bacteriological status and appearance can
be achieved if cattle are slaughtered under minimum stress
conditions followed by electrical stimulation, hot deboning
using muscle seaming, vacuum packing and quick chilling with
the meat ready to leave this part of the meat plant within 5
hours of sticking. For the lower valued parts of the carcass,
hot boned meat also has advantages if used in manufactured
products, and the possibility of upgrading the meat from these
parts of the carcass by reforming and extension not only with

textured vegetable proteins but also secondary animal proteins
(from mechanically deboned meat, blood, bones, etc.) is very
real.

By starting with leaner carcasses, it should be possible
to incorporate all the fat from the body into the meat products,
while at the same time effecting a reduction in the average
fatness of the meat and meat products consumed. Any disadvantage
which lean carcasses might have in terms of increased evaporat-
ive losses over fatter carcasses under conventional chilling
conditions in side form should be avoided by the hot boning and
vacuum packing operation.

After conditioning, the vacuum pack primals would be cut
into retail sized portions prepared for individual wrapping and
possibly freezing and, in addition to moulding, the cutting
process applied at each stage would take whatever steps were
necessary to produce fairly small, uniform sized pieces.

Some tentative estimates have been made of the improvement
in efficiency which might be achieved by adopting such new
technology as outlined above. These suggest an improvement of
as much as one third.

IMPLICATIONS FOR THE PRODUCER

In order to produce the maximum of lean meat from the
calves available with the greatest efficiency, the producer
must have a good feedback of carcass information (primarily
weight and estimated lean content) linked to price incentives
and penalties to encourage the necessary changes. It is this
requirement for the systematic communication of meat trade
requirements back to the producer in a clear and comparable
way that leads us to attach so much importance to carcass
classification schemes - against the wishes of a majority of
meat traders who fear their flexibility in trade will be
reduced. Classification information is valuable in giving the
producer guidance in the short term on when to slaughter in

order to obtain the best return, while in the longer term, classification data can assist him in assessing the merits of alternative production systems and in helping him to decide which breeds or crosses meet his targets (Meat and Livestock Commission, 1975_1). Using such information, we have experimented with simple indices of carcass quality (essentially the percentage achieving low fatness in conjunction with superior conformation) which demonstrate effectively differences between breeds and systems in the types of carcass produced (Meat and Livestock Commission, 1975_2). However, improved carcass quality is only one factor to be considered in the economics of production.

Much of the beef produced in Europe derives from the dairy herd and the principal milk producing breeds have become dual purpose. There has been a general interest in using sire breeds which will confer more rapid growth rates and later maturity to the progeny than sires of the dam breeds. In Britain, there are about twice as many dairy as beef cows and the scope for using such crossing sires is limited to beef cows and to dairy cows not needed to produce replacements for the dairy herd. The dairy herd is not static and the long term scope will clearly be influenced by milk production policy within the EEC.

Only in recent years has work been started to examine the overall efficiency of the different sire breeds. Work in Britain has concentrated on comparing breeds at the same level of fatness, and the evidence so far is that breeds of widely differing types (eg Aberdeen Angus and Charolais) have similar efficiencies of converting feed into meat. However, the comparisons have not yet been carried out with sufficient replications to be sure that real differences of 5 to 10% do not exist. Further, making the comparisons at the same level of fatness may not be revealing the true differences between breeds in their efficiency, since different breeds may not reach their optimum at the same level of fatness. In order to examine this, a serial slaughter programme is being introduced in the MLC sire breed comparison work.

If the later maturing breeds do prove to have an advantage
in food conversion, or even if the differences between breeds
in their efficiency of converting feed into meat do prove to be
insignificant, then their use would allow the calf costs to be
spread over a greater carcass weight and also the costs of
slaughtering and subsequent meat handling would be spread over
a greater weight of meat. However, these later maturing animals
eat more food in total and yield a slower return on capital
invested, so it is important when comparing the various breeds
to take account of all important economic factors. While the
later maturing animals may turn out to be better under good
grazing conditions, there may be a place for the earlier maturing breeds in hill and upland areas where there is a shorter
grazing season and where it may be difficult for the later maturing breeds to perform efficiently.

A factor which has an important bearing on the choice of
sire breed and which may well weigh against the larger breeds
is dystocia. Some of the later maturing breeds have on average
a markedly higher incidence of calving difficulties than, for
example, the Hereford or Aberdeen Angus. There may be important
differences between the later maturing breeds, but there are
also important differences between bulls within breeds. By
careful selection, therefore, it should be possible to minimise
calving difficulties.

A further problem is that some of the later maturing
breeds fail to colour mark (that is to produce crossbred calves
from Friesian cows with colour patterns distinctive to the sire
breed), but this would be less of a problem given improved calf
marketing arrangements.

As it turns out, the later maturing breeds tend to have
both good conformation and a better lean meat to bone ratio
than, say, the Friesian. While there is not a great deal of
variation between and within breeds in the way in which lean
meat is distributed throughout the body, some of the later
maturing breeds appear to have some advantage. For example, in

MLC studies covering a wide range of breed types, Limousin cross Friesian carcasses at one extreme had some 3% more of their lean meat in the higher priced cuts than purebred Ayrshires at the other, which represents a small but important advantage (Kempster et al., 1976$_1$). The apparent relationship sometimes seen between conformation and composition (ie lean meat to bone ratio at constant fatness) appears to be a between breed phenomenon. Within breeds, there does not appear to be a good relationship between conformation and either the lean meat to bone ratio or the percentage of lean meat in the higher priced cuts. However, at the same weight and fatness, good conformation carcasses will tend to have the meat more thickly disposed than those with poorer conformation, but the true economic significanc of this has yet to be established.

Marked differences have been found between breeds in fat distribution (Kempster et al., 1976$_2$). The beef breeds as opposed to the predominantly dairy breeds tend to have a more favourable distribution having more of their total fat deposited subcutaneously. This has a bearing on the ease with which fat can be trimmed in preparing retail cuts of meat since it is less easy to trim intermuscular than subcutaneous fat, and it may alsc have implications for the accuracy of predicting lean meat content when one has to rely on assessments of the intact carcass.

While there may be little financial incentive at present for producers to make greater use of the later maturing crossing sires, one step which could have an immediate impact on efficiency is to stop castrating male calves. Within the EEC, Britain, Eire and France appear to be the outposts of beef production from male cattle which have been castrated. In Britain, rather less than 5% of males are reared as bulls, and any growth in bull production will have to come in the short term from cattle reared in yards because of limitations which have been imposed on grazing for reasons of public safety. At the meat trade end, there are some fears about young bulls, but they are now finding a ready market where lean carcasses are sought, and it seems that careful handling before slaughter can minimise the risk of dark cutting meat.

An implication of the foregoing discussion for beef production in the future is that if efficient lean meat production from a variety of breeds and husbandry systems is accepted, then the industry will have to develop procedures for handling variations in weight, fatness and shape. Variations in weight and fatness will arise since some breeds may reach their optimum efficiency at a higher weight and fatness level than others. However, such variations should not present the industry with a problem given a more flexible and innovative approach to processing and marketing through to the retail display cabinet.

A period of generalised growth studies must now be followed by studies of breed differences in growth patterns in order to have a better understanding of the concept of 'optimum' slaughter, and in order to determine how the differences between breeds may be exploited in different production, rather than marketing situations.

NOTE ON OPTIMUM LEAN MEAT PRODUCTION

The paper argues that the main objective in beef production should be increasingly to maximise the efficiency of converting resources (particularly feed) into lean meat. Following from this, studies on beef growth and development should be designed to find those combinations of breeds (and possibly sexes) and production situations which maximise this efficiency. For each type of cattle, there is an optimum point of growth under a particular system of production at which the lean meat produced has least unit cost. This occurs when the cost of production, involving the cost of the calf (where appropriate) its feed, housing, etc, when calculated per unit of lean meat is least. The least cost lean meat will be associated with a certain weight of fat and bone, all three together determining the optimum carcass weight.

There is bound to be some fat associated with the lean meat produced during growth, and in assessing efficiency least fat must not be construed as meaning greatest efficiency in the sense of converting the available resources into lean meat. This is because there are some overhead costs which, relative to lean meat, are highest at the very beginning of the growth period. On the other hand, it is unlikely that the most efficient converters will be among the fattest cattle, partly because the weight of fat will be contributing to the total body weight which has to be maintained, and partly because deposition of fat demands some four times as much energy as lean meat per unit weight.

The calculation of optima in commercial terms will clearly be affected to a degree according to whether fat is given a valuation of nothing, or something representing an average processing value. It will also depend on the calf cost assumed. One may, for example, take a situation where the whole cost of feeding the cow is to be borne by the calf or, alternatively, one may take a lower calf cost to reflect the market value for calves from cows kept mainly for milk.

There is little information available for lean meat optima for cattle under current conditions. Serial slaughter studies have been carried out recently by Frood (1976), Kay (Report to the Meat and Livestock Commission, 1976) and Andersen (1975). Using these data and assuming a total calf cost of 16 000 MJ (for the situation where the whole calf and cost of feeding the cow is to be borne by the calf), the data of Kay suggest an optimum weight for Hereford x Friesian and Friesian steers of the order of 400 kg liveweight whereas in the data of Frood (1976) it would be somewhat lower. Both Frood's and Andersen's data suggest it could be approaching 450 kg for bulls From the dissection data in these trials, it is calculated that optimum slaughter points lie within the range 4 - 8% subcutaneou: fat in carcass.

Data from the above trials, other dissection studies, and the application of a beef carcass classification scheme in Britain, has enabled the estimate of waste fat production reported in the foregoing paper to be given.

REFERENCES

Andersen, H.R. 1975. The influence of slaughter weight and feeding level on
the growth, feed conversion, carcass composition and conformation of
bulls. 430. Beretning Fra Statens Husdyrbrugs Forsøg, København, 1975.

Bansback, R.J., 1977. United Kingdom meat comsumption in 1979, 1982 and
1985. Proceedings of Meat and Livestock Commission Symposium "Meat
demand and price forecasting", Wye, 1977.

Frood, I.J.M. 1976. An investigation into the effect of sex and plane of
nutrition on the growth performance and carcass quality of British
Friesian cattle for beef production. Ph.D. Thesis, University of
Reading, 1976.

Grünewald, L., 1972. Differences in livestock and meat marketing in EEC
countries. Proceedings of Meat and Livestock Commission Conference
"British meat and the Common Market", Harrogate, 1972.

Harrington, G. and Kempster, A.J., 1977. An examination of the fatness of
commercial British beef carcasses and estimates of waste fat production. Anim. Prod., 24, 152. (Abstr.). Copies of the full paper are
available from the authors on request.

Kempster, A.J., Cuthbertson, A. and Smith, R.J., 1976_1. Variation in lean
distribution among steer carcasses of different breeds and crosses.
J. agric. sci., 87, 533-542.

Kempster, A.J., Cuthbertson, A. and Harrington, G., 1976_2. Fat distribution
in steer carcasses of different breeds and crosses. 1. Distribution
between depots. Anim. Prod., 23, 25-34.

Liebler, H.E., 1977. The German pig meat trade - lessons for Britain.
Proceedings of Meat and Livestock Commission/Farmers Weekly Conference "Pig Meat - increasing the market share", Bristol, 1977.

Meat and Livestock Commission, 1975_1. Progress on beef carcass classification. Marketing and Meat Trade Technical Bull., No. 22.

Meat and Livestock Commission, 1975_2. Guidelines for beef carcass improvement. Marketing and Meat Trade Technical Bull., No. 18.

O.E.C.D. 1977_1. Some aspects of the beef chain in the Federal Republic of
Germany. Proceedings of O.E.C.D. Symposium "Towards a more efficient
beef chain", Paris, 1977.

O.E.C.D. 1977_2. The beef chain in France. Proceedings of O.E.C.D. Symposium
"Towards a more efficient beef chain", Paris, 1977.

Rhodes, D.N. 1977. Breed and eating quality. British Cattle Breeders'
Club Winter Conference, Cambridge, 1977.

EATING QUALITY OF BUFFALO AND BOVINE CALVES SLAUGHTERED AT 20 - 28 AND 36 WEEKS OF AGE

A. Borghese, A. Romita, S. Gigli, A.Di Giacomo
Istituto Sperimentale per la Zootecnia, Rome - Italy.

ABSTRACT

Some physical characteristics (colour, cooking losses and Warner Bratzler Shear force values) and taste panel assessment were determined on some muscles from 30 Friesian male calves and 30 water buffalo male calves reared in identical conditions of environment and feeding and slaughtered at 20, 28 and 36 weeks of age.

There is no significant difference between species for meat physical characteristics.

At panel taste, buffalo meat cooked in an open pan is significantly more tender and juicy only when taken from animals at 36 weeks.

Generally reflectance percentage, cooking losses tenderness and panel taste mean scores decrease as the age increases.

INTRODUCTION

People think that meat quality of water buffalo *(Bubalus bubalis L.)* is less good that bovine meat, having a poor flavour that the consumer does not like. However, this poor flavour is found in old animals' meat, and not in young ones according to Wilson (1961), Ragab et al. (1966), Ferrara et al.(1969), Joksimovic (1969), Charles and Johnson (1972), Cockrill (1974), Ognjanovic (1974), Borghese et al. (1976).

Therefore, we intend to study if, and at which age, considerable differences appear on physical characteristics (tenderness, colour and cooking losses) and taste characteristics (tenderness, flavour and juiciness) between the two species, reared in identical conditions of feeding and environment.

In the present paper we report data on the animals slaughtered up to 36 weeks of age.

MATERIAL AND METHODS

The trial was carried out on 30 bovine male calves and 30 water buffalo male calves reared under identical conditions of feeding and environment and slaughtered at 20 - 28 - 36 weeks of age.

A meat sample, by removing the *longissimus dorsi* muscle between the 12th and 13th rib, was employed to evaluate colour with colorimeter Manuflex INRA, at 525 mµ wave length. Afterwards the same sample was equally subdivided in two pieces and one was cooked at 120°C in boiling oil and the other at 180°C in an oven, until the internal temperature reached 70°C, to calculate cooking losses and to remove three cores (1" diameter) for Warner Bratzler Shear determinations.

A panel taste trial was also carried out: *semitendinosus* muscle was cooked in a pressure-cooker for 30 minutes and *semimembranosus* slices on an open pan for 2 minutes without sauces:

TABLE 1

MEAT PHYSICAL CHARACTERISTICS

	20 weeks		28 weeks		36 weeks	
	Calves Mean	Buffaloes Mean	Calves Mean	Buffaloes Mean	Calves Mean	Buffaloes Mean
Colour						
Reflectance percentage	41.6a	43.4a	42.0a	42.5a	32.2b	33.0b
Cooking losses:						
Boiling oil	34.1a	33.4a	21.6b	23.2b	22.3b	26.0b
Oven	19.9a	18.5a	15.1b	14.0b	12.4b	11.3b
Warner Bratzler values:						
Boiling oil	4.8a	4.0a	5.4ab	5.8ab	7.5b	7.5b
Oven	4.9a	4.1a	5.0a	5.5ab	6.4b	6.1b

Different letters on the same line mean significant differences for $P \leq 0.05$.

each component of a 15 members panel, rated different meat
samples to identify animal species and to evaluate tenderness,
flavour and juiciness on a nine point hedonic scale (Borghese
et al., 1976): 6 900 scores were carred out.

RESULTS

Meat physical characteristic

The mean values of meat colour, cooking losses and Warner
Bratzler Shear force are reported in Table 1.

For all the considered parameters there is no significant
difference between species at the same age.

Meat colour is similar at 20 and 28 weeks but it becomes
darker at the latter age ($P \leq 0.05$).

The variability of cooking losses and of tenderness values
is very high; the values tend to decrease with the age: the
significance of differences is also shown in Table 1.

Significant differences between cooking methods apppear at
all ages for cooking losses and only at 36 weeks of age as far as
Warner Bratzler values are concerned.

Panel taste

The percentage of judges who identified the difference
between the species in meat sample was 15 - 20% without differenc
between cooking methods.

Panel taste mean scores are reported in Table 2. Generally
the judges overscored water buffalo meat in a significant way
as far as tenderness and juiciness of meat at 36 weeks of age
cooked in an open pan.

As a general trend the mean scores decrease with increase
in age of the animals.

TABLE 2

PANEL TASTE MEAN SCORES

	Calves			Buffaloes		
	20 weeks	28 weeks	36 weeks	20 weeks	28 weeks	36 weeks
Pressure-cooker						
Tenderness	7.81a	7.82a	7.46b	7.97a	7.76b	7.52c
Flavour	7.28	7.27	7.30	7.49a	7.20b	7.35ab
Juiciness	5.47a	5.50a	6.07b	5.65a	5.45a	6.25b
Open pan						
Tenderness	7.37a	7.25a	6.31b	7.60a	7.34b	6.63c
Flavour	7.11a	6.39b	5.93c	7.12a	6.31b	6.03c
Juiciness	5.74a	5.37b	4.63c	5.77a	5.51a	4.88b

Different letters in the same line, for calves and buffaloes separately, mean significant differences for $P \leq 0.05$ among the ages.

Judges scored better (P≤ 0.05) meat cooked in pressure cookers for tenderness and flavour at all the ages, but for juiciness only at 36 weeks of age.

CONCLUSIONS

Water buffalo meat, on visual inspection by judges, was lighter than calf meat, and colorimeter confirmed this fact according to Matassino et al. (1976$_2$), even if differences were not significant; also Joksimovic and Ognjanovic (1977) did not find significant differences in myoglobin and total pigment content between the two species. Calf meat, according to most authors cited by Matassino et al. (1976$_1$) and water buffalo meat (Intrieri et al., 1972; Matassino et al., 1976$_2$) becomes darker at increasing ages of the animals.

Cooking losses also decrease with animal age and change with cooking method (Bouton and Harris, 1972; Shaffer et al., 1973; Bouton et al., 1976): in fact heat and water content transfer is much faster through boiling oil than through air in a closed oven, without significant difference between species.

Water buffalo meat tenderness at Warner Bratzler Shear and at panel taste judgment decreases as the age increases and calf meat shows a similar trend according to most authors (Dumont 1952; Tuma et al., 1963; Walter et al., 1963; Adams and Arthaud 1963; Webb et al., 1964; Romans et al., 1965; Boccard 1966; Sliger et al., 1966; Hunsley et al., 1971; Brekke and Wellington, 1972; Prost et al., 1975; Dutson et al., 1975, Reagan et al., 1976).

Flavour and juiciness mean scores also decrease with age (P≤ 0.05) except for juiciness with pressure cooking, which was better at 36 weeks of age, both for calves and buffaloes (P≤ 0.05), contrary to Hunsley et al.(1971), where taste panel scores for flavour, juiciness and tenderness at different ages were not significantly different.

In conclusion water buffalo meat had similar results to calf meat and on many occasions even better scores.

REFERENCES

Adams, C.H. and Arthaud, V.H. 1963. Influence of sex and age differences on tenderness of beef. J.Anim. Sci. 22, 1112.

Boccard, R. 1966. Age and meat production. Tierzucht. Zucht. Biol. 82, 271.

Borghese, A., Romita, A. and Gigli, S., 1976. Prove comparative fra vitelli bovini e bufalini allevati fino a 20 settimane di età. III. Caratteristiche fisiche ed organolettiche della carne. Ann. Ist. Sper Zoot, 9, 163-170.

Bouton, P.E. and Harris,P.V.,1972 The effects of cooking temperature and time and some mechanical properties of meat. J.Food. Sci.37, 140.

Bouton, P.E., Harris, P.V. and Shortose, W.R. 1976. Peak shear-force values obtained for veal muscle samples cooked at 50 and 60°C: influence of ageing. J. Food Sci. 41, 197-198.

Brekke, C.J. and Wellington, G.M. 1972. Effect of animal weight on palatability of veal roasts. J. Anim. Sci. 35, 937-940.

Charles, D.D. and Johnson, E.R., 1972. Some carcass characteristics of the Australian water buffalo in 'The 9th Conf. Austr. Soc. of Anim. Prod.'

Cockrill, W.R. 1974. Alternative livestock: with particular reference to the water buffalo (Bubalus bubalis): in Cole, D.J.A. and Lawrie, R.A. Meat 21 Easter school in Agricultural Science University of Nottingham Butterworth. 507-524.

Dumont, B. 1952. La tendreté de la viande. Ann. Zootech. 3, 71-95

Dutson, T.R., Smith, G.C., Hostetler, R.L. and Carpenter, Z.L. 1975. Postmortem carcass temperature and beef tenderness. J. Anim. Sci., 41, 289.

Ferrara, B., Minieri, L., De Franciscis, G. and Intrieri, F. 1969. La produzione della carne bufalina in Italia. Acta Med. Vet. 15, 313-366.

Hunsley, R.E., Vetter, R.L., Kline, E.A. and Burroughs, W. 1971. Effect of age and sex on quality, tenderness and collagen content of bovine longissimus muscle. J. Anim. Sci. 33, 933-938.

Intrieri, F., Zicarelli, L., Di Lella, T. and Rinaldi, G. 1972. Sualcune caratteristiche chimiche, fisiche e chimico-fisiche del muscolo Longissimus dorsi di vitellone bufalino. Acta. Med. Vet. 18, 78-87.

610

Joksimovic, J. 1969. Physical, chemical and structural characteristics of
     buffalo meat. J.Sci. Agric. Res. 22, 110-151.

Joksimovic, J. and Ognjanovic, A. 1977. Comparison of carcass yield carcass
     composition and quality characteristics of buffalo meat and beef. Meat
     Sci. 1. 105-110.

Matassino, D., Girolami, A., Colatruglio, P., Cosentino, E., Votino, D. and
     Bordi, A. 1976$_1$. Colour evaluation of muscles in 10 bovine crossbreeds.
     EEC Agric. Res. Seminar, Crit. Meth. Assess. Carcass Meat Charact. Beef
     Prod. Exper., Zeist (1975), EUR 5489,285-299.

Matassino, D., Romita, A., Girolami, A. and Cosentino, E. 1976$_2$. Studio
     comparativo fra bufali e bovini sulla qualità della carne. II. Colore
     della carne all'età di 20 settimane. Atti del II° Convegno Nazionale
     ASPA Bari, 387-392.

Ognjanovic, A. 1974. Meat and meat production, in 'The Husbandry and Health
     of the Domestic Buffalo'. Ed. W. Ross. Cockrill. F.A.O Roma.

Prost, E., Pelczynska, E. and Kotula, A.W. 1975. Quality characteristics
     of bovine meat. II. Beef tenderness in relation to individual muscles ,
     age and sex of animal and carcass quality grade. J. Anim. Sci, 41,
     541-547.

Ragab, M.T., Darwish M.Y.M. and Malek, A.G.A. 1966. Meat production from
     Egyptian buffaloes. II. J. Anim. Prod. UAR. 6, 31.

Reagan, J.O., Carpenter, Z.L. and Smith, G.C. 1976. Age-related traits
     affecting the tenderness of the bovine *longissimus* muscle. J.Anim. Sci.
     43, 1198-1205.

Romans, J.R., Toma, H.J. and Tucker, W.L. 1965. Influence of carcass
     maturity and marbling on the physical and chemical characteristics of
     beef. I. Palatability, fiber diameter and proximate analysis. J. Anim.
     Sci. 24, 681.

Shaffer, T.A., Harrison, D.L. and Anderson, L.L. 1973. Effects of end point
     and oven temperatures on beef roasts cooked in oven film bags and open
     pans. J.Food Sci. 38, 1205-1210.

Sliger, R.L., Ramsey, C.B., Cole, J.W. and Hobbs, G.S. 1966. The relationships
     of chronological and physiological age of beef females to carcass and
     palatability characteristics. J. Anim. Sci. 25, 255.

Tuma, H.J., Henrickson, R.L., Odell, G.V. and Stephens, D.F. 1963. Variation
     in the physical and chemical characteristics of the *longissimus dorsi*
     muscle from animals differing in age. J. Anim. Sci. 22, 354.

Walter, M.J., Goll, D.E., Anderson, L.P.and Kline, E.A. 1963. Effect of
    marbling and maturity on beef tenderness. J. Anim. Sci. 22, 1115.

Webb, N.B., Kahlenberg, O.J. and Naumann, A.D. 1964. Factors influencing
    beef tenderness. J. Anim. Sci. 23, 1027.

Wilson, P.N. 1961. Palatibility of water buffalo meat. J. Agric. Soc. Trin.
    61, 457-459-460.

DISCUSSION ON SECTION 3

R. Boccard *(France)*

The first question is to Dr. Buchter. She mentioned
values for good tenderness, for medium tenderness and for tough
meat in terms of 10 and 15 kg. I am a little afraid of these
values - what are the other units you used?

L. Buchter *(Denmark)*

I presented the values in order to make it possible for
you to interpret the data in the Table so that effects of sire,
ageing etc. could be compared. The values are relative to the
methods we use and to the figures given in the Table.

R. Boccard

What was the size of your core?

L. Buchter

2 cm^2, but again it depends upon the jaws you use.

R.B. Thiessen *(UK)*

I have a question for Mrs. Buchter. Professor Lawrie
indicated that there were a number of factors which contributed
to tenderness, and you have found sire differences. Could you
say what you feel these differences in tenderness might be due
to, in histological terms. How do you feel that you could
select sires, particularly at an early age, maybe using muscle
biopsies, to take advantage of this?

L. Buchter

You are touching down on to a very tricky area, but I
think that the differences found in young bulls come partly from
myofibrillar toughness, and may be different degrees of cold
shortening during the handling, even if you standardise it at a
very good treatment, but it also probably depends a lot on the
connective tissue. These things need more attention, but these
are just our practical results. As for how we could select for
this trait, it seems clear that it should be done at progeny

testing stations. At the stage where we are now we do make a
meat quality analysis on our progeny test for meat and by that
we can trace the sires which give extreme toughness. We have
had sires which have produced 18 young bulls and, even when the
meat was completely aged and every technological care was taken,
a value of 20 kg was obtained as an average of 18 young bulls;
and other sires give an average of 7 kg. So the whole range,
from tender to tough, is covered.

R.B. Thiessen

Is temperament involved here?

L. Buchter

We have sometimes asked the man at the progeny testing
station to record the temperament. However, these are very
subjective evaluations.

A. Cuthbertson *(UK)*

I would like to raise this question of eye muscle size
and tenderness which our French colleagues have been talking
about, suggesting that there may be a linkage due to the connective tissue. Have you any evidence that the differences between
the animals that you have observed might be a reflection of
differences of muscle size?

L. Buchter

Yes, there is a small amount of evidence in our cross
breeding experiments; in his thesis work Liboriussen found something which could sustain this idea. As far as I know, there
was no evidence in our dual purpose breeds.

D.E. Hood *(Ireland)*

A question to Lis Buchter also, in relation to the level
of intramuscular fat which she finds desirable from an eating
quality point of view. We have heard a lot about lean meat production and, clearly, from these results, it would appear that
it is desirable to have some fat present. In the figures which
refer to young bulls, do you consider an equivalent amount of

fat desirable in other types of beef? Secondly, what level of
fatness in the carcass corresponds to this level of intramuscular
fat?

L. Buchter

The evidence is from the same experiment that Liboriussen
spoke about yesterday. You have liveweights from 180 up to
540 kg and feeding levels between 100% and 55%. What I recommend
is close to 2.5% intramuscular fat. It is difficult to achieve
in fast growing young bulls. You have to operate the 100% feeding level in order to get up to that; at 85% it is already
lower. You ask how much fat this corresponds to in the carcass,
but I cannot specify this.

A.J.H. van Es *(Netherlands)*

Many people prefer the meat of deer and pheasant to that
of beef. I wonder why - is it tenderness, or flavour, or fat
content? Or something else?

H. de Boer *(Netherlands)*

It is expensive and so it may be associated with quality.
If there is a reason, I think it would be mainly the flavour,
but I may be mistaken.

L. Buchter

I think it is a flavour different from beef, not tenderness.

G. Torreele *(Belgium)*

Is there any evidence up to now that, at a similar level
of fatness, one progeny group has more intramuscular fat than
the other?

L. Buchter

There is a heritability coefficient of intramuscular fat
in the *longissimus dorsi* which is close to 0.4 which means that
different sires have a tendency towards depositing different
amounts of intramuscular fat.

G. Torreele

My question was this: whether this varies within levels
of extramuscular fat, in order to see whether little extramuscula
fat could be combined with a high content of intramuscular fat.

L. Buchter

It is possible - but I need the views of my colleagues on
this. However, the correlation coefficient between subcutaneous
and intramuscular fat in general is very high.

M.J. Clancy (Ireland)

I would like to ask Lis Buchter about the differing progeny. She said some sires produced young bulls whose meat was
tender as well as other sires producing young bulls with tough
meat. Is there a difference in the fat levels of these two
toughness types?

L. Buchter

No. These things go across levels of intramuscular and
subcutaneous fat, as we have seen in the crossbreeding experiment

R.G. Kauffman (USA)

We have had to rely a great deal on our animal geneticist
to produce lean versus fat kinds of carcasses. On the other hanc
I think we have been dependent a lot on our animal geneticists
to help us select for quality. I have had a great deal of
difficulty in sorting out quality and quantity but I wonder
whether indeed we have to rely on genetics so much for quality
now, which would relieve the burdens of those in animal breeding.
I would like to put the question in a general way about the role
which genetics plays in relation to environment in producing a
particular end quality product - and you can use tenderness as
your base if you like - and then tell me which of these variable
tend to affect tenderness most: chronological age of animal,
which I include as an environmental effect, not genetic; the
level of fatness in the muscle, as has been discussed; electrica
stimulation to prevent cold shortening; postmortem ageing;
mechanical tenderisation, including the meat grinder; other

processing techniques that make it into an emulsified product; chemical tenderisation such as Papain; and stress, which of course is related, as we know, to genetics. I am convinced that we are asking too much of our animal breeders to select animals which are not only lean, but also of a certain quality characteristic, when we can do so much with the environment.

L. Buchter

We are not asking too much from the geneticist because we have first controlled and optimised our technological treatment of the meat, so the last step is to be done via the genes. That is what we are lacking and in my paper I indicated what improvements can be expected.

R. Boccard

A point in relation to the different factors influencing tenderness. We know well the technological factors, but when we discuss the general improvement of meat production we have to look for biological variation, and to try to give the geneticists some tools for ultimate improvement. Of course we know it is possible to change tenderness considerably by technological means but it is questionable whether we can see the hamburger as the definitive solution to the tender meat problem!

H. de Boer

It is of prime importance to define what me are required to produce, and we must focus on that point as well.

D.N. Rhodes (UK)

We have a large amount of data collected over the last ten years which is derived from sets of animals such as Mr. Cuthbertson was describing to you which are of defined breeds, raised under definite conditions, slaughtered in groups with some sort of age or weight. If breed has anything to do with genetics then we can see whether correlations exist between breeds and eating quality. I do not have the data with me, but, in brief, one can say that we have no correlation whatsoever between what is generally understood to be high quality meat produced

from the Aberdeen Angus, the Devon, crosses of the Hereford, pure
and crossed on various other beef or dairy or dual purpose breeds
There is a tremendous variability in the eating quality of beef,
and looking at this bulk of data - which now includes some six
or seven hundred individual animals - we cannot pick out, as we
had expected to be able to do, a clear indication that if you
take such and such a breed you will get better meat. If we forget everything else except tenderness then you get a tremendous
range of tenderness within each breed group. Remember, these
data are not like Lis Buchter's; she includes a defined sex/age
group method of raising. I am talking about beef production in
a really very broad sense. My own conclusion would be that it
is all very unfortunate, but there is a source of variability in
toughness for which we just do not have any indication yet of
the real cause.

D. Lister (UK)

I just want to offer a little bit of information about the
problem which was bothering Dr. Torreele and Lis Buchter a little
while ago. It is quite well known that animals put their fat in
different parts of the body and in general if they deposit fat
subcutaneously then they also tend to deposit it intramuscularly.
The two go together and, if animals put their fat predominantly
as kidney knob and channel fat, then they are also apt to deposit
intermuscular fat predominantly. You can actually use these
characteristics to define populations.

If you add subcutaneous and intramuscular fat together and
divide it by intermuscular plus kidney knob and channel fat then
you can get a ranking according to what I call the 'fat partition
index' for want of better terminology. This says that, if you
have animals which are not going to be very fat in terms of subcutaneous tissues, they are likely to be at risk in terms of the
effect of intramuscular fat on eating quality. I suspect that
Lis Buchter's bulls are liable to fall into the region where
intramuscular fat can conceivably present a problem in terms of
eating quality. There seems to be a critical level of intramuscular fat which is necessary to enjoy the qualities of meat tc

the full and there is a risk in some members of the population
of their not actually obtaining that certain minimum threshold
value which will allow the consumer to enjoy the meat.

D.N. Rhodes

I think one should again make the point that there is a
big interaction when you are talking about texture or toughness,
and cooking methods and consumer preference; and you cannot stand
up in a meeting like this and refer to such and such a number.
The measurements of texture are only arbitrary numbers, they
depend upon the heating which has been given to the sample, and
the way the machinery is used to work it. They do not have any
objective significance. When cooked by a Frenchman or an Italian
the internal temperature may rise to 30°C only, and you are then
talking about a texture value of raw meat. Someone else may
refer to numbers obtained in samples with an internal temperature
of perhaps 74°C when all of the myofibrillar protein is denatured
and precipitated, the collagen itself has shrunk, most of it has
been gelatinised, the residual background texture under those
conditions is due to the highly cross-linked collagen and, moreover, we have vastly overcooked the precipitated proteins. You
just cannot make the statement, "The texture was so and so."
You have got to postulate everything else, and if you bring in
the last point, you have got to relate it to the consumer as well
because, if we give French or Italian rare steaks to most people
in England, they will politely ask you if you would not mind
cooking it for them.

H. de Boer

An important question is how far the different requirements
interfere with the efficiency of production and, in particular,
with the level of fatness in the carcasses which determines to a
great extent the efficiency of production. Mr. Cuthbertson was
saying that the level of fatness in general in British slaughter
cattle was about one-fifth of the amount of fat above the optimum,
which is not so very much. To what level of fatness does this
correspond, more or less?

G. Harrington *(UK)*

I realise that we are throwing around a concept which is as yet only poorly defined. It refers to populations of cattle in which we want to minimise the feed cost per unit of lean meat produced.

As was pointed out quite rightly, this calculation is only possible given an assumption of the maternal feed cost, or the energy cost of the calf. If we say the calf is free, then the efficiency deteriorates from day one. I do not think that is a reasonable assumption, even though dairy producers have to have a calf to get cows in milk. I think the economics of dairy production would be seriously hampered if they had to give their calves away, so the first problem we have is to make sensible statements about the energy cost of the calf in a mixed milk and beef system. We are very much open to suggestions on how this should be. We can make estimates of the energy cost of calves for their one third of our cows which are suckler cows, whose sole function in life is to produce a calf every year. We can make arbitrary decisions that half of that is a reasonable energy cost to put on a calf from the dairy herd, and we have made these calculations. Given these rather poor assumptions, we have looked at the available information which has now improved since this meeting, because we have more to go on now, to help us to see where optimum production in that sense lies with our current cattle so that we can estimate that it falls at a certain level of fatness in the spectrum.

We then look at the way cattle are now produced and we deduce that one fifth of the fat which is produced with our current cattle is in excess of the fatness which would be required if those cattle were killed at their optimum. This does not mean to say that we cannot improve cattle. We could sweep away the cattle we have and replace them with other breeds and with other sexes and have immediately a more efficient beef production system with a better efficiency of converting food to lean meat; and, if you compared that carcass framework with the one we have now, then far more than 20% of the fat we now produce would be said to be more than required. Our base line is a bad one. In fact, this calculation says that 6% of all the carcass weight we produce is fat in excess of the amount required for optimum lean production with our current cattle. These are rather difficult concepts. We are making some broad calculations with hopelessly inadequate data. There is a fundamental distinction between this approach which says, "Here is the feed, here are the calves: what is the optimum slaughter regime for the different breeds under the various feeding systems to minimise the feed cost of the lean meat being produced?" and the other method, which says this: "We have done some consumer research and we think the housewife wants this fat and this size joint". If we put it all together into a carcass specification and say that we need 300 kg carcasses with this fat and try to push our production into that framework, that, in many ways is the application of modern consumer marketing techniques to the beef industry. We feel that the primary factor we are dealing with is the cost of our raw material and I cannot conceive of alterations in consumer preferences in the next twenty, or thirty, or fifty years which would not be better dealt with by the meat technologist handling the product than by asking the animal production person to alter cattle and perhaps to move them away from the most efficient point of slaughter.

H. de Boer

I agree with you that the contribution to increasing efficiency in the whole process should not come only from the producers. We should expect accommodations from the other sides as well.

L. Buchter

This is just an apology to Dr. Kauffman for my rather abrupt reaction to his comments. I felt that this meeting was to discuss not what we could do in technology but rather what we could achieve in the growth and development fields. That is why I point to those possibilities that are within these limits and not to what could be done technologically because, as I also say in my paper, those muscles in which we are really looking

for the ultimate qualities or the best qualities in tenderness
and flavour are only those than can be used for steaks or roasts.
Therefore we must be aware that it is very difficult to achieve
better meat quality if the cost escalates also.

H. de Boer
        Thank you very much. I will now close this session.

SECTION 4

METHODS OF QUANTIFYING GROWTH AND DEVELOPMENT

Co-ordinator:

St. C.S. Taylor

METHODS OF QUANTIFYING GROWTH AND DEVELOPMENT: GENERAL REVIEW

St. C.S. Taylor

ARC Animal Breeding Research Organisation,
West Mains Road, Edinburgh EH9 3JQ, UK

ABSTRACT

A great variety of methods have been used to quantify growth and development. A number of these are described in outline. In each case some attempt is made to point to their advantages and their limitations and also to indicate inherent dangers in their usage and interpretation. Illustrative examples are given wherever possible.

The three major classifications of allometry, growth curve analysis, and nutritional models are introduced in anticipation of their elaboration in subsequent papers. Other topics included are graphical techniques and methods of scale transformation; exponential smoothing, growth rate curves, growth as a time series, autoregression, prediction of growth; dimensional analysis, food allometry, the relationship between growth curves and allometry, broken or segmental curves; stochastic models; linear and dynamic programming; biological control models; principal component analysis for size and shape; and the special problems associated with data obtained from serial slaughter experiments.

INTRODUCTION

Over the decades, quantifying growth has always been a
popular occupation. In biologically-orientated reviews of growth
Bertalanffy (1960) gives 273 references and Bailey and Zobrisky
(1968) give 187, while in an excellent mathematically-orientated
review, Kowalski and Guire (1974) give 220. It is hoped that
all references in this section on 'quantifying growth' will be
brought together and presented in some loosely classified form.

ALLOMETRY

The allometry equation has several outstanding properties.
1) It is effective in describing curvatures of the type that
occur during the main postnatal period of rapid growth, and
during various other limited periods of growth. 2) The growth
coefficient is simple to calculate, and the estimate is
usually fairly robust. 3) The growth coefficient is unaffected
by any scale transformation of either variable. For example,
the coefficients for carcass muscle in relation to carcass bone
are similar for cattle, sheep and pigs (Jackson, 1967). 4) The
equation has a widely accepted usage and apparently simple
interpretation.

The disadvantages lie in 1) Non-additivity of two or more
allometric equations, which make it conceptually illogical and
methodologically difficult. Is some new additive form possible?
2) Over-emphasis in the log-log plot on early points and a
packing together of late points, so that estimates usually
emphasise relative rates of growth at early stages while non-
linearity can be obscured at later stages. 3) For extended
periods of growth, the simple log-log plot nearly always becomes
curved. Thus the range over which any allometric relation holds
should always be clearly specified.

The extension to multivariate allometry holds out considerable promise. Combinations of traits may be found, for
example, that are invariant under nutritional changes. Recent
work in this field will be discussed by Dr. Seebeck.

GROWTH CURVES

Despite the profusion of possible growth equations, the
number of curves that have been widely used is quite limited.
Apart from polynomials, Brody's curve, the Gompertz, the
logistic, and the generalised logistic or Richards curve and
the Pearl-Verhulst possible account for 80 to 90% of all growth
equations used. Basic information on growth curves can be
found in Brody (1945), while an excellent summary of growth
equations used in practice (with useful extensions) has been
given by Grosenbaugh (1965).

Most commonly used growth equations are incapable of
describing growth over the whole range from embryo to adult.
As with the allometric equation, it is therefore important to
specify the range over which the growth equation being used has
an acceptable goodness to fit. For example, Brody's curve
gives a good description of the growth in body weight of
domesticated animals from the time they are about 30% mature
onwards.

The choice of growth curve can result in misleading
interpretations of data. For example, the Gompertz curve has
rapid final curvature when compared with Brody's curve. Hence
if a Gompertz curve is fitted, the data can be interpreted as
showing continued linear growth in the adult animal; whereas
if Brody's curve is fitted, it is no longer necessary to
propose slow linear growth late in the growing period.

Confused interpretations can also arise from over-
parameterisation. For example, linear body measurements in
growing cattle, when expressed as a percentage of their adult
value, clearly show that some body measurements are less mature
than other body measurements at every immature age. Body
measurements can be ranked and the ranking, with a few minor
exceptions, remains largely unchanged throughout growth.
Fitting Brody's curve to such data gives an exponential time
parameter for each body measurement which gives precisely the

same ranking for earliness of maturity as that found empirically
If the more complex generalised logistic is fitted to the same
data, it is found that the best overall fit for all body measurements is given when the curvature parameter is fixed at $m = \frac{1}{2}$.
Thus, Brody's curve with curvature $m = -1$ does not give the best
overall fit. Interpreting the time parameter when $m = \frac{1}{2}$ no
longer gives the order of maturing found empirically or with
Brody's simple curve. Somehow curvature or shape and rate of
maturing have become confounded, and the simple interpreted
result lost.

The statistical professionalism of Grizzle and Allen (1969)
among others in their use of polynomials for comparing growth
curves will be discussed by Professor Finney. Their analysis
apparently gives not much more than the answer 'Yes, the curves
are probably different' or 'No, they are probably not different';
the biology remains in the graphs of the different growth curves.

An interesting approach is one which will be described later
this morning by G. Torreele, who has related growth curve
parameters to body composition.

NUTRITIONAL MODELS

A simple and robust, conceptually acceptable model - even
with a relatively poor fit - is often to be preferred to a more
complex model that extrapolates to absurdity and includes
relationships known not to hold - even though it gives a much
better fit to given set of data for a specific situation.

In most models there is an absence of parameters affected
by nutritional history. It is usually assumed that growth is
entirely Markovian, that is, that the growing animal has no
memory of how it got to its present body size and composition.
The one and only aim of such a growth process is to guide the
system to a genetically pre-set target at maturity. This may
be predominantly true of animal growth. A Markovian process
may provide a reasonable first approximation. Nevertheless
there is ample evidence that subsequent growth depends on the
nutritional path followed at earlier stages.

There is an absence of models that incorporate adaptive control processes. As every biologist knows, these comprise a large part of every animal's growth. The exception, occasionally included in nutritional models, is a target mature weight.

There is also an absence of limits. At the most primitive level, there is the limit to final size and, as just mentioned, some models do include this. But important limits such as the limits to food intake or growth rate at different stages of growth (which vary depending on nutritional history) or limits to the rates of protein, fat and bone deposition are almost invariably absent. The outstanding exception is the model of Fawcett, and this, in my assessment, is its greatest single merit. It is an example of how a biologist trained as economist, and hence trained in linear programming, brings a new essential ingredient into growth models - because linear programming is primarily about limits. It is in keeping with the spirit of this seminar that Fawcett should spend less time on previous nutritional models and devote most of his time to familarising us with the properties of his own model.

There is an absence of **genetic** differences in the shadowy animals that appear in nutritional text books. For example, there is the famous 300 kg cow which may be a fully grown Dexter or a very immature Charolais. Genetic scaling, which is an essential ingredient in almost any good growth model, is discussed later.

System-analysis models of how an animal grows and when optimal slaughter should occur are frequently subject not only to many of the above criticisms but additionally to both the advantages and disadvantages of 'systems-analysis' models as a class.

Finally, there is little doubt that many more restricted feeding experiments are needed on different genotypes to provide data to test models of regulated growth.

GRAPHICAL METHODS

A remarkable degree of understanding of quantitative data
on growth and development can be achieved by simple graphical
representation. Even in the age of computers, there is nothing
unsophisticated about a graph, especially a computer-produced
one. There is no way of conveying so much quantitative
information almost instantly and with so little mental effort.
A table containing several classifications and many sequences
of 50 or 100 values can be absorbed at one graphic glance.

SCALE TRANSFORMATION

A simple but sometimes effective method of comparing
differences between mean growth curves was given by Rao (1958)
also by Wingerd (1970). Different treatment means are plotted
against the corresponding sequence of overall means. The result
is usually to produce straight lines if the treatments are not
excessively different from the mean curve. The statistical test
is then simple linear covariance analysis. As an example, the
method might be used to test for a sex difference in growth
curves.

LINEAR TRANSFORMATION OF GRAPHICAL AXES

For a description of this method of quantifying complex
changes in shape, see D'Arcy Thompson (1945).

GENETIC SCALE TRANSFORMATIONS

A criticism of most nutritional models is that they do not
distinguish between genotypes. To do this completely is, of
course, impossible. Major effects, however, can be attributed
to genetic differences in body size. Thus a 100 kg animal might
be a mature Oxford Down sheep, a Charolais bull calf 2 months
old, or a Dexter about 9 months old - that is, animals respective
100%, 10% and 30% mature. The nutritional requirements for
maintenance and growth are quite different in each case. Tables

for values genetically scaled by, say, mature body weight, would be considerably more relevant in describing the nutritional requirements of different genotypes. The general technique of genetic scaling for body size should in my opinion (even if in nobody else's) be incorporated into all growth analyses, growth equations and growth models. Some of the techniques have been described by Fitzhugh and Taylor (1971) and by Taylor (1965, 1978).

EXPONENTIAL SMOOTHING AND PREDICTION

Sometimes the purpose of a statistical analysis is to obtain a quantitative prediction of growth over the next time interval. Such a procedure is helpful when slaughter or backfat testing or the like is scheduled to occur at a fixed weight. A method of exponential smoothing has been well described by Hirschfeld (1970). The principle is widely used in many disciplines. The weighting given to any value decays exponentially the further distant the value is in the past: the more rapid the rate of decay, the more prediction depends on the immediate past and the less it depends on the distant past.

Another use for this same method is to smooth an irregular curve, and then it can be used bi-directionally. A typical use is smoothing notoriously erratic growth-rate curves.

GROWTH AS A TIME SERIES

The methods used in the statistical analysis of time series such as moving averages or serial correlations can be applied directly to growth curves. These are of special interest in mature animals which may show cyclic fluctuations in their body weight or food intake.

AUTOREGRESSION

Regressing values at each age on the value at the previous age provides a method for describing the linear growth process

for an individual. At each step, a random growth error occurs.
These errors become an established part of the animal's growth,
being slowly eroded as time elapses and the animal approaches
its 'target' body size and composition. This simple process
results in a Brody-type curve with correlated errors. More
complex curves, including periodic fluctuations, can be
described by multiple regression on values at two or more
previous ages. The method is useful for predicting growth a
few steps ahead. By calculating correlations between growth
errors at different ages any persisting or permanent effects
or early deviations on subsequent growth can be studied.

STOCHASTIC MODELS

Ultimately, all growth processes have a stochastic component. Comparison of growth in uniformly treated identical
twin cattle reveals inherent random errors or growth occurring,
persisting, and being eliminated with varying intensities as
different body parts mature.

Analysing growth by autoregression is equivalent to treating it as a linear process in the stochastic sense. Treating
growth as some other type of stochastic process might be equally
rewarding (eg Parks, 1973), Introducing a stochastic error,
however, can be a needless sophistication if the description
of the underlying growth process is inadequate.

DIMENSIONAL ANALYSIS

Under the title of the theory of biological similarities,
relationships have been derived that must hold because of
dimensional equivalence and the conservation laws of physics
(eg Derome, 1977). For example, heat loss is proportional to
surface area; surface area is proportional to the two-thirds
power of volume; density is independent of body size; hence heat
loss must be proportional to the two-thirds power of body
weight. This is only an approximation to the observed relationship. Where is the discrepancy? Dimensional analysis always

leads to relationships of allometric form, which should therefore be prevalent in biology.

FOOD ALLOMETRY

Body weight has been related to cumulated feed intake
(Roux, 1974; Hind, 1978). Surprisingly, an allometric relationship has been found to hold over a fairly extended part of the
growing period despite the relationship being conceptually
impossible both at birth and maturity.

GROWTH CURVES AND ALLOMETRY

No realistic equations for cumulated food intake and body
weight both as functions of time could be combined to give an
allometric equation. In general, allometry is inconsistent
with growth curves in time - except for certain limited classes.
For example, suppose two different variables, y and x, followed
a generalised Gompertz with the same power function of time but
with different time parameters. If the values of y and x at
maturity were y_M and x_M, and $y/y_M = u_y$ and $x/x_M = u_x$, then
$\ln(-\ln u_y)$ would be linearly related to $\ln(-\ln u_x)$. Although
the types of assumptions are the same, this is no longer
allometry. Such a relationship holds remarkably well in cattle
between height at withers which is early maturing and width at
hooks which is late maturing. Unlike the log-log plot for which
the points at the late ages are closely spaced and form a
hooked curve, the points are now linear and evenly spaced. The
significance of this example is 1) it shows that allometry breaks
down where one would expect it to, and 2) for any given relationship some pseudo-logical derivation can be concocted that produces
an acceptably linear relationship.

BROKEN OR SEGMENTAL CURVES

Fitting a sequence of straight lines to different segments
of an allometric curve or any other curve, is usually highly
artificial. A functional transformation eg from x to $x\ln(1-x)$

may make many of the original linear segments disappear and may
produce other apparently linear segments in the overall curve.

Again, this analytical device may be useful, but its
empirical usefulness should not in itself serve as a passport
into the realm of biological concepts.

LINEAR AND DYNAMIC PROGRAMMING

Although together because of their nominal similarity,
the objective is to indicate their total dissimilarity as methods
of analysing growth. Linear programming is a powerful technique
of optimisation when quantities are discrete and pre-set limits
must not be exceeded. For example, conformation scores progress
by integer steps; animals have finite upper limits to their
food intake; and so on. Thus Fawcett describes his model of
growth as based on recursive linear programming. An optimum
is found for each day's growth; the animal is allowed to grow
in just this way; thereafter a new optimum is found for the next
day's growth; and so on. Dynamic programming on the other hand
does not involve limits and need not involve optimisation; it
is a stepwise procedure for solving a complex set of differential
equations all of which describe continuous processes. Each
step is made sufficiently small for all changes to be treated
as linear. Both techniques are computer-orientated and powerful,
but in common with every other method of quantifying growth,
they are totally ineffective unless the underlying concepts are
biologically sound.

BIOLOGICAL CONTROL MODELS

As mentioned earlier, biologists know full well that animals
are systems in which adaptive control processes predominate.
Growing animals adapt to changes in food intake by speeding up
or slowing down their metabolic processes, by changing their
blood concentrations, and in the longer term their body
composition. Some of these adaptive changes occur quickly.
In cattle, adapting to a new diet may take days or weeks, while

adaptive changes in response to restricted food intake may take
months or years to occur. Models which include biological
control processes are uncommon but urgently need to be developed
beyond the early brave attempt by Weiss and Kavanau (1957).

PRINCIPAL COMPONENT ANALYSIS FOR SIZE AND SHAPE

Multivariate analysis in one or other of its many forms is
bound to be useful when large numbers of carcass components are
being examined. For example, when individual muscles are
dissected, an obvious question in comparing carcasses is to
ask whether most of the difference is due to a general uniform
increase in size or whether the distribution of muscles has
also changed. A good example of the technique of multivariate
analysis will be provided by McFie's paper.

SERIAL SLAUGHTER

Quantifying the growth and development of the carcass and
its components requires a fairly specialised kind of statistical
analysis. At each slaughter age, some observations continue
for some individuals but not for others. As I understand it
from Professor Finney a statistical methodoloy specific to this
situation has not been developed. I hope that the discussions
arising from this seminar will stimulate him and others to
direct statistical research into this difficult, challenging
and important area.

REFERENCES

Bailey, M.E. and Zobrisky, S.E. 1968. Changes in proteins during growth
and development of mammals. In Body Composition in animals and man.
pp. 87-125. National Academy of Science, Washington.

Bakker, H. 1974. Effect of selection for relative growth rate and body
weight of mice and rate, composition and efficiency of growth. Meded.
Landhouwhogeschool, Wageningen, 74-8.

Barton, A.D. and Laird, Anna, K. 1969. Analysis of allometric and non-
allometric differential growth. Growth, 33, 1-16.

Bellman, R. 1957. Dynamic Programming. New Jersey. Princeton Univ. Press.

Bertalanffy,L. von. 1960. Principles and theory of growth, Chap. 2. of
Fundamental Aspects of Normal and Malignant growth. Ed. W.W. Nowinski.
Amsterdam, Elsevier.

Breirem, K. and Homb, T. 1972. Energy requiréments for Growth. Chap. 8. in
Handbuck der Tiernährung Vol. II Hamburg and Berlin, Paul Parey.

Brody, S. 1945. Bioenergetics and Growth. New York and London. Reinhold.

Cartwright, T.C. 1974. Net effects of genetic variability on beef production
systems. Genetics, 78, 541-561.

Cock, A.G. 1966. Genetical aspects of metrical growth and form in farm
animals. Quart. Rev. Biol., 41, 131-190.

Dent, B. and Casey, H. 1967. Linear Programming in Animal Nutrition. London,
Crosby and Lockwood.

Derome, J.R. 1977. Biological similarity and group theory. J. theor. Biol.,
65, 369-378.

Fawcett, R.H. 1977. Nutritional models of growth. In the proceedings of
this seminar.

Fearn, T. 1977. A two-stage model for growth curves which leads to Rao's
covariance adjusted estimators. Biometrika, 64, 141-143.

Fitzhugh, H.A. and Taylor, St.C.S. 1971. Genetic analysis of degree of
maturity. J. anim. Sci., 33, 717-725.

Fitzhugh, H.A. Jr. 1976. Analysis of growth curves and strategies for
changing their shape. J. anim. Sci. 42, 1036-1051.

Gould, S.J. 1966. Allometry and size in ontogeny and phylogeny. Biol. Rev.,
41, 587-640.

Graham, N.McC., Black, J.L., Faighney, G.J. and Arnold, G.W. 1976.
Simulation of growth and production in sheep. Agricultural systems,
1, 113-138.

Graham, N.McC. and Searle, T.W. 1972. Growth in sheep. J. agric. Sci. (Camb.), 79, 383-389.

Grizzle, J.E. and Allen, D.M. 1969. Analysis of growth and dose response curves. Biometrics, 25, 357-381.

Grosenbaugh, L.R. 1965. Generalisation and reparameterisation of some sigmoid and other non-linear functions. Biometrics, 21, 708-714.

Grover, N.B., Bloch, M. and Gross, J. 1970. Computer analysis of growth of rats. Growth, 34, 145-152.

Gunther, B. 1975. Dimensional analysis and theory of biological similarity. Physiol. Rev., 55, 659-699.

Hind, E. in press. Efficiency of lean meat production by dairy steers. (unpublished).

Hirschfeld, W.J. 1970. Time series and exponential smoothing methods applied to the analysis and the prediction of growth. Growth, 34, 129-143.

Jackson, T.H. 1967. The allometric relationship between carcass muscle and carcass bone in Scottish Blackface sheep. Anim. Prod., 9, 531-533.

Kendell, M.G. and Stuart, A. 1966. The Advanced Theory of Statistics. Vol 3. London, Griffin and Co.

Kowalski, C.J. and Guire, K.E. 1974. Longitudinal data analysis. Growth, 38, 131-169.

Laird, A.K. 1966. Dynamics of embryonic growth. Growth, 30, 263-275.

Laird, A.K. 1966. Postnatal growth of birds and mammals. Growth, 30, 349-363.

Lofgreen, G.P. and Garret, W.N. 1968. A system for expressing net energy requirements and feed values for growing and finishing beef cattle. J. anim. Sci. 27, 793-806.

McFie, H.J.H. 1977. Quantifying breed differences in shape, In the proceedings of this seminar.

McMahon, T. 1973. Size and shape in biology. Science. 179. 1201-1204.

Parks, J.R. 1973. A stochastic model of animal growth. J. theoret. Biol., 42, 505-518.

Penrose, L.S. 1974. Some notes on discrimination: size and shape. Ann. Eugen. (London), 13, 228-237.

Rao, C.R. 1958. Some statistical methods for comparison of growth curves. Biometrics, 14, 1-16.

Roux, C.Z. 1974. The relationship between growth and feed intake. Agronomalia, 6, 49-52.

638

Roux, C.Z. 1976. A model for the description and regulation of growth and
    production. Agronomalia, 8, 83-94.

Russell, W.S. 1975. The growth of Ayrshire cattle: analysis of linear body
    measurements. Anim. Prod. 21, 217-226.

Schake, L.M. and Riggs, J.L. 1975. Calorific efficiency of beef production.
    J. anim. Sci., 40, 561-566.

Swartz, J. and Bremermann, H. 1975. Discussion of parameter estimation in
    biological modelling. J. mathematical Biol. 1, 241-257.

Taylor, St.C.S. 1965. A relation between mature body weight and time taken
    to mature in mammals. Anim. Prod., 7, 203 - 220.

Taylor, St.C.S. 1978. Genetic size-scaling rules and genetically standardised
    growth equations (unpublished).

Thompson, D'Arcy W. 1945. Essays on Growth and Form. Ed. W.E. Le Gros
    Clark and P.B. Medawar. Oxford University Press.

Timon, V.M. and Elsen, E.J. 1969. Comparison of growth curves of mice
    selected and unselected for postweaning gain. Theor. Appl. Genet.
    39, 345-351.

Toates, F.M. and Booth, D.A. 1974. Control of food intake by energy supply.
    Nature, 251, 710-711.

Weiss, P. and Kavanau, J. Lee. 1957. A model of growth and growth control
    in mathematical terms. J. gen. Physiol. 41, 1-47.

Wilkins, B.R. 1966. Basic mathematics of control. In Regulation and Control
    in Living Systems. pp. 12-58. London, Wiley and Sons.

Wilson, B.J. 1977. Growth curves, their analysis and use. In Growth and
    Poultry Meat Production. Ed. Boorman. K.N. and Wilson, B.J. Edinburgh,
    Brit. Poult. Sci. Ltd.

Wingerd, J. 1970. The relation of growth from birth to two years to sex,
    parental size and other factors, using Rao's method of transformed
    time scale. Human Biol., 42, 105-131.

Zotina, R.S. and Zotin, A.L. 1972. Towards a phenemenological theory of
    growth. J. theor. Biol., 35, 213-225.

A SURVEY OF ALLOMETRIC ANALYSIS

R.M. Seebeck

CSIRO Division of Animal Production, Tropical Cattle Research
Centre, P.O. Box 542, Rockhampton, Queensland 4700
Australia.

ABSTRACT

Allometric analysis is used to give a description of development such that differences in shape (or composition or function) are studied in relation to the size of the animal. In this framework differences in, for example composition, can simply be due to differences in size or due to other factors such as genotype or treatment.

When the ratio of the relative growth rates of the component parts is constant the simple allometric equation, $y = ax^b$, can be used, and differences between groups can be identified in terms of the constants in this equation. This equation has particular merits in its simplicity and easy statistical tests. Although the relationships may deviate from this simple model, extra variates may be included into the equation to account for such things as treatment group, curvilinearity and rate of growth affecting development.

When allometry is extended to a more realistic multivariate model, several approaches have been made, none of which is completely successful. The most useful approach involves a general concept of the relationship between a shape vector and particular size variables. It can be shown that the association between shape and size depends on 1) the particular set of attributes contributing to the shape vector and 2) the particular size variable. Although shape may be defined in several ways for a particular set of attributes, this does not affect the association. It is shown that by using the shape vector as dependent variables in a multi-variate least squares model, factors such as treatment group and rate of growth can be assayed for their effect on shape independent of the size variable.

1.   PURPOSES OF ALLOMETRIC ANALYSIS

In this survey, I will discuss allometry in terms of shape
as it was originally defined, although as Gould (1966) has
pointed out it applies to composition (as in the example I will
use later) and physiological function such as basal metabolism,
etc. The aim of allometric analysis is to make some
generalisations to assist in interpretation of shape differences
between animals. In particular, shape often differs due to
differences in size of the animals, and it is the particular
purpose of allometric analysis, on the one hand, to study the
relationship between shape and size and, on the other hand,
having removed those size - associated differences in shape, to
study other sources of variability of shape.

Commencing from physical principles, a general approach to
discussing biological similarity particularly in terms of
physiological function, using allometry, has been developed
(see review by Günther 1975), where similarity criteria are
divided into geometric and physical similarity, the latter
comprising dynamic, kinematic, hydrodynamic and thermic
similarity. While related to the present discussion, its
formalism puts it aside from my development of the subject.

The variations in size and shape to be studied can be
classified in terms of those due to growth of an animal (growth
allometry) and those that are associated with differences in
mature size (size allometry). These latter differences can be
due to differences related to evolution, or merely arising from
the static comparison of related forms at the adult stage. The
type of data obviously influences the purpose of the allometric
analysis. When studying growth allometry, observations may be
made on the same organism at different times (what Tanner 1962,
calls a longitudinal study). For some studies, however, growth
allometry is not studied on the one animal (for example when
the 'shape' measurements are destructive as in carcass
composition) so that inferences have to be made from observations
made on a set of animals varying in size (a cross-sectional

study). Sometimes, as in many zoological studies, this is done by a population sample at a particular time. In other studies a serial sampling technique is used, as is desirable in composition studies of farm animals (Seebeck 1968). In the latter case, a random allocation of animals to sampling time (or size) can be made, removing some forms of bias that arise in the other method. The inferences that can be made in either case cannot refer to the growth of the component parts of an individual but solely to the mean of a group.

Depending on the type of data, size allometry _versus_ growth allometry, longitudinal _versus_ cross-sectional, different mathematical models may be appropriate. In this survey, I commence with the simple equation and develop up to the multi-variate model with quite general concepts of size and shape analysis.

Unfortunately, a longitudinal study of growth allometry has mathematical difficulties due to correlated errors, which have not yet been resolved for general application. Thus the techniques to be discussed are mainly those for cross-sectional studies of either growth or size allometry.

2.   BIVARIATE ALLOMETRIC EQUATION

Huxley (1932) in his book 'Problems of Relative Growth', presented the model of a constant 'differential growth ratio'. If x_2 is a measure of organ size and x_1 is a measure of the size of the rest of the body, then allometry assumes that the ratio of the relative growth rates is a constant. Thus

$$\left(\frac{dx_2}{dt} \middle/ x_2 \right) \quad = \quad b \left(\frac{dx_1}{dt} \middle/ x_1 \right) \dots \quad \dots \quad \dots \quad (1)$$

and b is differential growth ratio. Integration of both sides of (1) yields

$$\log x_2 = b \log x_1 + \log a \qquad \dots \quad \dots \quad \dots \quad (2)$$

or $\qquad x_2 = ax_1^b$

Either of these two equations may be referred to as the 'allometric' equation and b as the allometric coefficient. However allometry need not necessarily be restricted to this type of equation - even a straight line with an intercept would express shape changes, but not with the functional implications of the allometric equation.

If $b = 1$, then the organ/body ratio x_2/x_1, is constant. The 'shape' is then constant, all points lying on the same ray emanating from the origin, with slope a. This condition is known as isometry. With values of b different from 1, the organ/body ratio increases ($b > 1$) or decreases ($b < 1$) with increasing body size.

One of the difficulties in application of the allometric equation to data is the treatment of error. The approach to estimating b, the differential growth ratio, will vary according to the assumptions made, and discussions may be found in Sprent (1966, 1968, 1972), Hopkins (1966), Jolicoeur and Mosimann (1968) Jolicoeur and Heusner (1971), as well as many references cited in these papers.

Use of the bivariate allometric equation (as well as the multivariate form which the above papers also discuss) to give a simple definition of shape changes associated with size may be justified if either (a) the data closely follow the allometric equation with little error, or (b) there is considerable biologically-uninteresting measurement error masking a single underlying relationship (as was assumed by Hopkins 1966).

There are many cases where there are deviations from the simple allometric equation which are biologically interesting. With the simple model, these can be estimated by adding extra terms to the allometric equation.

Consider an equation

$$\log y_{ij} = b_i \log x_{ij} + \log a_i + e_{ij} \quad \dots \quad \dots \quad \dots \quad (3)$$

where y_{ij} is the weight of the organ y of the jth animal of the ith breed, x_{ij} is the weight of the remainder of the body of the jth animal of the ith breed,

b_i is the regression of y on x for the ith breed,

a_i is a constant for the ith breed,

and e_{ij} is error, normally and independently distributed with a mean of O and variance of σ^2.

This is of course equivalent to equation (2), the allometry equation, with allowances for different differential growth ratios and different constants, and assuming that x and y are lognormal.

Such a model is useful in comparisons of different groups. If there are different b's between groups then these would reflect major differences in development between the groups - the ratios of the relative growth rates of the two parameters differ. If the value of b is approximately the same for each group, then differences in the values of a represent constant percentage differences in the weight of y at varying values of x.

The above model then assumes that, although the different sub-populations may have different allometric coefficients and constants, the ratio of relative growth rates is constant over the range of the data with each sub-population. In some cases this has been found not to be true and a $(\log x)^2$ term has been introduced to allow for curvilinearity. This destroys the simplicity of interpretation of allometry but may still have some use. Another possibility is the use of concomitant variables in the allometric equation to reduce variability about the relationship. One that could be used is the age at the time of measurement (which indicates in a general way the growth rate up to that time). This is relatively easy to interpret - a faster growing animal having a different ratio of specific growth rates may be explainable in terms of such things as different nutrient availabilities and mature sizes.

Seebeck (1968) extended the above approach to a multi-variate model with several dependent lognormal variates (eg weights of carcass components) on the single independent lognormal variate, for which appropriate statistical tests are available (eg Seal 1964). I have now shown that this is exact only under limited restrictions. These are that if we have, say, 3 carcass components, muscle, bone and fat any two of these can be used as the dependents to represent shape, <u>so long</u> as the independent variate representing size is the geometric mean of all the carcass components, ie $(\Pi X)^{1/3}$. In effect then if muscle and bone represent shape, its association with that particular size variable can be tested, as well as the effects of say genotype which can be represented by additional independent variates. This has some relationship to the methods of Mosimann which will be described in the next section. However he develops a representation of shape which can be related to size variables other than the geometric mean. Making use of my original method, I then go on to develop his concept further so that other factors such as genotype and growth rate can be tested for their effects on shape independently of a particular size variable.

3. MULTIVARIATE ALLOMETRY

As noted by Mosimann (1970) the first question to answer in studying the shape of animals is 'when do two animals have the same shape?' This question must be modified to be able to be answered - "when do two animals have the same shape with respect to a finite number of measurements?" The simple allometric equation can only answer it with respect to two measurements, in which in effect, one variable, the dependent, may be considered as shape and the other as size. However bivariate comparisons are very restricted. Even with three variables, two animals may have the same shape with respect to length, width and height but the addition of a fourth, such as head width, might reveal a shape difference not susceptible to definition by the first three measurements.

Sprent (1972) points out a mathematical difficulty associated with any generalisation to p dimensions of a linear relationship between logarithms. Does a straight line in two dimensions generalise to a straight line or to a plane in three dimensions? The mathematician may consider either a reasonable generalisation but this is of little help to a biologist for model building.

Jolicoeur (1963$_{1,2}$) did not assume a straight line relationship for the logarithms of the data but proposed that the hypothesis of isometry of the variables $x_1, \ldots, x_k$ could be tested by comparing the eigen vector of the first principal component of the sample covariance matrix of the logarithms of the x's with the theoretical eigen vector

$$(1/k)^{1/2} \ (1, \ldots, 1).$$

The hypothesis of isometry is then accepted or rejected depending on the agreement of the sample eigen vector with this theoretical one using a test with a multivariate normal assumption developed by Anderson (1963). It is equivalent to testing for a straight line generalisation if all other eigen values are zero. As Mosimann (1970) has pointed out, there are several limitations to Jolicoeur's approach. Firstly, it assumes that the size variable is $(\Pi X)^{1/k}$, and as has been shown by Mosimann and will be demonstrated here, this is only one of a large number of possible lognormal size variables. Also one could accept the possibility that the logarithms of shape variables are more variable than the logarithm of the size variable, in which case the isometry hypothesis is true if any population eigen vector is $(1/k)^{1/2} \ (1, \ldots, 1)$. Sprent (1972) has also pointed out that Jolicoeur's basic definition has the serious weakness that even in the bivariate case it is easy to concoct a covariance matrix with two eigen values not very different from each other and one of these having (1,1) as the associated eigen vector. This will occur with an elliptic but almost circular scatter of points on the logarithmic scale when the major axis is equally inclined to the variate axes.

With such a scatter we would not accept the hypothesis of simple allometry and thus the concept of isometry is unrealistic.

Hopkins (1966), following on from the work of Teissier (1955, 1960), has preferred a factor analytic approach, to that of principal components. This has a greater potential in that it allows more flexibility in the mathematical model, but there are serious interpretational difficulties due to non-uniqueness that have yet to be resolved (Sprent 1972).

Mosimann (1970; $1975_{1,2}$) has developed a more general concept of size and shape variables, following on from the definition of isometry as proposed by Jolicoeur ($1963_{1,2}$). I will try to summarise this work, of which a more full description, relatively easily understood by biologists, is given in Mosimann and James (1977).

Consider a vector of k measurements, all positive numbers expressed in the same units, say millimetres or kilograms. We denote these measurements by $(x_1,...,x_k) = x$ and call x a 'data' vector. The approach is to treat each data vector as a pair of variables: (1) a vector variable which identifies the positively directed ray from the origin on which $\underset{\sim}{x}$ lies (either a 'shape' vector or a 'size - ratio' vector), and (2) a scalar variable, G, (the 'size') which identifies how far from the origin x is found on that positively-directed ray.

If we have two animals such that

$$\underset{\sim}{x}_1 = (x_{11}, ...,x_{1k})$$

$\underset{\sim}{x}_2 = (x_{21}, ...,x_{2k})$, then these two individuals have the same shape with respect to these measurements if one vector is a scalar multiple of the other, that is, if

$$\underset{\sim}{x}_1 = a\underset{\sim}{x}_2$$

for some a > o. This definition reflects a common geometrical thinking since Euclid (eg with similar triangles) and the same concept of shape underlies the bivariate allometric equation.

The shapes can be named in a variety of ways. One is by the direction cosines of the rays. Another is by its proportion vector. That is, if we divide each element of x by Σx, we obtain the vector

$$\left(\frac{x_1}{\Sigma x}, \ldots, \frac{x_k}{\Sigma x}\right)$$

A third sample is a ratio vector. Take the last element x_k and divide it into each element of x obtaining

$$\left(\frac{x_1}{x_k}, \ldots, \frac{x_{k-1}}{x_k}, 1\right)$$

A **size variable** G (x): (1) must have as its value some positive number and (2) must exhibit the homogeneity property $(G(ax) = aG(x)$ for all positive x and all $a > 0$. In our definitions of shape we have in fact defined three size variables $(\Sigma x^2)^{1/2}$ (for direction cosines), Σx and x_k, and each has this homogeneity property.

Finally, we note another type of vector which may be used to name the positive rays. The data vector $x = (x_1, \ldots, x_k)$ can be regarded as having embedded in it a one-dimensional data vector $x_1 = (x_1)$, a two dimensional data vector $x_2 = (x_1, x_2)$ etc. We can then define a sequence of size variables for $1, \ldots, k$ dimensions respectively such as, for example, $x_1, (x_1 + x_2), \ldots, (x_1 + \ldots + x_k)$ or $x_1, x_2, \ldots, x_k$. Both these sequences are sequences of size variables, and in both G_1 is only a function of x_1, G_2 of (x_1, x_2) etc. When the sequence is defined in a special way so as to be 'regular' (c.f. Mosimann 1975[1,2]) then vectors of ratios of size variables

$$\left(\frac{G_2}{G_1}, \frac{G_3}{G_2}, \ldots, \frac{G_k}{G_{k-1}}\right)$$

name the positive rays and have certain advantages over shape-vectors in that it can be shown that $\frac{G_2}{G_1}$ alone determines the

2-dimensional rays of (x_1, x_2) and $\begin{pmatrix} \frac{G_2}{G_1} & \frac{G_3}{G_2} \end{pmatrix}$ alone determines the

3-dimensional rays of (x_1, x_2, x_3), etc.

The choice of size variable depends on both geometrical and biological considerations (and, as we have seen in the previous section and will enlarge later, on statistical limitations). In a given study more than one size variable is often of interest because different biological concepts are reflected by different size variables. As we will see, allometric relations with respect to different size variables need not be the same. Also relationships among size variables themselves may be of interest.

Up till now in this section we have represented each data vector by a pair of variables, a shape vector and a scalar size variable Now suppose that we have a sample of N animals and hence N data vectors. Each x is a value of the positive random vector $\underset{\sim}{X}$ which represents the population. For the time being, this random vector $\underset{\sim}{X}$ may have any distribution. Associated with it is the random shape vector $\underset{\sim}{Z}(\underset{\sim}{X})$ and a random size variable $G(\underset{\sim}{X})$. We can define 'isometry with respect to size G' as the statistical independence of the random shape vector $\underset{\sim}{Z}(\underset{\sim}{X})$ and size $G(\underset{\sim}{X})$. Thus if the ratio vector $\underset{\sim}{X}/X_k$ is independent of size $(\Pi X)^{1/k}$, the geometric mean, then we have isometry with respect to the geometric mean.

The use of these concepts gives rise to two theorems for which exact statements and proof are given elsewhere (Mosimann 1970, p.937; 1975_1, pp. 209-10).

A. If some size variable b is independent of any single shape vector (like $\underset{\sim}{X}/\Sigma X$, $\underset{\sim}{X}/X_k$, etc.) or any single regular size - ratio vector, then that size variable is independent of every shape (or regular size ratio) vector. Thus for the purposes of studying independence, any shape (or regular size - ratio) vector may be used to represent shape.

B. In the population, at most one size variable can be independent of shape (unless the other size variable divided by the first is a degenerate random variable, or shape has virtually no variation).

If we now assume that the random data vector $\underset{\sim}{X}$ has a multivariate lognormal distribution it can be shown that some shape vectors (those involving powers of ratios and products of the elements of X) themselves follow a multivariate lognormal distribution. This arises because they are the differences and means of the logarithms and therefore retain normality.

When the size variable as well as the shape vector is lognormal, then log size (eg $\bar{Y}$) and log shape (eg $Y_1 - Y_3$, $Y_2 - Y_3$, 0) are independent if and only if the population multiple correlation of $Y_1 - Y_3$, $Y_2 - Y_3$ with $\bar{Y}$ is zero since the combined vector $(Y_1 - Y_3, Y_2 - Y_3, \bar{Y})$ is a linear transformation of $\bar{Y}$ and hence multivariate normal. A test of the significance of the multiple correlation (R) is an exact test of isometry for that lognormal size variable. As has been shown previously, all shape variables are equivalent, so that R can be used to test indirectly the relationship of that size variable to a shape vector that is not itself lognormal (such as $X/\Sigma X$). Using multiple regression when log G is the dependent variable and $Y_1 - Y_k, \ldots, Y_{k-1} - Y_k$ are independent variables then the F-value = (MS due to regression)/(MS residual) gives the test that the population multiple correlation is zero, and hence that shape is independent of size. Although this procedure is suggested for ease of calculation and uses a regression programme, the test as used here does not involve the distinction of dependent and independent variates, and is not a conditional test. Perhaps more important than testing for the association of shape and size is a measurement of the degree of this association. It can be shown that the multiple correlation coefficient, from the regression model, can be used to measure the association of shape with size, with the restrictions as above on log-normality.

The general approach of Mosimann has its limitations. Firstly, it is particularly suited to size allometry, rather than growth allometry, or at least to cross-sectional studies as may be used in carcass composition experiments. Secondly, once having defined shape and size variables, Mosimann (eg James and Mosimann 1977) restricts analysis to their distribution properties (means, variances and covariances). No attempt is made, as in classical allometry, of summarising allometric relationships by a single coefficient reflecting the mean trend of shape with particular size variables.

However, it is possible to use the shape vector as a set of dependent variates in a multivariate least squares model and extend the approach I have used for the bivariate model in the last section, so that the effects of treatment group, concomitant variables, etc. can be assayed for their effects on shape independent of the particular size variable chosen. I will give an example in the next section.

The extension of this to longitudinal studies appears fraught with difficulties. Obviously the approach of, for example, Grizzle and Allen (1969) for growth curves provides one starting point but much theoretical and empirical work remains to be done. Another possibility is to describe development using the dynamic modelling approach (partial differential equations) instead of using least squares principles. But I do not know how to, for example, contrast groups of animals, using such models.

4. EXAMPLE OF APPLICATION OF BIVARIATE AND MULTIVARIATE ALLOMETRY

The example I will use and describe briefly is on carcass composition data from a serial slaughter experiment of three breeds of steers that were raised under field conditions in the sub-tropics. The full analysis will be reported elsewhere. Suffice it to say here that the three breeds (F_3 generation) are Africander cross (AX), Brahman cross (BX) and Hereford-Shorthorn

TABLE 1

ANALYSIS OF STEER CARCASS COMPOSITION DATA WITH THE ALLOMETRIC EQUATION.
VALUES EXPRESSED IN LOGARITHMS

Independent Variates	Dependent Variates					
	Bone	Muscle	I	Fat II	III	Fascia & Tendons
μ	1.2411	1.8535	1.3936	1.3903	1.3952	0.3945
Breed						
AX	0.0042	0.0165^a	-0.0453^a	-0.0416^a	-0.0484^a	0.0245
BX	0.0004	0.0013^b	-0.0166^a	-0.0030^{ab}	0.0042^{ab}	-0.0202
HS	0.0046	-0.0178^c	0.0619^b	0.0445^b	0.0443^b	-0.0043
TSW						
overall	0.539	0.676	2.226	1.801		0.516
AX	0.698^a	0.816^a	1.922^a	1.341^a		0.733^a
BX	0.292^b	0.445^b	2.924^b	2.332^b		-0.212^b
HS	0.627^{ab}	0.766^a	1.832^a	1.730^{ab}		1.026^{ab}
$(M*B*F*F\&T)^{1/4}$						
overall					2.142	
AX					1.739^a	
BX					2.995^b	
HS					1.692^a	
Age (per day)					0.00031	

^{abc} Values in the same group with different superscripts are significantly different ($P < 0.05$)

(HS). The definition of the breeds and of the slaughter and dissection procedures are described elsewhere (Seebeck 1973$_1$, 2, 3), although these are a separate group of cattle to that reported in the latter two papers.

In Table 1 bivariate allometric analyses (actually one dependent 'shape' variate at a time, with one independent 'size' variable, plus other independent variates to express deviations from allometry) are given. These describe the development of bone, muscle, fat and fascia and tendons dissected from one side of the dressed carcass, against total side weight (TSW), with modifications for breed at the geometric mean of TSW, and an interaction between breed and TSW so that separate relationships for breed could be derived and tested for a significant differences between them. These do show different slopes between the breeds for all components, and some differences at the mean TSW in the position of the lines. With these models, $\mu \cdot$ (ie $a + b\bar{X}$) is given instead of a, since with different sub-population slopes being fitted, the positions of the lines at the mean of the independent variates is more relevant than at the intercept (where the values of the independent covariates are zero).

In the case of fat, two extra models are given. In Model II, age in days is included as an additional covariate, which showed that, over all breeds, the slower growing animals (ie with a larger age) were more fat. In Model III, the geometric mean of the carcass components, $(M*B*F*F\&T)^{1/4}$ is used as an alternative size variate, and in this case there is not much difference from the results with TSW in Model I.

In Table 2, multivariate analysis of variance is used to analyse shape vectors against size for the carcass composition data (ignoring fascia and tendons), with statistical tests given as χ^2-tests or F - tests as appropriate (degrees of freedom in parentheses). In Models, I, II and IV, the geometric mean of the carcass components is used as the size variable and in Model III muscle weight is used. In each model a breed

TABLE 2

ANALYSIS OF CARCASS COMPOSITION DATA OF STEERS USING MULTIVARIATE ALLOMETRY

Independent Variate Group	Shape Vector $\equiv$ Dependent Variates			
	I M/B,F/M	II M/B,F/M	III M/B,F/M	IV M,B
Breed χ^2(4)	21.90***	20.19***	13.19*	20.19***
Breed* $(M*B*F)^{1/3}$ χ^2(4)	12.38*	12.56*		12.56*
$(M*B*F)^{1/3}$ F(2,39) F(2,38)	43.45***	13.95***		36.22***
Breed*(M) χ^2(4)	-	-	7.00	
(M) F(2,38)	-	-	5.68**	
Age F(2,38)		3.45*	31.20***	3.45*
R^2	0.843***	0.867***	0.811***	0.847***

    *   Significant (P < 0.05)

   **   Significant (P < 0.01)

  ***   Significant (P < 0.001)

TABLE 3

CONSTANTS AND SIGNIFICANCES OF DIFFERENCES IN ANALYSIS OF MULTIVARIATE
ALLOMETRY DATA EXPRESSED IN MODEL I OF TABLE 2. VALUES EXPRESSED IN
LOGARITHMS

Independent Variates	Dependent Variates	
	M/B	F/M
μ	0.6121	-0.4612
Breed		
AX	0.0132	-0.0543^a
BX	0.0013	-0.0099^a
HS	-0.0145	0.0642^b
$(M*B*F)^{1/3}$		
overall	0.104	1.417
AX	0.099	1.021^a
BX	0.099	2.168^b
HS	0.114	1.062^a

[a,b] Values in the same group with different superscripts are significantly
different (P < 0.05)

x size term is included, but it is not significant in Model III. In Model II, age at slaughter has a significant effect on shape, independent of size and this effect is more significant when size is measured by muscle weight in Model III. Model IV is the same as Model II except that muscle and bone weights are used as the shape vector instead of the size - ratio vector, muscle/bone and fat/muscle, as in the first three models. This shows that all tests, except for that of the size variable, are the same for the different shape vectors, when the size variable is the geometric mean of the components (they are not for other size variables).

Table 3 gives the constants and indications of significant differences for the analysis of the size - ratio vector given in Model I of Table 2. It shows that there is little variation in muscle/bone, either with size or between breeds, but that for fat/muscle, the ratio increases more rapidly in BX, while at the geometric mean of the size variable, HS has the highest fat/muscle ratio.

With this set of data, none of the size variables showed significant curvilinearity (size^2 terms were not significant), showing that the ratio of the relative growth rates of components was constant.

All statistical analyses were performed with the author's non-orthogonal analysis of variance and covariance programme, SYSNOVA, which can also handle multi-way factorial data and hierarchies within factors, with or without specified inter-actions.

REFERENCES

Anderson, T.W. 1963. Asymptotic theory for principal component analysis.
    Ann. math. Statist., 34, 122-48.

Gould, S.J. 1966. Allometry and size in ontogeny and phylogeny. Biol.
    Rev., 41, 587-640.

Grizzle, J.E. and Allen, D.M. 1969. Analysis of growth and dose response
    curves. Biometrics, 21, 708-14.

Günther, B. 1975. Dimensional analysis and theory of biological similarity.
    Physiol. Rev., 55, 659-99.

Hopkins, J.W. 1966. Some considerations in multivariate allometry.
    Biometrics, 22, 747-60.

Huxley, J.S. 1932. Problems of relative growth, London: Methuen and Co. Ltd.

James, F.C. and Mosimann, J.E. 1977. Exact statistical methods for
    studying size allometry. II. Variation in Florida red-winged black-
    birds (unpublished manuscript).

Jolicoeur, P. 1963_1: The multivariate generalisation of the allometry
    equation. Biometrics. 19, 497-77.

Jolicoeur, P. 1963_2. The degree of generality of robustness in Martes
    americana. Growth, 27, 1-27.

Jolicoeur, P. and Heusner, A.A. 1971. The allometry equation in the
    analysis of the standard oxygen consumption and body weight of the
    white rat. Biometrics, 27, 841-55.

Jolicoeur, P. and Mosimann, J.E. 1968. Intervalles de confiance pour la
    pente de l'axe majeur d'une distribution normale bidimensionelle.
    Biométr.-Praxim., 9, 121-39.

Mosimann, J.E. 1970. Size allometry: Size and shape variables with
    characterisations of the lognormal and generalised gamma distributions.
    J. Am. statist. Ass., 65, 930-45.

Mosimann, J.E. 1975_1 Statistical problems of size and shape. I. Biologica
    applications and basic theorems. In: Statistical Distributions in
    Scientific work, Vol. 2. Eds.: G.P. Patil, S. Kotz and J.K. Ord,
    Dordrecht-Holland: D. Reidel Publishing Company.

Mosimann, J.E. 1975_2. Statistical problems of size and shape. II.
    Characterisations of the lognormal, gamma and dirichlet distributions.
    In: Statistical Distributions in Scientific Work, Vol. 2. Eds. :
    G.P. Patil, S. Kotz and J.K. Ord, Dordrecht-Holland: D. Reidel
    Publishing Company.

Mosimann, J.E. and James, F.C. 1977. Exact statistical methods for studying size allometry. I. Theory and methods for size-shape analysis (unpublished manuscript).

Seal, H.L. 1964. Multivariate statistical analysis for biologists, London: Metheun and Co. Ltd.

Seebeck, R.M. 1968. Developmental studies of body composition. Anim. Breed. Abstr., 36, 167-81.

Seebeck, R.M. 1973_1. Sources of variation in the fertility of a herd of zebu x British cattle in Northern Australia. J. Agric. Sci. 81, 253-62

Seebeck, R.M. 1973_2. The effect of body-weight loss on the composition of Brahman cross and Africander cross steers. I. Empty body weight, dressed carcass weight and offal components. J. Agric. Sci., 80, 201-10.

Seebeck, R.M. 1973_3. The effect of body-weight loss on the composition of Brahman cross and Africander cross steers. II. Dissected components of the dressed carcass. J. Agric. Sci. 80, 411-23.

Sprent, P. 1966. A generalised least squares approach to linear functional relationships. Jl. R. Statist. Soc. B, 28, 278-97.

Sprent, P. 1968. Linear relationships in growth and size studies. Biometrics. 24, 639-56.

Sprent, P. 1972. The mathematics of size and shape. Biometrics, 28, 23-37.

Tanner, J.M. 1962. Growth at adolescence, 2nd Ed., Oxford: Blackwell.

Teissier, G. 1955. Sur la détermination de l'axe d'un nuage rectiligne de points. Biometrics, 11, 344-57.

Teissier, G. 1960. Relative growth. In: The physiology of crustacea, Vol. 1. Ed. : T.H. Waterman, New York: Academic Press.

GROWTH CURVES: THEIR NATURE, USES, AND ESTIMATION

D.J. Finney
Department of Statistics and ARC Unit of Statistics,
University of Edinburgh, UK

Any critical account of growth curves and associated
statistical techniques would be lengthy and necessarily highly
mathematical. I make no claim to familiarity with the extensive
literature; Kowalski and Guire (1974), in an excellent survey
without excessive mathematics, gave a set of 220 references.
This paper examines the logical structure that needs to be understood by biologists. The statistical theory of estimation and
techniques of computation, though important, are matters for
specialists and are less suitable for oral presentation. Here
attention is concentrated on the use to which curves are to be
put, the type of algebraic formulation appropriate to the curve
itself and to the variability of individual observations and to
different principles of estimation. Special emphasis is placed
on the logical distinction between the growth curve for an
individual animal and the mean for a population.

WHAT IS A GROWTH CURVE?

Initially I fix attention on a single variate, a property
of an animal that can be measured repeatedly during life. From
birth to death, the variate will follow a continuous curve,
although only isolated points are recorded. To a first approximation, the curve is smooth, steadily increasing to maturity,
and subsequently nearly constant or perhaps increasing or
decreasing slightly. Under closer examination, irregularities
emerge - small waves within each day imposed upon the general
trend, growth spurts, and ill-effects of short term illness.

If we average over animals within a population (eg over
cattle of one breed on a stated feeding plan), the irregularities
disappear because animals are out of phase with one another,
and a simpler curve appears. The distinction between individual

growth curves and the idealised mean curve for a population or
a treatment group is essential.

Any data available for study will consist of pairs of age
and measurement corresponding to one or more animals. Except
for errors due to inaccurate measuring instruments, which I
shall ignore, the observations are true values from the individual curves. Alternatively, they can be regarded as estimates
of points on a mean curve for a population of which animals
measured are representative. The isolated points on true
curves may be used either for interpolation in order to make
inferences about other ages for the same animals or for estimating properties of the mean curve. A special difficulty can
enter if slaughter at different ages brings heterogeneity of
age span: biases may enter because economic considerations
oblige investigators to slaughter when some arbitrary size
criteria are satisfied rather than at predetermined ages.

Under the head of "growth curves", some authors include
dependence of response to a drug on time since administration
and time trends in the number of individuals in a population.
I do not: I restrict myself to the growth of animals in any
quality thought worthy of measurement.

USES OF GROWTH CURVES

Any continous curve, however irregular its appearance, can
be represented by a mathematical function. The growth curve
for a single animal is an exceedingly complicated function; in
a practical problem, a rather gross approximation that takes
account of the general trend but ignores short-term fluctuations
and local irregularities would have to be used. The form of
a mean curve ought to be simpler but, as with most biological
phenomena, one is unlikely ever to have the exact mathematical
formulation and must therefore use approximations that are
plausible in respect of particular applications.

Growth curves are used for many purposes, among which are:

a)   to summarise descriptively how a measurement changes with time in an animal or a group of animals;

b)   to provide a norm with which individual animals may be compared, in order to judge whether they are exceptional or whether remedial action is needed;

c)   to compare the consequences of different feeding or other management treatments in experiments directed at improving the size or conformation of livestock;

d)   to estimate for individual animals either the value that a variate had at an age when no measurement was made or the age at which the variate achieved a specified value;

e)   to predict the value of a variate at some future date or, especially for slaughter purposes, to predict the ages at which individual animals will achieve specified measurements;

f)   to study in depth, as part of a fundamental research project, the growth of animals in relation to physiology and metabolic processes.

Doubtless this list could be extended. To use a type of curve appropriate to (f) in a situation better described by (a) would be foolish, and almost certainly impracticable because the data are inadequate. Needs such as (a) or (b) can be met by curves whose mathematical specifications bear little relation to underlying biology, whereas (f) must involve an attempt, however oversimplified, to describe biological processes by mathematical models.

FORMS OF CURVES

Usually a "family of curves", a mathematical function that can be modified to suit particular data by numerical changes in one or more parameters, will be adopted. For example, the straight line:

$$Y = \alpha + \beta x \qquad (3.1)$$

represents a dependence of a variate Y on age x in terms of
two parameters, α   and   β whose numerical values define the
position and slope of the line. The greater the number of
parameters the more extensive the family, and the greater is the
possibility of making it fit different needs. The next simplest
curves for statistical use are the polynomials:

$$Y = \beta_0 + \beta_1 x + \beta_2 x^2 + \beta_3 x^3 + \ldots . \qquad (3.2)$$

They are widely used, adaptable, and entirely satisfactory for
some applications; they rarely pretend to true mathematical
description of a biological process. Although general theory
teaches that any growth curve can be represented as closely as
may be desired by a polynomial equation, this is little practical
comfort if the approximation needs 15 parameters. Biologists do
not always appreciate the immense variety of mathematical functions available; with some ingenuity, any curve can be represented by functions that look very different but that are indistinguishable numerically or visually in practical applications.

A larger family of curves retains linearity in the parameters;
it replaces each power of x in equation (3.2) by any function
of x that does not in itself introduce additional parameters,
for example

$$Y = \beta_0 + \beta_1/x + \beta_2 \log(x+5) + \beta_3 \sin\left(\frac{\pi x}{12}\right) \qquad (3.3)$$

These are almost as easy to handle, numerically and statistically,
as the polynomials, but are seldom more realistic. Together
with the polynomials, they may suffice empirically in some uses
of growth curves, but not over a wide range.

For any variate closely related to the general size of an
animal, the growth curve is likely to have three major phases,
an initial slow but steady acceleration, a middle period of
rapid growth that may look approximately linear, and a closing
phase in which growth slows towards the mature level. In

studies of meat production, the third stage may be of lesser
interest; its omission may permit simpler mathematical
formulations to be used. Polynomials are notoriously bad at
representing curves that flatten to the horizontal at one or
both ends: an adequate polynomial requires far more parameters
than non-linear alternatives. Sometimes the appearance of a
curve can be simplified by transformation of scales, for example
by drawing it in terms of logarithm of age instead of age or
logarithm of weight instead of weight. This pictorial gain is
irrelevant to fundamental discussion and I shall not discuss
it further.

At least four parameters are essential to a function with
the desired features. The widely used equation (in terms of
log x a logistic)

$$Y = \delta - \frac{\delta - \gamma}{1 + \alpha x^{\beta}} . \qquad (3.4)$$

gives a sigmoidal curve between a minimum of γ at age zero
and an approach to a maximum at δ as age advances. Other
functions of very different algebraic appearance can be almost
indistinguishable in the fitting of data. Equation (3.4) and
others have a symmetry in relation to logarithm of age that may
not be appropriate, but that can be removed only at the cost
of including at least five parameters, for example by

$$Y = \delta - \frac{\delta - \gamma}{(1 + \alpha x^{\beta})^{\theta}} . \qquad (3.5)$$

Grosenbaugh (1965) has listed other possibilities. Similarly,
any need to allow for a complexity of growth curve beyond a
steady increase to maximum slope and thereafter a steady decrease in slope would demand extra parameters. Another
important type of equation can be written

$$Y = \alpha + \beta(1 - e^{-\gamma x}) , \qquad (3.6)$$

only three parameters but no approach to the horizontal when
x is small.

Curves for other classes of measurement (eg biochemical
properties of the animal) may show entirely different trends
with time. We can readily devise equations to represent curves
with any qualitatively stated features, but to estimate satisfactorily more than 4 parameters for a curve representing
biological phenomena requires very good and extensive data.
An excess of parameters permits widely different numerical
values to be equally successful in fitting the data, and therefore gives standard errors of individual estimates so large as
to prevent useful inferences. Adoption of a many-parameter
equation when no more than a straight line is needed can be
as misleading as adoption of an inadequate polynomial where
the depth of an investigation should be matched by a curve
approximating more closely to truth.

MULTIVARIATE RECORDS AND CONSISTENCY

Animal growth studies will usually include measurement of
several variates, some recorded frequently on each animal,
others only at wide intervals. Many variates (eg weights of
internal organs or of joints of meat) can be measured only once
per animal; little can be said about individual curves for
these, but data from animals slaughtered at different ages may
permit a mean curve to be estimated.

Multivariate data give much information about patterns of
growth, but vastly more complicated statistical methods are
needed. In addition to well-explored problems of multivariate
statistical analysis, questions of internal consistency may
arise. For example, assumption of any equation other than (3.1)
for two different variates is likely to be inconsistent with
assumption of the same form for a variate that is essentially the
sum of the first two. A ratio of variates introduces greater
troubles. Over a short period, to describe the growth of two
length measurements and the growth of their ratio by straight
lines may be adequate; in a fundamental study of growth over

a prolonged period, the inconsistency would be serious.
Allometric relations represent one attempt to overcome this
difficulty, but I think they cannot completely succeed.

VARIANCE STRUCTURE

Data for growth curve studies will usually consist of
measurements made at the same ages in each of several animals;
successive ages need not be equally spaced though this is common.
As explained in Section 1, the measurements give points almost
exactly on the individual response curves. However, a measurement on any animal will deviate from the mean at that age by a
statistical "error" and the errors at different ages will be
correlated: the existence of some correlation is obvious, for
an animal that is much above average at one age must still be
above average a short time later, though the correlation may
be smaller after a longer interval.

The position is different and simpler if each animal is
recorded only once. An experiment may be planned in terms of
serial slaughter, with members of a group of identically treated
animals chosen at random for slaughter and measurement at
predetermined ages, a procedure that is unavoidable for some
variates. This removes all correlations and makes all measurements statistically independent. Individual growth curves
cannot be discussed, but a mean curve for the group can still
be estimated, albeit with less precision than when each animal
is measured at several ages. Statistical analysis can still be
considered a special case of that outlined below, but is so
much more simple as to need no further mention here.

FORMAL SPECIFICATION

We can now state a general formulation. The value of a
single variate for animal i at time x can be written as
y_i , where

$$y_i = f(x) + \varepsilon_i(x) \quad ; \qquad (6.1)$$

in this equation, $f(x)$ is the mean growth function for the
population, a function such as is found on the right hand side
of equations (3.1) - (3.6), and $\varepsilon_i(x)$ is the deviation of the
individual curve at time x . Taking expectations over the
population,

$$E\{\varepsilon_i(x)\} = 0 \qquad\qquad (6.2)$$

(by definition) and

$$E\left(\{\varepsilon_i(x)\}^2\right) = V(x) \quad , \qquad\qquad (6.3)$$

the variance expressed as a function of x . One more function,
$\rho(x_p, x_q)$, expresses the correlation between errors at times
x_p, x_q :

$$E\left[\varepsilon(x_p)\ \varepsilon(x_q)\right] = \rho(x_p, x_q)\{V(x_p)V(x_q)\}^{\frac{1}{2}} \quad , \quad (6.4)$$

where necessarily (when x_p, x_q are equal)

$$\rho(x_p, x_p) = 1 \quad ; \qquad\qquad (6.5)$$

commonly ρ declines as the interval between x_p, x_q increases.

This is too complicated for most purposes: to estimate
variance and covariance parameters as well as those of $f(x)$
would need an enormous body of data. If interest is restricted
to a short time interval, even such simplifications as

$$V(x) = \text{constant} \qquad\qquad (6.6)$$

and

$$\rho(x_p, x_q) = \text{constant} \qquad\qquad (6.7)$$

or

$$\rho(x_p, x_q) = \omega^{\left|x_p - x_q\right|} \qquad (6.8)$$

for a constant ω may be acceptable. For a theoretical study, as distinct from analysis of a particular experiment, relating $V(x)$ to the expectation $f(x)$ rather than directly to the time scale x may be preferable; even an approximation such as

$$V(x) = \lambda f(x) \qquad (6.9)$$

may suffice. Experience must teach whether conclusions are sensitive to changes in the variance and correlation assumptions.

An entirely different approach is possible. Suppose that the growth curve for animal i is

$$y_i = g_i(x) \quad , \qquad (6.10)$$

a member of a family in which the k parameters take the values $\theta_{i1}, \theta_{i2}, \ldots, \theta_{ik}$. In so far as the function is an exact representation of the individual curve, the k parameters completely identify the curve and variation between individuals can be discussed in terms of a multivariate distribution of the θ's. Such "regression models of the second kind" (Sprent, 1969) are attractive for small k; they have been used especially where $g(x)$ could be taken as polynomial and equal spacing of x facilitated the definition of the θ's as coefficients of orthogonal components. However, the complexity of individual curves (noted in Section 1) raises questions about the adequacy of $g_i(x)$ as a true curve unless k is very large. The curve for animal i is unique, so that oscillations about a simple smooth curve cannot logically be ascribed to statistical error; nor can the consequences of describing a growth curve by, say, a 4-parameter logistic when it should be a 10-parameter function of another kind be easily judged. What meaning do the θ's have if they are not the

"true" parameters? Whether the function be correct or not,
mean values of the θ's need not correspond at all closely with
a reasonable mean curve.

The objections to representing each individual growth curve
by a mathematical function of much simpler type (and one with
few parameters) seem philosophically important, especially as
statistical criteria for the adequacy of approximation are not
appropriate. The practice is common; used carefully, it should
be satisfactory, but further theoretical and empirical study is
desirable. Possibly a synthesis of the two approaches will help,
though this may encounter the logical flaws and the statistical
difficulties of both.

ESTIMATION

If no more is required than a descriptive summary of growth,
any simple fitting of curves should suffice and standard polynomial regression may be a good choice. No reference to
orthogonal polynomials is needed; although available tables
make them easily used when ages at measurement are equally spaced,
and they can be constructed for unequal spacing, this device for
expediting computation is not of fundamental importance. Without
a much deeper quantitative understanding of growth processes,
bias is inevitable if withdrawal from observation is dependent
upon growth, for example if fast-growing animals are slaughtered
when young, or animals that do not thrive are culled.

Similarly for establishing a normal pattern of growth, to
smooth individual curves by adopting a simple function is harm-
less. A function linear in all parameters is perhaps desirable,
in order that the curve for mean values of the variate shall be
identical with the curve based on means of the parameters. As
long as no extrapolation is involved, polynomials may again be
adequate.

Estimating the position of an individual growth curve elsewhere than at a date of observation can be a very different

task. If observations are frequent, the problem is primarily
one of interpolation from neighbouring observations. Since the
observations are error-free points on a complicated curve, the
question is mathematical rather than statistical; judgement is
needed in choosing between a few points that represent well local
"waves" in the curve and many points that allow greater flexibility in representing long-term trends. Linear interpolation
between observations on either side of the point of interest may
suffice, or 3-5 observations might be employed if the curve is
evidently fairly smooth. If animals are observed less frequently
and not all at the same ages - for example, some at 6, 12, 18,
24, ... weeks of age and others are 3, 9, 15, 21, ... weeks -
interpolatory estimation for one animal may gain in accuracy
from information given by other animals. I see interesting and
complicated possibilities here, but do not know whether they
have been studied.

Extrapolatory estimation, or prediction beyond the latest
observation, is more difficult, important though it may be for
policy and planning. Whether a curve is drawn through observations for an individual or fitted to means, extrapolation can be
dangerously misleading, especially with a polynomial curve which
may turn suddenly as soon as it goes beyond the points used to
construct it. Even an estimate of equation (3.4) may predict
the approach to an asymptote quite incorrectly. For the
individual, autoregressive time series techniques are an alternative (Hirschfeld, 1970). The measurement at any time may be
regarded as comprising the sum of a function of time (the trend),
a function of the earlier measurements on the same individual
(possibly only the most recent being needed), and a random
deviation. The distinctive feature is that dependence on previous observations allows chance variations to be incorporated
into the future growth of the individual. If the process can
describe well the true pattern of growth, it may offer good
prospects for prediction; however, a long sequence of observations is needed in order to estimate the parameters, and to
verify from known measurements the quality of "prediction".

The method is little use if each animal has been measured only a few times. Like other extrapolation procedures, even when applicable it offers no safeguard against unrecognised changes in conditions. It may prove more successful for skeletal measurements than for gross weight or for a measurement that involves subjective assessment. Autoregressive procedures have not been outstandingly successful for prediction in other fields; can any calculations based on monthly or weekly weights of an animal predict at all closely what further growth will occur in the next 10 weeks, or how many weeks will be needed for an increase of 20 kg?

There remain problems of experiments on feeding and management. I assume an experiment to have a sound treatment structure (perhaps simply a control compared with one or more carefully specified dietary regimes), animals of the same age randomly divided between the treatments, and all animals weighed (or otherwise measured) at t selected ages. Commonly three or four ages of measurement will be used, but the number can be larger. Wishart's (1938, 1939) approach, extended by many people, is today particularly associated with Elston and Grizzle (1962), Grizzle and Allen (1969), Potthoff and Roy (1964), and Rao (1958, 1965). Treatments can be compared by analysis of any combination of weights calculated for each animal, for example final weight or total of all weights. Naturally one wants to draw inferences in terms of a minimal set of meaningful combinations. An obvious step is to find for each animal the unique polynomial of lowest degree that fits exactly the t times of observation; this can be expressed in terms of orthogonal components, so that for any animal it is

$$Y = u_0 + u_1 \gamma_1(x) + u_2 \gamma_2(x) + \ldots + u_{t-1} \gamma_{t-1}(x) \quad , \quad (7.1)$$

where $\gamma_i(x)$ is a polynomial of degree i in x and $\gamma_i(x)$, $\gamma_j(x)$ are orthogonal for every j $\neq$ i. The coefficients $u_0, u_1, \ldots, u_{t-1}$ are then an alternative representation of the individual weights $y_1, y_2, \ldots, y_t$ in the sense that either set of quantities determines the other uniquely, but they are likely to be a more effective summary: u_θ is an average weight, u_1 a conventionalised measure of linear increase, and so on. Comparisons between treatments, and between animals within treatments, can now be made in terms of the u_i. If the growth curve is simple, possibly only the first two or three coefficients carry useful information and treatment differences can be expressed solely in terms of these. If extremes of age are included, the facts remain but the transformation to the u_i may be far less satisfactory.

Often the precision of treatment comparisons can be increased by a covariance analysis on one or more variates <u>known</u> to be unaffected by the treatments: for example, weights or other measurements of animals taken before randomisation, or a characteristic that is genetically determined even though not manifested until after treatments are applied. If the true growth curves were known to be quadratic, third and higher degree polynomial coefficients would be independent of treatment and therefore candidates for use as concomitants (as first suggested by Leech and Healy, 1959). The improvement of precision by covariance analysis underlies the papers to which I have referred, technicalities of this recovery of information accounting for most of the algebraic complexity.

The theory is sound; the papers are excellent (the differences between the authors are of secondary importance); the computations can readily be systematised for routine handling. One thing needed is more experience of whether in practice the potentialities for higher precision are fulfilled. Even more important questions relate to the legitimacy of the method. That a response curve is truly quadratic or cubic will never be known. If a covariance analysis uses a concomitant affected by the applied treatments, the adjustments to means will bias comparisons; mere absence of "statistically significant" differences between treatments is not usually deemed sufficient evidence that no treatment effects exist. Also, the method takes no account of the logical requirement that weights of an individual animal lie exactly on its growth curve, which may therefore be more complicated in form than the mean curve. The method deserves trial, but should not be accepted unreservedly as a safe guide to inference.

I have no time to discuss non-polynomial curves. Any equation linear in its parameters is easily estimated, and even the ideas of Rao and others might be generalised, but such equations are not likely to be very useful. Non-linearity introduces new problems. Under fairly standard assumptions of normality of distribution, and with some knowledge of the variance and correlation structure discussed in Section 6, parameters can be estimated by least squares; the computations, formidable some years ago, present no difficulty to a computer provided with a good optimisation programme. Treatments may still be compared in terms of the parameter estimates, but I doubt whether additional information can be recovered by an analysis generalised from that for polynomials.

CONCLUSION

I am conscious that I have presented a first draft for an introduction to a treatise on growth curves, not a new synthesis or a guide to the statistical analysis of certain types of data. You may judge my offering unsatisfactory and disappointing: I believe the subject to be more complex and more subtle than is generally recognised. To argue about the merits of Rao or Grizzle and Allen may involve heavy mathematics, but the problems are purely technical and potential gains in precision may be small. Biologist and statistician need deeper understanding of the underlying nature of the curves, the aims of particular studies, and the types of interpretation that analysis can support. Beyond these lie questions, until now scarcely touched, of optimal experimental design for stated objectives, taking account of constraints such as the impossibility of slaughtering one animal at two different ages, and possibly requirements that the sale value of animals and carcasses shall not be disregarded.

REFERENCES

Elston, R.C. and Grizzle, J.E. (1972). Estimation of time-response curves and their confidence bands. Biometrics, 18, 148-159.

Grizzle, J.E. and Allen, D.M. (1969). Analysis of growth and dose response curves. Biometrics, 25, 357-381.

Grosenbaugh, L.R. (1965). Generalisation and reparameterisation of some sigmoid and other non-linear functions. Biometrics, 21, 708-714.

Hirschfeld, W.J. (1970). Time series and exponential smoothing methods applied to the analysis and prediction of growth. Growth, 34, 129-143

Kowalski, C.J. and Guire, K.E. (1974). Londitudinal data analysis. Growth, 38, 131-169.

Leech, F.B. and Healy, M.J.R. (1959). The analysis of experiments on growth rate. Biometrics, 15, 98-106.

Potthoff, R.F. and Roy, S.N. (1964). A generalised multivariate analysis of variance model useful especially for growth curve problems. Biometrika, 51, 313-326.

Rao, C.R. (1958). Some statistical methods for the comparison of growth curves. Biometrics, 14, 1-17.

Rao, C.R. (1965). The theory of least squares when the parameters are stochastic and its application to the analysis of growth curves. Biometrika, 52, 447-458.

Sprent, P. (1967). Estimation of the mean growth curves for groups of organisms. Journal of Theoretical Biology, 17, 159-173.

Sprent, P. (1969). Models in Regression and Related Topics, London: Methuen and Co. Ltd.

Wishart, J. (1938). Growth-rate determinations in nutrition studies with the bacon pig, and their analysis. Biometrika, 30, 16-28.

Wishart, J. (1939). Statistical treatment of animal experiments. Journal of the Royal Statistical Society, suppl. 6, 1-22.

NUTRITIONAL MODELS OF GROWTH

R.H. Fawcett

School of Agriculture, University of Edinburgh
West Mains Road, Edinburgh EH9 3JG: Scotland

ABSTRACT

A critique of models specifying energy and protein requirements for growth is presented together with a methodology for linking requirements for growth to parameters expressing limits to growth and its composition. This provides a basis for definition of optimal paths to a specific carcass composition.

INTRODUCTION

Nutritional models of growth attempt to replace the
classical growth curve by a set of input/output relationships
describing how input per unit time is converted to output per
unit time. Whole families of growth curves can be produced
by manipulating input per unit time between the limits of zero
and some finite maximum and integrating over time.

Assessment of any model should commence by asking how well
does the model achieve its stated purpose. The critique
presented here is based upon a set of criteria relating to the
design of economically efficient systems of meat production.

The general requirements of any model are that it should
be

a) comprehensible,

b) easy to use,

c) reasonably accurate.

The specific requirements of models for the design of
animal production systems are that they should be adequate for
response to changes in

a) environmental temperature,

b) composition of diets,

c) growth rates and carcass quality

and if possible be incorporated into some form of mathematical
optimising procedure.

The search for the appropriate functional form of the
growth curve led to the development of complex mathematical
models of growth which included some useful biological principles

Weiss and Kavanau (1957) use a set of differential
equations to define growth as the net balance of mass produced

over mass destroyed and terminal size is represented as a
stationary equilibrium between incremental and decremental
components. Growth regulation occurs automatically by a
negative feed back mechanism and provides an adequate
description of chick growth to maturity.

A more recent approach is to be found in Parks (1970[1,2,3]) where a systematic attempt is made to relate growth to
food consumption, energy and protein.

Roux (1974), (1976) has followed the same growth as Parks
and has described the growth process as a single dimensional
stationary gaussian markov process. Such models are important
from the point of view of ad libitum growth and simple control
mechanisms but require a sound knowledge of differential
equation systems and markov processes before they are
understood.

A related set of models have been derived by workers
involved in nutritional advice on how to feed farm livestock
and are reviewed by Moe and Tyrell (1971). These models are
based solely on the energy requirements for maintenance and
growth. Differences between them lie in scientific interpretation rather than empirical validity. They are reasonably
successful in predicting the consequences of a given quantity
of a diet to a given weight of animal but require a cumbersome
set of adjustments to the availibility of ME, in response to
feeding level, and the energy requirements for growth in response
to increased growth rate.

The models are based on empirical work which validates
the system of calculation. Computer models have been built
around ARC (1965), Lofgreen and Garret, (1968) and Nehring et
al, (1971). Adaptation of ARC (1965) into linear programming
form was affected by McHardy (1965). Simplification of ARC
(1965) is to be found in Harkins et al, (1974) and Newbery
(1974) who integrates the implied growth equations in order to

establish input/output relationships for beef cattle on large
scale experimental trials.

A critical reappraisal of ARC (1965) is to be found in
Webster et al. (1974) suggesting that fasting metabolism is
not the appropriate basis from which to calculate energy
requirements for production.

Graham et al. (1976) considers these models to be a gross
over simplification of animal, diet, environment interactions
and too narrowly constrained in their area of application.
Graham's group have developed a computer model predicting

1) daily balances of energy and nitrogen,

2) daily deposits of wool, conceptus, milk and body tissues,

3) daily changes in empty live weight.

Whilst effective use of an input/output system does not
require any knowledge of what goes on inside a 'black box',
the time has come wherein our knowledge of the underlying
physiological functions enables us to replace one large box by
several smaller boxes with some increase in accuracy and better
understanding of the metabolic processes governing growth
response to nutrient input.

The reasons why foods should have differential efficiencies
of use for maintenance and growth must be exposed. A possible
solution to this riddle may be found by simulation of daily
nitrogen and protein balance. Whittemore's concept of protein
corrected maintenance is entirely consistent with the comments
of Webster et al (1974). It should be pointed out that
Kielanowski (1966) paved the way by demonstrating evidence for
'relatively' stable energy costs of protein and energy deposition
Kotarbinska (1969) provides the means of converting protein
and lipids into live body gain. This work forms the basis of
the Whittemore, Fawcett models (1974), (1975), (1976) and the
model of Monteiro (1975).

Monteiro establishes a functional relationship between food intake, weight gain and body composition based on the differential energy requirements of fat, fat free tissue and gut fill. An allometric function relating changes in fat free tissue to changes in body weight was used to extend the model into situations where direct measures of body composition are not available. Monteiro reported that the bulk of the differences between Jersey and Friesian could be accounted for by differences in composition of weight gain. It is reasonable to assume that large and small types function with different efficiency at equal live weights, and that differential efficiencies over the same age interval correspond to breed characteristics of size, degree of maturity and appetite.

BUILDING A MODEL

Feed input is one of the major determinants of the profitability of animal production systems. Graham et al. (1976) state simply that efficient use of feedstuffs requires knowledge of 3 items,

1)   an animal's requirements for nutrients,

2)   biological value of feeds,

3)   some method of using the information to reach a specific objective.

The model of Whittemore and Fawcett (1976) can be developed to provide all three within a linear programming framework.

Growth as a net balance

The concept of turnover of body mass (Weiss and Kavanau) or energy mass is rejected in favour of a model of protein turnover since it is assumed that the recovery of useful energy from catabolism of lipid is 100% efficient, whereas on deamination of protein there is a significant loss of energy as heat production.

Partitioning of nutrients

The idea of a metabolic objective function has been proposed by MacClean (1977) but provides no practical advice.

Models which require more than one product from metabolism run into this problem. The models cited by Moe and Tyrell all work on the principle of maintenance first, production later. Clearly such models can never deal with turnover, pregnancy or lactation. Defining maintenance by weight stasis assumes that body fat is not being catabolised to fuel protein synthesis. At zero nitrogen retention it is thought that 5% of the protein mass is broken down and re-synthesised involving considerable heat production. A model based on a constant energy requirement for synthesis must reduce estimates of maintenance by the amount of heat produced from turnover at zero retention. This is the basis of Whittemore correction. Maintenance is then a residual energy requirement for other body work.

Whether or not there is energy demand for cold thermogenesis depends upon the rate of heat production. Animals growing quickly have lower critical temperatures. Energy for cold thermogenesis can be supplied firstly by depressing the daily rate of lipid deposition until the minimum lipid protein ratio in the gain is obtained thereafter both lipid and protein deposition are reduced in the same ratio until sufficient heat output is generated. This is easily incorporated into a linear programming framework (Whittemore and Fawcett 1975). The relationship between critical temperature and daily heat production is independent of body weight. Marginal increments in daily heat production required to establish equilibrium with the environment are proportional to metabolic body weight. Thus smaller pigs tend to a higher heat production per kg metabolic body weight and require higher environmental temperatures to avoid body losses to cold thermogenesis.

The Requirements for nutrients

The model partitions the digestible energy and protein
components of the feed intake into live body gain, urinary
loss and heat loss by means of simultaneous nitrogen and energy
balance, and incorporates two fundamental assumptions.

Firstly, there is a finite maximum daily rate of protein
retention hence the daily rate of retention Pr is less than or
equal to the finite maximum P̂r. Secondly, there is a minimum
lipid: protein ratio Ỹ in the daily gain so that the amount of
lipid retained must be greater than or equal to a multiple Ỹ of
the amount of protein retained. The energy requirement for
lipid production is assumed to be constant at 0.0535 MJ/gm of
which 0.0393 MJ is retained. The energy requirement for protein
synthesis is constant at 0.0309 MJ/gm of which 0.0236 MJ is
retained. This gives rise to a variable cost of protein
deposition since protein retained is the net result between
breakdown and synthesis. The digestible protein (P) is split
into two components that which can be used for new synthesis
(Pn) and that which is to be deaminated (P-Pn). Protein
retention (Pr) can be defined as total synthesis (Px) less
breakdown (Pb).

$$Pr = P\bar{x}-Pb \qquad (1)$$

Total synthesis can be regarded as the sum of new-synthesis
and resynthesis.

$$Px = Pn + (1 - \emptyset) Pb \qquad (2)$$

Recovery from the amino pool of the body protein involves
endogenous losses ∅Pb.

$$(P-Pn) + \emptyset = P-Pr \qquad (3)$$

The total amount deaminated and then excreted is given by
the necessary condition to balance the utilisation of protein

mass and only 0.0115 MJ/gm is recovered as available energy. Whittemore and Fawcett (1976) showed that (Z) the ratio of protein retained : total protein synthesis could be expressed as a diminishing function of the degree of maturity.

$$\frac{Pr}{Px} = Z = 0.23 \quad \frac{(P\hat{t} - Pt)}{(P\hat{t})} \tag{4}$$

(Where Pt is a current protein mass and $P\hat{t}$ the mature protein mass). If Z is a proportion of protein retained from the total synthesis Px, then (1-Z) Px is the amount of protein which is broken down. Substituting for Pb in equation 2.

$$Px = Pn + (1-\emptyset)(1-Z) Px \tag{5}$$

With $\emptyset$ constant at 6% we derive the result

$$K = \frac{Pn}{Px} = (0.06 + 0.94 Z) \tag{6}$$

If a chemical value (V) is ascribed to the protein $(0 < V < 1)$ the amount of protein available for new-synthesis can be regarded as $PV \geqslant Pn$. If Pn/Px = K and Pr/Px = Z then Pr/Pn = Z/K from which can be deduced that $Pr \leqslant \frac{ZV}{K}$ P. Clearly biological values should only be determined in situations of protein limited growth in the presence of adequate energy when this restraint is binding. The biological efficiency of conversion of digested protein into retained protein is then given by ZV/K and is dependent partly upon the quality of that protein symbolised by V and partly upon the state of the animal Z/K the limit to the biological efficiency of conversion (Table 1, column 4).

Value of Feeds

Chemical value, V, is related to the amino acid profile of the proteins in the animal's body but distorted by the differentia efficiency of utilisation of certain amino groups. Oser (1959) developed a concept of essential amino acid index and Bender (197

TABLE 1

EFFICIENCIES OF PROTEIN TURNOVER

1	2	3	4
	Z	K	Z/K
% Degree of Maturity	Protein retained / total synthesis	new synthesis / total synthesis	Retention / new synthesis
0	0.230	0.276	0.833
20	0.184	0.233	0.790
40	0.138	0.199	0.693
60	0.092	0.146	0.630
80	0.046	0.103	0.446
100	0	0.060	0

a chemical score in order to predict the biological value of proteins, but neither method was entirely satisfactory due to the effects of level of feeding and age of the animal subjects involved in the bioassay. Substitution possibilities exist between amino acids, methionine and cystine, arginine and tryptophan, phenylalanine and tyrosine, thus the preferred amino acid profile for chemical value should be expressed in terms of independent groups consisting of one or more amino acids. Such a profile would be considered correctly adjusted when the addition of one member of each group in isolation could not increase nitrogen retention. Dietary response to a specific amino acid will indicate that only that essential amino acid group is limiting. Limiting amino acids can be identified by comparison against the preferred amino acid profile, Table 2 column 7. The chemical value ascribed to barley of 0.457 is the minimum value obtained from the division of the percentage of each amino acid group in barley protein by the percentage of that amino acid in the required profile. The number of grammes of protein of chemical value unity which can be obtained from a kilogramme of feed ingredient is obtained by multiplying the number of grammes of digestible protein per kilogramme by the chemical value. This is listed for each of the limiting amino acid groups in Table 3.

TABLE 2

CHEMICAL VALUES AS DETERMINED BY LIMITING AMINO ACIDS

Amino Acid	Barley		Soya		Fish		Required	Ref
	1 %AA	2 1/7	3 %AA	4 3/7	5 %AA	6 5/7	7 %AA	Number
Histidine	2.1	0.84	2.6	1.04	2.3	0.92*	2.5	1
Isoleucine	4.9	1.225	5.0	1.25	4.9	1.225	4.0	2
Leucine	6.1	0.76	7.6	0.95	7.6	0.95	8.0	3
Lysine	3.2	0.457*	6.2	0.88	7.4	1.06	7.0	4
Meth + cystine Tryosine	3.6	0.90	2.8	0.70*	4.0	1	4.0	5
Penylalanine	9.3	1.33	8.6	1.23	7.9	1.13	7.0	6
Threonine	3.5	0.875	4.2	1.05	4.6	1.15	4.0	7
Tryptophan	1.4	1.4	1.4	1.4	1.4	1.4	1.0	8
Valine	5.0	5.0	5.0	5.0	5.6	5.6	1.0	9
g/kg Protein	77		410		580			

* Limiting Amino Acid Group

TABLE 3

GRAMMES OF CHEMICAL VALUE UNITY SUPPLIED BY 1KG FEED FOR EACH POSSIBLY LIMITING AMINO ACID

Limiting Amino Acid ref.		Barley	Soya	Fish
Histidine	1	64.7	426.4	533.6
Lysine	4	35.2	360.8	614.8
Meth & Cys	5	69.3	287.0	580.0

For the purposes of computerised diet formulation chemical value must be a consistent pre-determinable attribute of the feeding stuff. The values contained in Table 3 have this important attribute which biological values cannot have since they are specific to the age of the animals used on the bio-assay.

The digestible energy values of feeds for monogastric animals does not appear to be a serious source of error and the published values have been used from Whittemore and Elsley (1975).

Methodology and Objectives

Kotarbinska (1969) tabulates the ratio of fat free body mass to protein mass in the whole body and the ratio of fat free gain to protein gain over live weight ranges. From the initial condition of live weight and protein mass the lipid and protein content of pigs over the live weight range 20 - 100 kg can be established and is described by $L = 0.67P^{1.36}$. A function of this nature may be unique for ad libitum fed pigs but can be manipulated by controlling intake. Kotarbinska's fat free ratios are specific to the growth path of the experimental animals because the whole of the fat free gain is ascribed to protein gain.

Fatty tissue is known to contain about 10% water. If (1.1 * lipid gain) is subtracted from live body gain a corrected fat free gain independent of lipid is obtained. Division of corrected fat free gain by protein gain yields a corrected fat free ratio (x 1) and provides a base for a more general relationship between live weight gain ΔW and protein (Pr) and lipid (Lr) in the gain.

$$\Delta W = x_1 \, Pr + 1.1 \, Lr$$

$$\text{Where } x_1 = 3.772 \left(\frac{Pt}{Pt}\right)^{-.11256}$$

Similarly a corrected fat free ratio in the whole body can be derived

$$Y = 4.251 \left(\frac{Pt}{Pt}\right)^{-.11386}$$

These ratios are expressed as functions of the degree of maturity rather than live weight in order to incorporate the effect of different genotypes having different protein masses at the same live weight.

Grading schemes for pigs can be based upon the P_2 back fat measurement. The statistical estimation problem of the model builder is that from a knowledge of protein mass, lipid mass and live weight he needs to predict carcass weight and grade. Alternatively commencing with carcass weight and grade he needs to predict live weight, lipid mass and protein mass. A slightly more devious route has had to be used in practice based on published data relating carcass dissection to chemical analysis which cites the 'C' measurement rather than P_2. Dr. Kempster of MLC has provided an equation transforming C to P_2. Dissected fat can be predicted from live weight and C pooling evidence from several sources and lipid mass can be directly estimated from dissected fat. This provides the basis of fat free live weight (FFB).

Protein mass P_t can now be estimated from the fat free live weight divided by the corrected fat free ratio in the whole body. The generalised equation can be derived FFB = 4.251 $Pt^{0.88614} Pt^{0.11386}$ which can be solved for Pt and may well be an adequate description of the growth and development of an animal.

The pathway is now complete in that the lipid and protein content of any increment in live weight from weaning to a fixed slaughter point can be defined.

Eg given the initial conditions:-

1)   final carcass weight 77 kg, or live weight 100 kg.

2)   target P_2 23 mm,

3)   initial weight and composition 20 kg live, 3 kg protein, 3 kg lipid .

The growth path of Kotarbinska's pigs can be simulated by a stepwise ascent of a compositional function linking the initial and final composition, thus defining the lipid protein ratio associated with daily increments in protein mass. A simplified linear programming matrix expressing the daily equilibrium in activity rates is presented in Table 4. This represents a linear set of difference equations the coefficients and restraints of which can be updated daily.

Growth curve analysts, Zotin and Zotina (1972), Parks and Roux envisage growth as a stationary markov process in that the genetic characteristics of the animal are expected to bring it to a predetermined stationary state via a predictable route under voluntary feed intake.

The present methods of utilising nutritional models of growth requires preselection of an average growth rate. The linear programming model can be used to explore the limits to the growth and intermediate body composition of an animal when provided with three essential pieces of information;

1)   the limit to daily feed intake,

2)   the limit to daily rate of protein deposition,

3)   the minimum lipid : protein ratio in the gain.

The effect of variation on the target 'P_2' backfat measurement on the expected yield and composition of bacon pigs at 100 kg live weight is shown in Table 5.

It is now a feasible assignment to identify not only the most profitable slaughter point but the diet and feeding scale required to achieve that objective.

A recursive linear programming algorithm can be used to define the least cost diet and feed consumption required to grow a pig to any particular live weight and grade within the limits imposed by its genetic potential.

The target P_2 and live weight require a pig to contain 15 kg protein and 27 kg lipid. Tests of the model are consistent

TABLE 4

LINEAR PROGRAMMING TABLEAU SIMPLIFIED FOR A LEAST COST GROWTH PATH

Growth rate	Protein retention	Lipid retention	Protein intake	Barley	Soya	Fish	Tallow	Restraints
1	$-x_1$	-1.1						≤0 Growth rate
	1							≤Pr̂ Rate of protein deposition
	Y	-1						=0 Composition of gain
	x_2			-35.2	-363	-613		≤0 Lysine
	x_2			-69.3	-287	-580		≤0 Meth + Cyst
	x_2			-64.7	-426	-534		≤0 Histidine
	x_3	.0535	.0242	-12.7	-15	-15.1	-29	=-EM Energy Balance
			-1	77	410	580		=0 Protein Balance
				1	1	1		≤F Air dry intake
+P				$-P_B$	$-P_S$	$-P_F$	$-P_t$	= Max

x_1 = ratio of fat free body gain to protein gain

Y = Lipid protein ratio in the gain

x_2 = (K/Z) efficiency of protein turnover

x_3 = energy cost of protein retention

P = value of live weight gain

P_B, P_S, P_F, P_t, = prices of feed ingredients

Pr̂ = maximum rate of protein deposition

EM = residual energy requirement

F = maximum air dry intake

TABLE 5

THE EFFECT OF TARGET 'P$_2$' ON YIELD AND COMPOSITION OF BACON PIGS AT
100 KG LIVE WEIGHT

Target P$_2$	Estimated Carcass Wt. kg	Lipid kg	FFB kg	Protein kg
17	74.0	19.5	80.5	16.7
18	74.5	20.8	79.2	16.5
19	75.0	22.0	78.0	16.2
20	75.5	23.3	76.7	15.9
21	76.0	24.5	75.5	15.7
22	76.5	25.8	74.2	15.4
*23	77.0	27.0	73.0	15.2
24	77.5	28.3	71.7	14.9

with the three experiments of Fowler and Livingstone (1971) in
that the total energy and protein requirements for this lipid
and protein growth are independent of the route chosen provided
that it is effected in the minimum possible time. This requires
that the daily rate of protein deposition is maximised.

CONCLUSION

The recursive linear programming model is in fact a Markov
process in disguise and does not require any knowledge of
differential equations and can satisfy all the criteria of a
model which could be useful in the design of animal production
systems capable of incorporating aspects of environment, genetic
variation, diet composition and economic objectives. The
unsolved problem is what governs feed intake. There is the
possibility of feed back control limiting the rate of heat
production in any particular environment.

Lack of objective criteria on which to define appetite
means that the model cannot fully explore the full range of
possible diets. Currently a maximum ME intake or air-dry weight

intake has to be used.

There is no reason why similar models could not be
developed for ruminant animals provided that a systematic
relationship can be expressed between chemical composition and
a grading scheme which will permit estimation of the expected
value of the final carcass.

REFERENCES

ARC (1965) Nutrient requirements of farm livestock, No.2. Ruminants
    Agricultural Research Council, London.

Bender, A.E. 1973. Chemical scores and availability of amino acids in
    Protein in Human nutrition, ed. J.W.G. Porter and K.A. Rolls, Academic
    Press, London.

Fowler, V.R. and Livingstone, R.M. 1971. Anim. Prod. 13, 59-60

Harkins, J. Edwards, R.A. and McDonald, P. 1964. Anim. Prod. 141-148.

Kielanowski, J. 1966. Anim. Prod. 8, 121-128.

Kotarbinska, M. 1969. Chemical composition of pigs, Badania nad przemiana
    energii n rosnacych swin. Pub. Warsaw Inst. Animal Physiol. No. 239.

Lofgreen, G.P. and Garret, W.N. 1968. J. Anim. Sci. 27, 793-806.

MacClean, F.I. 1977. J. Theor. Biol. 65, 513-522.

McHardy, 1965. An investigation of the application of programming techniques
    to farm management problems. Ph. D. Thesis, Edinburgh University.

MLC 1973. Progress pig carcass classification, Dec. 1973. Tech. Bull. No.
    10. Meat and Livestock Commission, Bletchley.

MLC 1975. Commercial product evaluation report, April 1975. Meat and
    Livestock Commission, Bletchley.

Moe, P.W. and Tyrell, H.F. 1973. J. Anim. Sci. 37, 183-189.

Monteiro, L.S. 1975. Anim. Prod. 20, 315-335.

Nehring, K., Beyer, M. and Hoffman, B., 1971. Futtermitteltabellenwerk, VEB,
    Deutscher Landwirtschaftsverlag, Berlin.

Newbery, D.M.G. 1974. J. Agric. Sci. 82, 1-10.

Oser, B.L. 1959. An integrated amino acid index for the production of
    biological values of proteins. In Protein and Amino Acid Nutrition,
    ed. A.A. Albanese. Academic Press, London.

Parks, J.R. 1970. 1) Am. J. Physiol., 219, 833-836.
                  2) Am. J. Physiol., 219, 837-839.
                  3) Am. J. Physiol., 219, 840-843.

Roux, C.Z. 1974. Agroanimalia 6, 49-52.
           1976. Agroanimalia 8, 83-94.

Webster, A.J.F. Brookway, J.M. and Smith, J.S.1974. Anim. Prod. 19, 127-139.

Weiss, P. and Kavanau, J.L. 1957. J. Gen. Physiol., 41, 1-47.

Whittemore, C.T.W. and Elsley, F.W.H. 1976. Practical Pig Nutrition.
    Farming Press, Ipswich.

690

Whittemore, C.T.W. and Fawcett, R.H. 1976.   Anim. Prod. 22, 87.

1975.   Model Pig University of Edinburgh

1974.   Anim. Prod. 19, 221-231.

Zotina, R.S. and Zotin, A.I. 1972.   J. Theor Biol. 35, 213-225.

QUANTIFYING BREED DIFFERENCE IN SHAPE

H.J.H. Macfie

Meat Research Institute, Langford, Bristol, United Kingdom

ABSTRACT

The definition and estimation of shape using multivariate statistical techniques that operate on all characters simultaneously is discussed. The basic principle is to find linear combinations that are correlated with growth and then the remaining variation is taken to represent shape variation. The procedure used depends on whether concomitant variables exist that can together be taken to represent growth. An example is given in which the muscle distribution of three breeds of cattle is compared.

INTRODUCTION

The aim of this paper is to show how differences in shape among populations containing a growth effect may be displayed and quantified using multivariate techniques.

Definition of shape

Shape may be defined in a variety of ways and using a variety of measurements, but the principle in each case is the same; having defined a suitable measure of size, a new measure that is in some sense independent of size is defined to be a measure of shape. For example if size is defined as body weight then the vector of ratios of parts to the whole is a shape vector. This topic has been well discussed by Mosimann (1970) and Sprent (1972).

Growth and shape estimation

The concept of designating one variable to represent size and adjusting all other variables to a constant value of size is particularly useful in a part - whole study such as the examination of differences in muscle distribution or body composition, as was done by Mukhoty and Berg (1971). If, however, skeletal measurements are included it becomes much more difficult to define a suitable size variable and one solution in studies that contain a large growth effect is to perform a principal components analysis, and obtain that linear combination of all the variables that shows the maximum variation among the individuals. In many cases the coefficients of this linear combination, termed the first principal component (PC), are all positive and nearly equal, and may be taken to represent a linear function of size. Rao (1964) and Jolicoeur $(1963_1, 1963_2)$.

The second PC is that linear combination of the variables that displays the maximum variation among the individuals not accounted for by the first PC. Subsequent PCs are defined similarly until all the variation has been accounted for. In geometrical terms each PC represents a plane in the multidimensi(

variable space defined by a set of perpendicular axes - one for each variable. Each PC plane is perpendicular to every other and thus if one plane represents size all other PCs display variation that is independent of size and may be designated as shape variation. This approach was used by Brown et al. (1972) in a study of the variation in 10 body measurements on Hereford and Angus bulls at 4, 8 and 12 months of age. The first PC, accounting for 56 to 68% of the total variation, gave nearly equal weighting to the 10 measurements and was taken to be a linear function of size, while the second PC, accounting for 10% of the total variation, contrasted tall, narrow animals with short, wide bodied animals.

Growth invariance

A major problem in the interpretation of shape vectors, obtained using principal components, is that although the vectors are independent of size they may still contain a growth effect. To avoid this Brown et al. (1972) performed a separate analysis on each age group. However in many experiments the changes in shape with growth are of very great interest - also the number of animals in one age group may be too small to perform a separate principal components analysis. Furthermore, once growth related shape vectors have been identified, we may examine shape vectors that are growth invariant. These vectors reflect shape differences among animals that were present throughout the age range under study, and if all the animals are from similar breed, sex and nutritional backgrounds, these vectors quantify biological variation in shape.

The identification of growth related shape vectors is only possible when concomitant variables are available that can together be taken to characterise size. Any PC showing a significant correlation with the concomitant variables and also being orthogonal to the first PC (size function) can be ascribed as a growth related shape vector. Alternatively it is possible to use canonical correlations to find successive linear combinations of the variables that are most highly correlated

694

with linear combinations of the concomitant variables. An
example of this approach is given in this paper.

Estimation of shape differences

In the previous section we have considered the estimation
of shape vectors within a particular population. In the case
where it is required to estimate the shape differences <u>between</u>
populations the procedure is to remove the variation due to
size and growth using principal components, or canonical correlations if concomitant variables are present. The remaining
variation is taken to be shape variation. Successive linear
combinations, called canonical variates, Rao (1952), that display
the maximum ratio of between to within population variation are
then obtained and as these variates are uncorrelated within
populations a 2 dimensional display of the differences may be
obtained by plotting the population means relative to the
first 2 canonical variates axes.

The mathematical theory to perform a canonical variates
analysis in growth invariant space was developed by Burnaby
(1966) and Gower (1976). An example of its application when
there are no concomitant variables is given by Reyment and
Banfield (1976). In this paper an example is given of the case
where concomitant variables are present.

The motivating problem in this application is the comparison
of the muscle distribution of three breeds of cattle. Thus the
term 'shape' relates specifically to the components of the
musculature and excludes other important components of overall
shape such as skeletal structure and fat. However the principles
discussed in the foregoing sections are just as relevant to
muscle distribution as to overall shape.

MATERIAL AND METHODS

Experimental material

Three breeds used in the experiment were Hereford (17)
Hereford x British Friesian crosses (19) and British Friesian

(19). Steers from each breed were reared under normal commercial conditions and where possible two animals from each breed were slaughtered and dissected at 0, 6, 12, 18 and 24 months of age. Full details of the experiment are contained in Owers et al. (1968). For this study the data consisted of the weights of the nine major muscle groups depicted in Figure 1.

Statistical Methods

To stabilise the variation it was necessary to transform all the weights to log base 10. Total muscle weight (TMW) was taken to represent size and two forms of allometry were used to relate y - the weight of a muscle group - to x - the total muscle weight. These were

1) $\log y = a + k \log x$

2) $\log y = a + k \log x + 1 (\log x)^2$.

The second form expresses a curvilinear relationship between log y and log x and was found to give a significant improvement in fit for 6 of the 9 muscle groups. To allow for this curvilinear growth a canonical correlations analysis between the 9 muscle groups (log base 10) and log TMW and $(\log TMW)^2$ was performed separately on each breed. The resulting linear combinations were tested for parallelism using methods developed by MacFie (1977). Finally a canonical variates analysis on both the uncorrected and growth invariant data was performed.

RESULTS

Canonical correlations

The coefficients of the two pairs of linear combinations $\underline{L}$ and $\underline{M}$ of the muscle groups and log TMW, $(\log TMW)^2$ respectively are shown in Table 1 for each breed separately, for the pooled within breed and overall.

Comparing the different analyses it can be seen that the contrasts expressed by the canonical vectors are the same in each case. The difference in size of the concomitant canonical

coefficients between the separate breed and pooled/overall is
due to a scaling constraint and is unimportant. Testing the
breed vectors for parallelism revealed no significant difference,
MacFie (1977), and so the pooled within breed pairs were used
to estimate the growth effect.

Before discussing the breed comparisons it is worth
examining the coefficients of the canonical correlation vector
pairs.

In each of the analyses the first vector has a very small
coefficient for the quadratic term, implying that this vector
pair represents a linear growth effect. This is supported by
the fact that all the coefficients of the muscle groups are
positive.

The effect expressed by the second vector of coefficients
for the concomitant variables is a contrast between the linear
and quadratic terms. This contrast will increase with log TMW
and appears to be expressing the curvilinearity observed at the
higher weights. This is represented in the associated vector
of muscle groups by a contrast between muscle groups 1 and 2
(large positive) and 5 and 9 (large negative). Using the well
known classification scheme of Butterfield and Berg (1966) these
groups would all have been diphasic with impetus high-low for 1
and 2 and low-high for 5 and 9. This vector is thus a contrast
between regions of musculature that increased or decreased their
proportions during growth.

Canonical variates analyses

The two linear combinations of the muscle groups that
display the maximum variation between the 3 vectors of breed
means before the data was corrected for growth, are shown in
Table 2. Significance testing, Bartlett (1947), indicated that
only the first canonical vector displays a significant proportion (86.3%) of the variation between breed means. This is
shown clearly in Figure 2 where the breed means are plotted

1. **Muscles of the Proximal part of the Pelvic Limb.**

2. **Muscles of the Distal part of the Pelvic Limb.**

3. **Muscles of the Sub-Lumbar Region.**

4. **Muscles of the Abdominal Wall.**

5. **Muscles of the Proximal part of the Thoracic Limb.**

6. **Muscles of the Distal part of the Thoracic Limb.**

7. **Muscles connecting the Thoracic Limb to the Neck.**

8. **Muscles of the Dorso-Spinal and Complexus Regions.**

9. **Intrinsic Muscles of the Neck, Thorax and Rest.**

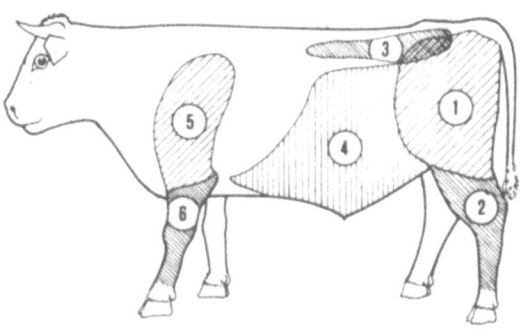

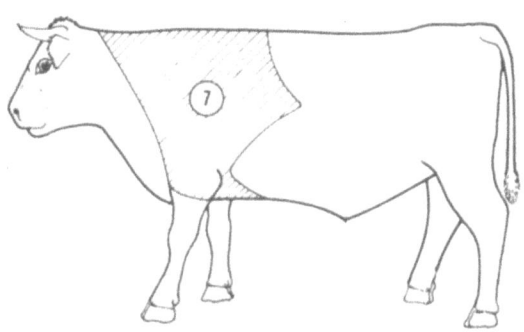

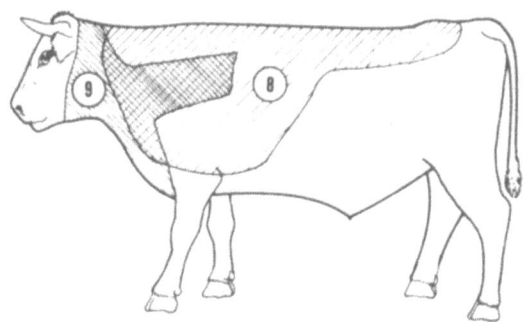

Fig. 1.

TABLE 1

SUMMARY OF CANONICAL CORRELATIONS

Breed	Vector Pair	$\log$ TMW	$\frac{M}{2}(\log^2 TMW)$	$\underline{L}$ Muscle Group									Canonical Correlations
				1	2	3	4	5	6	7	8	9	
Hereford	1	0.615	-0.001	0.186	0.045	0.014	0.055	0.072	0.009	0.058	0.102	0.070	1.00
	2	17.966	-2.083	-7.588	-2.677	0.521	-1.627	6.676	0.361	1.133	0.460	2.445	0.90
Hereford x Friesian	1	0.642	-0.003	0.198	0.038	0.021	0.055	0.066	0.008	0.053	0.099	0.069	1.00
	2	18.868	-2.150	-6.119	-2.473	0.985	-0.196	4.355	1.102	-1.791	3.858	-0.079	0.84
Friesians	1	0.553	0.000	0.174	0.036	0.007	0.056	0.063	0.034	0.035	0.085	0.067	1.00
	2	14.992	-1.728	-5.377	-0.206	1.107	-1.358	3.350	1.962	0.645	-0.734	1.269	0.89
Overall CSSP matrix	1	0.342	-0.001	0.106	0.023	0.008	0.031	0.042	0.011	0.028	0.051	0.040	1.00
	2	9.541	-1.097	-3.020	-0.941	0.303	-0.915	1.796	1.508	0.383	0.367	0.683	0.82
Pooled within breed CSSP matrix	1	0.347	-0.001	0.107	0.024	0.008	0.030	0.042	0.010	0.028	0.052	0.040	1.00
	2	9.635	-1.108	-3.219	-0.743	0.241	-0.987	1.725	1.336	0.547	0.514	0.739	0.82

TABLE 2

CANONICAL VARIATES ANALYSIS: ROOTS, VECTOR COEFFICIENTS AND MEANS FOR
UNADJUSTED DATA

	Non zero roots			
	1	Sig	2	Sig
	1.0240	**	0.1619	NS
% of variance	86.3		13.7	

	Vector coefficients	
Muscle group	1	2
1	24.33	-0.79
2	-30.90	1.51
3	4.52	-18.03
4	13.00	12.68
5	7.67	-21.14
6	39.44	35.74
7	-21.61	7.60
8	-20.24	2.69
9	-12.65	-18.39
Constant term	5.58	-6.33

	Breed means	
	1	2
Hereford	-0.97	0.44
H x F	-0.46	-0.51
Friesian	1.32	0.11

700

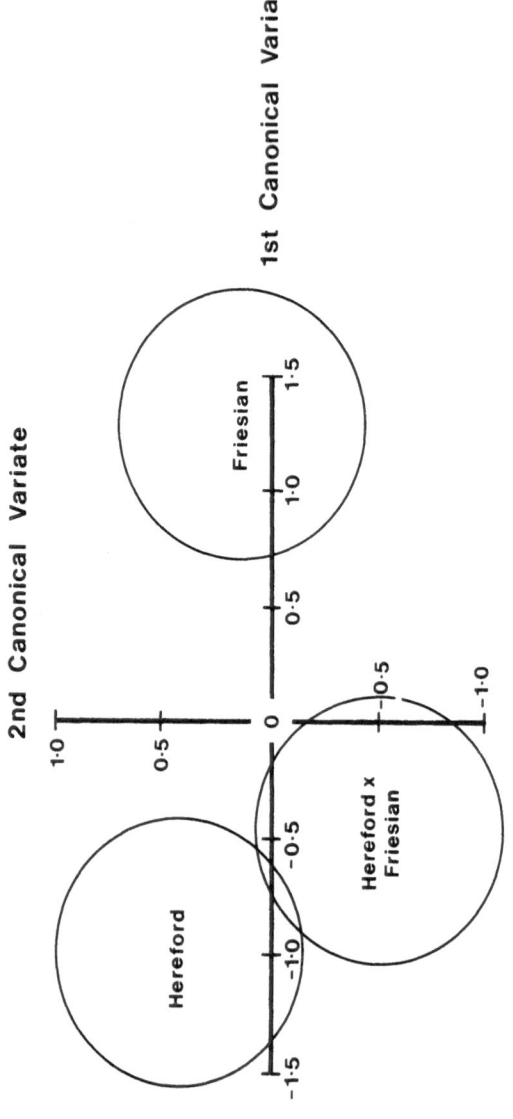

Fig. 2.

relative to the first two canonical variates. The 95% confidence regions for the means are also shown.

This plot indicates that the first canonical variate accounts for differences between the Friesians and the other two groups. Examining the coefficients of this variate indicates a contrast between muscle groups 1 and 6 (positive coefficients) with muscle groups 2, 7, 8 and 9. The positions of these groups, shown in Figure 1, suggested two anatomical contrasts at work: the first between the upper and lower section of the rear leg, the second between the lower front leg and the neck and back groups.

A similar analysis on the data adjusted for growth appears in Table 3 and Figure 3 shows the corresponding plot. It is clear that all significant breed differences are removed by projecting the data into growth invariant space. It is concluded that the muscle distribution of the three breeds did not differ when the growth effect was removed.

DISCUSSION

The fact that there is no difference in muscle distribution between these groups once growth related differences have been removed may not surprise many workers in this field. Clearly the next step is to include breeds that are quite different visually, into the analysis. A similar examination using skeletal measurements is also planned.

TABLE 3

CANONICAL VARIATES ANALYSIS: ROOTS, VECTOR COEFFICIENTS FOR DATA
ADJUSTED FOR GROWTH

	Non zero roots			
	1	Sig	2	Sig
	0.0485	**	0.0074	NS
% of variance	86.8		13.2	

	Vector coefficients	
Muscle group	1	2
1	2.54	0.01
2	-3.53	-0.19
3	1.07	-3.23
4	2.09	1.41
5	-0.11	-1.58
6	6.40	3.82
7	-2.09	0.41
8	-3.76	1.83
9	-1.97	-2.42

Constant term

	Breed means	
	1	2
Hereford	-0.1168	0.0566
H X F	-0.0691	-0.0633
Friesian	0.1736	0.0127

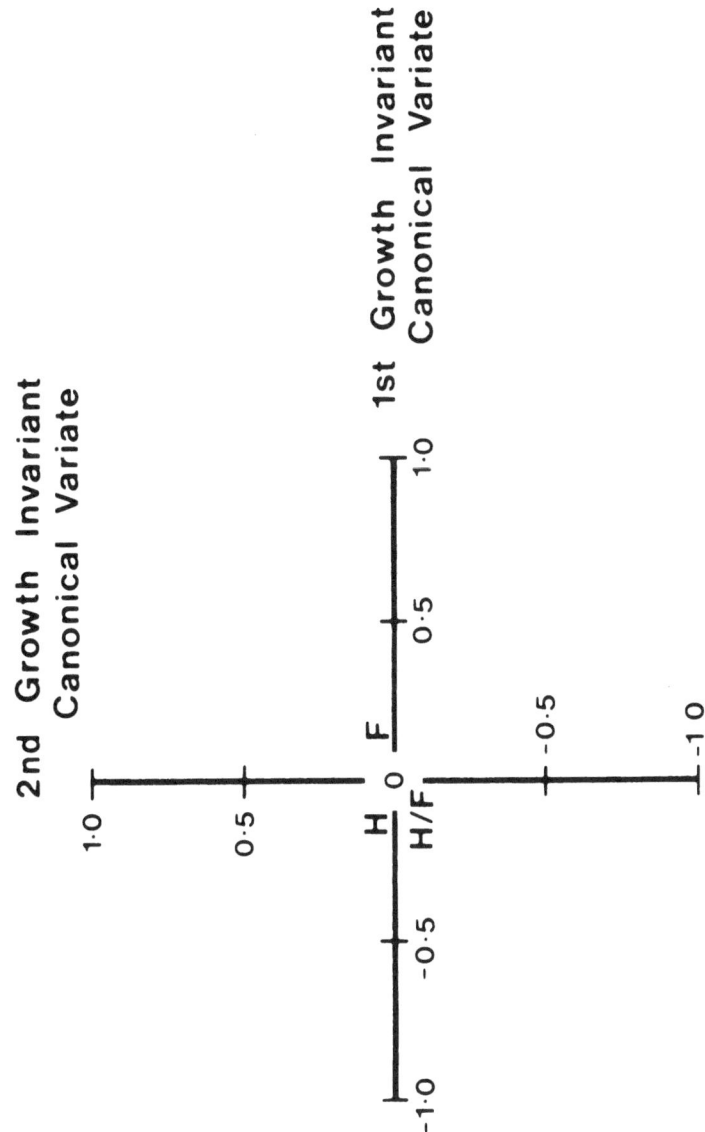

GROWTH INVARIANT CANONICAL VARIATES ANALYSIS

Fig. 3.

704

REFERENCES

Bartlett, M.S. 1947. Multivariate analysis. J. Royal Stat. Soc. 9, 176-197.

Barnaby, T.P. 1966. Growth invariant discriminant functions and
    generalised distances. Biometrics. $\underline{22}$, 96-110.

Butterfield, R.M. and Berg, R.T. 1966. A classification of bovine muscles
    based on their relative growth patterns. Res. Vet. Sci. $\underline{7}$, 326-332.

Gower, J.C. 1976. Growth-free canonical variates and generalised inverses.
    Bull. Geol. Inst. Univ. Uppsala, N.S. $\underline{7}$, 1-10.

Jolicoeur, P. 1963_1. The multivariate generalisation of the allometry
    equation. Biometrics. $\underline{19}$, 497-499.

Jolicoeur, P. 1963_2. The degree of robustness in Martes americana.Growth,
    $\underline{27}$, 1-27.

MacFie, H.J.H. 1977. Univariate and multivariate statistical methods –
    investigation of growth and shape. Ph D. Thesis. University of Bath.

Mosimann, J.E. 1970. Size Allometry: Size and shape variables with
    characterisations of the normal and log normal distributions. J.
    Americ. Stat. Assoc. 930-945.

Mukhoty, H. and Berg, R.T. 1971. Influence of breed and sex on the
    allometric growth patterns of major bovine tissues. Anim. Prod. $\underline{13}$,
    219-227.

Owers, A.C., Pomeroy, R.M., Williams, D.R. and Scott, B.M. 1968. Major
    beef research project. Smithfield Club and Agricultural Research
    Council.

Rao, C.R. 1952. Advanced statistical methods in biometric research. New
    York: Wiley.

Rao, C.R. 1964. The use and interpretation of principal components
    analysis in applied research. Sankhya $\underline{26}$, 329-358.

Reyment, R.A. and Banfield, C.F. 1976. Growth-free canonical variates
    applied to fossil foraminifers. Bull. Geol. Inst. Univ. Uppsala. N.S.
    $\underline{7}$, 11-21.

Sprent, P. 1972. The mathematics of size and shape. Biometrics $\underline{28}$, 23-37.

AN APPROACH TO THE COMPARISON OF GROWTH CURVES OF DUTCH FRIESIAN, BRITISH FRIESIAN AND HOLSTEIN FRIESIAN COWS.

H. Bakker and W.J. Koops

Department of Animal Science of the Agricultural
University Wageningen, The Netherlands.

ABSTRACT

   In this paper methods are described to select a mathematical function to fit the growth curve of cattle and to adjust for gestation and lactation effects on body weight. The method of choosing the growth function is based on the comparison of residual variances after fitting the Richards function with substitution of different values of n. The Richards function is a general formula for a number of well known growth functions. The von Bertalanffy function ($n = 1/3$) gave the smallest residual variance in our data. Adjustment for gestation and lactation effects were made by multiplication of the von Bertalanffy function by two separate functions describing both effects respectively. Residual variation was decreased considerably by the adjustments. The formulae derived were illustrated in samples of different subgroups of a crossbreeding experiment of Dutch Friesian, British Friesian and Holstein Friesian cattle.

INTRODUCTION

In most European countries beef production is mainly
dependent on dual purpose cattle populations. Consequently,
beef production ability (growth rate, feed conversion, slaughter
quality) should be emphasised in the breeding goal in addition
to milk production ability. Optimal slaughter weight and
slaughter category (eg veal calves, slaughter bulls, steers,
etc) are determined by the growth curve characteristics, mature
size and rate of maturity.

Selection for growth characteristics can be done after
fitting growth functions for all breeding animals. However,
in the female population the growth curve is influenced by
gestation and subsequent lactations from circa 18 months of
age on. Thus fitting the growth function should be based on
data prior to the first gestation or adjustments for gestation
and lactation is necessary.

In this paper will be discussed:

A method to select a mathematical function to fit the growth
curve

A method to adjust for gestation and lactation by means
of the growth function

An application of both methods by comparison of growth
curves of different groups in a crossbreeding experiment
of Dutch Friesian, British Friesian and Holstein Friesian
cattle.

These three items of the paper will be discussed separately.

SELECTION OF THE GROWTH FUNCTION

Mathematical functions to describe growth curves can be
distinguished in functions which take in account certain
fundamental postulates about the growth process (theoretical
functions) and functions which lack these theories of growth
as a background (Richards, 1959, Parks, 1971). The latter

functions are often third or fourth degree polynomials. The
choice between so-called theoretical functions and polynomials
depends on the purpose of these functions. If the parameters
are for biological interpretation theoretical functions are
more suitable than polynomials.

As we want to use the parameters of the function for
biological interpretation theoretical functions will be discussed
only.

The most familiar theoretical functions can be derived from
the Richards formula (Richards 1959).

$$\hat{Y}_t^n = A^n - (A^n - Y_o^n) \, e^{-kt}$$

In which

$\hat{y}_t$ = estimated body weight at age t

A = mature weight

y_o = weight at t=0 (= birth weight)

k = rate constant, which determines the spread of the curve
along the time axis

n = weight exponent constant

The well known growth functions (Brody, von Bertalanffy,
Gompertz, and Logistic) can be found by substituting different
values for the parameter n.

n = 1 : $\hat{y} = A - (A-y_o)e^{-kt}$ (Brody function)

n = 1/3 : $\hat{y}^{\frac{1}{3}} = A^{\frac{1}{3}} - (A^{\frac{1}{3}} - y_o^{\frac{1}{3}})e^{-kt}$ (von Bertalanffy function)

n → 0 : $\ln\hat{y} = A - (\ln A - \ln y_o)e^{-kt}$ (Gompertz function)

n = -1 : $1/\hat{y} = 1/A - (1/A - 1/y_o)e^{-kt}$ (Logistic function)

Variation of the parameter n has also consequences for the
fraction of mature size A at which the point of inflection is
estimated.

Weight at point of inflexion is $\hat{y}_i = (\frac{1/n - 1}{1/n})^{1/n}$. A

So n = 1   gives $\hat{y}_i$ = oA

    n = 1/3   gives $\hat{y}_i$ = .30A

    n → 0   gives $\hat{y}_i$ = .37A

    n = -1   gives $\hat{y}_i$ = .50A

In theory the best function to describe the growth curve is the four parameter Richards function by means of an iterative procedure. In practice however, fitting the Richards function is complicated by the number of parameters in the iteration and lack of convergence of the n parameter. In addition wrong estimations of n occur when local minima in the residual variance are met during the iteration.

An alternative method is to fit the Richards function to a sample of the total data set, using a number of alternative values of n. The best function can be selected based on the smallest residual variance.

This method is demonstrated in two sub samples from the base population of a crossbreeding experiment. The base population of this experiment consisted of 120 female calves in each of two breeding districts in the Netherlands. Description of this experiment was presented by Politiek (1974). We used data obtained from five animals out of each of the two sets of 120 female calves in the base population. Body weight was estimated at birth and taken monthly up to circa 600 days of age and 0, 50 and 100 days after each calving. In total 27 body weights data were available from each cow.

Iteration of the growth function was done by substituting 8 different values for n in the Richards function (eg -1, -1/3, 1/4, 1/3, 1/2, 3/4, 1, 1 1/4).

Results for the 10 cows are presented in Figure 1. It follows clearly from this figure that fitting the von Bertalanffy function gives lowest residual variances compared to the other

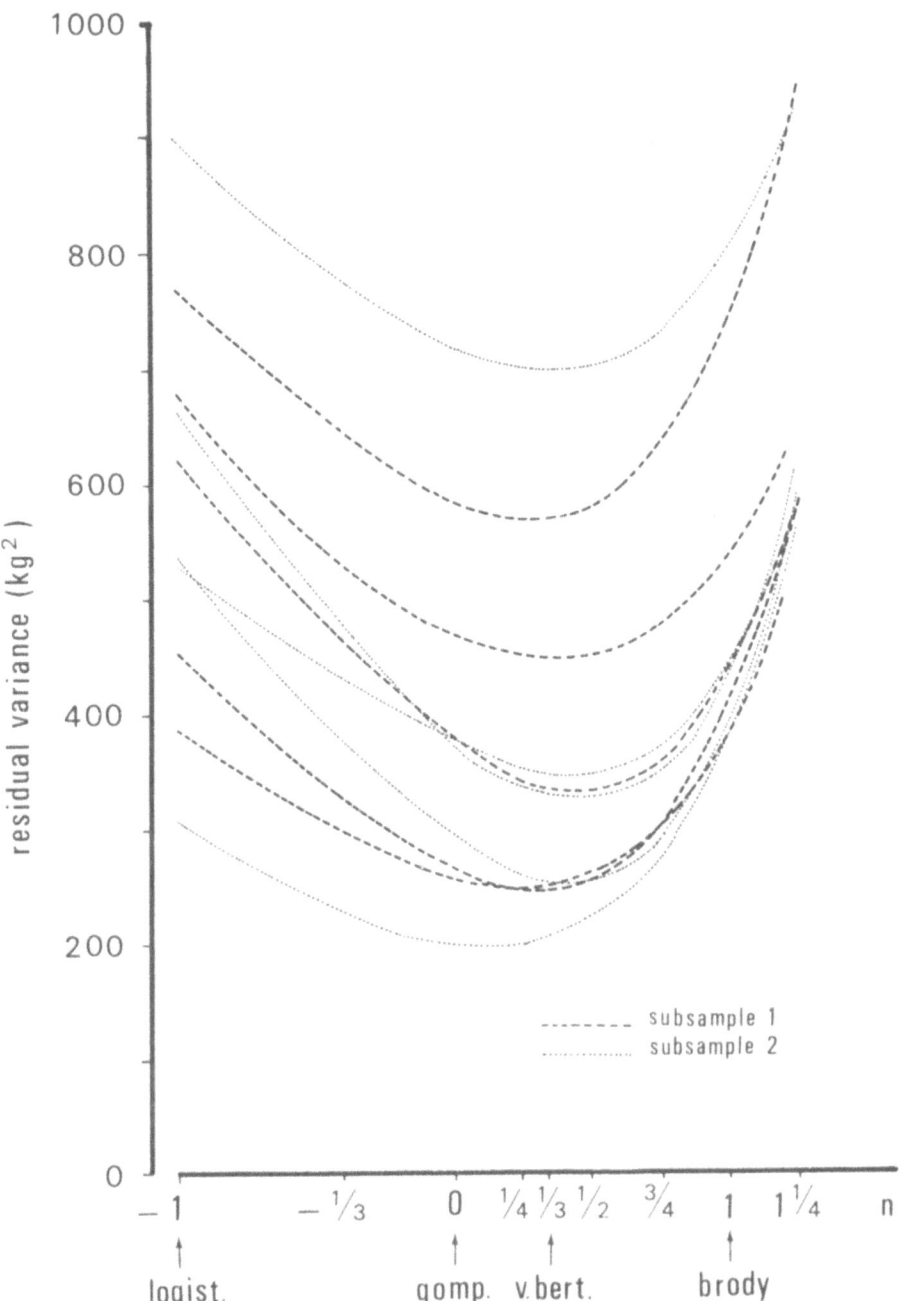

Fig. 1. Residual variances when different values of n are substitued in the Richards function to describe the growth curve of 10 cows.

common growth functions. This was both when residual variances
were expressed on a linear body weight scale or on the transforme
scale specified by the value of n. It was concluded from these
results that the von Bertalanffy function was the best function
to describe the growth pattern of these cows. So this function
was used in further research.

ADJUSTMENTS FOR GESTATION AND LACTATION

When body weight data of a cow are collected while the
cow is lactating or pregnant adjustments for these effects are
necessary when a growth function is fitted. In our case we
used two separate functions to adjust the von Bertalanffy functior

Adjustment for the effect of gestation:

$$f(x_d) = e^{k2 \frac{x_d}{D}}$$

in which $f(x_d)$ = effect of gestation on body weight

x_d = number of days pregnant

D = total gestation length (assumed to the 278 days)

k_2 = correction parameter for gestation. Roughly $k_2 \cdot y \approx$ total
weight of calf, placenta, etc.

Adjustment for the effect of lactation:

$$f(x_k) = e^{-k_3 \frac{x_k}{K}} e^{\frac{-x_k}{K}}$$

in which $f(x_k)$ = effect of lactation on body weight

x_k = number of days in lactation

K = number of days in lactation when the effect on body
weight is maximal (assumed to be 150 days)

k_3 = correction parameter for lactation $k_3 \cdot e^{-1} \approx$ maximum
percent body weight decrease due to lactation. $k_3 \cdot 0.37$.
$y \approx$ maximum body weight loss due to lactation.

The adjustment of the growth function is obtained by multiplication of the von Bertalanffy function by $f(x_d)$ and $f(x_k)$.

So $\hat{y} = f(x_L) \cdot f(x_d) \cdot f(x_k)$

in which $\hat{y}$ = estimated body weight

$f(x_L)$ = von Bertalanffy function as a function of age x_L

or

$$\hat{y} = A\left[1 - \left\{1 - (\frac{y_o}{A})^{1/3}\right\} e^{-k_1 x_L}\right]^3 \cdot e^{k_2 \frac{x_d}{278}} \cdot e^{k_3 \frac{x_k}{150}} \cdot e^{\frac{-x_k}{150}}$$

This function was fitted to a data set obtained from the experiment of Stegenga, Vos and Vos (1968). They used mono-zygous twins to study the effect of age at first parturition on growth and development of Dutch MRY cattle. Data of 5 pairs out of 10 with the largest number of body weight data were used in our investigations. Body weights were taken monthly from circa 500 days to circa 1800 days of age. In a number of cases the following day at the same time repeated measurements were taken too. Information on age at first calving, number of gestations is presented in Table 1. Results of fitting the von Bertalanffy function with or without adjustments and the effects on residual variances are given in this Table too. It follows from this Table that R^2 after fitting the von Bertalanffy function and adjustment for gestation and lactation is high. The effect of adjustment is also clearly demonstrated by comparison of the residual variation when the von Bertalanffy function is fitted with adjustment and when no adjustment is made. Quite often the residual variation is halved by the adjustment. The S_{duplo} is indicating that a considerable amount (circa 1/4 - 1/3) of the variation after fitting the function with adjustments is dependent on measurement errors.

It can be calculated from the data in Table 1, that the average k_2 = 11.1% and the average value of $k_3 \cdot 0.37 = 7.6\%$. Assuming a body weight of a cow of 600 kg it can be concluded that the total fluctuation due to lactation and gestation is 45 + 65 = 110 kg.

TABLE 1

GROWTH CURVE PARAMETERS AFTER FITTING THE VON BERTALANFFY FUNCTION TO BODY WEIGHT DATA OF 5 MONOZYGOUS MRY TWINS AND THE EFFECTS OF ADJUSTMENT FOR GESTATION AND LACTATION ON RESIDUAL VARIATION.

Twin pair	Animal	Age at first calving (days)	Number of pregnancies	$A+S_A$ kg [1)	$k_1+S_{k_1}$ *100 [1)	$k_2+S_{k_2}$ [1)	$k_3+S_{k_3}$ [1)	R^2	S (vB+adj) kg [2)	S (vB) [3)	S total [4)	S duplo [5)	Number of obs.
1	1	735	4	650+15.7	.24+.01	.145+.018	.311+.063	.934	22.9	46.4	86.1	4.4(17)	46
	2	928	4	632+12.2	.26+.01	.133+.014	.204+.044	.966	17.9	33.0	94.2	5.8(20)	52
2	3	718	4	601+14.1	.23+.01	.122+.015	.448+.057	.970	17.4	32.9	97.3	5.6(24)	60
	4	855	4	598+15.0	.26+.01	.097+.018	.154+.062	.959	22.2	28.8	106.3	6.7(22)	60
3	5	787	4	536+14.6	.26+.01	.099+.018	.321+.065	.962	17.9	26.4	88.9	4.7(24)	51
	6	1015	3	542+20.6	.26+.01	.072+0.26	.163+.084	.965	19.4	23.6	100.5	4.8(25)	51
4	7	753	4	552+17.0	.25+.01	.152+.017	.317+.067	.978	16.3	32.4	105.4	3.7(26)	52
	8	1090	3	553+13.4	.27+.01	.151+.016	.085+.051	.985	15.6	29.0	122.5	6.4(25)	52
5	9	727	2	624+11.0	.24+.01	.078+.013	.003+.043	.990	15.1	19.1	147.6	5.0(26)	54
	10	1085	2	659+19.8	.22+.01	.068+.020	.052+.065	.987	18.0	21.2	155.6	5.9(26)	54

1) Asymptotic standard errors

2) Residual variation when the von Bertalanffy function is fitted with adjustments for gestation and lactation

3) Residual variation when the von Bertalanffy function is fitted without adjustments

4) Variation in the raw body weight data

5) Variation calculated from duplo observations

The application of the formula is demonstrated in Figure 2 for cows 1 and 2 from Table 1. It can be concluded from Table 1 and Figure 2 that the adjustment methods contribute considerably to the description of growth curves of cows in lactation or gestation. The adjustment method makes it possible to compare growth curve parameters of cows when data were obtained at different stages of lactation and/or gestation.

APPLICATION OF FITTING THE GROWTH FUNCTION WITH ADJUSTMENTS

The methods to select a growth function and to adjust for gestation and lactation were developed to describe growth curves of different subgroups of a crossbreeding experiment of Dutch Friesians, British Friesians and Holstein Friesians (Politiek 1974). As an illustration the results of fitting the von Bertalanffy function with adjustment for lactation are given here for samples of 7 animals out of each of the two base population groups (Politiek 1974) and the three F_1 populations. Data collection was as described in section 'Selection of the growth function'. As no body weight data were collected after the first stage (1 or 2 months) of gestation no adjustment for gestation was necessary. So x_d was assumed to be zero. Results are presented in Table 2.

TABLE 2

FITTING THE GROWTH FUNCTION WITH ADJUSTMENT FOR LACTATION TO GROWTH CURVES OF COWS IN DIFFERENT SUBGROUPS OF A CROSSBREEDING EXPERIMENT[1]

Subgroup	no of animals	no of obs/anim.	A	y_o	k_1 *1000	k_3 *100	S_{res}	R^2
Base 1	7	27	617	43.1	2.61	42.3	12.7	.994
Base 2	7	27	606	37.6	2.88	39.9	12.8	.995
F_1 (DF*DF)	7	27	592	41.6	2.81	33.8	15.2	.993
F_2 (HF*DF)	7	26	611	44.5	2.83	42.1	16.1	.994
F_3 (BF*DF)	7	27	622	42.3	2.91	37.6	14.9	.992

[1] Results are mean values of animals per subgroup.

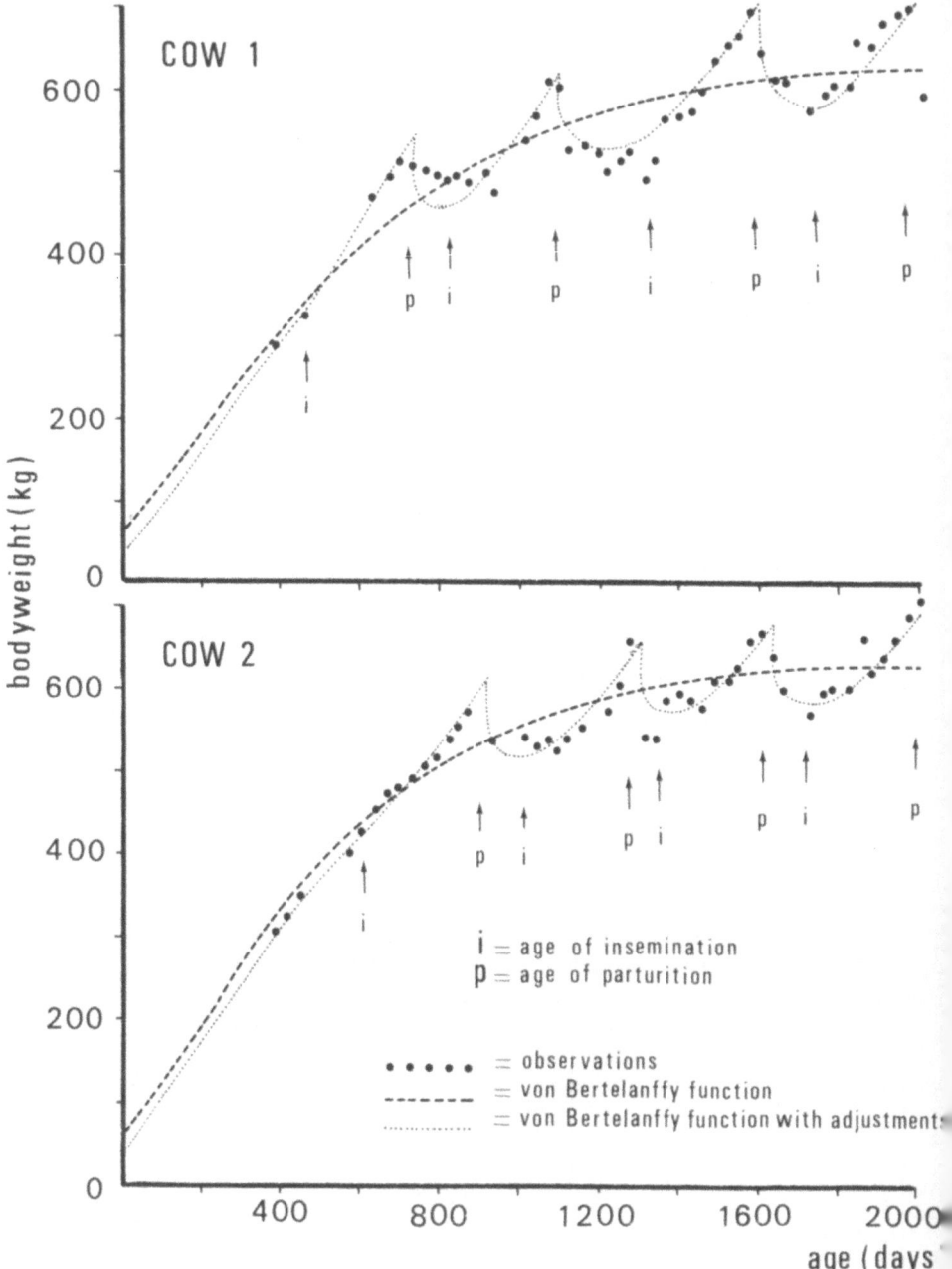

Fig. 2. Fitting the von Bertalanffy function with or without adjustments for gestation and lactation to growth curves of two cows from Table 1.

It follows from Table 2 that the fit, expressed in R^2, is very good in all cases. Residual variation is circa 15 kg. No further statistical analysis or interpretation will be given to the results in these small samples. Results of the growth curve analysis of animals in the crossbreeding experiment will be reported elsewhere.

REFERENCES

Parks, J.R. 1971. Phenomenology of animal growth. N.C. State Univ. Raleigh.

Politiek, R.D., 1974. The comparison of Friesians from different origin. Z. Tierzüchtg. Züchtgsbiol. 91, 1-10.

Richards, F.J. 1959. A flexible growth function for empirical use. J. Exp. Botany 10, 290-300.

Stegenga, Th., Vos, M.P.M. and Vos, H. 1968. De invloed van de leeftijd bij het eerste afkalven op groei en ontwikkeling bij het Nederlandse roodbonte vee. Landbouwk. Tijdschr. 80, 87-95.

THE RELATIONSHIP BETWEEN GROWTH CURVE PARAMETERS AND CARCASS COMPOSITION

G. Torreele

Department of Animal Breeding, Faculty of ·Agricultural Sciences, State University - Ghent Belgium

ABSTRACT

Different growth models are fitted to a series of individual (age, live weight) - sets from intensively fattened young bulls. Growth curve parameters and derived growth characteristics are analysed together with data relative to the slaughter value of the animals (carcass composition, feed intake, carcass measurements). Growth characteristics obtained following the different models are compared; their relationship with carcass composition is dealt with.

Growth cannot be studied properly without referring to a nutritional background. It is postulated that, if an animal is fed ad libitum a nutritionally adequate diet, the individual will determine not only his own ration (distribution of quantity of diet in time) but also his own growth curve (distribution of quantity of, say, live weight, in time). Both distributions, the intake and the growth curve are strict individual features of the animal and the diet. In this context, it is legitimate to study growth as a function of time as the independent variable.

The shape of a growth curve is described by the rate of change at any reasonable time. Besides, in post-natal growth, points of major importance are birth weight, weight and age at the (major) inflection point and weight and age when growth ceases (maturity). Mathematical growth curves of biological interest are such that parameters relate (in a more or less complicated way) to these data. Also, the variation in the growth curves encountered is described by the joint distribution of these points.

Apart from cows and the older bulls necessary for reproduction, most cattle are slaughtered before say 50% of their expected post-natal growth curve has been lived. Furthermore, most young bull performance tests, (the progeny tests for meat production and the majority of the feeding trials), are concentrated within the first 3, 10 to 15 or 18 months of post-natal life, irrespective of constant live weight, constant age or some fixed degree of finish being considered. In all such cases, the special problem is that, depending upon the individual, this period (say birth - 500 kg live-weight) covers the whole (or at least the major part) of the pre-inflection growth phase and none or only a limited part of the post-inflection period. The inflection point, if there is any, is difficult to determine and it is even more difficult to obtain satisfactory information about weight and age at maturity. It is thought that a solution to these problems may be reached by fitting one (or more) functions to the observations, thus, by the nature of the function and of the data it is possible and justifiable to derive estimates of the unknowns. However, unless it can be verified on complete growth curves it will remain unknown if the derived growth constants are facts of nature or mathematical artifacts of the data analysis.

In agriculture, we are not only interested in growth as such. One of the objectives is concerned with the costs of production, and another with the value of the product. Thus, we will consider growth and intake parameters as possible sources of information on slaughter value. In this contribution, only growth characteristics as possible indicators of carcass value will be discussed. It may be questioned if live-weight, being the sum of muscle, fat, bone and offals, can be used at all as an indicator of, for example, carcass composition. On the other hand it is thought that live-weight, rate of feed intake, growth rate, degree of maturity, bring some information as to the live-weight increase at a particular life stage. Also some information as to muscle proportions and to meat quality (succulence, taste, tenderness, etc) can be gained from growth data.

One of the results of a progeny test conducted by Martin at Gent is a set of 842 individual live-weight - age relationships together with the individual feed intake data and a series of measurements related to slaughter value (dressing percentage, carcass composition, linear measurements). (*) Growth and intake data relate to the period 10 days old - 450 kg live-weight. Slaughter value is measured at 450 kg. The idea is to investigate a possible relation between growth curve parameters based on this growth range and measurements of slaughter value, especially carcass composition.

There is a great number of curves of biological interest; the curves we will take into consideration are:

a: a simple logistic function expressed as a hyperbolic tangent (Hotelling, Peil, Richards).

$$f(1) = W = a_1 + a_2 \tanh a_3(t - a_4)$$

(*) 4 breeds, 108 sires (27, 25, 28, 28) and 842 progeny (205, 215, 210, 212)

b: a general logistic function expressed as a modified hyper-
    bolic tangent

$$f(2) = W \, a_1 + a_2 \tanh a_3 (\log t - a_4)$$

c: Richards' function

$$f(3) = W = A(1 - be^{-kt})^M$$

d: a function proposed by Lehmann

$$f(4) = W = e^{(a/k) - 1/(ke^{k(t - c)})}$$

e: a function derived by Janoschek

$$f(5) = W = b + a(1 - e^{-(kt)^P})$$

f: the integrated form of the Heinz' function (Sayag)

$$f(6) = W = B + (A/(k_2 - k_1)) \, (-e^{-k_1 t}/k_1 + e^{-k_2 t}/k_2)$$

For complete mathematical details concerning the growth
curves cited, we refer to the corresponding literature given
at the end of the paper.

The first curve, apart from giving the best fit to the
observations, is also a simple and good illustration of what
is meant by a curve of biological interest. For this function
we have:

- $W = a_1 + a_2 \tanh a_3 (t - a_4)$
- $dW/dt = K(U-W)(W-L)$ with K a proportionality factor, U
  the upper asymptote, L the lower asymptote. $K = (a_3/a_2)$,
  $U = (a_1 + a_2)$, $L = (a_1 - a_2)$
- $W_{birth} = a_1 - a_2 \tanh a_3 a_4$
- $W_{inflection} = a_1$
- $t_{inflection} = a_4$
- $dW/dt_{inflection} = a_2 a_3$

Analogous details can be found for the other functions
mentioned above. All six functions were fitted to a group of
215 animals belonging to the East Flemish breed. Fitting followed a modified Marquard - procedure (Ottoy, Mercy). As only

TABLE 1

MEAN VALUES AND HERITABILITIES FOR THE GROWTH CURVE PARAMETERS FOLLOWING $f(1)$*.

GROWTH MODEL: $f(1) + W = a_1 + a_2 \tanh a_3 (t - a_4)$

		Breed 1 (n = 183)	Breed 2 (n = 192)	Breed 3 (n = 185)	Breed 4 (n = 198)
$a_1 \pm s_{a_1}$	A	301 ± 3.0	301 ± 3.2	306 ± 3.1	300 ± 3.0
$a_2 \pm s_{a_2}$	B	361 ± 6.5	363 ± 5.9	370 ± 6.9	357 ± 5.5
$a_3 \pm s_{a_3}$ (* 10^{-4})	C	43 ± 0.8	44 ± 0.7	41 ± 0.7	38 ± 0.7
$a_4 \pm s_{a_4}$	D	232 ± 2.7	213 ± 2.4	242 ± 2.4	225 ± 2.3
$W_{birth} \pm s_{W_b}$	E	39.0 ± 0.5	40.5 ± 0.5	37.4 ± 0.4	39.3 ± 0.5
$W_{infl} \pm s_{W_i}$	F	297 ± 3.0	301 ± 3.2	306 ± 3.1	300 ± 3.0
450/U	G	0.700 ± 0.008	0.708 ± 0.008	0.681 ± 0.008	0.687 ± 0.007
$T_{infl} \pm s_{T_i}$	H	232 ± 2.7	224 ± 2.4	242 ± 2.4	233 ± 2.4
$U \pm s_U$	J	662 ± 8.4	660 ± 8.2	677 ± 9.3	662 ± 7.7
$dw/dt_i \pm s_{dw/dt_i}$	K	1.475 ± 0.011	1.474 ± 0.010	1.443 ± 0.009	1.439 ± 0.011
$h^2 \pm s_{h^2}$	A	0.76 ± 0.30	0.28 ± 0.22	0.002 ± 0.18	0.23 ± 0.21
id.	B	1.13 ± 0.33	0.36 ± 0.23	0.10 ± 0.20	0.00 ± 0.11
id.	C	0.98 ± 0.32	0.42 ± 0.25	0.49 ± 0.27	0.16 ± 0.20
id.	D	1.13 ± 0.33	0.24 ± 0.21	0.21 ± 0.22	0.28 ± 0.21
id.	E	0.80 ± 0.31	0.51 ± 0.26	0.07 ± 0.19	0.20 ± 0.21
id.	F	0.76 ± 0.30	0.28 ± 0.22	0.02 ± 0.18	0.23 ± 0.21
id.	G	0.80 ± 0.31	0.34 ± 0.23	0.16 ± 0.21	0.06 ± 0.18
id.	H	1.13 ± 0.33	0.24 ± 0.21	0.21 ± 0.22	0.28 ± 0.22
id.	J	0.94 ± 0.32	0.28 ± 0.22	0.07 ± 0.19	0.03 ± 0.17
id.	K	1.05 ± 0.33	0.55 ± 0.27	0.35 ± 0.24	0.47 ± 0.25

* The number of progeny-groups is respecively : breed 1 : 27; breed 2 : 25; breed 3 : 28; breed 4:28.

TABLE 2

MEAN VALUES AND HERITABILITIES FOR THE GROWTH CURVE PARAMETERS FOLLOWING $f(4)$*.

GROWTH MODEL:. $f(4) = W = e^{a/k - 1/ke(t-c)}$

		Breed (1) (n = 192)	Breed (2) (n = 205)	Breed (3) (n = 196)	Breed (4) (n = 205)
$a \pm s_a$ (×10⁻³)	A'	31 ± 0.3	33 ± 0.3	31 ± 0.2	32 ± 0.2
$k \pm s_k$ (×10⁻⁴)	B'	48 ± 0.5	49 ± 0.5	47 ± 0.4	48 ± 0.4
$c \pm s_c$	C'	908 ± 12	868 ± 10	918 ± 10	906 ± 10
$W_{birth} \pm s_{W_b}$	D'	41.0 ± 0.44	38.5 ± 0.38	36.7 ± 0.31	41.0 ± 0.38
$W_{infl} \pm s_{W_i}$	E'	303 ± 3.7	302 ± 3.4	304 ± 3.7	299 ± 3.5
$T_{infl} \pm s_{T_i}$	F'	234 ± 3.1	226 ± 2.5	240 ± 2.8	232 ± 2.7
$dw/dt_i \pm s_{dw/dt_i}$	G'	1.427 ± 0.009	1.459 ± 0.008	1.398 ± 0.008	1.351 ± 0.008
$h^2 \pm s_{h^2}$	A'	1.10 ± 0.30	0.26 ± 0.21	0.38 ± 0.24	0.45 ± 0.25
id.	B'	1.03 ± 0.32	0.29 ± 0.21	0.37 ± 0.24	0.44 ± 0.24
id.	C'	1.11 ± 0.33	0.20 ± 0.19	0.26 ± 0.22	0.38 ± 0.23
id.	D'	0.97 ± 0.32	0.46 ± 0.24	0.17 ± 0.20	0.57 ± 0.26
id.	E'	0.70 ± 0.29	0.41 ± 0.23	0.24 ± 0.22	0.38 ± 0.23
id.	F'	1.10 ± 0.30	0.48 ± 0.25	0.54 ± 0.26	0.40 ± 0.24
id.	G'	0.80 ± 0.30	0.54 ± 0.26	0.00 ± 0.17	0.48 ± 0.25

Note: The model of Lehmann is such that there is a fixed relation between live-weight at inflection and the upper asymptotic weight:

$$W_i = 0.3679 \, U. \quad U = e^{a/k}. \quad W_{infl} = e^{(a-k)/k}; \quad U/W_i = e = 2.718..$$

* The number of progeny groups is the same as in Table 1, however, the number of progeny is not the same.

TABLE 3

MEAN VALUES $\pm$ THEIR STANDARD ERRORS AND THE HERITABILITIES FOR SOME CHARACTERISTICS OF SLAUGHTER VALUE

		Breed 1 (n = 205)	Breed 2 (n = 215)	Breed 3 (n = 210)	Breed 4 (n = 212)
%M	A "	64.5 ± 0.26	64.1 ± 0.22	62.5 ± 0.23	66.8 ± 0.22
%F	B "	20.3 ± 0.26	21.0 ± 0.23	22.9 ± 0.23	18.8 ± 0.22
%B	C "	15.2 ± 0.06	14.9 ± 0.05	14.6 ± 0.05	14.4 ± 0.06
M/B	D "	4.28 ± 0.03	4.32 ± 0.02	4.30 ± 0.01	4.65 ± 0.02
S	E "	60.9 ± 0.12	60.9 ± 0.11	60.6 ± 0.10	62.3 ± 0.13
MPC	F "	0.385 ± 0.002	0.391 ± 0.002	0.362 ± 0.002	0.425 ± 0.002
h^2	A "	0.24	0.15	0.45	0.99
"	B "	0.40	0.17	0.49	0.84
"	C "	0.67	0.63	0.83	0.60
"	D "	0.24	0.41	0.58	1.00
"	E "	0.44	0.72	0.87	1.04
"	F "	0.22	0.17	0.49	1.23

%M = percentage of muscle in the carcass

%F = percentage of fat + tendon in the carcass

%B = percentage of bone + cartillage in the carcass

M/B = the proportion muscle to bone in the carcass

S = dressing percentage = weight of cold carcass divided by live weight

MPC = meat production coefficient = product of %M by S.

f(1) and f(4) were reasonably satisfactory (especially with
regard to the estimation of the upper asymptote and of weight
at inflection *), only these functions were used as models for
the 842 animals and further submitted to variance-covariance
analysis. Table 1 summarises some of the results obtained with
function 1; similar data for f(2) are tabulated in Table 2;
Table 3 contains the observations as to slaughter value while
Table 4 gives an idea about the relationship between the growth
data calculated for the different functions. Finally, Table
5 resumes the findings with regard to the relationship growth
data-carcass composition for function 1.

The Tables 1, 2 and 3 need little comment as they are
intended to sketch the 4 populations with regard to their
potential for growth and meat production. For most characters
considered there are definite differences between the 4 breeds.
Within each breed, for most of the data calculated or observed,
there are significant differences between the progeny groups.
The within breed variance for all the characters considered is
appreciable.

From Table 4 it is seen that, depending upon the character
considered, the 6 models give medium to high correlated estimates. The correlations are worst for U, the upper asymptote,
and best for T_i, the age at the inflection point. However,
this is a partial result and must be confirmed for the totality
of observations in that breed and for the other breeds.

Table 5 resumes what we were looking for, at first glance,
the results, although a number of them are highly significant,
are rather disappointing. The growth curve parameters A, B, C,
D do not deliver any substantial information as to carcass
composition; on the other side, the correlations with dressing
percentage are astonishingly high. Dressing percentage is also
correlated with all the other calculated growth characteristics.
Age at 450 kg shows no relation to carcass value; total intake

* Actually, having W_i greater than 450 kg and/or U larger than 2000 kg
  resulted in the rejection of the fit.

is highly correlated with T_{450} (0.75 on the average) and is also correlated with carcass composition (and not with dressing percentage).

TABLE 4

RELATIONSHIPS BETWEEN GROWTH DATA FOR DIFFERENT GROWTH MODELS (*)

4a:	Mean values and st.deviations for birth weight, weight and age at inflection and the calculated upper asymptote. (n:43)			
	W_{birth}	W_{infl}	T_{infl}	U
f(1)	41.7; 7.1	271.0; 15.8	201.4; 18.7	574 ; 41
f(2)	51.9; 6.8	248.7; 31.6	185.3; 27.6	1006 ; 238
f(3)	56.1; 7.5	260.0; 15.6	191.2; 16.0	772 ; 54
f(4)	39.2; 5.1	269.8; 31.1	200.6; 24.3	733 ; 84
f(5)	51.1; 6.5	255.1; 25.8	189.8; 28.6	676 ; 110
f(6)	54.7; 6.7	265.9; 41.5	198.8; 37..3	1017 ; 226

4b:	Correlations between calculated upper asymptotes (upper triangle) and between estimated weight at inflection (lower triangle) (n = 43)					
	f(1)	f(2)	f(3)	f(4)	f(5)	f(6)
f(1)	1.00	0.90	0.18	0.32	0.93	0.50
f(2)	0.95	1.00	0.21	0.48	0.99	0.40
f(3)	0.84	0.77	1.00	0.37	0.21	0.31
f(4)	0.88	0.89	0.78	1.00	0.44	0.49
f(5)	0.97	0.99	0.80	0.91	1.00	0.42
f(6)	0.57	0.55	0.63	0.71	0.63	1.00

4c:	Correlations between estimated ages at the inflection points (upper triangle) and the correlations between the estimated birth weights (lower triangle) (n = 43)					
	f(1)	f(2)	f(3)	f(4)	f(5)	f(6)
f(1)	1.00	0.92	0.94	0.92	0.96	0.75
f(2)	0.90	1.00	0.87	0.93	0.99	0.70
f(3)	0.53	0.52	1.00	0.92	0.92	0.86
f(4)	0.66	0.57	0.27	1.00	0.95	0.79
f(5)	0.97	0.95	0.55	0.61	1.00	0.77
f(6)	0.85	0.86	0.48	0.64	0.90	1.00

(*) This is a partial result, based on the 110 animals of breed 2; this gives 43 sets having 'reasonable' data for all six functions considered.

TABLE 5

PHENOTYPIC CORRELATIONS BETWEEN GROWTH CURVE PARAMETERS AND SLAUGHTER VALUE:
FUNCTION 1, MEAN OF THE 4 CORRELATION COEFFICIENTS

*)	A" %M	B" %F	C" %B	D" M/B	E" S	F" MPC
A	0.08	0.02	0.04	0.03	−0.12	0.02
B	0.06	−0.07	0.06	0.00	−0.13	−0.01
C	−0.04	0.06	0.08	0.03	0.17	0.04
D	0.06	−0.07	0.02	0.03	−0.13	0.00
E	0.12	−0.15	0.10	0.00	0.05	0.12
F	0.08	−0.09	0.04	0.02	−0.12	0.02
G	−0.07	0.09	0.07	0.00	0.17	0.01
H	0.06	−0.07	0.02	0.03	0.13	0.00
J	0:07	−0.08	0.05	0.01	−0.13	0.00
K	0.01	0.00	−0.06	0.05	0.06	0.05
T_{450}	−0.05	0.04	0.00	0.05	0.02	−0.04
In- take	−0.31	0.32	−0.08	0.14	0.00	0.26
$W_{i/u}$	−0.01	0.03	−0.05	0.06	0.11	0.04

Note: T_{450} = time needed to attain 450 kg (at birth, T = 0)

Intake = total quantity of fodder units consumed from birth to 450 kg

A, A" etc, cfr Tables 1 and 3

Significant values of r at = 0.05, 0.01, 0.0027 and 0.001 for resp.
700 and 900 degr. of freedom: 700   0.07   0.10   0.11   0.12
900   0.06   0.09   0.10   0.11

*) See Table 1.

At the beginning of the paper, we expressed our belief as
to the possibility of fitting one or more functions to incomplete
growth data and, from the growth curve parameters estimated,
make some valuable predictions with regard to slaughter value
especially carcass composition. The results obtained are such
that, although not destroying all hope in that direction, we
surely must temper our expectations for approaches of this
kind. Perhaps also the available observations (ie the reach
birth - 450 kg) and/or the functions used predetermined the results obtained. As a global result from this work we can state
that the estimates of the growth parameters and the derived
growth constants depended upon:

- the individual,

- for some individuals, depended upon the model fitted

- for some individuals and some models, depended upon the method of fitting (for example, a normal Marquard procedure versus the modified version)

- for some individuals, some models and some method of fitting, depended upon the accuracy used in the approach to iterative solution (cfr remark above)

- for some ....., depended upon the starting values.

The fitting of individual growth curves to a (moderate) large number of individuals reveals the existence of pronounced interaction between model and individual and consequently none of the 6 models can be used as a general one.

Perhaps a word about the relation between the growth curve parameters; a, k, and c of the Lehmann - function were all three highly correlated (say $r = 0.98$) while a_1, a_2, a_3 and a_4 showed high to moderate correlation: 12: (0.50, 0.57; 0.69, 0.64) 13: (-0.41, -0.46, -0.59, -0.40), 14: (0.89, 0.90, 0.90, 0.88), 23: (-0.88, -0.88, -0.86, -0.87), 24: (0.47, 0.48, 0.66, 0.60) and 34: (-0.50, -0.49, -0.61, -0.48).

ACKNOWLEDGEMENT

The author is indebted to Dr. J.P. Ottoy (*) and to Ir A. Calus (**) for helpful advice and practical aid in the elaboration of the analysis.

(*) Seminarie voor Toegepaste Wiskunde en Biometrie, Faculteit Landbouw, RUG (Dir. Prof. Ir G. Vansteenkiste)

(**) Bureau voor Biometrie, IWONL, Coupure 433, Gent. (Dir. Prof. Ir G. Vansteenkiste en Prof. Dr Ir A. Rotti).

728

REFERENCES

Bertalanffy, L. von, 1951. Theoretische Biologie II. A. Franke AG Verlag,
   Bern.

Brody, S., 1945. Bioenergetics and growth. Reinhold.

Brown, J.E., Fitzhugh, H.A. and Cartwright, T.C. 1976. A comparison of
   nonlinear models for describing weight-age relationships in cattle.
   Jnl Anim Sci. 42,4.

Fitzhugh, H.A. 1976. Analysis of growth curves and strategies for altering
   their shape. Jnl Anim Sci, 42,4.

Fitzhugh, H.A. and Taylor, St. C.S. 1971. Genetic analysis of degree of
   maturity. Jnl Anim Sci, 33,4.'

Harrison, G.A. Weiner, J.S., Tanner, J.M. and Barnicot, N.A. 1970. Biologie
   van de mens, I, II, Het Spectrum, Utrecht, Antwerpen.

Hotelling, H., 1927. Cited by J.F. Richards.

Janoschek, A., 1957. Das reaktionskinetische Grundgesetz und seine
   Beziehungen zum Wachstums-und Ertragsgesetz. Stat. Vierteljahresschrift,
   10, 1-2.

Landsberg, J.J. 1977. Some useful equations for biological studies. Expl.
   Agric. 13,.

Lehmann, R. 1975. Mathematische Grundlagen zur Analyse des Wachstums von
   landwirtschaftlichen Nutztieren. Archiv Tierzucht, 18, 3.

Ottoy, J.P. and Mercy, J. 1978. To be published.

Peil, J. 1971. Optimale Anpassung nichtlinearer Modellensätze bei der
   Auswertung biologischer Experimente. Konferenz 'Mathematische
   Optimierung bei der Modellierung biologischer systeme', Jean.

Rasch, D. 1965. Biometrische Methoden zur Beschreibung des Wachstums.
   Archiv Tierzucht, 8.

Richards, J.F. 1970. The quantitative analysis of growth, in 'Plant
   physiology' edited by Steward, F.C. New York, Academic press.

Rottiers, A. 1976. Verband tussen lichaamsmaten van een kalf en zijn latere
   waarde als slachtdier. Werk van einde studiën. Fac. Landbouw. RUG.

Sayag, G.J. 1965. Mathématiques appliquees aux sciences physiques et
   naturelles, Vol I, Editions DOIN, Paris.

Tanner, J.M. and Taylor, G.R. 1970. De groei. Perscombinatie N.V. Amsterdam.

Taylor, St. C.S. and Fitzhugh, H.A. 1971. Genetic relationships between
   mature weight and time taken to mature within a breed. Jnl Anim Sci.
   33, 4.

Torreele, G. and Ottoy, J.P. 1973. De gewichtstoename van intensief gemeste vleesstiertjes en de Weinbach - groeicurve, Landbouwtijdschrift, 2.

Vercruysse, A. 1977. Over de groei bij intensief gemeste vleesstiertjes. Werk van einde studiën, Fac. Landbouw, RUG.

Zotina, R.S., and Zotin, A.I. 1972. Towards a phenomenological theory of growth. J. Theor. Biol. 35.

St.C.S. Taylor *(UK)*

I would suggest that we end this session by asking one or two people to give their reactions to this morning's talks.

R.G. Kauffman *(USA)*

Very quickly, I would say that they were very informative and, I thought, very appropriate to tie in the rest of our carcass types of work. I can say very confidently, and I am sure it relates to others, that I did not understand everything which was said, but it should be helpful to us in understanding what we are doing in the future.

R.W. Pomeroy *(UK)*

The only thing which has really occurred to me this morning is that you could very well write a book which would put a lot of this mathematics into a language which the run-of-the-mill biologist can easily understand. That would be one of the most valuable things you could possibly do in the rest of your research career.

V.R. Fowler *(UK)*

My reaction to what I have heard this morning is that, after all the mathematical processes have been gone through, the conclusion is exactly what I would have made had I looked at the growth curve by eye in the first place. I can see that it is necessary to add a sort of moral tone to growth analysis, to have a great many very elegant equations and so on, but whether our conclusions at the end of the day actually change is still unclear to me.

D.J. Finney *(UK)*

Perhaps a little lightheartedly at this time of morning, I will make one or two comments. The first thing I noted down was that extrapolation is always dangerous. Yet it is the price of living. We inevitably extrapolate from our past lives in judging what we can do and should do tomorrow, and we do it at the

peril of our lives, because the sun may not rise tomorrow. I
think we have to keep that tension in mind all the time, but we
can never be certain that extrapolation, within any curve provided for you by any mathematician, statistician, or calculated
by yourself, will be correct, and if it is not, it will extrapolate and lead you into trouble.

With regard to this question of serial slaughter, where
the Chairman, in response to a chat he had with me yesterday,
corrected his final paragraph, I think that in a sense it still
gives a slight misunderstanding, at least of my own views. The
point I want to make is that the easy part of serial slaughter
records is where you have what I would regard as a true planned
experiment. The number of animals is taken, it is planned from
the start that some of these, chosen at random or chosen in an
objective manner, are going to be slaughtered at 40 weeks, and
some are going to be slaughtered at 50 weeks, and some at 60
weeks and so on. You then have a series of cross-sections of
your population from which you can synthesise information on a
mean pattern of growth, though, as several people have insisted
this morning, you cannot synthesise the curve for the individual
animal. There is no difficulty about this, although of course
you are getting much less information on the growth curves than
if you can measure each animal at many different times, and if
you have them all going through for the whole period. The difficulty which I mentioned to Dr. Taylor some months ago now, is
that if you have your observations tied up with commercial requirements - for example that you are obliged to slaughter
animals when they reach approximately a certain weight, or perhaps you are going to cull certain animals because you discover
fairly early in life that they are not progressing very favour-
ably - you then have a very much biased selection of animals
for elimination during the course of your observations, and
you have an inherently far more difficult problem in deciding
how to synthesise your data in order to produce a fair representation of an average growth curve. That, I think, is the problem
here. With the other type of data, where you simply have observations on animals chosen at random and chopped off at various

stages, there are computational difficulties, but nothing of
very deep interest.

In fitting polynomials and all growth curves I think we
are apt to forget that we should not really be treating the
individual observations as independent or equally correlated
with one another. If we have 20 successive observations at one
month intervals on an animal it really is not fair to use just
simple least squares with any equation in order to estimate its
parameters. It may work satisfactorily, but do remember that
adjacent observations are going to be more closely correlated
than observations a very long way apart, and this can introduce
misleading features into the fitting of a curve. I do not know
what the answer is. It is a very real problem, but do not be
surprised if occasionally you get odd results.

J. Kempster *(UK)*

May I make one point on Dr. MacFie's paper, because it
ties in with something from yesterday. It is just an example of
a difference that I do not think is going to be eliminated by
doing what Dr. MacFie said is projecting the data into growth
invariant space. This is from the first part of the paper which
I gave yesterday, and it gives the bone weight of the limb joints
as a percentage of total bone weight with total bone weight constant. You can see that there is quite a nice growth relationship. As you would expect, the later maturing breeds, the larger
breeds, will have more bone in the limb joints than in other
joints, because bone is early maturing. There is one breed
which stands out quite differently, and that is the South Devon.
I do not think that is going to be pulled in by the methodology
which he used, so here is an example. I think that in some of
Dr. Bech Andersen's data there are breeds which showed the same
sort of pattern, eg the Chianina and the Limousin.

J. Martin *(Belgium)*

When introducing this seminar, I was convinced having
looked through the papers, that we would be taking an important
step in relating the background of many problems we are confronted

with. I think we have been successful in this respect.

St.C.S. Taylor

    I have one brief comment to make before closing. About thirty years ago a little book used to be carried around by everyone, "The Rubaiyat of Omar Khayyam", and there is a verse from it that refers to the attitude of one or two people:

> Myself when young did eagerly frequent
> Doctor and Saint, and heard great argument
> About it and about: but evermore
> Came out by this same door as in I went.

Thank you all very much.

SECTION 5

SUMMARY AND NEEDS FOR FUTURE RESEARCH

Chairman:

H.J. Oslage

RECAPITULATION AND OUTLINE FOR THE FUTURE

H.J. Oslage *(West Germany)*

Ladies and gentlemen, we now start our last session of this seminar. The target is to summarise the results of the meeting as a whole and to get an outlook for future research though, with the large number of outstanding papers we have listened to, and also their specialisation and diversity within the framework of beef production, this may be a somewhat difficult task.

I very much hope that the discussions, both within and outside this room, will have contributed to a better understanding of the papers and may also have helped to clarify different, even opposing, standpoints, showing the way to new areas of future work.

We will now listen to short summaries by the co-ordinators, taken in order of the sections.

Summary by R.W. Pomeroy *(UK)* - Growth, general

The first session was a very short one, in which I looked at one or two of the major milestones in the study of growth and development, and pointed out some conflicts between the work of the Hammond school in the 1930s, and the later work of Butterfield around 1960. I suppose by now these conflicts are well known, but they are of such economic importance that it is surprising that it took so long to resolve them. If you look over the history of work in this field, and I suppose this is true of many other branches of science, there is, roughly speaking, a twenty to twenty-five year cycle in which important pieces of research are repeated.

On the Hammond side, there was the work of Walters in 1910 and Hammond about 1935; and before Walters, in about 1860 there was the Lawes and Gilbert work. There is an interesting gap of about fifty years between Lawes and Gilbert, and Walters and company, and it would make quite an interesting research study to see what happened in about, say, 1875. As Dr. Taylor said

earlier, Brody was apparently using the bivariate allometry
equation about 1910, Huxley either originated or revived it in
1932, and Butterfield revived it again around 1960.

We are in this situation where we are in danger of continually re-inventing the wheel, and I think it is probably time that
we reorientated some of our efforts and thinking in these directions.

Dr. Kauffman talked about some of the biological inter-
relationships in carcass work, and gave a very comprehensive
description of them, and in addition, he made one or two important points, particularly about the importance of <u>definition</u>.
Within an English-speaking country such as the United States,
perhaps definition is not too important, but in continental
Europe it most certainly is. Those of us who have worked on
international committees know very well that we spend a very
long time in unproductive argument, only to realise towards the
end of it that we are really talking about quite different things
I think that we have learned to define our terms of reference in
our definitions before we start, but it is a very salutary lesson
that, when you have more than one nationality involved in any
sort of discussion, it really is very important to define your
terminology at the outset.

Another point which emerges in this sort of work is the
importance of the sample. There has been quite a good deal of
work done where a very heterogeneous collection of animals -
heterogeneous as to breed, sex, level of nutrition and so on -
have been treated as though they were a growth series. Not only
have they been heterogeneous as to breed, but also with different
numbers of breeds, and different ages and weights. This is
absolute nonsense.

I realise that we are not in the spacious age of the 1920s
and the 1930s, when one could devote a fair amount of space to
writing up one's experiments, but nevertheless, I think that it
would be very useful in the literature if people would describe

what they did, how they did it and why they did it in a great
deal more detail than they do, even if it meant cutting down the
enormously long discussions which people tend to put at the end
of their papers. If the experiment has been well devised and
well designed and described, very few people in the field need
that discussion. I think it would be a very salutary exercise
if the editors of journals insisted on discussions being cut
down to a few paragraphs and a really adequate description of the
materials and methods, the nature of the animals and so on being
put in at the beginning.

Dr. Kauffman also made another important point. I may mis-
quote him a little but what I think he was saying was that we are
doing more and more experiments to find out what happens when
animals grow and develop. We have devoted very little attention
to why changes come about and how they come about. For the
future this is probably where we ought to put our effort - into
finding out why things happen and how they happen. If the distribution of individual muscles within the musculature is as
constant as it is alleged to be in animals of very different
shapes and so on, how is this harmony of muscle distribution
regulated? We know, I think, pretty well what happens nowadays;
what we want to know is why it happens and how it is brought
about.

Summary by R. Boccard *(France)* - Muscle and bone growth

The second section was devoted to muscle and bone growth
and brought up the non-variation of muscle distribution by any
such factors as breed, nutrition or conformation. In fact, when
we consider this aspect, and the first approaches to this problem
twenty years ago - or even at the beginning of the century - the
relative constancy of the muscle had already been noted. At
that time it was considered as an aspect of the general homeostas
or of an equilibrium of the animals, and was called 'anatomical
harmony'. This constancy is certainly statistically established
especially when animals are considered at the same carcass weight
and near the same state of fatness.

But we have to consider carefully all the details and at
the individual level we do find variation. In my opinion this
variation is worthy of attention. However, a new approach is
needed to turn the variation to good account. At the moment we
do not have adequate standardisation of methods - and I am thinking of dissection procedures here - but this is not the moment
to go into this question. There is a special group that has been
set up to examine the problem. Work on individual variations of
muscle weight should involve various types and it may be the
beginning of a new approach to this variation and its utilisation
in the selection of meat animals.

With regard to other characteristics, we see variations
which are important in the area of utilisation, although we are
unable to explain why the variation is there. This understanding
is necessary for the advancement of progress. Variations in the
muscle to bone ratio in relation to the variation of the thickness of muscle are for the moment far from being fully understood
and, owing to the influence which they exert over quality and
quantity of meat, new experiments need to be conducted to further
our knowledge in this field. These experiments will need better
definitions of the material and methods used, especially if the
data are to be made widely accessible. The instruments and
chemical methods employed should be standardised so that a well

established common basis is set up. The use of objective
measurements is another important target to aim for. Mathematical
expressions are adequate tools for expressing the continuous
variation of conformation, for instance. The well-defined methods
for measuring the histological structure of muscle are promising
achievements. The variation of meat structure is certainly a
large field where much work is still to be done.

Everyone here is working towards the improvement of meat
production by increasing the efficiency and the quantity of the
lean. This increase may be obtained by changing the number of
fibres, by changing the feeding system and by other means. However, we also have to look at the quality of the muscles we produce and we must not forget that collagen plays an important role
in this. Quality is something that should be emphasised in future
work.

The organoleptic characteristics of meat have to a large
extent escaped the traps of food scientists. But we do know how
some subtle metabolic and biochemical changes inside the muscle
fibre can induce considerable variations at the organoleptic
level. Consequently, as Dr. Lawrie has said, we have to consider
characteristics even at a level which is difficult to reach.
Only in this way will the quality be assessed so that we see, if
not improvement, then at least the maintenance of this quality.
To achieve this aim, a large amount of basic material is necessary
which could also become available from the sets of animals which
are dissected as part of the many current experiments. It is a
waste not to use this meat for principal determinations at the
chemical level.

There is great potential in the field of biological variation. Exploration of diverging types of cattle may provide a
broad variation of the structure and also of the biochemical
characteristics of the muscle. We can enlarge the range of
variation of the characteristic variables by the use of special
animals, which do not belong to the Bos family but to *Bos indicus*
or buffalo, for example, or even game, in order to experience a

wider variation in the lean/fat ratio. In addition, we can also use some specific types which we already have in hand - for instance, the double-muscled animals. I think that these will provide good starting points for new approaches to beef productio and possibly for new seminars also.

Altogether, we must also not forget the importance of the cow as a meat producer. More than one third of our beef originates from cows and perhaps we have paid too much regard to bull meat in recent years. We have to fill the gaps in our knowledge in this field of the meat from culled cows.

All this adds up to a sizeable programme of work. I expect that quantitative aspects will continue to receive prime attention, but I would, however, like to stress this: let us not forget the importance of quality.

Summary by P.L. Bergström *(Netherlands)* - Fatty tissue growth

In the session on fatty tissue growth, we had three main
papers and three short ones. In those of Mr. Williams and of
Dr. Hanrahan et al, fat distribution was the main subject to be
discussed.

It was pointed out in Mr. Williams's paper that the fat
distribution is different between beef and dairy types as far as
the main fat depots are concerned - subcutaneous, intermuscular,
kidney and channel fat. It was demonstrated that the subcutaneous fat as well as the kidney knob and channel fat, has to be
taken into account in the assessment of the overall fatness.
Within the subcutaneous and intermuscular fat, the differences
in fat distribution proved to be only small. Besides this,
some effects of the fattening system on the fat distribution
were mentioned.

In the paper of Dr. Hanrahan et al, referring to lambs it
was demonstrated that there exist breed differences in fat distribution which emphasise the risk of taking sample joints or
smaller or single depots to predict the carcass fat depots.

Two other papers were particularly concerned with the growth
of fat cells. One was prepared by Dr. Enser and Dr. Wood; and
the other by Mrs. Schön. We may accept that the increase in
weight of the fatty tissue is due primarily to an increase in
size of individual fat cells, whereas the number of fat cells
during the post-natal phase of growth does not increase to an
important extent as far as we can see. With regard to differences
in size of the fat cell, Mrs. Schön demonstrated differences
between the fat depots as influenced by category, fattening and
age.

During the discussion period we had an interesting contribution from Dr. Robelin of France, who demonstrated that, in
his studies with Charolais and Friesian young bulls, differences
existed in the increase of the fat cell diameter between the fat
depots, and that these effects seemed to be in agreement with

the expected rate of maturing of the given types. It was pointed
out by Dr. Enser that, because of the difficulties of measuring
the number of fat cells caused by the fact that the preadipocytes cannot be observed, it is desirable to collect more
data from mature animals.

Two other papers, those of Dr. Leat and Dr. Borghese et al.
and the paper of Dr. Hanrahan and his co-authors discussed the
fatty acid composition of the fatty tissue. The fatty acid
composition is influenced by age, stage of fattening, anatomical
location and diet, whereas Dr. Hanrahan demonstrated breed influences in fatty acid composition of the fat depots in lambs
and Dr. Borghese and his co-workers demonstrated influences of
the species between bovine and water buffaloes.

The fatty tissue distribution is of great importance for the
study of carcass composition, for in my opinion it is the main
source of difficulties involved in the use of more simplified
dissection methods such as standardised methods derived from
practice, and sample joints. In this respect I think that further
research is necessary to verify whether the principles pointed
out in this session are valid for other breeds and types as well
because up to now most of the research has been done on British
breeds, in general fattened up to a higher degree of fatness
than is common on the Continent.

The papers concerned with fat cell growth and fatty acid
composition have, in my opinion, given a very valuable contribution to a better understanding of the fundamental aspects of
the development and properties of fatty tissue. As a matter of
fact, we could have expected that the fatty acid composition,
which is a topic at the moment, would be introduced into the
discussion. It was pointed out that it is mainly the total
quantity of fat intake in human nutrition which causes the major
problems, but, as the papers demonstrated, there can at least be
an influence on the fatty acid composition of fat deposited in
the fatty tissues.

Summary by C. Béranger *(France)* - Growth and nutritional efficiency

The papers and discussion, on efficiency and growth, and interaction between genotype and nutrition and hormones and sex and so forth, have confirmed that there is a large variation existing between genotypes in growth capacity in relation to mature weight, but also within the same mature weight. Such differences appear also in dressing percentage, and therefore in carcass weight, and in carcass composition, resulting from differential growth of tissues and of fat, protein and water. We have also seen that there is variation in voluntary intake and that most of the variations between breeds are related to protein or muscle growth potential; the feed efficiency variations are related to both factors. When these differences seem to be well established between breeds, the next step should be to quantify them, especially their magnitude, and to express them correctly using different methods of comparison. It is also important to have a better understanding of the mechanisms involved in such variations.

Some aspects have been developed. First, we now have a better knowledge of the utilisation of energy and protein by the ruminant, and we can explain some variation by this nutrition aspect, especially in the case of protein synthesis. Some variation also appears, and can be explained, in maintenance requirements and in hormonal and metabolic profile, especially between breeds of dairy or beef types. Variation in protein synthesis, degradation and accretion are certainly of great importance, but knowledge about this is only beginning to be developed.

Secondly, we have new evidence of the importance of interaction between genotype and environment. This has been obtained in several experiments we have reviewed. First the level of energy feeding is the most important factor which is being studied in relation to the different genotypes, but also some environmental conditions and the adaptation of cattle to different conditions and different systems of production. The relationships between these different reactions of cattle to different levels of feeding, and their protein growth potential have been clearly demonstrated by many data. We already know that we have to fit the diet to the potential of the animals, and to adapt slaughter weight to breed and environment conditions to obtain the optimum carcass composition and also to obtain the optimum utilisation of feeds with a high efficiency..

So feed requirements have to be calculated for different breeds and different types of production. For example, for energy and protein we need different types of cattle; in France at Theix we are distinguishing different types of bulls: early maturing or late maturing and intermediate - steers fattened at a different age and weight, heifers of one or two years - all characterised by different systems of production and types and shape of growth curve.

We also have to compare and select animals in different conditions, essentially at different levels of feeding, if we want to adapt to the different conditions.

Many points will have to be studied in more detail in the future. First, it would appear that muscle and fat characteristics and their variation with the genotype and the level of muscle growth rate have to be more thoroughly studied as well as their consequence in meat quality and utilisation. One aspect i to ascertain the optimum level of muscle growth rate and the optimum level of protein deposition, and how the intramuscular fat could develop in relation to the growth, genotype and level of feeding. Secondly, the mechanisms involved in genetics and nutrition interaction have to be more precisely studied. The role of hormones and the possibility of modifying their action would certainly be a profitable area of future research.

We will have to define the optimal level of energy for maximum efficiency for the different genotypes and for the different systems of production. We know that certain differenc exist, but we have to quantify this optimum level for each type of cattle. We also have to determine more precisely the optimum level of protein supply and in particular that of amino acid supply at duodenum level for maximum utilisation of protein growth potential.

All these factors have their influence on selection, especially in the interaction between selection for milk production and for beef. Is it possible to select for a high milk yield, high appetite, high fat deposition and mobilisation, and simultaneously for a sufficient protein growth potential to produce meat? What will be the use of cross-breeding to remove the diverging effect of selection on different criteria which are opposite in dairy and beef types?

Many problems have been raised by this session, stressing the need for further studies on basic mechanisms and on selection processes, taking into account the importance of interactions with nutrition.

748

Summary by H. de Boer *(The Netherlands)* - Beef production and the
customer's requirements

The last short session is still quite fresh and there has
been hardly any time to reconsider different aspects of it.

We have tried to give some answers in this seminar on growth
to the question of <u>what</u> we should grow, especially in view of
considering improved efficiency in the whole chain of production
and the further stages in the chain which lead up to the consumer. In principle, the consumer should give the answer, as
he is the one who has the requirement. His wishes are trans-
lated through intermediate stages between the producer and the
consumer himself. This transformation is far from adequate,
the more so because the intermediate stages themselves play a
very important role. Both Dr. Buchter and Mr. Cuthbertson dealt
with this aspect and with possible improvements to which technological developments may contribute considerably.

During the seminar and also in this last session we looked
mainly at systematic beef production - predominantly by means of
fattening young bulls. This is one stage in the whole process
of cattle production; as a part of its own it is working within
the margin between the calf and the final product. This sets
conditions to possibilities and also to aspects of efficiency
to which nutritionists and geneticists could contribute. The
essential points remain, to define more precisely what should be
produced and next how to meet the requirements. In the short
term this may mean to make the best use of existing feeds and
breeds; in the long term expected future requirements may be
more influential. This last aspect is of specific importance to
the geneticists. However, I do not think we could yet come to
any firm conclusions or make any recommendations in this respect.

Apart from the influence of tradition, considerations of
economy and convenience will influence future consumer requirements, providing further scope for increased efficiency in beef
production. Many alternative developments remain possible,
however, and it will be necessary always to be alert to changes,
and to re-evaluate continuously the progress of rationalisation.

Summary by St. C.S. Taylor *(UK)* - Quantifying growth

Mr. Chairman and colleagues: I have no special right to say so, but I was personally delighted with the papers that were given in the section on methods of quantifying growth - their incisiveness, their variety, their commonsense.

The specific methods described are summarised in the general review and in the abstracts of the papers, and it seems unnecessary to repeat them here.

I would point out that the word 'growth' must now be interpreted in its widest sense. Growth no longer is meant to refer only to growth of body weight, as has tended to become a habit. Growth should include growth of body fat and its curve, growth of lean and its curve and, most importantly, growth or increase in amount of food consumed. Thus, when talking of a mathematical or statistical analysis of growth curves, we must automatically include all these variables in our thinking.

Although the papers were intentionally restricted to mathematical analysis, the methods described will have little value in quantifying growth if the experimental data are unsuitable. Hence the section on quantifying growth' should (in theory) have a large corresponding section on experimental design and objectives.

Possibly the most important general comment about the various methods of analysis proposed was made by Professor Finney. He said that results must be suspect when different methods of analysis lead to different conclusions. Conversely, if results are to have credibility, different methods of analysis should all lead to the same conclusion.

Finally, since limits to growth still tend to be largely ignored, despite their fundamental importance, acquiring greater familiarity with linear programming and how this technique can handle limits might be a worthwhile recommendation. Thank you.

H.J. Oslage *(West Germany)*

I would like to thank all the co-ordinators. I appreciate their strict discipline which allows us time for discussion.

I suggest we now start the discussion, keeping in mind the target of outlining focal points for future research as we have asked for; we will do this according to the sequence of our programme.

The first section dealt with Patterns of Growth and Development of Muscle, Bone and Fatty Tissues. I think we may start to discuss these three sub-sections and the summaries by Dr. Pomeroy, Mr. Boccard and Mr. Bergström.

The discussion is now open.

A.J.H. van Es *(Netherlands)*

I think that part of the problem in making meat from dual-purpose breeds lay in predicting the lean of a given animal. I wonder if it could be done in such a way that measurements of length and maybe width were taken in some simple way from the animal at a young age and that these were extrapolated to, let us say, mature bone measurements, and from that a prediction of the mature weight could possibly be made. Together with some figure of maturity, we might derive the percentage of that which we would like to have at slaughter.

G. Torreele *(Belgium)*

I think that what you are asking for is impossible, except if we had some way of fitting a good growth curve and extrapolating from that curve the estimated weight of meat. Also, the degree of fatness interferes, because this is a matter of intake as well.

A.J.H. van Es

The farmer likes to know if this animal has a high lean potential when it is young, because there is quite a large degree of variability within dual purpose breeds.

G. Torreele

We made a study of all the possible measurements of a calf. We took 800 calves at 10 days old, and took 17 measurements from each and related them to the later potential for beef production. I cannot remember any of them being important.

H.J. Langholz *(West Germany)*

I would just like to add that, by AI, we may derive at least one half of our prediction of fattening performance from the sire whose measurements etc. could be assessed.

J. Martin *(Belgium)*

I wonder if we cannot do something in this way about the length of gestation. We have more work to do on prenatal growth.

R.W. Pomeroy *(UK)*

Extending what Professor Martin has just said, this prenatal and immediately postnatal period is a very much neglected area in growth. After all, if you think of an animal slaughtered at, say, 18 months of age, the prenatal period is half of that postnatal bit which we are all looking at, and the postnatal period is tending to get shorter all the time so, if there is any impact of the prenatal growth on the postnatal growth, we ought to be looking at it. It has been an almost entirely neglected area.

J.H. Oslage

The comments on this have been more or less concerned with the possibility of evaluation of the development of an animal at a very early stage. I would like to see whether you have any additional comments on the summaries of the co-ordinators.

J.D. Wood (UK)

I would like to ask a question of Mr. Boccard and Mr. Bergström. I was under the impression from Mr. Bergström's statement that the variation in muscle weight distribution due to sex, breed, weight or fatness was very small. Yet Mr. Boccar said in his summary that there was need for more work in this area. As someone who is involved in work in this field, I would really like this to be discussed a bit further because I am a little confused.

R. Boccard (France)

Generally speaking, when we consider the big cuts we have a very real equilibrium between animals, as I have said, when we consider them at the same weight and fatness. But when you analyse the distribution of muscle very carefully, you can find some variation in the weight distribution of the muscles. I think that, even though this variation is small, and furthermore not attractive at the commercial level, I feel that the genetici has to exploit this small variation in order to further, by small stages, the possibility of improvement, because it is certain that genetic change can be achieved. Perhaps we have to look for these variations in specific cases. That is why I proposed the study of some extremes which nature offers - for example, the double muscled animal is a type which we could well use to learn more of the mechanism involved in this variation.

P.L. Bergström (Netherlands)

Although the group means are hardly different in muscle weight distribution, as far as breed, types, quality-classes, and so on are concerned, we find within each group an important variation. This variation seems to be of different origin - in part genetic. It is apparent that we cannot observe, either in the living animal or in the carcass, this individual variation. This is quite logical, because the morphological aspects of an animal are of quite a different nature. At the moment I could not say whether we can make any use of that for, when we look at the data in our group of very good quality cows, we find both

animals with a favourable, and with an unfavourable muscle weight
distribution, but there is very nearly the same range in the very
poor quality classes. However, if we simply conclude that, because of the very small practical differences between a group
means, we can leave out this aspect, then we are making a mistake.
There is still a part of the variation which we cannot yet classify, though I cannot give an explanation for this. I think this
corresponds with what Dr. Boccard has said.

R.M. Seebeck *(Australia)*

The problem, as Dr. Taylor has just pointed out to me, is
that people have been saying the same thing about muscle weight
distribution fairly often now - that you cannot get muscle
weight distribution differences. It all started with Butterfield
with a very heterogeneous set of data, and he did not find muscle
weight distribution between his particular breeds. There has
been plenty of work since which in fact contradicts that, not
only in cattle but in other species. I have data, and I am sure
you can get data from France (a lot of it is unpublished). It
is my fault that I have not published a lot of these differences
and in my case the biggest difference is between some peculiar
breeds where we found some quite significant differences.
Certainly, I can get a relationship between muscle weight distribution and growth rate, that the faster growing animals are
the ones with the better muscle weight distribution; I can get
sire differences; we can get muscle weight differences in pigs
with Richman and Berg's data, and so on. These differences may
not be very great but I think they are there, and it is a matter
of trying to do something about them once we have satisfied
other criteria. Maybe it is not economic, maybe there are other
things we should be working on before that, but I think that, if
we repeat often enough that we do not get muscle weight distribution differences, there is a danger that everyone will come to
believe it, and it certainly is not true.

H. Bakker *(Netherlands)*

I would like to comment on the suggestion of Dr. Boccard
that there are possibilities to select for the muscle weight dis-

tribution. I think there are a few factors necessary to be able
to select for muscle weight distribution. You need to be able to
recognise the trait, the trait needs to be heritable, and you
need to make a decision on how much emphasis to put on to the
trait in comparison with all the other traits you wish to select
for. Answering the question, I would say that, with the state
of information we have right now, I am not too sure that we
should select for muscle weight distribution.

J. Kempster (UK)

The last speaker said exactly what I was going to say. I
think we have to be very cautious that we do not get this whole
thing out of perspective. I think we should look at other aspect
of animal production, where we can see quite clear responses to
selection - for example in milk production, or in pig production.
Both of these have fairly clear, often single, objectives, and
in selection we obtain very clear responses. But here we are
talking about a character such as lean distribution, which in
overall economic terms is one-sixteenth as important as fat distribution, and one-fourth as important as lean/bone ratio,
certainly in the national British population; and, while we
continue to talk about these small amounts of variation, and
worry about whether these should go into selection indexes, we
are not going to make responses in the main characteristics of
importance - rate of lean tissue gain within breed, lean tissue
food conversion efficiency within breed.

G. Torreele (Belgium)

Perhaps I can add something to this. First of all, let me
try to quantify that variation in muscle distribution and give
some examples with regard to young bulls on the blackboard (not
recorded). In addition, it should be stressed that conformation
has a lot more to do with carcass length than with muscle distribution, as is illustrated by the concept of blockiness (weight
divided by length).

P.L. Bergström

Dr. Seebeck said that the whole thing was started by Butterfield, and that everyone was inclined to follow him. I think that is quite correct - and that is exactly what I did not want to do. On the other hand, Dr. Kempster said that it is of very little economic importance, and that is certainly true as well. At the moment we are between these two extremes. It is a difficult decision to make - whether to stop the whole business or to try and go a little further and at least get a better idea.

J.H. Oslage

I think we will allow the co-ordinator to have the last word on this question, and continue to a discussion of the second section and Dr. Béranger's summary.

H.J. Langholz

Dr. Béranger, I wonder why you did not stress the problem of the differences between breeds concerning feed intake a bit more, and our need to study this in relation to the energy supply in our different feeding systems. I feel, as a geneticist, that the normally high correlation we are calculating between daily gain and food conversion is very much dependent on the energy we are supplying to the cattle and which depends on the variation in the feed intake. If, for example, you have a very high nutritional supply, the feed intake variation comes into the picture and brings this correlation to zero; and I expect that when we are on a very low plane of nutrition the same will apply. This gives us geneticists a much better understanding of this relationship, and I think we should stress, even more than you did in your summary, that more attention needs to be paid to this special aspect. Maybe you would care to comment on that?

C. Béranger (France)

I agree with you, and if I have not brought this subject out in my summary it is because this aspect of intake came out in the discussion, but was not well established in many of the papers. Most of them were related to growth, growth efficiency, but the importance of variation in feed intake in relation to growth has not been so well explained. I agree very much that this needs to be an important part of any future research.

A.J.H. van Es

I think that a point was made by Dr. Fowler, and later by someone else, concerning the optimal slaughter time. We could not get a satisfactory answer on how to find the theoretically optimal slaughter point, and I think this could be stressed a little more.

V.R. Fowler *(UK)*

I am surprised that we have not heard more from worried geneticists about how to put all this interesting information together. Dr. Kempster, for example, spoke with great authority about how important three or four different aspects of production were, but in doing so he had to decide what the conditions of production are going to be, in order to say how important these things are. For example, if an animal is fed on concentrates, the answer may be different from that for an animal fed on roughages. The sort of thing which Lis Buchter talks about might be very important in Denmark but it might be very unimportant in the United Kingdom. I think it would be a very useful exercise at some future date for someone to try to put together the relative importance of all these different component which can be manipulated genetically, and all the different components which can be manipulated by the system of production - for example the killing weight - and all those different things which can be manipulated by the Rhodes approach, the processing aspects. All these compartments need considering separately and then finally they need putting together.

J.H. Oslage

As there are no other comments on this section, or on section 3 which dealt with the Possibilities for the Improvement of Beef Production in relation to the consumer's and customer's requirements, we will go on to discuss section 4 - Methods of Quantifying Growth and Development.

H.J. Langholz

I want to add one point, Dr. Taylor, to your conclusions.
I think we should lay a little more stress on the importance of
this kind of study when we are looking for breeding goals. For
example, we should include these kinds of studies in comparing
breeds, as Dr. Bakker did in his paper. Another application
concerns the procedures we use in performance testing in the
dual purpose breeds. What are the consequences for the mature
cow when we select our young bulls just according to growth on
a very intensive diet? I think the studies here from Belgium
are very important and should be followed up by others in other
countries where the available data is plentiful. That is just a
comment I wanted to add to your remarks.

FINAL CONSIDERATIONS by

J.H. Oslage

If there are no further comments this is the right time
for the Chairman to give the final word to this last section and
to give a short comment on the co-ordinators' summaries and on
our last discussion, I propose that we start with the question
of nutritional efficiency in growth and the effects of genotype,
sex and hormones and their interaction, as in section 2. This is
not, as you may think, because I am a nutritionist and specially
qualified in this field, but because I feel there is some
logical justification, since nutrition and the processes of
metabolism really form the basis of production. As pointed out
by Dr. Béranger, the papers in this section, and the discussions
also, have outlined the pressing need for more and better data
on the determining parameters of the growth process. We especially need more understanding of the maximum capacity for protein
accretion, and of the composition of the gain during the growth
period, which means the prediction of retained energy at a given
liveweight gain, or for a given breed under different conditions
of nutrition. Data of this kind have to be more complete and
taken to a great depth by further investigations into the
physiological and biochemical background. We should, as Dr.

Pomeroy outlined, not only be able to answer the question, "What happens?", but also to know <u>why</u> it happens and <u>how</u> it happens. So we should strengthen our efforts to get better information on what we call maintenance requirements, on the nitrogen or protein metabolism, on the energetic efficiency of the synthesis of the organic substance and the metabolic turnover, just to mention a few topics we have discussed during the past days.

I would like to say a word to those in the field of endo-crinology. As hormones regulate and control all metabolic pro-cesses, research in this field gets more and more important. In the special topic of this meeting on beef production, even such a current problem as the regulation of feed intake seems to somebody to be a subject needing more, and intensified, re-search. Experiments on these topics should be carried out, as has been outlined by some co-ordinators, with the most important different breeds in Europe. Considering the different conditions in the European countries, it might also be valuable to include the different categories, although the superiority of bulls for producing meat has again been underlined at this meeting. Re-garding the diversity of conditions in Europe as far as breed, feeding conditions, management systems, market requirements and so on are concerned, it would be very profitable if experiments on the above-mentioned problems could be carried out in closer collaboration, or as a concerted action between the research institutes of the different countries. More data and knowledge in this direction will enable us to get better estimations of the nutritional requirements of growing cattle under the differ-ent conditions of production, and this, as everyone agrees, is the most important basis for an economical beef production. Improved lean production is a fundamental conditio sine qua non, as has been pointed out several times during our meeting.

Data of the kind that I am talking about are at the same time essential for animating any model calculations, or for giv-ing a real sense of practical value for such calculations. Such models of biological processes or, to be more in line with the topic of our meeting, models of growth and development, are of

increasing value and necessity to predict the input/output relationships under different and changing conditions. This
meeting may stimulate people concerned in this field to continue
their work in close contact with biologists and in a biologically
meaningful way.

Turning now to our first section on patterns of growth and
development of muscle, bone and fatty tissues, I believe that
some of these points are also valid for this field, especially
concerning the importance of research on more basic questions
in physiology and biochemistry. Being a layman in the field of
meat quality, I have the impression that this area is especially
difficult and complex as it is not only influenced by the
normal factors such as animals and environmental factors which
we production people have to contend with, but the question of
meat quality is additionally influenced by a series of 'post
mortem' factors. Without a proper understanding of the physiological and biochemical processes we are unlikely to get any
remarkable progress in this field.

An interesting aspect of the reports and of our discussion
seems to be the results reported on muscle distribution. The
constancy of a single muscle in relation to the total mass of
muscles is somewhat contradictory to the widespread effort of
the breeders to increase some special parts of high value of
of the carcasses. It therefore seems to be in the spirit of
this meeting and its discussions to clarify the opposite meanings
and doubts in this field by further research.

Another field of great interest and common importance which
has arisen in this section is the question of fat distribution
and its relation to lean, depending on breed , age and nutrition
in regard to the amount as well as to the distribution in the
body. The reported results, as well as the discussions, have
pointed out that future research in this direction should be
based more on anatomical criteria than on the so-called 'usual'
cuts because the latter have a very limited comparability.

This leads me to another point, which will be the last of my comments but which seems to me to be very important. According to Dr. Pomeroy's comments I would like to stress particularly the overall necessity of two points. The first is a better and clearer definition of the technical terminology we use - and, if possible, the use of the same definition of a particular term in the institutes in the various countries. In any country we will always find people who want to create new terms, which of course are always better in their understanding. I am not against improvements in our special terminology but, on the other hand, I am not in favour of misunderstandings arising through the use of different terms or definitions for the same thing.

The second point concerns the question of the necessary experimental design for researching a given problem. Everyone knows that in most cases there are several possible methods of designing and carrying out an experiment. But each kind of experiment has its special minimal requirements to be fulfilled. This starts with the number of animals or replicates, their sex, age or weight, experimental period, and includes feeding and environmental conditions, going through to the kind and accuracy of sampling, chemical or physical methods of investigation, and many other things.

I would like to close my remarks with a proposal that we should establish some general rules, recommendations and minimum planning requirements for various kinds of animal experiments. This, together with a commonly agreed definition of terms, would be of considerable assistance and would make our research results more readily comparable and thus of greater value.

That, ladies and gentlemen, ends my short summary, and with this I will close the session.

CLOSING REMARKS

by

H. de Boer

During this seminar we have been dealing with a complex matter, and the programme has been somewhat overcrowded. Although different aspects have been summarised during this last session, it seems necessary to reconsider the matter as a whole in order to get the best benefit out of it.

The way in which this was done for the Theix colloquium on genotype-nutrition interactions appealed to me and we might try to achieve a similar synthesis on this multidisciplinary seminar. In this context, an early appearance of the printed proceedings will be of importance. The preparation has been given into professional hands this time and I have confidence that Janssen Services will manage this aspect.

Now that two of the three years' CEC beef production research programme have passed it seems logical that the seminars develop a more interdisciplinary character. In order to be able to cover such a broad field we have had to call on the help of many different scientists in the planning and realisation of this conference. I would like to express sincere thanks to all of them, and also to the local organiser, Professor Martin and the Research Centre for Beef Production of Melle, which we associate with the remarkable social highlights of this Ghent seminar.

LIST OF PARTICIPANTS

BELGIUM

Professor Dr. J. Martin

Bosstraat 1
B - 9230 - MELLE

Dr. G. Torreele

Leerstoel voor Veeteelt RUG
Coupoure 533
9000 GHENT

Ir. R. Verbeke

Studiecentrum Rundvleesproduktie,
Bosstraat 1
B - 9230 - MELLE

DENMARK

Mrs. Lis Buchter

Danish Meat Research Institute,
Maglegardsvej 2
Postbox 57
ROSKILDE

Dr. B. Bech Andersen

National Institute of Animal Science
Rolighhedsvej 25
1958 - COPENHAGEN

Dr. T. Liboriussen

National Institute of Animal Science
Rolighedsvej 25
1958 - COPENHAGEN

Mr. K. Kousgaard

Danish Meat Research Institute
Maglegardsvej 2
Postbox 57
ROSKILDE

FRANCE

Mr. C. Béranger

INRA
Laboratoire: Production de la Viande,
Theix
63110 - BEAUMONT

Mr. R. Boccard

INRA
Station de Recherches sur la Viande,
Theix
63110 - BEAUMONT

Mr. B. Bonaiti

Département de Genetique Animale
CNRZ
Domaine de Vilvert
78350 - JOUY-en-JOSAS

Mr. B.L. Dumont

Laboratoire de Recherches sur la Viande
de l'INRA
CNRZ
78350 - JOUY- en -JOSAS

Mr. Y. Geay

INRA
Station de Recherches, Production de
la Viande,
Theix,
63110 BEAUMONT

Mr. R. Jarrige

INRA
Elevage
Theix
63110 BEAUMONT

Mr. J. Robelin

INRA
Station de Recherches, Production de
la Viande,
Theix,
63110 BEAUMONT

IRELAND

Dr. M.J. Clancy

The Agricultural Institute
Dunsinea Research Centre
Castleknock
Co. DUBLIN

Dr. F.J. Harte

The Agricultural Institute
Grange,
Dunsany,
Co. MEATH

Dr. D.E. Hood

Meat Research Department
The Agricultural Institute
Dunsinea
Castleknock
Co. DUBLIN

Dr. J.P. Hanrahan

The Agricultural Institute
Belclare,
Tuam
Co. GALWAY

ITALY

Dr. A. Borghese

Istituto Sperimentale per la Zootecnia
Via Onofrio Panvinio 11
00162 ROME

Dr. E. Cosentino

Istituto di Produzione Animale dell'
Università di Napoli,
Facoltà di Agraria
80055 - PORTICI

Dr. A. Romita

Istituto Sperimentale per la Zootecnia
Via Onofrio Panvinio 11
00162 - ROME

764

NETHERLANDS

Dr. H. Bakker

Afd. Veeteeltwetenschap Landbouwhoge-
school
Duivendaal 5
Postbus 338
WAGENINGEN

Dr. P.L. Bergstrom

IVO 'Schoonord'
Driebergseweg 10-d
ZEIST

Ir. H. de Boer

IVO 'Schoonord'
Driebergseweg 10-d
ZEIST

Dr. A.J.H. van Es

Instituut voor Veevoedingsonderzoek
Runderweg 6
LELYSTAD

UNITED KINGDOM

Mr. A. Cuthbertson

Meat & Livestock Commission
PO Box 44
Queensway House,
Bletchley
MILTON KEYNES MK2 2EF

Dr. M.B. Enser

ARC Meat Research Institute
Langford
Bristol BS18 7DY

Professor D. J. Finney

Department of Statistics & Agricultural
Research Council Unit of Statistics,
University of Edinburgh
James Clerk Maxwell Building,
King's Buildings,
Mayfield Road,
EDINBURGH EH9 3JZ

Mr. R. Fawcett

Edinburgh School of Agriculture
King's Buildings
West Mains Road,
EDINBURGH EH9 3JZ

Dr. V.R. Fowler

The Rowett Research Institute
Greenburn Road,
Bucksburn
ABERDEEN AB2 9SB

Mr. G. Harrington

Meat & Livestock Commission
PO Box 44
Queensway House
Bletchley
MILTON KEYNES MK2 2EF

Miss E. Hind	ARC Animal Breeding Research Organisation King's Buildings West Mains Road, EDINBURGH EH9 3JQ
Dr. W.M.F. Leat	ARC Institute of Animal Physiology Babraham CAMBRIDGE, CB2 4AT
Dr. H.J.H. MacFie	ARC Meat Research Institute Langford BRISTOL BS18 7DY
Professor R.A. Lawrie	University of Nottingham Department of Applied Biochemistry & Nutrition School of Agriculture Sutton Bonnington LOUGHBOROUGH LE12 5RD
Dr. D. Lister	ARC Meat Research Institute Langford BRISTOL BS18 7DY
Dr. R.W. Pomeroy	ARC Meat Research Institute Langford BRISTOL BS18 7DY
Dr. D.N. Rhodes	ARC Meat Research Institute Langford BRISTOL BS18 7DY
Dr. St. C.S. Taylor	ARC Animal Breeding Research Organisation King's Buildings West Mains Road, EDINBURGH EH9 3JQ
Mr. D.R. Williams	ARC Meat Research Institute Langford BRISTOL BS18 7DY
Dr. J.D. Wood	ARC Meat Research Institute Langford BRISTOL BS18 7DY

UNITED STATES OF AMERICA

Dr. R,G. Kaufmann	Department of Meat & Animal Science 1805 Linden Drive University of Wisconsin MADISON, Wisconsin 53706

766

Dr. R. Daenicke

Institut für Tierernährung der Forschungsanstalt für Landwirtschaft, Völkenrode, Bundesalle 50 3301 BRAUNSCHWEIG

Professor Dr. E. Farries

Institut für Tierzucht und Tierverhaltung der Forschungsanstalt für Landwirtschaft, Mariensee 3057 NEUSTADT 1

Professor Dr. F.W. Huth

Institut für Tierzucht und Tierverhaltung der Forschungsanstalt für Landwirtschaft, Mariensee 3057 NEUSTADT 1

Dr. S. Kögel

Bayerische Landesanstalt für Tierzucht Prof. Dürrwächter-Platz 1, 8011 GRUB

Professor Dr. K.J. Langholz

Institut für Tierzucht und Haustiergenetik, Albrecht Thaer Weg 1 GOTTINGEN

Professor Dr. J.H. Oslage

Institut für Tierernahrung der Forschungsanstalt für Landwirtschaft, Völkenrode Bundesalle 50 3301 BRAUNSCHWEIG

Dr. W. Pabst

Institut für Tierzucht und Haustiergenetik, Albrecht Thaer Weg 1 GOTTINGEN

Professor Dr. K. Rohr

Institut für Tierernahrung der Forschungsanstalt für Landwirtschaft, Völkenrode Bundesalle 50 3301 BRAUNSCHWEIG

Professor Dr. L. Schön

Bundesanstalt für Fleischforschung Oskar van Millerstrasse 20 KULMBACH

AUSTRALIA

Dr. R:M: Seebeck

CSIRO Division of Animal Production Tropical Cattle Research Centre PO Box 542 ROCKHAMPTON, Queensland 4700

EEC - ADMINISTRATION

Mr. P. L'Hermite

Commission of the European Communities
DG VI-E-4
Loi 84 - 8/24
200, rue de la Loi
B - 1049 - BRUSSELS
Belgium

Mr. R. Kuyl

Commission of the European Communities
DG VI-E-4
Loi 84 - 8/24
200, rue de la Loi
B - 1049 - BRUSSELS
Belgium

Mr. G.J. Breslin

Commission of the European Communities
DG XIII
Batiment Jean Monnet,
Plateau du Kirchberg
LUXEMBOURG
Luxembourg

RECORDING PERSONNEL

Mr. S.E.W. Hallam

Janssen Services
14, The Quay,
Lower Thames Street,
LONDON, EC3R 6BU, UK

Mrs. V. Johnson

Janssen Services
14, The Quay,
Lower Thames Street,
LONDON, EC3R 6BU, UK

Mr. R. Rice

Janssen Services
14, The Quay,
Lower Thames Street,
LONDON, EC3R 6BU, UK